East Texas Baptist College
Library
Marshall, Texas

Viruses of Vertebrates

Viruses of Vertebrates has become the standard work of reference on the properties and characteristics of viruses affecting man, domestic animals, birds and other vertebrates. In this new edition systematic arrangement of the contents reflects developments in classification, and unclassified viruses are now relatively few. Some rearrangement has also been possible in the enlargement of generic descriptions of morphology and chemical and physical properties so that facts are not repeated in the descriptions of individual viruses.

Painstaking study and condensation of essential information have been necessary to keep the extent of the new edition within reasonable bounds, and the authors have achieved a clear and readable prose style coupled with a very considerable factual content. The continuing increase in the use of laboratory animals and tissue cultures, the development of electron microscopy, and recognition of the importance of the zoonoses, all increase the likelihood of workers being faced with a new or unfamiliar virus. The virologist, the student and researcher in allied fields of microbiology will find this edition of 'Andrewes and Pereira' an invaluable addition to their library shelves.

VIRUSES *of* VERTEBRATES

Sir CHRISTOPHER ANDREWES

M.D., F.R.C.P., F.R.S.

Late Wellcome Trust Fellow, Late Deputy-Director
National Institute for Medical Research
Mill Hill, London

and

H. G. PEREIRA

M.D.

National Institute for Medical Research
Mill Hill, London

THIRD EDITION

The Williams & Wilkins Company
Baltimore

SANS TACHE

7 & 8 Henrietta Street, London WC2 8QE
A division of Crowell Collier and Macmillan Publishers Ltd.

First Edition 1964
Second Edition 1967
Third Edition 1972

ISBN 0 7020 0429 4

Published in the United States of America by
The Williams & Wilkins Company, Baltimore

Made and printed in Great Britain by
William Clowes & Sons, Limited, London, Beccles and Colchester

Contents

Contents

Preface to the Third Edition

Since 1967, when the Second Edition of this book was published, it has become steadily clearer that most viruses can be logically placed into genera. These genera have morphological, chemical and physical characters in common, so that it is justifiable to deal with these in a generic description and to avoid repeating the facts under accounts of individual viruses. The reader will therefore have to seek more information under generic descriptions at the beginning of each chapter. The International Committee for the Nomenclature of Viruses has erected a number of new genera, so that there are eleven more here than in the Second Edition of this book. Further a move has been made towards creating higher taxa: two families, Picornaviridae and Papovaviridae, with three and two genera respectively have been created. A third family, Togaviridae, proposed by an Arbovirus Study Group, although not officially approved by the ICNV, will be adopted to cover the genera Alphavirus and Flavovirus.

The wholly unclassified viruses have become steadily fewer; they are now dealt with in one chapter of 15 pages, compared with six chapters of 72 pages in the First Edition and five chapters of 46 pages in the Second Edition.

To avoid overloading the book it has been necessary to be very selective in the matter of references. We have therefore included those likely to be most helpful, either because they afford the fullest account of the facts, or because they describe recent work not referred to in other books or reviews. We have accordingly frequently omitted reference to the first reported description of a new finding, though a few classical papers are mentioned.

C. H. ANDREWES

March 1972 H. G. PEREIRA

Preface to the First Edition

There are books on the principles of virology and books on virus diseases, especially those of man. This volume deals rather with the viruses themselves, their properties and their relations to one another. Their pathogenic effects are, in principle, considered only so far as is necessary to identify a virus and to indicate its importance. In practice an account of the disease-producing powers of viruses will be found to occupy considerable space, for it is largely by their deeds that we know them.

Viruses attacking only insects are not considered nor are plant viruses. The book does, however, deal not only with viruses of man but also with those affecting other vertebrates. The wider view is necessary if one is to consider viruses from the point of view of taxonomy. Moreover, with the widespread use of various laboratory animals and of tissue cultures derived from various species, workers are bound to encounter viruses of unknown origin. The importance of the zoonoses—diseases of other animals transmissible to man—also is increasingly recognized.

The book will, it is hoped, serve three purposes. By the orderly arrangement of the known facts about viruses it should help the enquirer to discover rapidly whether this or that is known about a particular virus, and guide him to fuller sources of information on the matter. Secondly it is intended to help him, as Bergey's manual helps the bacteriologist, to identify an unknown virus which he may encounter. Thirdly, there is, I hope, a long-term scientific justification for the book. We do not know enough to classify all viruses in an orderly manner. A partial attempt to do so is made in Parts I and II by arranging into groups such viruses as seem ripe for such an attempt. I have tried to avoid what I feel is a pit-fall, the proposal of a classification in advance of adequate knowledge: so Part III contains the viruses which we cannot yet classify rationally, arranged according to the species they attack. It is hoped that as knowledge grows, one may be able, in possible future editions, to promote more and more viruses from Part III to a more satisfactory status in Parts I and II.

Finally I apologize, particularly to many veterinary friends, for the presumptuous attempt of an individual virologist to deal with all viruses of vertebrates.

January 1964 C. H. ANDREWES

Abbreviations

The following conventional abbreviations have been used:

BUDR = 5-bromo-2′-deoxyuridine
CAM = chorio-allantoic membrane
CE = chick embryo
CF(T) = complement fixation (test)
CNS = central nervous system
CPR = cytopathic effect
CSF = cerebrospinal fluid
DNA = deoxyribonucleic acid
EM = electron microscope (or microscopic)
EMC = encephalomyocarditis
FUDR = 5-fluoro-2′-deoxyuridine

$$\text{GC content} = \text{moles}\ \frac{\text{guanine} + \text{cytosine}}{\text{total nucleotides}} \times 100$$

HA = hæmagglutination
HAI = hæmagglutination-inhibition
IB = inclusion body
IC = intracerebral
IM = intramuscular
IN = intranasal
IP = intraperitoneal
IUDR = 5-iodo-2′-deoxyuridine
IV = intravenous
MK = monkey kidney
MW = molecular weight
NDV = Newcastle disease virus
RBC = red blood cell
RDE = receptor destroying enzyme
RNA = ribonucleic acid
SV = simian virus
TC = tissue culture
VN(T) = virus neutralization (test)
UV = ultraviolet

TABLE 1

PROVISIONAL VIRUS CLASSIFICATION

Nucleic acid	RNA				DNA		
Symmetry	Cubical		Helical	Uncertain	Cubical		Uncertain
Presence of outer membrane	0 (ether-stable)	+ (ether-labile)	+ (ether-labile)	+ (ether-labile)	0 (ether-stable or -labile)	+ (ether-labile)	+ (ether-stability varies within the group)
	Picornaviridae Reovirus	Togaviridae	Orthomyxovirus Paramyxovirus Rhabdovirus	Leukovirus Arenavirus Coronavirus	Parvovirus Papovaviridae Adenovirus Iridovirus	Herpesvirus	Poxvirus

PART I

RNA Viruses

1

Picornaviridae

Review: Plummer (1965).

This has been adopted as a family name to include the 3 genera enterovirus, rhinovirus and calicivirus. The family is characterized as follows: virions consist of a naked capsid with icosahedral symmetry, containing 20 to 30 per cent single-stranded RNA with a molecular weight of $2{\cdot}5 \times 10^6$ to $2{\cdot}8 \times 10^6$ daltons. Over-all diameter 30 to 40 nm. Ether-resistant. Viral replication not dependent on cellular DNA function. Viral synthesis and maturation in the cytoplasm.

ENTEROVIRUS

This genus includes viruses inhabiting preferentially the intestinal tract of vertebrate hosts, such as polioviruses, coxsackieviruses echoviruses, and a number of bovine, porcine, murine, avian and other viruses. Besides the habitat, the main character distinguishing these viruses from those of other genera of Picornaviridae is their stability at acid pH. Cationic stabilization to thermal inactivation (Wallis & Melnick, 1962) and inhibition of viral replication by 2-(α-hydroxybenzyl) benzimidazole (Eggers & Tamm, 1961) have also been suggested as characters peculiar to the genus, but exceptional behaviour is observed with some members (see Plummer, 1965).

Morphology. Electron microscopic examination of poliovirus (Dales et al., 1965; Mattern & Daniel, 1956; Horne & Nagington, 1959), coxsackieviruses A (Mattern & DuBuy, 1956) and B (Morgan et al., 1959) and echoviruses (Duffy et al., 1962; Rifkind et al., 1961; Jamison, 1969) reveals approximately spherical naked virions with uniform diameter variously estimated between 20 and 30 nm and a nucleoid 6 to 20 nm across. Variations in size have been reported by Jamison & Mayor (1966) who describe poliovirus 1 and echovirus 19 as being larger (mean diameter 23·5 nm) than other echoviruses and coxsackie B2 (mean diameter 21·0 nm). Finch & Klug (1959) suggested on the basis of X-ray diffraction studies that polioviruses have icosahedral symmetry with 60 protein sub-units 6 to 6·5 nm across forming a shell around an RNA core. Icosahedral capsids made up of 32 (Mayor, 1964; Jamison, 1969) or 42 (Agrawal, 1966) capsomeres have been suggested on the basis of negative contrast pictures of poliovirus.

The morphogenesis of poliovirus (Mattern & Daniel, 1956; Dales et al., 1965), coxsackievirus (Stuart & Fogh, 1961) and echovirus (Rifkind et al., 1961; Duffy et al., 1962) takes place mainly in the cytoplasm although nuclear and nucleolar participation in viral synthesis has been suggested. Virions are often seen in association with small cytoplasmic vesicles or with fibrillar structures. Crystalline arrays of virions are often seen. The role of cytoplasmic membranes in poliovirus biosynthesis has been described by Caliguri & Tamm (1970).

Similar morphological features have been described for enteroviruses of murine (Leyon, 1951; Dales & Franklin, 1962), bovine (Polson & Kipps, 1965; Gralheer et al., 1965), avian (Richter et al., 1964) and porcine (Singh et al., 1961; Meyer et al., 1964) origins although it has been suggested that the last are slightly but significantly larger (diameter 32 to 36 nm) than human enteroviruses. McFerran et al. (1971) found no morphological differences between enteroviruses of porcine, bovine and ovine origins, all measuring 28 $\pm$ 3 nm in diameter.

Chemical composition. Infectious single-stranded RNA has been obtained from human (Colter et al., 1957a; Sprunt et al., 1959), porcine (Brown & Stewart, 1960), murine (Colter et al., 1957b; Ada & Anderson, 1959) and avian (Vindel, 1963) enteroviruses. An RNA content of 22 to 30 per cent has been estimated for polioviruses (Schwerdt & Schaffer, 1955) and 21 to 31·7 per cent for murine enteroviruses (Rueckert & Schäfer, 1965; Faulkner et al., 1961; Scraba et al., 1967; Burness, 1970). Recent MW estimates of the RNA of human (Tannock et al., 1970; Granboulan & Girard, 1969) and murine (Burness, 1970) enteroviruses give values of $2{\cdot}5 \times 10^6$ to $2{\cdot}7 \times 10^6$ which are higher than the previously accepted value of 2×10^6.

The base composition of the RNAs of representative enteroviruses is shown in Table 2. A study of polynucleotide sequences by annealing experiments revealed considerable areas of homology in RNAs of poliovirus types 1, 2 and 3 (Young et al., 1968).

Analysis of the protein composition of polioviruses (Maizel & Summers, 1968), coxsackieviruses (Kiehn & Holland, 1970), murine (Burness & Walter, 1967; O'Callahan et al., 1970) and bovine (Johnson & Martin, 1970) enteroviruses reveals 3 major structural polypeptides with molecular weights ranging from 24,000 to 35,000, one minor structural polypeptide in some strains and several nonstructural components, the largest being a precursor cleaved into the structural polypeptides on assembly into virions (Jacobson & Baltimore, 1970).

Physico-chemical characters. Physical properties of representative enteroviruses are shown in Table 3. Enteroviruses are resistant to lipid solvents, survive well at −76°C in 50 per cent glycerol in the cold, but on the whole not very easily preserved by lyophilization. Most of the infectivity may be lost through this procedure, although the remaining fraction may survive well. Freeze-drying is withstood better in the

TABLE 2
BASE COMPOSITION OF ENTEROVIRUS RNAs

Virus	*Base composition (per cent)*				*Reference*
	G	*A*	*C*	*U*	
Poliovirus 1	24·7	28·5	21·7	25·0	Schaffer et al., 1960
Poliovirus 2	23·2	28·7	22·5	25·5	
Poliovirus 3	24·0	27·5	21·5	25·7	
Coxsackie A9	27·7	27·0	20·7	24·7	Mattern, 1962
Coxsackie A10	28·3	27·3	21·0	24·8	
Columbia SK (ME)	23·7	25·1	24·2	26·9	Rueckert & Schäfer, 1965
EMC	23·7	27·4	23·5	25·6	Faulkner et al., 1961
EMC	24·4	25·8	25·1	24·8	Burness, 1970
Bovine-VG-5-27	22·6	29·3	26·6	21·6	Martin et al., 1970
Bovine-VP-7-19	22·5	31·3	24·0	22·5	
Bovine-VC-65-182	23·1	31·0	25·0	20·5	

presence of 5 per cent glucose and 5 per cent dextran (Tyrrell & Ridgewell, 1965).

Other physico-chemical properties will be mentioned in the description of each member of the genus.

Antigenic structure. Over 60 antigenic types have been recognized among human enteroviruses. These are divided into 4 subgroups: (International Enterovirus Study Group, 1963): polioviruses (3 serotypes), coxsackieviruses A (24 serotypes), coxsackieviruses B (6 serotypes) and echoviruses (34 serotypes). It has been suggested that new human enteroviruses should be numbered sequentially from 68, irrespective of subgroups (Rosen et al., 1970).

Cultivation. Most, although not all, grow in tissue culture causing a type of cellular damage which is fairly characteristic. Cytological and

TABLE 3

BIOPHYSICAL PROPERTIES OF SOME ENTEROVIRUSES (INFECTIOUS VIRIONS)

Virus	*Particle weight (daltons)*	*Sedimentation coefficient* ($\times 10^{-13}$)	*Diffusion coefficient* (cm^2/sec)	*Frictional ratio*	*Water of hydration (g of water/ g dry virus)*	*Partial specific volume (ml/g)*	*Buoyant density (in CsCl or* C_5CO_4*) (g/ml)*	*Reference*
Poliovirus 1	$6{\cdot}8 \times 10^6$	160			0·3		1·34	Schwerdt & Schaffer, 1955
Poliovirus 2	$6{\cdot}8 \times 10^6$	158			0·28	0·64	1·34	Schaffer & Schwerdt, 1959
Poliovirus 3	$6{\cdot}4 \times 10^6$	157			0·37			Mattern, 1962
Echovirus 19		157					1·34	Fabiyi et al., 1964
EMC	$8{\cdot}5 \times 10^6$	162·3	$1{\cdot}44 \times 10^{-7}$	1·13	0·29	0·687	1·35	Burness & Clothier, 1970 Goodhart, 1965
Mengo	$8{\cdot}32 \times 10^6$	151	$1{\cdot}44 \times 10^{-7}$	1·1	0·23		1·32	Scraba et al., 1967
Teschen							1·34	Warrington, 1967
Bovine		156					1·34	Polson & Kipps, 1965

biosynthetic alterations associated with the growth of enteroviruses have been reviewed by Godman (1966). Virus is adsorbed to lipoprotein cell receptors present only in susceptible cells (Holland & McLaren, 1961). Cellular resistance to infection may be overcome by using infectious RNA rather than virus (Holland et al., 1959). Mechanisms of viral penetration and uncoating are unknown. Soon after infection the synthesis of cellular nucleic acids and proteins is inhibited. Virus replication is independent of cellular DNA function (Simon, 1961; Reich et al., 1961). Replication of poliovirus (Holland & Bassett, 1964; Crocker et al., 1964, coxsackievirus (Mattern & Chi, 1962), echovirus (Godman et al., 1964) and murine enteroviruses (Hausen, 1962; Franklin & Rosner, 1962) is entirely cytoplasmic. RNA replicative forms have been demonstrated in cells infected with poliovirus (Baltimore et al., 1964; Pons, 1964), EMC virus (Montagnier & Sanders, 1963) and murine encephalomyelitis virus (Hausen, 1965). Virus induced RNA-dependent RNA polymerases have been described during the replication of poliovirus (Baltimore & Franklin, 1963; Holland & Bassett, 1964), mengovirus (Baltimore & Franklin, 1963) and EMC (Horton et al., 1964). Synthesis of early and late proteins leads to the formation of 'procapsids' containing large polypeptides which are cleaved during the process of assembly with RNA (Jacobson & Baltimore, 1968). *In vitro* assembly of poliovirus has been described by Phillips (1969).

Habitat. As the name implies, enteroviruses are found primarily as inhabitants of the intestinal tract, particularly of young hosts. They commonly cause no illness, but may spread from the gut and cause destructive lesions in the central nervous system.

HUMAN ENTEROVIRUSES

Poliomyelitis

Synonyms: Acute anterior poliomyelitis. Infantile paralysis. Heine-Medin disease. *Poliovirus hominis.*

Reviews: Bodian & Horstmann (1965). Schaffer & Schwerdt (1959) (Physical properties).

Physico-chemical characters. Survives well at −20°C, and for 8 years at −70°C. As with many other picornaviruses, not readily preserved by freeze-drying, readily inactivated by heat—even 50°C for 30 minutes; but results vary according to strain and suspending medium; 30 minutes at 60°C seems always effective. The curve of inactivation by 0·1 per cent formaldehyde has been studied by several workers.

The effect of many other chemicals has been reported; oxidizing agents seem particularly effective: data are reviewed by Gard (1955). UV radiation and high-speed electrons have also been used for inactivating without destroying antigenicity. A radio-sensitive target with an estimated MW of $2{\cdot}23 \times 10^6$ daltons has been described (Ohlbaum et al., 1970).

Antigenic properties. There are 3 serological types, I, II and III, virtually distinct in neutralization tests, but showing a little overlap in cross-immunity experiments and rather more in the CF test. Of 4 antigens separated by density gradient sedimentation, 2, called C and D, were studied in complement fixation and gel-diffusion tests by Le Bouvier et al. (1957; 1959). D was type specific and was degraded by heat and other treatment to C, a smaller particle showing more cross-reactivity and probably devoid of nucleic acid. Very good CF antigens have been made from brains of baby mice, to which some strains have been adapted (Casals et al., 1951; Selzer & van den Ende, 1956). Microflocculation (Smith et al., 1956), micro-precipitin (Eggers & Sabin, 1961), and micro-ouchterlony (Selzer, 1962) tests have been described, giving mainly type-specific reactions.

Within the 3 main types, minor antigenic variants occur (Wecker, 1960; Plotkin et al., 1961; Gard, 1960). These may be important in epidemiological studies especially in association with trials of attenuated virus; the antigenic characters are stable in culture but apparently less so in infected people or experimental animals.

Cultivation. *In fertile eggs.* A type II virus was adapted to growth in 6- to 9-day-old chick embryos by Roca-Garcia et al. (1952), using yolk-sac inoculation or (Cabasso et al., 1952) allantoic inoculation. Lesions of the nervous system, sometimes fatal, were produced (Love & Roca-Garcia, 1955).

In tissue culture the virus has been grown in many tissues of primates since the report by Enders, Weller & Robbins (1949). Cultures of kidney, from man, rhesus, patas and other monkey species, monkey fibroblasts, human amnion, leucocytes, HeLa and other human cell lines have been widely used. Growth in nonprimate tissues has mainly been unsuccessful. Enders et al. (1967), however, obtained growth of type I in chick embryo and hamster cells, when cell-fusion, induced by Sendai virus, permitted entry of virus into cells. Curiously, the mouse-adapted strains have not been grown in mouse tissues.

Viruses can be attenuated by growth in eggs or in culture and used for immunization. Plaques are formed on monolayers of suitable primate cells. Several 'marker' characters have been described, depending on the ability of virus to grow under differing conditions; these are useful for identifying particular strains, especially attenuated ones. These include the T (temperature) marker (Lwoff & Lwoff, 1960), the d (pH) marker (Dulbecco & Vogt, 1958), MS (monkey stable cell line) and the A markers (stability in cations) (Melnick, 1962) and the intratypic antigenic structure (McBride, 1959; Gard, 1960). Attempts to map the poliovirus genome by recombination and/or complementation between temperature sensitive or other mutants have been described (Cooper, 1968; Bengtsson, 1968).

Pathogenicity. The viruses infect human pharynx and alimentary tract, reaching Peyer's patches and lymph-nodes. Thence they may, rather transiently, pass into the blood-stream and reach the nervous system, spreading further along nerve pathways. Invasion of the CNS is, however, exceptional. Two rather different views are contrasted by Bodian & Horstmann (1965). According to their view, the primary site of multiplication is in lymphoid tissues while according to Sabin (1956) it is in oropharyngeal and intestinal mucosa. Infection in man may be wholly inapparent. It may be a nondescript febrile illness, often called abortive poliomyelitis, or it may involve the nervous system causing nonparalytic and paralytic disease: paralysis usually affects the limbs, but can involve the cranial nerves; an encephalitis affecting higher centres is rare. The incubation period is from 4 to 35 days with an average of 10 days.

Experimentally. A similar disease can be induced in many primates after inoculation by various routes, but most readily by IC or IN inoculation. Chimpanzees and cynomolgus monkeys are readily infected *per os*, and chimpanzees, unlike lower primates, can readily be made into symptomless intestinal carriers. Type II was adapted by Armstrong (1939) to cotton-rats and, later, mice. Subsequently types I and III were also adapted to mice by intraspinal injection (Li & Habel, 1951; Li & Schaeffer, 1953). In contrast to the situation with many other viruses, suckling mice are less readily infected than adults are, but adaptation of type II to sucklings has been achieved (Casals et al., 1951) and subsequently also type I (Selzer & Butchart, 1959). Hamsters have also been infected.

Pathological lesions. The most important lesions are usually in the spinal cord particularly the anterior horns, where neurones are destroyed; cellular infiltration with mononuclear cells follows in the wake of this damage. Other parts of the CNS are involved to a varying extent. Eosinophilic inclusions are found in cytoplasm of infected cells (Beale et al., 1956), but are not very characteristic. Very small intranuclear inclusions are also described (Hurst, 1931). Myocarditis may occur in man and in experimental animals.

Ecology. In countries with lower standards of hygiene, virus excreted in faeces is spread very readily, so that children are almost universally infected at an early age. The proportion of paralytic cases is very low, but paralysis is chiefly seen in children and the disease there deserves its name of 'infantile paralysis'. There is evidence that strains of low virulence have selective advantage under poor social conditions and compete successfully with virulent strains. With better hygiene the virus spreads less readily and infections occur later in life, when paralysis is apt to be more serious. Virus has been recovered from flies in the field (*Lucilia* & *Phormia*); attempts to prove multiplication in flies have been suggestive but unconvincing. There has been much argument as to the relative importance of the oropharynx and intestine in spreading infection in man. Contact infection may take place amongst monkeys in the laboratory.

Control. This is not the place to review the great successes in preventing poliomyelitis with inactivated, formalinized virus (cf. Salk, 1959) or with the attenuated viruses first tested by Koprowski et al. (1952) and later developed by Sabin (1959). Use of the last seems to hold most promise of the ultimate elimination of the virus (Sabin, 1962). Cockburn & Drozdov (1970) have contrasted the dramatic fall in incidence in temperate countries following vaccination with the increase in the tropics; here, vaccination, as currently carried out, has been less successful.

Coxsackieviruses

Reviews: Dalldorf & Melnick (1965). Tobin (1953). Dalldorf (1955). Löffler (1958).

Named after a village in New York state where an outbreak occurred. These viruses are divided into two groups, A and B, on the basis of biological characters. They are similar in physico-chemical and other fundamental properties, and these will be considered for the two groups together, with a bracketed (A) or (B) to indicate for which viruses particular information is available.

Physico-chemical characters. Stable between pH 2·3 and 9·4 for 1 day, or between pH 4·0 and 8 for 7 days (Robinson, 1950). Resistant to 5 per cent lysol and 70 per cent ethanol, but inactivated by 0·1 N HCl or 0·3 per cent formaldehyde (Kaplan & Melnick, 1952). Coxsackie B but not A viruses, in general, are inhibited by 2(α-hydroxylbenzyl)-benzimidazole (Eggers & Tamm, 1961a, b).

Haemagglutination. Type A7 agglutinates those fowl RBCs which are sensitive to vaccinia haemagglutinin (Grist, 1960); A21 and B3 will agglutinate human O cells. The haemagglutinins apparently form part of the virus particle, except in the case of A7 (Williamson & Grist, 1965).

Antigenic properties. There are numerous serological types in each group; a list of prototype strains described up to 1965 has been published (Dalldorf & Melnick, 1965). The 24 A types and 6 B types may be distinguished by neutralization tests carried out in suckling mice or in tissue culture, by gel diffusion or by complement fixation. Cross-reactions in the CF test between different types are not very troublesome; however, sera of convalescent patients may show heterotypic responses. This may be due to presence of an antigen common to a number of types (Beeman & Huebner, 1952). Minor antigenic crossings may occur between some coxsackie A and some echo viruses. Cross-reactions among A types were studied in detail by Kamitsuka et al. (1965), and those among B types by Wenner et al (1965). Schmidt et al. (1963) separated group and specific antigens from coxsackie B and A9 viruses. Frommhagen (1965) described C and D antigens similar to those described for poliomyelitis (p. 8).

Cultivation. *In fertile eggs.* A few isolates of several A types have been adapted to grow in yolk sacs or on the CAM (Huebner et al., 1950; Godenne & Curnen, 1952); skeletal muscle of infected embryos may almost disappear.

In tissue cultures of monkey kidneys, B types grow readily. A large number of B strains fail to produce cytopathic changes in human amnion or diploid cells, but have been grown occasionally in other tissues of man and various monkey species; also in cultures of hamster kidney (Barron & Karzon, 1959), pig kidney (Guérin & Guérin, 1957), calf and lamb (Lenahan & Wenner, 1960) and mouse interscapular fat (Stulberg et al., 1952). CPE similar to that of poliovirus. Although several serotypes of A viruses grow readily in primate, mouse and other cultures, the majority do not readily do so; ability to grow varies from one isolate to another (Dalldorf, 1957).

When we come to pathogenicity, the A and B coxsackies have to be considered separately. Table 4 is modified from that published by Tobin (1953).

Coxsackie A Viruses

Pathogenicity. May be present in stools, especially in children in summer months; isolated more frequently when diarrhoea is present, but not necessarily causative. Frequently associated with herpangina, a short febrile illness with sore throat, in which are seen small papules or vesicles around the fauces, soon breaking down into shallow ulcers. Aseptic meningitis may also be caused, particularly by type A9. One type, A21, or Coe Virus, may give rise to the picture of a common cold (see p. 13). Infection of laboratory staff working with A virus is not infrequent.

TABLE 4

THE PATHOGENICITY OF A AND B COXSACKIES

Characteristic	Group A	Group B
Optimum age of mice for inoculation	48 hours	24 hours
Incubation period in suckling mice	2–6 days	4–12 days
Type of paralysis	Flaccid	Spastic
Myositis in suckling mice	Generalized	Focal or absent
Panniculitis in suckling mice	Absent	Marked
Encephalitis in suckling mice	Absent	Often present
Pancreatitis in weaned mice	Absent	Often present
Growth in primate tissue cultures	Exceptional	Usually positive
Chief clinical picture produced in man	Herpangina	Epidemic myalgia, aseptic meningitis and myocarditis in infants

Experimentally the viruses characteristically infect suckling mice, producing paralysis as a result of acute necrotic myositis, infrequently also myocarditis or hepatitis. Further details are shown in Table 4. Suckling mice are readily infected intramuscularly or by any other route, ground-up whole carcasses being used for passage. The viruses occasionally require adaptation to produce typical symptoms. The virus multiplies after injection into baby rabbits; myositis is produced, but there are usually no obvious symptoms (Bell & Hadlow, 1959). Verlinde & Versteeg (1958) produced pneumonia by inoculating pigs; and Verlinde & Ting (1954) produced fatal myositis in suckling ferrets. Lepine et al. (1952) found the merion, a North African rodent, very susceptible, new-born hamsters less so. Some strains show neuropathogenicity for monkeys (Dalldorf, 1957).

Pathological lesions of muscles of baby mice consist of an eosinophilic hyaline necrosis.

Characteristics of particular coxsackie A viruses

A7. This type can cause paralytic disease in monkeys, cotton-rats and new-born mice (Grist & Roberts, 1966). Russian workers have suggested that it be called a fourth type of poliovirus as it has been recovered from paralytic cases in man (Voroshilova & Chumakov, 1959).

A9 is in some ways intermediate between coxsackie- and echoviruses. It resembles the echos in being inhibited by 2(α-hydroxybenzyl)benzimidazole (Eggers & Tamm, 1961a, b). It has also been particularly associated with aseptic meningitis sometimes with an exanthem, and it shows some antigenic relation with A23–echo 9; it also produces CPE in rhesus kidney tissue cultures, and causes myocarditis in mice (Lerner, 1965). A14 also is more neurotropic for monkeys than most As. A16 has been associated in Canada (Robinson et al., 1958) and elsewhere with a febrile illness with vesicular rash: 'Hand, foot and mouth disease'; and so has A5.

A21 is serologically identical with Coe virus (Lennette et al., 1958) which produces common-cold-like symptoms, particularly amongst recruits. It is more commonly recovered from pharynx than faeces. It can be grown in embryo rabbit kidney, HeLa and other human cells and has a low pathogenicity for suckling mice. Its effects when given intranasally to volunteers have been described by Parsons et al. (1960), and its method of spread by Buckland et al. (1965). One strain caused tracheo-bronchitis or bronchopneumonia when given by fine spray (Couch et al., 1965). Recently isolated strains agglutinate human O RBCs.

A23 has been considered to be identical with echo 9 (Sickles et al., 1959) and will be considered under that heading (p. 17).

Coxsackie B Viruses

Pathogenicity. Inapparent infections are common. Particularly associated with epidemic pleurodynia or myalgia (Bornholm disease); in this condition fever is associated with severe thoracic or abdominal pain. Orchitis occurs as a complication. Aseptic meningitis may be caused, more commonly than with coxsackie As. A serious, often fatal, myocarditis due to coxsackie Bs occurs in new-born infants (Gear, 1958). There is evidence suggesting that congenital heart disease may be associated with maternal infections by coxsackie B viruses especially

B3 and B4 (Brown & Evans, 1967). These 2 types pass the placental barrier and cause many abortions in mice (Selzer, 1969). Myocarditis and pericarditis appear much more rarely in older children or adults. Upper respiratory involvement may be the only sign of infection, especially with B3 and B5.

Experimentally. Suckling mice can be infected by injection by various routes, but preferably IC. The lesions are predominantly in the brain. In contrast to the flaccid paralysis caused by the A viruses, B-infected ones show spasticity or spastic paralysis and tremors. Many strains produce histological lesions, and virus can be recovered, yet no illness is produced. Some B4 strains do not even multiply (Köppel, 1963). Necrosis of muscles, if it occurs, is focal. Myocarditis, hepatitis, parotitis and pancreatitis occur, particularly in older mice. Necrosis in the interscapular brown fat pad may be visible with the naked eye and is generally the most prominent lesion microscopically. Mice recovering from encephalitis may develop cerebral cysts.

Baby hamsters may also be infected and myocarditis has been produced by inoculating young cynomolgus monkeys (Lou et al., 1961). Gravid mice are more susceptible than normal. B viruses, as well as As, have caused infections in laboratory workers.

The lesions in brain and elsewhere are mainly those of acute cell-necrosis. Lepine et al. (1952) found degenerative and inflammatory lesions in muscle removed by biopsy from two human infections.

Ecology. (A and B strains.) Viruses are widespread, usually in the summer and autumn months in the alimentary tracts especially of children and of those living under poor social-economic conditions. Their presence in the gut is probably transient—a week or less. Virus passes out in the urine and faeces and has been recovered from sewage and from flies. Infections causing myocarditis in infants have been contracted from mothers. Viruses closely related or identical to types B1, B3 and B5 have been isolated from nose, throat and rectal swabs of dogs (Lundgren et al., 1968). A viruses have also been isolated from mosquitoes.

Echoviruses

Echo stands for 'enteric cytopathic human orphan' (orphan implies lack of association with disease). However, many viruses in the group are now known to cause aseptic meningitis and other troubles; so the O of echo is not always relevant. Typically they are less neuropathogenic than poliovirus, do not produce disease in suckling mice as do coxsackies nor cause common colds as do rhinoviruses; they produce a

CPE in rhesus kidney cultures. Some are intermediate in character between typical echos and other picornaviruses.

Reviews: Committee (1957). Melnick (1965). Müller (1961). Wenner & Behbehani (1968).

Physico-chemical characters. Different serotypes and even strains of one type differ in their rate of inactivation at 37°, 22°, 4° and −20°C. In general, they are inactivated in 30 minutes at 65°C, survive well at −70°C, but lose much activity on drying. Echo 1, at least, is very resistant to the photodynamic action of toluidine blue (Hiatt et al., 1960).

Haemagglutination. Human 0 RBCs are agglutinated by many strains of Types 3, 6, 7, 10, 11, 12, 13, 19 (Goldfield et al., 1957; Lahelle, 1958). A temperature of 4° and pH 6·3–6·6 are optimal. Simon & Domök (1963) separated agglutinating and nonagglutinating particles from echo 6. The haemagglutinin is not separable from the virus particle. Receptor-destroying enzyme does not affect agglutinability of cells, nor, in contrast to reovirus haemagglutinins, does chloroform (Leers, 1969).

Antigenic properties. Neutralization tests in tissue culture and complement-fixation tests separate the echoviruses into different serological types. A list of prototype strains described up to 1965 has been published (Melnick, 1965). The haemagglutination inhibition test is less specific. Thirty-four serotypes were recognised in 1970. Some types, e.g. echo 6, exist in 2 phases, one reacting poorly with homologous antibodies (Karzon et al., 1959). Cross-neutralization tests against prototype viruses may be 'one-way', so that new isolates may mistakenly be thought to represent new types. Cross-reactions may occur in the neutralization of CF test between types 12 or 13 and 18 and between 22 and 23. Type 8 is now considered as a subtype of type 1. There is evidence of a slight antigenic relation between echoviruses and poliovirus (echo 6) or coxsackies (see echo 9, p. 17). Collaborative studies of antisera against prototype viruses have been carried out on an international basis and the results summarized by Hampil & Melnick (1968) and Melnick & Hampil (1970). Lim & Benyesh-Melnick (1960) have proposed a scheme for testing with pools of antisera designed to facilitate the rapid typing of unidentified strains. The effect of heat on the antigenicity of echo 6 has been investigated by Forsgren (1966) with results similar to those obtained with other picornaviruses.

Cultivation in fertile eggs is not reported. Although most strains grow well, with characteristic CPE, in cultures of rhesus, cynomolgus or cercopithecus kidneys, some strains grow better in human amnion cell cultures. Types 7, 8 and 12, but not the others reported on, grow well in kidney cultures from *Erythrocebus patas* monkeys (Hsiung & Melnick, 1957). Plaques formed on monolayers in culture are of different sizes and appearance and are not necessarily always the same for a particular type. Many strains have been adapted to grow in HeLa and other lines of human cells, also (echo 4 and 9) in calf or pig cells (Lenahan & Wenner, 1960). Growth may be poorer when the pH of the medium becomes more acid, in contrast to what occurs with rhinoviruses (Barron & Karzon, 1957).

Habitat. In intestinal tracts of many persons, especially children, in the summer and under poor social-economic conditions. Most isolations of virus have been from stools or rectal swabs, some from throat or CSF. Strains isolated from dog faeces were echo 6 or very closely related to it (Pindak & Clapper, 1964; Lundgren et al., 1970).

Pathogenicity. Most infections are inapparent. Many of the serotypes, however, have caused fever accompanied by signs of aseptic meningitis, sometimes with an exanthem. Exceptionally there may be paresis, usually transient (types 9 and 16). The same serotypes have been associated with mild gastro-enteritis. Features of infection with some of the more important types are recorded below. Some types appear to cause mild respiratory disease. The association of various serotypes with aseptic meningitis, encephalitis, exanthems and other clinical features has been tabulated by Kibrick (1964) and by Wenner & Behbehani (1968).

Experimentally, echoviruses have rarely caused disease in experimental animals. Some epidemic strains of echo 9 have been adapted to suckling mice in which they have behaved like coxsackie A viruses and have also caused glomerulonephritis. It has been possible in culture to segregate strains virulent or avirulent for mice (Margalith et al., 1967). A strain of echo 6 has produced infections like those associated with coxsackie B viruses (Vasilenko & Atsev, 1965). Focal lesions have been found in the CNS of occasional monkeys inoculated in the brain or cord with types 1, 2, 3, 4, 6, 7, 8, 9, 10, 12, 14 and 17; at times such lesions have appeared following intramuscular injections (Wenner & Chin, 1957). They have consisted of focal infiltration, less frequently neuronal destruction. Symptoms in the monkey are commonly absent, but may occur after intraspinal injection (Lou & Wenner, 1962).

Characteristics of particular echoviruses

Echovirus 4 has caused at least 2 considerable outbreaks of aseptic meningitis, 70 per cent of the patients having also gastro-intestinal symptoms (Chin et al., 1957; Johnsson, 1957).

Echovirus 6 has also caused a number of similar outbreaks of aseptic meningitis in children and adults; gastro-intestinal disturbance was, however, less common (Kibrick et al., 1957). Localized muscle weaknesses have occurred and there were maculo-papular rashes in one outbreak. This type has rarely been isolated apart from association with disease.

Echovirus 8 was recovered from patients with respiratory and intestinal symptoms (Rosen et al., 1958).

Echovirus 9 was the cause of a widespread epidemic of aseptic meningitis occurring all over Europe in 1955 and 1956 and reaching North America in 1957. In many but not all outbreaks there was a maculo-papular rash. Familial infections were common. In contrast to the prototype echovirus 9 virus, many epidemic strains could be adapted to cause coxsackie A-like disease in suckling mice (Tyrrell et al., 1958). In other respects, also, the virus was intermediate between typical coxsackie and typical echoviruses and Tyrrell et al. (1958) reported some antigenic relation with coxsackie A9. It has been proposed (Sickles et al., 1959) to rename it coxsackie A23; it behaves, however, like an echovirus by Eggers & Tamm's (1961a) benzimidazole test. It can be grown in hamster-kidney cultures (Barron & Karzon, 1957).

Echovirus 10 has now been removed from the echoviruses and is considered under Reovirus (p. 52).

Echovirus 11 under the name of U virus was associated by Philipson & Wesslén (1958) with respiratory illness in children; it has also caused rashes. In experiments on volunteers with this strain Buckland et al. (1959) produced no typical colds but rather gastro-intestinal disturbances.

Echovirus 16 caused the so-called 'Boston exanthem' (Neva & Enders, 1954). There were a very few cases of aseptic meningitis; many had a maculo-papular rash, not appearing till the fever was over.

Echovirus 18 was recovered from faeces of numerous infants under 1 year old during an outbreak of diarrhoea in 1955 (Ramos-Alvarez, 1957) and it turned up in another outbreak in 1956. Evidence that it caused the diarrhoea was fairly good, but implication of other echo-viruses as causative agents of infantile diarrhoea is still debatable. Unlike coxsackieviruses (Ramoz-Alvarez & Sabin, 1956), they do undoubtedly turn up more frequently in infants than in older children.

Echovirus 20, first referred to as JVI, was recovered by Cramblett et al. (1958) from stools of children with fever, coryza and watery diarrhoea. Adult volunteers were infected with the virus by Buckland et al. (1961); 27 of 43 developed symptoms, chiefly constitutional, 2 had symptoms like those of a cold while 8 had abdominal symptoms.

Echovirus 22 and 23 behaved unlike other echos and resembled coxsackie A viruses in their reaction to 2(α-hydroxylbenzyl) benzimidazole (Eggers & Tamm, 1961a). Nuclear changes produced in infected cultures also differed from those of other echoviruses (Shaver et al., 1958). Echo 22 has been associated with outbreaks of respiratory disease, with some pneumonia, among premature infants (Berkovich & Pangan, 1968).

Echovirus 25 has produced pharyngitis, cervical adenitis and fever in volunteers (Kasel et al., 1965); also rashes in infants.

Echovirus 28 (JH or 2060 virus) is now included with the rhinoviruses (see p. 30).

Ecology. This is as much as had been described for coxsackieviruses (p. 14). Though the distribution of the viruses is world-wide, all reported epidemics have occurred in the temperate zones, beginning in summer or autumn months. Sporadic isolations have been made throughout the year.

SIMIAN ENTEROVIRUSES

Synonym: ECMO (enteric cytopathogenic monkey orphan).

Review: Kalter (1960). Hull (1968).

The extensive use of monkeys for work on poliomyelitis and other viruses has led to the discovery of many 'orphan' viruses present in monkey kidneys or stools. These have been given numbers in an SV (simian virus) series by Hull (1968) and his collaborators. Hull (1968) lists 14 strains identified as enteroviruses and 5 strains considered to be picornaviruses but possibly not enteroviruses. Heberling & Cheever (1965) characterized 15 strains as typical enteroviruses, i.e. small RNA viruses, ether- and acid-resistant and stabilized by molar magnesium chloride. These 15 strains were divided into 4 groups according to type of CPE and plaque formation in monkey-kidney cultures. Like coxsackie A viruses, they are but little inhibited by HBB (2(α-hydroxybenzyl) benzimidazole); on the other hand only 2 infected suckling mice, when they produced lesions like those caused by coxsackie B viruses. Several antigenic types could be distinguished, but a definitive classification into distinct serotypes must await further study. Most strains have been isolated from intestinal contents of rhesus or cynomolgus monkeys but Hull's strains SV4 and SV28 have been found mostly as latent viruses in monkey-kidney cultures (Hull, 1968). Isolates from baboons have been described (Fuentes-Marins et al., 1963).

BOVINE ENTEROVIRUSES

Synonym: ECBO viruses (enteric cytopathic bovine orphan).

Review: Kalter (1960).

Haemagglutination. 5/11 strains tested by Moscovici & Maisel (1958) agglutinated bovine RBCs at 5–8°C, not at room temperature. Three of them also clumped guinea-pig cells. Elution was rather rapid. Strains isolated by Inaba et al. (1959) agglutinated horse and sheep RBCs at 4°. Viruses of one serotype agglutinate rhesus RBCs (La Placa et al., 1963). Agglutination of sheep RBC is described by Nath et al. (1970).

Antigenic properties. Neutralization tests in tissue culture reveal that there are numerous serotypes. McFerran (1958) found 3 types in Northern Ireland, one corresponding to an American type. Klein (quoted by Kalter, 1960) separated 6 types. On the other hand 8 viruses isolated by Kunin & Minuse (1958) were all alike. La Placa et al. (1965) compared 22 prototype strains from various parts of the world using immune sera prepared in chickens. They put the viruses into 2 main groups. The 11 in the first group were closely related, but had broadly reacting and highly specific variants. The 11 in the second group were less homogeneous. Growth of group 2 but not of group 1 strains is inhibited by 2-(α-hydroxybenzyl) benzimidazole (Portolani et al., 1968). Barya et al. (1967) divided 13 strains into 4 serotypes on the basis of neutralization kinetics. Possible antigenic relationships between human and bovine enteroviruses are suggested by the neutralization and precipitation of human strains by bovine sera and vice versa (McFerran, 1962; Kanamitsu et al., 1967; Urasawa et al., 1968). Some bovine enteroviruses are said to share an antigen with poliomyelitis virus type 2 (Nath et al., 1971).

Cultivation. *In fertile eggs.* Kunin & Minuse's (1958) strain could be isolated by amniotic inoculation and also produced pocks on the CAM. According to Liess & Hopken (1964) strains have differing ability to grow in eggs; 9 of 14 strains grew after amniotic inoculation, smaller numbers when injection was on the CAM, into allantoic cavity or yolk sac.

In tissue culture of calf-kidney cells, the viruses produce CPE, and at least those of Moll & Finlayson (1957) and Moll & Davis (1959) grow also in cell cultures from pig, guinea-pig, rhesus and patas monkeys and man, but not in HeLa cells nor cat cells and poorly in those from swine, dog and rabbit. Several other workers have isolated these viruses in monkey-kidney cultures. Formation of plaques of several types permits the separation of different agents.

Pathogenicity. Not known to be naturally pathogenic for cattle, but one strain caused bloody diarrhoea in colostrum-deprived calves. It also led to abortion in pregnant guinea-pigs (Bögel & Mussgay, 1960). The Kunin strain produced lesions in suckling mice and hamsters like those caused by coxsackie A viruses (see p. 12).

Ecology. Viruses have been isolated most frequently from younger calves.

PORCINE ENTEROVIRUSES

Synonyms: Infectious porcine encephalomyelitis or poliomyelitis (and similar names). Teschen disease. Talfan disease. Ansteckende Schweinelahmung. ECSO viruses (enteric cytopathic swine orphan).

Reviews: Mayr (1962). (Teschen disease.) Kalter (1960).

Haemagglutination has been looked for but not as yet demonstrated.

Antigenic properties. Ten serological groups are distinguished by Alexander & Betts (1967). Strains associated with Teschen and Talfan disease were placed into 3 subtypes within the first serotype (Darbyshire & Dawson, 1963). Several other classification schemes have been proposed by workers in different countries (e.g. Dunne et al., 1967; McConnell et al., 1968; Morimoto, et al., 1968), but comprehensive comparative studies are awaited. Intermediary antigenic types have been described (Christov, 1966).

Cultivation. All strains can be readily cultivated in pig-kidney cells causing CPE similar to that of other enteroviruses and producing plaques.

Distribution. Teschen and Talfan diseases are widespread in Europe, but have only caused serious losses in Eastern Europe, especially Poland and Czechoslovakia, also in Madagascar (Lepine & Atanasiu, 1950). Mild strains are probably present in Canada (Richards & Savan, 1960; Greig et al., 1962), the United States (Koestner et al.,

1962) and Australia. Other porcine enteroviruses have world-wide distribution.

Pathogenicity. Teschen and Talfan diseases may occur sporadically or in outbreaks. In the severe form, with a mortality averaging 70 per cent, there is a mild fever, soon followed by convulsions, prostration, stiffness and paralysis. Residual paralysis occurs in animals which survive. With milder strains, infection may be inapparent. Mild British strains are more apt to cause ataxia than paralysis (Done, 1961). In Czechoslovakia wild boars are affected. The disease has been transmitted to young pigs by IC, IN, IM or IP inoculation and by feeding, but not by SC inoculation. Viraemia occurs early in the incubation period, which varies between 4 and 28 days and may precede paralysis by 10 to 12 days (Horstmann, 1952). Other species are insusceptible—except for *Potamochaerus*, the wild Madagascan boar.

Other porcine enteroviruses have been isolated from outbreaks of enteritis, there is little evidence that these viruses are naturally pathogenic for pigs. Beran et al. (1960) did, however, produce fatal infections with diarrhoea and some nervous symptoms in new-born pigs deprived of colostrum. Diffuse pneumonitis (Meyer et al., 1966) and cardiac lesions (Long et al., 1969) have also been observed in inoculated pathogen-free piglets. Dunne et al. (1965) described an epizootic, resulting from a group of viruses in this category, causing reproductive failure characterized by infertility and stillbirths. The viruses are non-pathogenic for other species also, including suckling mice. Nardelli et al. (1968) observed an infection in pigs caused by a porcine enterovirus and clinically resembling foot-and-mouth disease. The virus caused fatal paralysis in day-old mice, but failed to infect cattle or other species.

Pathological lesions in Talfan and Teschen diseases are more widespread than in human poliomyelitis, as there is a diffuse encephalomyelitis. Cytoplasmic eosinophilic masses occur in nerve cells. Myocardial lesions are reported.

Ecology. As with enteroviruses of man, the virus is normally a harmless inhabitant of the intestinal tract; it is probably only virulent strains which cause epizootics. Infected animals liberate virus both in mouth secretions and in stools. Beran et al. (1958) isolated viruses with increasing frequency in pigs up to the age of 9–10 weeks. Moscovici et al. (1959) did not obtain their virus from rectal swabs of adult pigs. Beran et al. (1958) made no isolations during the winter months and suggested that virus-shedding was seasonal. Some virus was excreted throughout a month (Wenner et al., 1960).

Control. Effective vaccines against Teschen disease have been made, particularly with virus cultivated in tissues. Virus may be given SC either living, attenuated in culture, or inactivated by formaldehyde. Both types gave 80–86 per cent protection (Mayr & Correns, 1959; Mayr, 1962).

MURINE ENTEROVIRUSES

Encephalomyocarditis Virus (EMC)

Synonyms: Well-known strains are Col-SK, MM and Mengo; these are alike antigenically, but show certain differences in biological behaviour (Craighead, 1965a).

Reviews: Jungeblut (1958).

Physico-chemical characters. Unusually heat-stable. In 0·1 per cent bovine albumin it was inactivated in 30 minutes at 60° but not at 56°; but in 20 per cent monkey sera it survived 30 minues at 80°C and 96°C for 20 minutes. Rather resistant to phenol, formalin and ethanol. Formalin 0·5 per cent inactivated in 6 hours, but not in 1 hour (Dick, 1948). When held in distilled water or in molar NaCl at 37°C it was much more resistant than strains (FA and GD VII) of mouse encephalomyelitis (Speir et al., 1962). It was also fairly stable when held 2 hours at 37° between pH 3 and pH 9 (Speir et al., 1962). Colter et al. (1964) describe variants of Mengo virus unstable below pH 6·8.

Haemagglutination. Shown by Hallauer (1951) to agglutinate sheep RBCs in the cold, being adsorbed and then eluted at higher temperatures. Craighead (1965b) found that the effective temperature range varied from one strain to another. RDE destroys the receptors on the sheep cells. Some strains also agglutinate human O cells, and Craighead & Shelokov (1962) isolated a stable agglutinin active for human and guinea pig cells. Haemagglutination is inhibited by specific antisera. Haemagglutinin was lost after 30 passes in a mouse ascites tumour (Furusawa & Cutting, 1962).

Antigenic properties. The various strains (MM, Mengo, Col-SK, etc.) are identical when compared by neutralization, CF or haemagglutinin-inhibition tests (Warren et al., 1949; Dick, 1949). Neutralizing antibodies have been found in the sera of wild rats and other species; haemagglutinin inhibitors found in various animal sera may be non-specific substances rather than true antibodies.

Cultivation. *In fertile eggs*, the virus will grow when inoculated on the CAM or into the allantoic or amniotic cavities or yolk sac. Infected embryos usually die, particularly if incubated at 35°C (Schultz & Enright, 1946). Maximum titres are obtained in 48 hours. Embryos show widespread degenerative changes and haemorrhages (Dick, 1950).

In tissue culture it grows in embryonic and other tissues of chick, mouse, man, monkey, hamster and cattle; also in HeLa and other human cell-lines and in certain mouse tumours, though not all strains do equally well in all these tissues. The cytopathic effects (Barski & Lamy, 1955) differ from those of poliomyelitis and related viruses; there are retraction of nucleus and cytoplasm and nuclear fragmentation resulting in a basophilic mass. Cytoplasmic inclusions are not seen. With the aid of appropriate buffers, plaque development on monolayers is obtained (Bellett, 1960). Ellem & Colter (1961) isolated variants from plaques of 3 types: one of these formed crescentic plaques, following the lines of cell architecture.

Habitat. Mice are the commonest natural hosts but the virus has also been isolated from the blood and from stools of human beings; from captive chimpanzee, gibbon, monkeys of several genera, a wild mongoose, pigs and several kinds of rodents—rats, a water-rat (*Hydromys*), cotton-rats and a squirrel. Serological evidence suggests that other rodents are probably infected in nature.

Pathogenicity. Infection is probably inapparent in rodents infected in nature. Virus was, however, isolated from the faeces of a red squirrel in an area where these animals were dying with paraplegia (Vizoso et al., 1964). Not all the viruses isolated from human sources were necessarily the cause of disease. It seems, however, that infection can occasionally cause febrile illness in man with symptoms of CNS involvement and lymphocytic pleocytosis in the cerebrospinal fluid (Dick et al., 1948; Gajdusek, 1955). Myocarditis in man is not recorded. The Vilyuisk virus isolated from cases of human encephalomyelitis in Siberia was shown to be related to EMC (Casals, 1963; Petrov, 1970). In captive chimpanzees (Helwig & Schmidt, 1945), monkeys and pigs (Murnane et al., 1960) illness has been accompanied by myocarditis, often fatal. Experimentally, the virus will infect a wide host range. Mice and hamsters commonly die with CNS involvement after inoculation by any route. Some strains after IC passage produced necrosis of pancreas, parotid and lacrimal glands (Craighead, 1965a); diabetes may follow (Craighead & McLane, 1968). Guinea-pigs and rabbits are more resistant, but the former develop myocarditis or become paralysed. Baby rats die after inoculation, but in adults there is a chronic inapparent infection. Inoculated wild African rats and mongooses have died, apparently of myocarditis (Kilham et al., 1956). A species of jerboa (*Meriones* *eybicus*) is highly susceptible to experimental infection (Dickinson & Griffiths, 1966). Fatal encephalitis is produced by inoculating monkeys; different species show differing susceptibilities to various virus strains. Several species have contracted infection *per os*. The pathogenesis of experimental infection in mice has been described (Dickinson & Griffiths, 1966; Friedman & Maenza, 1968).

By passage in mice in different ways, virus was adapted to cause predominantly lesions of nervous system or of muscles (Kuwata, 1956–57).

Pathological lesions in infected animals are those of encephalitis with scattered focal necrosis; or, particularly near the site of intramuscular inoculation, necrosis of muscle fibres with inflammation and oedema. Similar necrotic lesions occur in the myocardium.

Ecology. The natural reservoir is probably in rodents, other species contracting infection from them. Cage infection has been recorded amongst cotton-rats and mice. Virus is present in urine and faeces. Though it has been recovered from wild-caught mosquitoes (*Taeniorrhynchus*) in Uganda, there is no evidence that insects can act as vectors.

Control. A formolized virus vaccine protected mice experimentally (Kauffmann & Frantzen, 1948); so did that attenuated by passage in a mouse tumour (Furusawa & Cutting, 1962).

Mouse Encephalomyelitis

Synonym: Mouse poliomyelitis.

Description: von Magnus et al. (1955).

Several viruses are included under this name. On the one hand are those related to the TO (Theiler original) strain; on the other are the FA, GD VII and related strains. The last have biological, but not antigenic, characters intermediate between TO virus and the EMC viruses (cf. p. 22). The mouse encephalomyelitis (ME) strain studied by Franklin et al. (1959) and others is in fact EMC. The FA and GD VII viruses were originally thought to be virulent variants of TO; this now seems more than doubtful.

TO Strain

Some statements in the literature about this virus refer in fact to the other strains.

Haemagglutination. Unlike GD VII, does not agglutinate human RBCs.

Antigenic properties. Can be studied by neutralization tests in mice or, more readily, in tissue culture (Pette, 1959). Gard (1956–57) has described an immuno-inactivation test more sensitive than conventional tests. Several workers report some cross-neutralization between the TO and GD VII strains (Olitsky, 1945; Shaw, 1956); there are also said to be minor antigenic differences amongst TO-like viruses.

Cultivation. The virus usually fails to grow in eggs, but has been adapted by Shaw (1953) to grow in yolk sacs and in 'Maitland-type' chick-embryo tissue culture. It will grow (Falke, 1957) in mouse-kidney cultures, and in these, after adaptation, cytopathic effects appear.

Pathogenicity. Normally an inapparent infection occurring in the intestines of laboratory mice all over the world. Similar viruses have been recovered from wild mice, cotton-rats and kangaroo rats (*Dipodomys*), but there is a possibility that these had been infected from laboratory mice. The virus produces flaccid paralysis usually of hind limbs in a very small proportion of infected mice—a few amongst thousands. It is not necessarily fatal. Exceptionally, a number of paralytic cases may occur together (Gard, 1944).

Experimentally, response to inoculation is irregular; the incubation period after IC inoculation is 12–29 days. Very young mice may die without definite symptoms, older ones may have progressive flaccid paralysis, and still older ones an inapparent infection. Several workers have obtained mouse colonies free from infection (von Magnus & von Magnus, 1949), and mice from such colonies are more regularly susceptible. In them, virus multiplies at the site of an intramuscular inoculation and may also cause paralysis when given *per os* or by other routes (von Magnus, 1951; Dean, 1951). Golden hamsters have also been infected, but symptoms may be absent unless mice are interpolated in the transmission series (Dean & Dalldorf, 1948).

Pathological lesions in the CNS resemble those of poliomyelitis in man. There is destruction of anterior horn cells especially of the cord, with associated inflammatory reaction.

Ecology. New-born and suckling mice enjoy passive protection from maternal antibody. Soon after weaning they are found to be carriers and become increasingly resistant to infection. The carrier rate becomes progressively less as they grow older; at 7½ months it was only 54 per cent and fell further thereafter.

FA and GD VII Strains

One spontaneous occurrence of disease caused by an FA strain is recorded; otherwise these viruses have come to light after inoculation of mice with materials such as other viruses. It appears (Smithburn, 1952) that several ostensibly new arthropod-borne viruses do in fact belong in this group, having presumably come from the mice used for inoculation.

Physico-chemical characters. The effects of physical and chemical agents are fully reviewed by Gard (1955). The half-life at 50°C is 8·5 minutes. It survives better when suspended in milk. The virus is stable between pH 3 and pH 10; resistant to ether, chloroform and detergents and fairly resistant to phenol. Hypochlorite inactivates quickly in the absence of organic matter.

Haemagglutination. The GD VII strains agglutinate human O cells at 0°C, being eluted at higher temperatures (Lahelle & Horsfall, 1949). After trypsin treatment, virus preparations agglutinate at room temperature also (Morris, 1952).

Antigenic properties. Positive CF tests are reported but neutralization tests have been chiefly used. As already mentioned, several authors report some antigenic cross-reactions between TO and GD VII; GD VII and FA are also antigenically related.

Cultivation. The FA virus has been cultivated on CAM, yolk sac and allantoic cavity of fertile eggs (Riordan & Sá-Fleitas, 1947). GD VII has also been grown in eggs; also in tissue cultures of mouse brain (Parker & Hollender, 1945). Both strains have been cultivated in chick- and mouse-embryo tissue cultures and in cultures of mouse-tumour cells, GD VII more readily (Shean & Schultz, 1950). GD VII will grow in hamster (BHK21) cells and, very imperfectly, in HeLa cells (Sturman & Tamm, 1969).

Pathogenicity. When first isolated GD VII caused mostly flaccid paralysis, on passage generalized encephalitis. FA has always caused widespread encephalitis, flaccid paralyses only being seen after IP inoculation. Neither virus regularly appears in faeces. Intramuscular inoculation of either strain produces local myositis and GD VII at least apparently multiplied in the muscle (Rustigian & Pappenheimer, 1949). Some strains infect cotton-rats (Committee, 1948).

Pathological lesions. The changes due to diffuse encephalitis do not recall those of poliomyelitis in man closely, as do those of the TO virus.

Ecology. Presumably exist as latent infections of numerous mouse stocks, occasionally activated by various injections (Melnick & Riordan, 1947). The viruses are not normally contagious amongst mice. Thompson et al., however (1951), report an outbreak in young mice in a colony; a virus like GD VII was isolated. A virus antigenically related to GD VII has been isolated from rats (McConnel et al., 1964).

Control. Mandel (1951) reports that a carbohydrate with inhibiting properties for GD VII (but not for FA or TO) can be extracted from intestines of adult mice and guinea-pigs.

ENTEROVIRUSES OF OTHER MAMMALS

A picornavirus has been isolated from an aborted foal (Böhm, 1964); it was said to be stable over a wide range of pH. Enteroviruses have also been recovered from sheep (McFerran et al., 1969).

For picornaviruses isolated from dogs, see p. 14; for those from cats see p. 40.

AVIAN ENTEROVIRUSES

Apart from a number of strains isolated from apparently normal animals (Burke et al., 1959; Taylor & Calnek, 1962), avian enteroviruses have been associated with avian encephalomyelitis and with hepatitis in ducks and turkeys.

Avian encephalomyelitis

Synonym: Epidemic tremor.

Reviews: Jungherr & Minard (1942). Calnek et al. (1961)—immunization.

Physico-chemical characters. The virus is stabilized against inactivation at 56° by $MgCl_2$. Its rate of inactivation at other temperatures and in the presence of various diluents is described by von Bülow (1964).

Antigenic properties. Neutralizing antibodies, most readily titrated in tissue culture, are present in sera of recovered birds (Hwang et al., 1959). CF antigen made from brains of infected chicks may be useful in diagnosis (Sato et al., 1969). Antigenically similar viruses were recovered from rectal swabs from normal chicks (Burke et al., 1959). Other enteroviruses, falling into 15 serotypes by neutralization tests in tissue culture, have been recovered from normal chicks, especially those 4 to 12 weeks old (Taylor & Calnek, 1962). Avian encephalomyelitis virus thus appears, like Teschen disease virus among porcine ones, to be a particularly neurotropic member of the avian enteroviruses.

Cultivation. *In fertile eggs:* grows after inoculation into amnion, allantoic or yolk sac; infected embryos are atrophied and show neuronal degeneration, often hydrocephalus (Casorso & Jungherr, 1959). Some have encountered erratic results, attributable to using eggs from an infected flock, antibodies having passed from an immune hen into the yolk. Some strains unadapted to vigorous growth in eggs, may be titrated after extraction of virus from the chicks which hatch out from inoculated eggs (von Bülow, 1965).

Some strains of the virus grow also in chick-embryo *tissue cultures*, better in monolayers than in whole minced embryo (Hwang et al., 1959), also in cultures of monkey kidney (Halpin, 1966).

Distribution. Formerly thought to be confined to the Western hemisphere, but now recognized in other continents including Australia.

Pathogenicity. Characteristic symptoms in chicks are ataxia, followed by tremors of head and neck, later somnolence and death. Mortality up to 50 per cent, average 10 per cent. The disease occurs only in young chicks, especially those 2 to 3 weeks old; also in pheasant and guinea-fowl chicks. Cataracts and other ocular lesions may occur as sequelae (Peckham, 1957).

It may be readily transmitted IC; other routes give erratic results. The incubation period is from 6 to 10 days. Experimental infection of young ducks, turkeys and pigeons is possible, but mammals are insusceptible. No viraemia has been detected. Virus is shed in the faeces; also in the eggs of laying birds.

Pathological lesions are those of neuronal degeneration especially of Purkinje cells, with masses of mononuclear cells around bloodvessels. There is hyperplasia also of lymphatic tissue elsewhere.

Ecology. Readily transmitted by contact. Commoner in winter and spring: arthropod transmission is unlikely. Circumstantial evidence suggests that transmission through the egg is probable.

Control. Schaaf & Lamoreux (1955) suggest the immunization of older birds with live virus into the wing webs so that their offspring are resistant. Later, Schaaf (1959) suggested a vaccine made from brain tissue, virus having been inactivated by β-propiolactone; this has given promising results (MacLeod, 1965). The possible usefulness for vaccination of several attenuated strains is discussed by von Bülow (1964). Antibody made from the yolks of eggs laid by hyperimmune ducks is of value (Rispens, 1969)

Duck Hepatitis

Review: Asplin (1961).

Haemagglutination has been looked for without success, but Taylor & Hanson (1967) have described an indirect test using tannic-acid-treated sheep RBCs.

Antigenic properties. Neutralizing antibodies are present in sera of survivors. There are 2 serotypes (Asplin, 1965). Hyperimmune serum confers passive immunity. In gel-diffusion tests antigen from liver suspensions or tissue cultures gave 2 lines of precipitate with immune duck sera.

Cultivation. The virus can be cultivated in fertile hens' eggs, killing most of the embryos in 4 days. Virus is present in allantoic fluid. It becomes attenuated for ducklings by egg-passage and can be used for immunization. Virulence returns after passage through ducks (Asplin, 1958). In cultures of chick-embryo tissues the virus multiplies, but without CPE (Pollard & Starr, 1959). Fitzgerald et al (1963) observed cell necrosis in monolayers of duck-embryo kidney cultures.

Distribution. Recorded from many European countries including the U.S.S.R.; also from Egypt, Southeast Asia, the United States and Canada.

Pathogenicity. The virus attacks ducklings 3 days to 3 weeks old, mortality being very high. Ducklings may die within an hour of the onset of symptoms. Older ducks are immune, but may harbour virus. Other birds and mammals are resistant.

Pathological lesions are mainly those of haemorrhage, necrosis in the liver, and oedema. Histologically hepatic cell necrosis and proliferation of bile ducts with cellular infiltration are seen: perivascular cuffing may be seen in the CNS (Hanson, 1958).

Ecology. Virus is excreted in the faeces of recovered or latently infected birds and infection may well be by this route, but spread of infection is erratic.

Control. Ducklings hatched from eggs laid by immunized birds resist infection (Asplin, 1956), or they may be immunized with virus attenuated in chick embryos, a virus-infected needle being thrust through the foot-web. Hyperimmune sera may be used for passive protection.

Other Avian Hepatitis Viruses

Tzianabos (1966) has described a virus causing hepatitis in turkeys. It was serologically related to duck hepatitis virus, but was infectious only for turkeys, though it multiplied in 6–7 day-old fowl embryo. A virus killing goslings with hepatitis (Wachnik & Nowacki, 1962) may or may not be related.

RHINOVIRUS

This genus includes viruses inhabiting primarily the respiratory tract of mammalian hosts. They are acid-labile, and some have low-ceiling temperature of growth. The name was originally used for human strains associated with common colds, but has been extended to viruses from other hosts, including foot-and-mouth-disease virus.

HUMAN RHINOVIRUSES

Synonyms: Common cold viruses; Coryza viruses.

Reviews: Andrewes (1962; 1966). Andrewes & Tyrrell (1965). Gwaltney & Jordan (1964). Tyrrell & Chanock (1963) (description of the group).

Morphology. Virions are 20 to 30 nm in diameter (Chapple & Harris, 1966; McGregor et al., 1966), unenveloped, with icosahedral symmetry. A triangulation number identical to that of enteroviruses (T = 3) has been suggested (McGregor & Mayor, 1968). Crystalline virus aggregates are seen in the cytoplasm of infected cells, especially when grown in excess of magnesium ions (Blough et al., 1969).

Chemical composition. Demonstration of RNAse-sensitive infectious nucleic acid indicates that the viral genome consists of single-stranded RNA (Dimmock, 1966). This is confirmed by sedimentation coefficients at different ionic strengths and by noncomplementary base composition (Brown et al., 1970). Virions contain 29·8 per cent RNA with high (35 per cent) content of adenilic acid and MW of $2{\cdot}1 \times 10^6$ daltons (McGregor & Mayor, 1971). Brown et al. (1970), on the other hand, estimated a molecular weight of $2{\cdot}4 \times 10^6$ to $2{\cdot}8 \times 10^6$, i.e. of the same order as that of the RNAs of enteroviruses. Five structural polypeptides identified by polyacrylamide gel electrophoresis (Medappa et al., 1971).

Physico-chemical characters. Infective virus has a buoyant density in CsCl of 1·4 (Chapple & Harris, 1966; McGregor et al., 1966) contrasting with other picornaviruses for which the corresponding value is nearer 1·3. Virions were found by McGregor & Mayor (1971) to have partial specific volume of 0·682, sedimentation coefficient of 158S, diffusion coefficient of $1{\cdot}7 \times 10^7$ cm²/sec and MW of $7{\cdot}1 \times 10^6$ daltons. Survive well at −76°; more readily preserved by freeze-drying than many other picornaviruses. Ether-stable. Differ from enteroviruses in greater acid-lability, being quickly inactivated at pH 5·3; also in not being effectively stabilized against inactivation at 50° by 1M $MgCl_2$ (Dimmock & Tyrrell, 1964; Ketler et al., 1962). 2(α-hydroxybenzyl) benzimidazole, which inactivates coxsackie B and most echoviruses, fails to do so with rhinoviruses or coxsackie As (Eggers & Tamm, 1961a). 'Common Cold Virus' dried in air does not survive well (Lovelock et al., 1952). Rhinoviruses can be preserved by freeze-drying from a medium containing 10 per cent dextran and 10 per cent glucose (Tyrrell & Ridgewell, 1965).

Haemagglutination. None reported. Drescher & Tyrrell (1969) observed that the spontaneous aggregation of trypsin-treated human RBCs in certain solutions was inhibited by rhinoviruses: this inhibition was neutralized by specific antisera.

Antigenic properties. Neutralization tests can be carried out in tissue culture by plaque inhibition and other methods. There are numerous serotypes—at least 4 or 5 of the strains cultivable in rhesus kidney cultures (M strains), and altogether probably over 100. The first 55 to be fully compared have been given numbers by Kapikian et al. (1967). A further 34 serotypes have since been officially recognized (Kapikian et al., 1971). One M strain, B632, shows antigenic overlap with a virus which has been classified as echo 28 (p. 18), but has greater claims to be considered an atypical rhinovirus. Fenters et al. (1966) found low levels of cross-reactions between some serotypes in sera of immunized calves: Stott et al. (1969), however, reported that serological responses in man to several rhinoviruses were highly specific. Chapple et al. (1967), studying complement-fixing antibodies, found no rises in titre following infection or vaccination. There is correlation between level of neutralizing antibody to a serotype of rhinovirus and resistance to that serotype (Bynoe et al., 1961). Several studies suggest, however, that the level of IgA in nasal secretions may be more important than antibody titres in serum (Douglas et al., 1967; Perkins et al., 1969).

Cultivation. Several claims to have cultivated Common Cold viruses in *fertile eggs* remain unconfirmed.

Rhinoviruses can be grown with varying success in *cultures* of *primate tissues*. It is preferable for all of them, and essential for most, at least on first isolation, to cultivate at lower temperatures (about 33°) and lower pH than is conventional for virus culture and to maintain good oxygenation by rotation (Tyrrell et al., 1960). The M strains grow and produce CPE in rhesus kidney cultures as well as those of human embryonic kidney and the cancer lines HeLa and HEp-2, and other human cells (Taylor-Robinson et al., 1962). The majority of strains (H strains) are best isolated in human embryonic kidney, but many of these grow also in lines of diploid cells from human embryonic lung and some in KB and HeLa cells (from human cancer) (Hayflick & Moorhead, 1961; Taylor-Robinson et al., 1962). Conant & Hamparian (1968) could adapt all the 68 strains they tested to growth, with CPE, in a line of HeLa cells, while Douglas et al. (1966) adapted even H strains to monkey-kidney cells. The DC strain now known as type 9, the one first cultivated (Andrewes et al., 1953), in human embryonic lung, could subsequently only be isolated in diploid lines, though still later adapted to grow in other cultures (Tyrrell et al., 1962). Some rhinoviruses which cannot be grown by the methods just described can be isolated with the help of an improved method of organ culture (Hoorn & Tyrrell, 1965). Some strains can then be adapted to grow by more conventional methods; others have been grown only in organ cultures of nasal or tracheal epithelium of which the ciliated epithelium remains intact and active. In one series 75 per cent of colds yielded a virus (Tyrrell & Bynoe, 1966). A simplified organ culture technique in test-tubes was described by Harnett & Hooper (1968). Cytopathic effects resemble those caused by enteroviruses. 'Microplaques' produced on cell sheets may be counted for purposes of assay (Parsons & Tyrrell, 1961), and macroscopic plaques may also be obtained (Porterfield, 1962). Plaque-formation is better in the presence of increased magnesium (Fiala & Kenny, 1967). Stott & Tyrrell (1968) have described a colour or micrometabolic-inhibition test for rapid titration of viruses and antisera.

Distribution. World-wide except in small isolated communities.

Pathogenicity. Rhinoviruses produce typical common colds, especially in adults, with profuse nasal discharge often preceded by sore throat and followed by cough; they are usually afebrile. The incubation period is usually about 48 hours, but some strains take 3 or 4 days to produce symptoms. Secondary bacterial invaders may cause sinusitis and other complications. Rhinoviruses are frequently isolated during exacerbations of illness from patients with chronic bronchitis.

Experimentally washings or cultures produce the typical disease when given intranasally to volunteers. The colds occurring naturally or experimentally induced in chimpanzees (Dochez et al., 1930) are probably, but not certainly, caused by rhinoviruses. Chimpanzees naturally or deliberately infected with rhinoviruses have had only subclinical infections (Dick, 1968). Gibbons could also be infected (Pinto & Haff, 1969), but not patas monkeys.

Ecology. Virus is present in the nose and throat, not, or rarely, in faeces; and infection is via the respiratory tract. Cate et al. (1967) confirmed that rhinoviruses rarely survived passage through the gut, but it appeared that something other than gastric acidity inactivated them. Colds are commoner in winter, but reasons for seasonal variation are obscure (Andrewes, 1964). An alleged effect of chilling in inducing colds has not been experimentally demonstrable (Andrewes 1950; Dowling et al., 1958). Repeated colds in individuals can be at least partly explained by the existence of numerous serotypes. There appears to be a transient nonspecific immunity after a cold (Fleet et al., 1965). Epidemiology in relation to prevalence of various serotypes is discussed by Hamparian et al. (1964), Dick et al. (1967) and Gwaltney et al. (1968).

Control. No effective virus vaccines are available, though experiments suggest that specific immunity to a particular strain may be produced (Price, 1957; Doggett et al., 1963; Committee, 1965). Bacterial vaccines, directed against secondary invaders, are widely used, but are of doubtful value. Attempts to control by air-hygiene are unpromising. Claims for the value of ascorbic acid were not confirmed in a careful trial.

EQUINE RHINOVIRUSES

Review: Burrows (1970).

Physico-chemical characters. Buoyant density in CsCl 1·45 g/ml —greater than that of human rhinoviruses. Base composition also different (Rowlands et al., 1971).

Antigenic properties. Two serotypes occur (Ditchfield & Macpherson, 1965); type 1 appears to be the commoner and 60 per cent or more of adult horses have neutralizing antibodies against it.

Cultivation. Growth with CPE occurs in tissues of horse, monkey, man, rabbit and other species. The special requirements of low bicarbonate and low temperature necessary for many human rhinoviruses are not needed for equine strains.

Distribution. Recorded from Britain, Germany, Canada and the United States, probably world-wide.

Pathogenicity. The natural disease resembles the common cold of man with rhinitis and pharyngitis, but there is more tendency to viraemia accompanied by fever (Plummer, 1962; Ditchfield & Macpherson, 1965; Wilson et al., 1965). The incubation period after experimental infection was 3–7 days. Rabbits, guinea-pigs, monkeys and man can be infected experimentally; a human subject developed fever and viraemia (Plummer, 1963).

Ecology. Spreads readily in stables and probably not only by direct contact but also through contamination of 'stable furnishings'. In this respect and in persistence of virus in pharyngeal secretions for up to a month, its biological behaviour resembles that of foot-and-mouth disease (Plummer & Kerry, 1962).

FOOT-AND-MOUTH DISEASE

Synonym: Aphthous fever.

Review: Brooksby (1958; 1967). Bachrach (1968).

Morphology and development. Most workers agree that the diameter is between 20 and 25 nm and that it is roughly spherical or hexagonal. Electron-microscopy reveals a granular structure and a central region of lower density (Bradish et al., 1960). There are said to be capsomeres rather larger than those of other picornaviruses and consisting of short, hollow cylinders (Schulze & Gralheer, 1964). The granular appearance may be due to the 7 nm noninfectious component association with antigenic reactivity. Using a stroboscopic technique, Breese et al. (1965) obtained pictures more consistent with a rhombic triacontrahedron with 32 surface units than with a 42-unit icosahedron. Studies with fluorescent antibodies (Mussgay, 1958) indicate that viral development takes place in the cytoplasm.

Chemical composition. The virion contains 31·5 per cent RNA, the remainder being protein (Bachrach et al., 1964). RNA extracted from virus preparations with phenol is infective. MW of RNA estimated as $2{\cdot}8 \times 10^6$ (Wild & Brown, 1970). The base composition (molar fractions) of viral RNA is $G = 0{\cdot}24$; $A = 0{\cdot}26$; $C = 0{\cdot}28$; $U = 0{\cdot}22$ (Bachrach et al., 1964). Virus-specific RNAs with different properties, some thought to represent replicative forms, have been detected in infected cells (Arlinghaus et al., 1969; Wild & Brown, 1970).

Van de Woude & Bachrach (1968) presented evidence of a single structural polypeptide, but Wild et al. (1970) suggest that there may be 5 major and 1 minor polypeptides in the virion.

Physico-chemical characters. Virions have a sedimentation coefficient of 150S and a buoyant density of CsCl of 1·40 (Trautman & Breese, 1962; Wild & Brown, 1967). Virus is stable at pH 7·4–7·6, readily destroyed on the acid side of pH 6, but more stable again in the region of pH 3. Because of its lability at pH 5, Plummer (1965) groups it with the rhinoviruses as an 'acid-labile picornavirus'. Different strains or even clones of given strains may vary in acid sensitivity (Asso et al., 1966) and in thermostability (Ahl, 1968). The virus is always inactivated at its isoelectric point, which varies with ionic strength (Van de Woude, 1967). Sensitivity to X-radiation is described by Bonet-Maury & Frilley (1946). Heat inactivation is promoted by molar $MgCl_2$ (Wittmann, 1967). It may survive for several weeks on hay and other materials (Bedson et al., 1927). Disinfection is commonly carried out with 2 per cent commercial NaOH or KOH or 4 per cent Na_2CO_3, but the virus is even more sensitive to acids. In carcasses and meat, rigor mortis is accompanied by acid formation which destroys the virus in muscles. In glandular organs and bone marrow, on the other hand, less acid is formed and virus may persist for long periods. For preparing vaccines 0·05 per cent formaldehyde acting at room temperature for 4 days is effective (Sellers et al., 1959). The use of hydroxylamine and several other chemical inactivants is described by Fellowes (1966; 1967). Details of inactivation by formaldehyde, pH and temperature are given by Bachrach et al. (1957), while Brown et al. (1963) describe the effects of these and other agents on the various properties of the virus.

Haemagglutination. Haemagglutinins have not been satisfactorily demonstrated.

Antigenic properties. Neutralization tests can be carried out in guinea pigs or, more accurately, in tissue culture. Detailed techniques are described by Brooksby (1949), and he also (1952) describes the complement-fixation techniques applicable to this infection. There are 7 serological types known as A and O (Vallee), C (Waldmann), S.A.T. 1, S.A.T. 2, S.A.T. 3 (South Africa) and Asia 1. They can be distinguished by various serological tests, though with CF there are various degrees of overlapping. Within the main types there are variants and subtypes, differing sufficiently to affect the results of vaccination and to confuse diagnosis. Including subtypes, 40 distinct antigenic strains can now be recognized (Brooksby, 1967). Davie (1962, 1964) describes classification of subtypes. The Institute at Pirbright, Surrey, is recognized by F.A.O. as a World Reference Laboratory for typing strains. Freshly isolated virus is apt to be antigenically unstable, but reports of transformation of one major type into another are now discredited. Several antigens corresponding to virions, viral subunits, internal and possibly nonstructural components have been described (Brown & Crick, 1958; Cartwright, 1962; Cowan & Graves, 1966; Rowlands et al., 1969). Evidence of 3 distinct antigenic specificities on the surface of the virion has been presented by Brown & Smale (1970).

Cultivation. *In fertile eggs.* Virus was adapted to growth in eggs by Skinner (1954) who inoculated the embryos IV and incubated them at 35°C.

In tissue culture it grows in embryonic guinea-pig tissues (Maitland & Maitland, 1931) and in bovine, porcine and ovine kidney. Mouse- and egg-adapted strains do not do as well as others (Sellers et al., 1959). Cytopathic effects are much more marked in porcine than in bovine tissues, and it is possible to titrate virus in them by enumerating plaques (Sellers et al., 1959) or by a colour test depending upon changes in pH (Martin & Chapman, 1961). The course of development of virus in cultures of cattle-tongue epithelium was described by Cartwright et al. (1957). Changes produced in such cultures are described by Pay (1957): cells growing in membrane-like sheets rapidly rounded up and nuclei became pycnotic; in cells with a polygonal type of growth, however, though virus multiplied, CPE was minimal. A cell-line derived from baby hamster kidney has been used with success (Mowat & Chapman, 1962). Cultures of calf thyroid are particularly sensitive and yield virus in exceptionally high titre (Snowdon, 1966). Crick et al. (1966) have described a plaque technique for titrating infectious RNA.

Distribution. It has been possible to keep Australia, New Zealand, the United States and Canada free from the disease. Infection is endemic in continental Europe, Asia, Africa and South America. The O, A and C serotypes are widespread; Asia 1 has been found only in Asiatic countries. The S.A.T. 1, 2 and 3 strains were confined to Africa until, at the end of 1961, S.A.T. 1 spread to the Middle East, reaching European Turkey in 1962.

Pathogenicity. The disease in naturally affected cattle is usually not fatal, but leads to serious loss of condition and consequent economic loss. There occur fever and vesicular eruptions on mouth, tongue, muzzle, hooves, udder, sometimes elsewhere; the lesions lead to salivation and lameness. Secondary bacterial infections of denuded areas commonly occur, especially on the feet. Myocardial damage is severe in some outbreaks and may be fatal to calves. Lesions in pigs are similar; lameness is a prominent symptom. Sheep and goats are less severely affected. Deer, antelopes, other ruminants and elephants may also be infected. The incubation period after contact is 2 to 4 days or less. Some strains in nature infect pigs better than cattle (Brooksby, 1950). Hedgehogs may contract infection naturally and are infectious for other species (McLauchlan & Henderson, 1947).

Experimentally infected cattle show lesions in 10–12 hours at the site of inoculation, which is usually the tongue. Generalization to the feet and elsewhere follows in 2 to 4 days.

Guinea-pigs are susceptible to inoculation into the foot-pad and were for long the animal chiefly used experimentally; adaptation of strains is necessary before they grow well. Techniques for quantitative studies in cattle and guinea-pigs are described in a monograph by Henderson (1949). Suckling mice were shown (Skinner, 1951) to be highly susceptible; they develop spastic paraplegia after IP inoculation, and virus passed in them will later infect older mice also. Some neurotropic strains have been developed by IC passage in weaned mice (Hofmann, 1941). Strains attenuated for cattle by passage through mice or chick embryos may still be pathogenic for young guinea-pigs when inoculated IM (Skinner et al., 1962). Hamsters are very susceptible up to 60 days old (Eunes, 1955). Newborn rabbits may also be infected. Birds are fairly resistant, but Skinner (1959) infected fowls and other poultry: local vesicular lesions and occasional generalization occurred. A very few cases of a mild infection in human beings are reported; there were local vesicles, usually on the hands.

Pathological lesions. The vesicular lesions are associated with swelling, ballooning degeneration and necrosis of cells in the lower layer of epithelium. The lesions in guinea-pigs have been described by Platt (1958); pancreatitis was common. Platt (1956) also reported on those in 7-day mice; these showed widespread necrosis of skeletal muscles with some myocardial necrosis. In 3–4-week-old mice, there was more myocardial and often pancreatic necrosis. In older mice there was frequently damage to suprascapular fat-pads.

Ecology. The disease is extremely contagious. Infective saliva and excreta contaminate the environment, so that animals are very readily infected either directly or indirectly. Several recent writers have brought forward evidence of the effect of weather on spread; this is apt to be down-wind and precipitation of suspended particles by rain may be important. Sellers & Parker (1969) have studied the shedding of infective aerosols by various species and persistence of virus in the air. In Britain where the disease is not endemic, virus has repeatedly been introduced from South America, sometimes through uncooked garbage fed to pigs. There is circumstantial evidence that it may also reach Britain through the agency of migrating birds, particularly starlings (Wilson & Matheson, 1952). Virus has been isolated from the urine of recovered cattle during several months (Waldmann et al., 1931). Burrows (1966) has described persistence of virus for months in the pharynx of recovered cattle. Sheep (Burrows, 1968) and brown rats (Capel-Edwards, 1970) may be chronic carriers, and, in Botswana, virus has been recovered from clinically normal herds of cattle (Hedger, 1968); but it is not certain that carriers do transmit infection.

Control. Where the disease is not endemic a policy of quarantine, slaughter and disinfection of infected premises has proved efficient and best on the whole economically. For endemic areas vaccination is widely used; formolized vaccine adsorbed on to alum has been mainly used. Acetylethyleneimine seems to be as effective and perhaps more certain in its action (Brown et al., 1963). The inclusion of saponin increases potency (Hyslop & Morrow, 1969). Vaccine was previously made from epithelium of tongues of infected cattle, more recently from bovine epithelium infected and grown in tissue culture. Still more recently living virus has been attenuated by passage either in adult mice in tissue culture or in eggs (Skinner, 1960), and these attenuated strains have given encouraging results when tested in the field in South and East Africa (Mowat & Prydie, 1962; Martin & Edwards, 1965; Mowat et al., 1969).

OTHER BOVINE RHINOVIRUSES

Review: Bögel (1968).

Antigenic properties. The few strains yet isolated are serologically similar.

Cultivation. Resembles human rhinoviruses in growing best at 33° and in a slightly acid medium. Growth, with CPE, has been seen only in calf-kidney cells (Bögel & Böhm, 1962). One strain would, on first isolation, grow only in organ cultures (Reed et al., 1971).

Pathogenicity even for colostrum-free calves, is minimal; at most some increased nasal discharge is seen. Virus has been recovered only from nasal secretions. No evidence of pathogenicity for other species including man and new-born mice. Association with acute respiratory disease in cattle has been described by Rosenquist (1971).

CALICIVIRUS

This genus is distinguished from others of the picornaviridae family mainly by morphological characters: virions are slightly larger (35 to 40 nm in diameter), and the surface of capsids seen by negative contrast has characteristic dark areas interpreted as hollows penetrated by the negative stain giving an appearance of cups from which the generic name is derived. Stability at pH 5 but not at pH 3 and lack of stabilization by 1 M $MgCl_2$ are other properties of the genus (Wawrzkiewicz et al., 1968). The genus includes vesicular exanthema virus of pigs and a number of feline viruses formerly referred to as rhinoviruses.

VESICULAR EXANTHEMA
(of pigs)

Reviews: Madin & Traum (1955). Hagan (1958). Bankowski (1965).

Morphology and development. Virions are 35 to 40 nm in diameter (Oglesby, 1965; Wawrzkiewicz et al., 1968) with typical surface structure sometimes giving rise to the description of surface spikes (Zee et al., 1968a) or hollows (Wawrzkiewicz et al., 1968). In thin sections of infected cells, virions are seen in cytoplasmic cisternae and in crystalline arrays with a centre-to-centre spacing of 38 nm (Zee et al., 1968a, b). There is no evidence of nuclear participation in viral synthesis.

Chemical composition. Infectious single-stranded RNA obtained from virions has a sedimentation coefficient of about 37S and contains 29·0 per cent adenilic acid, 25·1 per cent cytidilic acid, 20·6 per cent guanilic acid and 25·3 per cent uridylic acid (Wawrzkiewicz et al., 1968). Virions contain 90 per cent protein and 20 per cent single-stranded RNA with MW 2×10^6 daltons (Oglesby et al., 1971).

Physico-chemical characters. Virions have a sedimentation coefficient of 160 to 170S and a buoyant density in CsCl of 1·36 to 1·377 g/ml (Oglesby, 1965; Wawrzkiewicz et al., 1968). Thermal inactivation not stabilized by 1 M $MgCl_2$ (Oglesby, 1965; Zee & Hackett, 1967). Resistant to ether, chloroform, phospholipase, 1 per cent deoxycholate and 0·3 per cent tween 80 (Wawrzkiewicz, et al., 1968). Two per cent NaOH is recommended for killing (Mott et al., 1953).

Haemagglutination. Tests against cells of many species have all been negative.

Antigenic properties. Up to 13 serological types have been separated by neutralization, complement-fixation or agar-gel diffusion tests; they have been identified by lettering: A, B, C, etc. Until recently only the B type had been identified outside California, but 2 new ones were reported from New Jersey in 1956.

Cultivation. The viruses grow with varying ease in kidney, skin, or embryonic tissue cultures of swine, horse, dog and cat producing CPE (Madin, 1958). On monolayers of pig-kidney plaques of 2 sizes develop, virus from larger plaques being more virulent for swine (McClain et al., 1958). Cellular changes are described by Zee et al. (1967).

Distribution. The virus was first recognized in California in 1932 and did not spread thence until 1952 when it suddenly invaded most of the United States. By slaughter and other means it is thought to have been eliminated by 1958.

Pathogenicity. Only pigs are naturally infected. It is important mainly because of the close similarity to foot-and-mouth disease. It is much milder, but may be fatal in suckling pigs. Vesicles appear on snouts, tongue, feet and teats. There are fever, loss of weight and difficulty in walking.

Experimentally the incubation period after intradermal inoculation usually about 48 hours. Some strains will produce local lesions when inoculated on to the tongues of horses and dogs and, very exceptionally, guinea-pigs. Madin & Traum (1953) infected hamsters. Virus is present in all tissues of infected pigs with traces in urine and faeces.

Ecology. The virus passes by contact amongst pigs, but much less readily than does foot-and-mouth disease. Infection is probably spread largely through uncooked garbage. Bankowski (1965) has described the sudden appearance, relatively brief prevalence and swift eradication of this disease.

Control is by quarantine and slaughter and by enforcement of laws on cooking garbage. A formolized vaccine has been found effective experimentally (Madin & Traum, 1953).

FELINE CALICIVIRUSES

Morphology. Virions are 30 to 40 nm in diameter showing surface structure consistent with 32 morphological units penetrated by stain

(Zwillenberg & Bürki, 1966; Almeida et al., 1968; Studdert et al., 1970).

Chemical composition. Infectious single-stranded RNA has been extracted from virions (Adldinger et al., 1969).

Physico-chemical characters. Virions have a buoyant density of 1·37 to 1·39 g/ml in CsCl (Studdert et al., 1970). Infectivity resistant to lipid solvents. Labile at pH 3; not stabilized by 1 M $MgCl_2$ (Crandell, 1967; Studdert et al., 1970).

Antigenic properties. Several different serotypes have been distinguished (Bürki, 1965; Studdert et al., 1970). It is possible that immunity is short-lived (Prydie, 1966).

Cultivation. Grow in cultures of feline cells and more readily in stationary cultures than do rhinoviruses (Crandell, 1967).

Pathogenicity. Most strains have been isolated from respiratory infections, but there may be some with more affinity for the digestive tract (Bürki, 1966). When given to cats by aerosol, one strain caused rhinitis, conjunctivitis, oral ulcerations and bronchopneumonia (Holzinger & Kahn, 1970).

Ecology. Infections most often seen in catteries.

PROBABLE PICORNAVIRUS

NODAMURA VIRUS

This virus, isolated from *Culex tritaeniorhynchus* in Japan (Scherer & Hurlbut, 1967; Scherer et al., 1968), has the properties of a picornavirus. Diameter 28 nm (Murphy et al., 1970). Inactivation by sodium desoxycholate variable. The virus will multiply in mosquitoes and other insects and causes fatal paralysis in suckling mice. It has not been grown in tissue cultures or eggs. Serological studies suggest that pigs are the natural hosts.

REFERENCES

Ada, G. L., & Anderson, S. G. (1959) *Aust. J. Sci.*, **21**, 259.
Adldinger, H. K., Lee, K. M., & Gillespie, J. H. (1969) *Arch. ges. Virusforsch.*, **28**, 245.
Agrawal, H. O. (1966) *Arch. ges. Virusforsch.*, **19**, 365.
Ahl, R. (1968) *Arch. ges. Virusforsch.*, **24**, 361.
Alexander, T. J. L., & Betts, A. O. (1967) *Res. vet. Sci.*, **8**, 330.
Almeida, J. D., Waterson, A. P., Prydie, J., & Fletcher, E. W. (1968) *Arch. ges. Virusforsch.* **25**, 105.

Andrewes, C. H. (1950) *New Engl. J. Med.*, **242**, 235.
Andrewes, C. H. (1962) *Jl. R. Inst. publ. Hlth*, **25**, 31, 55, and 79.
Andrewes, C. H. (1964) *Science*, **146**, 1274.
Andrewes, C. H. (1966) *Ann. Rev Med.*, **17**, 361.
Andrewes, C. H., Chaproniere, D. M., Gompels, A. E. H., Pereira, H. G., & Roden, A. T. (1953) *Lancet*, **2**, 546.
Andrewes, C. H., & Tyrrell, D. A. J. (1965) *Viral and rickettsial diseases of man.* Eds.: Horsfall & Tamm. 4th Ed. p. 546. Philadelphia: Lippincott.
Arlinghaus, R. B., Kaczmarczyk, W., & Polatnick, J. (1969) *J. Virol.* **4**, 712.
Armstrong, C. (1939) *Publ. Hlth. Rep. Wash.*, **54**, 1719 and 2302.
Asso, J., Aynaud, J.-M., & Portalier, R. (1966) *C.r. hebd. Séanc. Acad. Sci., Paris*, **262D**, 585.
Asplin, F. D. (1956) *Vet. Rec.*, **68**, 412.
Asplin, F. D. (1958) *Vet. Rec.*, **70**, 1226.
Asplin, F. D. (1961) *Bull. Off. int. Épizoot.*, **56**, 793.
Asplin, F. D. (1965) *Vet. Rec.*, **77**, 1529.
Bachrach, H. L. (1968) *Ann. Rev. Microbiol.*, **22**, 201.
Bachrach, H. L., Breese, S. S., Callis, J. J., Hess, W. R., & Patty, R. E. (1957) *Proc. Soc. exp. Biol. Med., U.S.A.*, **95**, 147.
Bachrach, H. L., Trautman, R., & Breese, S. S. (1964) *Am. J. vet. Res.*, **25**, 333.
Baltimore, D., Becker, Y., & Darnell, J. E. (1964) *Science*, **143**, 1034.
Baltimore, D., & Franklin, R. M. (1963) *Cold Spring Harb. Symp. quant. Biol.*, **28**, 105.
Bankowski, R. A. (1965) *Adv. vet. Sci.*, **10**, 23.
Barron, A. L., & Karzon, D. T. (1957) *Proc. Soc. exp. Biol. Med., U.S.A.*, **94**, 393.
Barron, A. L., & Karzon, D. T. (1959) *Proc. Soc. exp. Biol. Med., U.S.A.*, **100**, 316.
Barski, G., & Lamy, M. (1955) *Annls Inst. Pasteur, Paris*, **89**, 318.
Barya, M. A., Moll, T., & Mattson, D. E. (1967) *Am. J. vet. Res.*, **28**, 1283.
Beale, A. J., Stevens, P. F., Davis, N., Stackiw, W., & Rhodes, A. J. (1956) *Can. J. Microbiol.*, **2**, 298.
Bedson, S. P., Maitland, H. M., & Burbury, Y. M. (1927) *J. comp. Path.*, **40**, 15.
Beeman, E. A., & Huebner, R. J. (1952) *J. Immunol.*, **68**, 663.
Bell, J. F., & Hadlow, W. J. (1959) *J. infect. Dis.*, **105**, 54.
Bellett, A. J. D. (1960) *Virology*, **10**, 285
Bengtsson, S. (1968) *Acta path. microbiol. scand.*, **73**, 592.
Beran, G. W., Wenner, H. A., Werder, A. A., & Underdahl, N. R. (1960) *Am. J. vet. Res.*, **21**, 723.
Beran, G. W., Werder, A. A., & Wenner, H. A. (1958) *Am. J. vet. Res.*, **19**, 545.
Berkovich, S., & Pangan, J. (1968) *Bull. N.Y. Acad. Med.*, **44**, 377.
Bough, H. A., Tiffany, J. M., Gordon, G., & Fiala, M. (1969) *Virology*, **38**, 694.
Bodian, D., & Horstmann, D. M. (1965) *Viral and rickettsial infections of man*, Eds.: Horsfall & Tamm. 4th Ed. p. 430. Philadelphia: Lippincott.
Bögel, K. (1968) *J. Am. vet. Med. Ass.*, **152**, 780.
Bögel, K., & Böhm, H. (1962) *Zentbl. Bakt. ParasitKde, I. Abt. Orig.*, **187**, 2.
Bögel, K., & Mussgay, M. (1960) *Zentbl. VetMed.*, **7**, 534.
Böhm, H. O. (1964) *Zentbl. VetMed.*, **11B**, 240.
Bonet-Maury, P., & Frilley, M. (1946) *Annls Inst. Pasteur, Paris*, **72**, 432.
Bradish, C. J., Henderson, W. M., Kirkham, J. B. (1960) *J. gen. Microbiol.*, **22**, 379.
Breese, S. S., Trautman, R., & Bachrach, H. L. (1965) *Science*, **150**, 1303.
Brooksby, J. B. (1949) *Spec. Rep. Ser. agric. Res. Coun. Lond.*, no. 9 (H.M. Stationery Office, London).
Brooksby, J. B. (1950) *J. Hyg. Camb.*, **48**, 184.
Brooksby, J. B. (1952) *Spec. Rep. Ser. agric. Res. Coun., Lond.*, no. 12 (H.M. Stationery Office, London).
Brooksby, J. B. (1958) *Adv. Virus Res.* **5**, 1.
Brooksby, J. B. (1967) *Nature, Lond.*, **213**, 120.
Brown, F., & Crick, J. (1958) *Virology*, **5**, 133.

Brown, F., Hyslop, N. St. G., Crick, J., & Morrow, A. W. (1963) *J. Hyg., Camb.*, **61**, 337.
Brown, F., Newman, J. F. E., & Stott, E. J. (1970) *J. gen. Virol.*, **8**, 145.
Brown, F., & Smale, C. J. (1970) *J. gen. Virol.*, **7**, 115.
Brown, G. C., & Evans, T. N. (1967) *J. Am. med. Ass.*, **199**, 183.
Buckland, F. E., Bynoe, M. L., Philipson, L., & Tyrrell, D. A. J. (1959) *J. Hyg., Camb.*, **57**, 274.
Buckland, F. E., Bynoe, M. L., Rosen, L., & Tyrrell, D. A. J. (1961) *Br. med. J.*, **1**, 397.
Buckland, F. E., Bynoe, M. L., & Tyrrell, D. A. J. (1965) *J. Hyg., Camb.*, **33**, 327.
Bülow, V. von (1964) *Zentbl. VetMed.*, **11B**, 674.
Bülow, V. Von. (1965) *Zentbl. VetMed.*, **12B**, 298.
Burke, C. N., Luginbuhl, R. E., & Jungherr, E. L. (1959) *Avian Dis.*, **3**, 412.
Bürki, F. (1965) *Arch. ges. Virusforsch.*, **15**, 690.
Bürki, F. (1966) *Zentbl. Bakt. ParasitKde, I. Abt. Orig.*, **200**, 281.
Burness, A. T. H. (1970) *J. gen. Virol.*, **6**, 373.
Burness, A. T. H., & Clothier, F. W. (1970) *J. gen. Virol.*, **6**, 381.
Burness, A. T. H., & Walter, D. S. (1967) *Nature, Lond.*, **215**, 1350.
Burrows, R. (1966) *J. Hyg., Camb.*, **64**, 81.
Burrows, R. (1968) *J. Hyg., Camb.*, **66**, 633.
Burrows, R. (1970) *Proceedings of the Second International Conference on Infectious Diseases, Paris.* pp. 154–64. Basel: Karger.
Bynoe, M. L., Hobson, D., Horner, J., Kipps, A., Schild, G. G., & Tyrrell, D. A. J. (1961), *Lancet*, **1**, 1194.
Cabasso, V. J., Stebbins, M. R., Dutcher, R. M., Moyer, A. W., & Cox, H. R. (1952) *Proc. Soc. exp. Biol. Med., U.S.A.*, **81**, 525.
Caligiuri, L. A., & Tamm, I. (1970) *Virology*, **42**, 100.
Calnek, B. W., Luginbuhl, R. E., McKercher, P. D., & van Roekel, H. (1961) *Avian Dis.*, **5**, 456.
Capel-Edwards, M. (1970) *J. comp. Path.*, **80**, 543.
Cartwright, B. (1962) *J. Immunol.*, **88**, 128.
Cartwright, S. F., Pay, T. W. F., & Henderson, W. M. (1957) *J. gen. Microbiol.*, **16**, 730.
Casals, J. (1963) *Nature, Lond.*, **200**, 329.
Casals, J., Olitsky, P. K., & Anslow, R. O. (1951) *J. exp. Med.*, **94**, 123.
Casorso, D. R., & Jungherr, E. L. (1959) *Am. J. vet. Res.*, **20**, 547.
Cate, T. R., Douglas, R. G., Johnson, K. M., Couch, R. B., & Knight, V. (1967) *Proc. Soc. exp. Biol. Med., U.S.A.*, **124**, 1290.
Chapple, P. J., & Harris, W. J. (1966) *Nature, Lond.*, **209**, 790.
Chapple, P. J., Head, B., & Tyrrell, D. A. J. (1967) *Arch. ges. Virusforsch.*, **21**, 123.
Chin, T. D. Y., Beran, G. W., & Wenner, H. A. (1957) *Am. J. Hyg.*, **66**, 76.
Christov, S. (1966) *Zentbl. VetMed.*, **13B**, 31.
Cockburn, W. C., & Drozdov, S. G. (1970) *Bull. Wld Hlth Org.*, **42**, 405.
Colter, J. S., Bird, H. H., & Brown, R. A. (1957b) *Nature, Lond.*, **179**, 859.
Colter, J. S., Bird, H. B., Moyer, A. W., & Brown, R. A. (1957a) *Virology*, **4**, 522.
Colter, J. S., Davies, M. A., & Campbell, J. B. (1964) *Virology*, **34**, 474.
Committee on Common Cold Vaccines (1965) *Br. med. J.*, **i**, 1344.
Committee on Enteroviruses (1957) *Am. J. publ. Hlth*, **47**, 1556.
Committee on Nomenclature of National Foundation for Infantile Paralysis (1948) *Science*, **108**, 701.
Conent, R. M., & Hamparian, V. V. (1968) *J. Immunol.*, **100**, 107.
Cooper, P. D. (1968) *Virology*, **35**, 584.
Couch, R. B., Cate, T. R., Gerone, P. J., Fleet, W. F., Lang, D. J., Griffith, W. R., & Knight, V. (1965) *J. clin. Invest.*, **44**, 535.
Cowan, K. M., & Graves, J. H. (1966) *Virology*, **30**, 528.
Craighead, J. E. (1965a) *Nature, Lond.*, **207**, 1268.
Craighead, J. E. (1965b) *Proc. Soc. exp. Biol., Med., U.S.A.*, **119**, 408.
Craighead, J. E., & Shelokov, A. (1961) *Proc. Soc. exp. Biol. Med., U.S.A.*, **108**, 823.

Craighead, J. E., & McLane, M. F. (1968) *Science*, **162**, 913.
Cramblett, H. G., Rosen, L., Parrott, R. H., Bell, J. A., Huebner, R. J., & McCullough, N. B. (1958) *Pediatrics*, **21**, 168.
Crandell, R. A. (1967) *Proc. Soc. exp. Biol. Med., U.S.A.*, **126**, 240.
Crick, J., Lebedev, A. I., Stewart, D. L., & Brown, F. (1966) *J. gen. Microbiol.*, **43**, 59.
Crocker, T. T., Pfendt, E., & Spendlove, R. (1964) *Science*, **145**, 401.
Dales, S., Eggers, H. J., Tamm, I., & Palade, G. E. (1965) *Virology*, **26**, 379.
Dales, S., & Franklin, R. M. (1962) *J. Cell Biol.*, **14**, 281.
Dalldorf, G. (1955) *Ann. Rev. Microbiol.*, **9**, 277.
Dalldorf, G. (1957) *J. exp. Med.*, **106**, 69.
Dalldorf, G., & Melnick, J. L. (1965) *Viral and rickettsial diseases of man.* Eds.: Horsfall & Tamm. 4th Ed. p. 474. Philadelphia: Lippincott.
Darbyshire, J. H., & Dawson, P. S. (1963) *Res. vet. Sci.*, **4**, 48.
Davie, J. (1962) *Bull. Off. int. Épizoot.*, **57**, 962.
Davie, J. (1964) *J. Hyg., Camb.*, **62**, 401.
Dean, D. J. (1951) *J. Immunol.*, **66**, 347.
Dean, D. J., & Dalldorf, G. (1948) *J. exp. Med.*, **88**, 645.
Dick, E. C., Blumer, C. R., & Evans, A. S. (1967) *Am. J. Epidemiol.*, **86**, 386.
Dick, E. C., & Dick, C. R. (1968) *Am. J. Epidemiol.*, **88**, 267.
Dick, G. W. A. (1948) *Br. J. exp. Path.*, **29**, 559.
Dick, G. W. A. (1949) *J. Immunol.*, **62**, 375.
Dick, G. W. A. (1950) *Proc. Soc. exp. Biol. Med., U.S.A.*, **73**, 77.
Dick, G. W. A., Smithburn, K. C., & Haddow, A. J. (1948) *Br. J. exp. Path.*, **29**, 547.
Dickinson, L., & Griffiths, A. J. (1966) *Br. J. exp. Path.*, **47**, 35.
Dimmock, N. J. (1966) *Nature, Lond.*, **209**, 792.
Dimmock, N. J., & Tyrrell, D. A. J. (1964) *Br. J. exp. Path.*, **45**, 271.
Ditchfield, J., & Macpherson, L. W. (1965) *Cornell vet.*, **55**, 181.
Dochez, A. R., Shibley, G. S., & Mills, K. C. (1930) *J. exp. Med.*, **52**, 701.
Doggett, J., Bynoe, M. L., & Tyrrell, D. A. J. (1963) *Br. med. J.*, **1**, 34.
Done, J. T. (1961) *Bull. Off. int. Épizoot.*, **56**, 117.
Douglas, R. G., Jr., Cate, T. R., & Couch, R. B. (1966) *Proc. Soc. exp. Biol. Med., U.S.A.* **123**, 238.
Douglas, R. G., Rossen, R. D., Butler, W. T., & Couch, R. B. (1967) *J. Immunol.*, **99**, 297.
Dowling, H. F., Jackson, G. G., Spiesman, I. G., & Inoue, T. (1958) *Am. J. Hyg.*, **68**, 59.
Drescher, J., & Tyrrell, D. A. J. (1969) *Am. J. Epidemiol.*, **89**, 98.
Duffy, P. E., Bell, A., & Menefee, M. G. (1962) *Virology*, **16**, 350.
Dulbecco, R., & Vogt, M. (1958) *Virology*, **5**, 220.
Dunne, H. W., Gobble, J. L., Hokanson, J. F., Kradel, D. C., & Bubash, G. R. (1965), *Am. J. vet. Res.*, **26**, 1284.
Dunne, H. W., Kradel, D. C., Clark, C. D., Bubash, G. R., & Ammerman, E. (1967) *Am. J. vet. Res.*, **28**, 557.
Eggers, H. J., & Sabin, A. B. (1961) *Arch. ges. Virusforsch.*, **11**, 120 & 152.
Eggers, H. J., & Tamm, I. (1961a) *J. exp. Med.*, **113**, 657.
Eggers, H. J., & Tamm, I. (1961b) *Virology*, **13**, 545.
Ellem, K. A. O., & Colter, J. S. (1961) *Virology*, **15**, 340.
Enders, J. F., Weller, T., & Robbins, F. (1949) *Science*, **109**, 85.
Enders, J. F., Holloway, A., & Grogan, E. A. (1967) *Proc. natn. Acad. Sci. U.S.A.*, **57**, 637.
Eunes, E. S. (1955) *Bull. Off. int. Épizoot.*, **43**, 756.
Fabiyi, A., Engler, R., & Martin, D. C. (1964) *Arch. ges. Virusforsch.*, **14**, 621.
Falke, D. (1957) *Z. Hyg. InfektKrankh.*, **143**, 645.
Faulkner, P., Martin, E. M., Sved, S., Valentine, R. C., & Work, T. S. (1961) *Biochem. J.*, **80**, 597.
Fellowes, O. N. (1966) *J. Immunol.*, **96**, 772.
Fellowes, O. N. (1967) *J. Immunol.*, **99**, 508.
Fenters, J. D., Gill, S. S., Holper, J. C., & Marquis, G. S. (1966) *Am. J. Epidemiol.*, **84**, 10.
Fiala, M., & Kenny, G. E. (1966) *J. Bact.*, **92**, 1710.

Finch, J. T., & Klug, A. (1959) *Nature, Lond.*, **183**, 1709.
Fitzgerald, J. E., Hanson, L. E., & Wingard, M. (1963) *Proc. Soc. exp. Biol. Med., U.S.A.*, **114**, 814.
Fleet, W. F., Couch, R. B., Cate, T. R., & Knight, V. (1965) *Am. J. Epidemiol.*, **82**, 185.
Forsgren, M. (1966) *Acta path. microbiol. scand.*, **66**, 262.
Franklin, R. M., Wecker, E., & Henry, C. (1959) *Virology*, **7**, 220.
Franklin, R. M., & Rosner, J. (1962) *Biochim. biophys. Acta*, **55**, 240.
Friedman, R., & Maenza, R. M. (1968) *J. Infect. Dis.*, **118**, 125.
Frommhagen, L. H. (1965) *J. Immunol.*, **95**, 818.
Fuentes-Marins, R., Rodriguez, A. R., Kalter, S. S., Hellman, A., & Crandell, R. A. (1963) *J. Bact.*, **85**, 1045.
Furusawa, E., & Cutting, W. (1962) *Proc. Soc. exp. Biol. Med., U.S.A.*, **109**, 417.
Gajdusek, C., & Rogers, N. G. (1955) *Pediatrics*, **16**, 819.
Gard, S. (1944) *Yale J. Biol. Med.*, **16**, 467.
Gard, S. (1955) *Wld Hlth Org. Monogr. Ser.*, **26**, 215.
Gard, S. (1956–57) *Arch. ges. Virusforsch.*, **7**, 449.
Gard, S. (1960) *Bull. Wld Hlth Org.*, **22**, 235.
Gear, J. H. S. (1958) *Progr. med. Virology*, **1**, 106.
Godenne, M. O., & Curnen, E. C. (1952) *Proc. Soc. exp. Biol. Med., U.S.A.*, **81**, 81.
Godman, G. C. (1966) *Int. Rev. exp. Path.*, **5**, 67.
Godman, G. C., Rifkind, R. A., Page, R. B., Howe, C., & Rose, H. M. (1964) *Am. J. Path.*, **44**, 215.
Goldfield, M., Srihongse, S., & Fox, J. P. (1957) *Proc. Soc. exp. Biol. Med., U.S.A.* **96**, 788.
Goodheart, C. R. (1965) *Virology*, **26**, 466.
Gralheer, H., Hahnefeld, H., Hahnefeld, E., Schulze, P., & Hantschel, H. (1965) *Arch. exp. VetMed.*, **19**, 171.
Granboulan, N., & Girard, M. (1969) *J. Virol.*, **4**, 475.
Greig, A. S., Mitchell, D., Corner, A. H., Bannister, G. L., Meads, E. B., & Julian, R. J. (1962) *Can. J. comp. Med.*, **26**, 49.
Grist, N. R. (1960) *Lancet*, **1**, 1054.
Grist, N. R., & Roberts, G. B. S. (1966) *Arch. ges. Virusforsch.*, **19**, 454.
Guérin, I. F., & Guérin, M. M. (1957) *Proc. Soc. exp. Biol. Med., U.S.A.*, **96**, 322.
Gwaltney, J. M., & Jordan, W. S. (1964) *Bact. Rev.*, **28**, 409.
Gwaltney, J. M., Hendley, J. O., Simon, G., & Jordan, W. S. (1968) *Am. J. Epidemiol.*, **87**, 158.
Hagan, W. A. (1958) *Ann. Rev. Microbiol.*, **12**, 140.
Hallauer, C. (1951) *Arch. ges. Virusforsch.*, **4**, 224.
Halpin, F. B. (1966) *Nature, Lond.*, **209**, 429.
Hamparian, V. V., Leagus, M. B., Hilleman, M. R., & Stokes, J. (1964) *Proc. Soc. exp. Biol. Med., U.S.A.*, **117**, 469.
Hampil, B., & Melnick, J. L. (1968) *Bull. Wld Hlth Org.*, **38**, 577.
Hanson, L. E. (1958) *Am. J. vet. Res.*, **19**, 712.
Harnett, G. B., & Hooper, W. L. (1968) *Lancet*, **1**, 339.
Hausen, H. (1962) *Z. Naturf.*, **17b**, 158.
Hausen, P. (1965) *Virology*, **25**, 523.
Hayflick, L., & Moorhead, P. S. (1961) *Exp. Cell Res.*, **25**, 585.
Heberling, R. L., & Cheever, F. S. (1965) *Am. J. Epidemiol.*, **81**, 106.
Hedger, R. S. (1968) *J. Hyg., Camb.*, **66**, 27.
Helwig, F. C., & Schmidt, E. C. H. (1945) *Science*, **102**, 31.
Henderson, W. M. (1949) *Spec. Rep. Ser. agric. Res. Coun., Lond.*, no. 8 (H.M. Stationery Office, London).
Hiatt, C. W., Kaufman, E., Helprin, J. J., & Baron, S. (1960) *J. Immunol.*, **84**, 480.
Hofmann, W. (1941) *Zentbl. Bakt. ParasitKde, I. Abt. Orig.*, **148**, 69.
Holland, J. J., & Bassett, D. W. (1964) *Virology*, **23**, 164.
Holland, J. J., McLaren, L. C., & Syverton, J. T. (1959) *Proc. Soc. exp. Biol. Med., U.S.A.*, **100**, 843.

Holland, J. J., & McClaren, L. C. (1961) *J. exp. Med.*, **114**, 161.
Holzinger, E. A., & Kahn, D. E. (1970) *Am. J. vet. Res.*, **31**, 1623.
Hoorn, B., & Tyrrell, D. A. J. (1965) *Br. J. exp. Path.*, **46**, 109.
Horne, R. W., & Nagington, J. (1959) *J. molec. Biol.*, **1**, 333.
Horstmann, D. M. (1952) *J. Immunol.*, **69**, 379.
Horton, E., Lin, S., Dalgarno, L., Martin, E. M., & Work, T. S. (1964) *Nature, Lond.*, **204**, 247.
Hsiung, G. D., & Melnick, J. L. (1957) *J. Immunol.*, **78**, 137.
Huebner, R. J., Ransom, S. E., & Beeman, E. A. (1950) *Publ Hlth Rep., Wash.*, **65**, 803.
Hull, R. N. (1968) *Virology monographs*, **2**. Eds.: Gard, Hallauer, & Meyer. Vienna New York: Springer-Verlag.
Hurst, E. W. (1931) *J. Path. Bact.*, **34**, 331.
Hwang, J., Luginbuhl, R. E., & Jungherr, E. L. (1959) *Proc. Soc. exp. Biol. Med., U.S.A.*, **102**, 429.
Hyslop, N. St. G., & Morrow, A. W. (1969) *Res. vet. Sci.*, **10**, 109.
Inaba, Y., Omori, T., Kodama, M., Ishii, S., & Matumoto, M. (1959) *C. r. Séanc. Soc. Biol., Paris*, **153**, 1653.
International Enterovirus Study Group (1963) *Virology*, **19**, 114.
Jacobson, M. F., & Baltimore, D. (1968) *J. molec. Biol.*, **33**, 369.
Jacobson, M. F., & Baltimore, D. (1970) *J. molec. Biol.*, **49**, 657.
Jamison, R. M. (1969) *J. Virol.*, **4**, 904.
Jamison, R. M., & Mayor, H. D. (1966) *J. Bact.*, **91**, 1971.
Johnsson, T. (1957) *Lancet*, **1**, 590.
Johnston, M. D., & Martin, S. J. (1970) *Biochem. J.*, **119**, 17P.
Jungeblut, C. W. (1958) *Handbuch der Virusforsch*, **4**, 459. Vienna: Springer.
Jungherr, E. L., & Minard, E. L. (1944) *Am. J. vet. Res.*, **5**, 125.
Kalter, S. S. (1960) *Bull. Wld Hlth Org.*, **22**, 319.
Kamitsuka, P. S., Lou, T. Y., Fabiyi, A., & Wenner, H. A. (1965) *Am. J. Epidemiol.*, **81**, 283.
Kanamitsu, M., Hashimoto, N., Urasawa, S., & Shiba, S. (1967) *Jap. J. med. Sci. Biol.*, **20**, 471.
Kapikian, A. Z. & colleagues (1967) *Nature, Lond.*, **213**, 761.
Kapikian, A. Z. & 15 others (1971) *Virology*, **43**, 524.
Kaplan, A. S., & Melnick, J. L. (1952) *Am. J. publ. Hlth*, **42**, 525.
Kappel, M. van (1963) *Z. Hyg. u. Grenz.*, **9**, 235.
Karzon, D. T., Pollock, B. F., & Barron, A. L. (1959) *Virology*, **9**, 564.
Kasel, J. A., Rosen, L., Loda, F., & Fleet, W. (1965) *Proc. Soc. exp. Biol. Med., U.S.A.*, **118**, 381.
Kauffmann, F., & Frantzen, A. (1948) *Acta path. microbiol. scand.*, **25**, 356.
Kelly, D. F. (1964) *Res. vet. Sci.*, **5**, 56.
Ketler, A., Hamparian, W., & Hilleman, M. R. (1962) *Proc. Soc. exp. Biol. Med., U.S.A.*, **110**, 821.
Kibrick, S. (1964) *Progr. med. Virol.*, **6**, 27.
Kibrick, S., Meléndez, L., & Enders, J. F. (1957) *Ann. N.Y. Acad. Sci.*, **67**, 311.
Kiehn, E. D., & Holland, J. J. (1970) *J. Virol.*, **5**, 358.
Kilham, L., Mason, P., & Davies, J. N. P. (1956) *Am. J. trop. Med.*, **5**, 647.
Koestner, A., Long, J. F., & Kasza, L. (1962) *J. Am. vet. med. Ass.*, **140**, 811.
Koprowski, H., Jervis, G. A., & Norton, T. W. (1952) *Am. J. Hyg.*, **55**, 108.
Kunin, C. M., & Minuse, E. (1958), *J. Immunol.*, **80**, 1.
Kuwata, T. (1957) *Arch. ges. Virusforsch.*, **7**, 333.
Lahelle, O. (1958) *Acta path. microbiol. scand.*, **44**, 413.
Lahelle, O., & Horsfall, F. L. (1949) *Proc. Soc. exp. Biol. Med., U.S.A.*, **71**, 713.
La Placa, M., Portolani, M., & Lamieri, C. (1963) *Arch. ges. Virusforsch.*, **13**, 587.
La Placa, M., Portolani, M., & Lamieri, C. (1965) *Arch. ges. Virusforsch.*, **17**, 98.
Le Bouvier, G. L., Schwerdt, C. E., & Schaffer, F. L. (1957) *Virology*, **4**, 590.
Le Bouvier, G. L. (1959) *Br. J. exp. Path.*, **40**, 452.

Leers, W.-D. (1969) *Arch. ges. Virusforsch.*, **28**, 116.
Lenahan, M. F., & Wenner, H. A. (1960) *J. infect. Dis.*, **107**, 203.
Lennette, E. H., Fox, V. L., Schmidt, N. J., & Culver, J. O. (1958) *Am. J. Hyg.*, **68**, 272.
Lepine, P., & Atanasiu, P. (1950) *Annls Inst. Pasteur, Paris*, **79**, 3.
Lepine, P., Desse, G., & Sautter, V. (1952) *Bull. Acad. natn. Méd., Paris*, **136**, 66.
Lerner, A. M. (1965) *Progr. med. Virol.*, **7**, 97.
Leyon, H. (1951) *Exp. Cell Res.*, **2**, 207.
Li, C. P., & Habel, K. (1951) *Proc. Soc. exp. Biol. Med., U.S.A.*, **78**, 233.
Li, C. P., & Schaeffer, M. (1953) *Proc. Soc. exp. Biol. Med., U.S.A.*, **82**, 477.
Liess, B., & Höpken, W. (1964) *Zentbl. Bakt. ParasitKde, I. Abt. Orig.*, **194**, 26.
Lim, K. A., & Benyesh-Melnick, M. (1960) *J. Immunol.*, **84**, 309.
Löffler, H. (1958) *Handbuch der Virusforsch.*, **4**, 631. Vienna: Springer.
Long, J. F., Kaszal, L., & Koestner, A. (1969) *J. infect. Dis.*, **120**, 245.
Lou, T. Y., & Wenner, H. A. (1962) *Arch. ges. Virusforsch.*, **12**, 241 and 303.
Lou, T. Y., Wenner, H. A., & Kamitsuka, P. S. (1960) *Arch. ges. Virusforsch.*, **10**, 451.
Love, R., & Roca-Garcia, M. (1955) *Am. J. Path.*, **31**, 901.
Lovelock, J. E., Porterfield, J. S., Roden, A. T., Sommerville, T., & Andrewes, C. H. (1952) *Lancet*, **2**, 657.
Lundgren, D. L., Meade, G. H., & Clapper, W. E. (1970) *Tex. Rep. Biol. Med.*, **28**, 49.
Lundgren, O. L., Clapper, W. E., & Sanchez, A. (1968) *Proc. Soc. exp. Biol. Med., U.S.A.*, **128**, 463.
Lwoff, A., & Lwoff, M. (1960) *Annls Inst. Pasteur, Paris*, **98**, 173.
McBride, W. D. (1959) *Virology*, **7**, 45.
McClain, M. E., Hackett, A. J., & Madin, S. H. (1958) *Science*, **127**, 1391.
McConnell, S. J., Huxsoll, D. L., Garner, F. M., Spertzel, R. O., Warner, A. R., & Yager, R. H. (1964) *Proc. Soc. exp. Biol. Med., U.S.A.*, **115**, 362.
McConnell, S., Spertzel, R. O., & Shively, J. N. (1968) *Am. J. vet. Res.*, **29**, 245.
Macferran, J. B. (1958) *Vet. Rec.*, **70**, 999.
McFerran, J. B. (1962) *J. path. Bact.*, **83**, 73 and 83.
McFerran, J. B., Clarke, J. K., & Connor, T. J. (1971) *J. gen. Virol.*, **10**, 279.
McFerran, J. B., Nelson, R., McCracken, J. M., & Ross, J. G. (1969) *Nature, Lond.*, **221**, 194.
McGregor, S., & Mayor, H. D. (1968) *J. Virol.*, **2**, 149.
McGregor, S., & Mayor, H. D. (1971) *J. Virol.*, **7**, 41.
McGregor, S., Phillips, C. A., & Mayor, H. D. (1966) *Proc. Soc. exp. Biol. Med., U.S.A.*, **122**, 118.
McLauchlan, J. D., & Henderson, W. M. (1947) *J. Hyg., Camb.*, **45**, 474.
Macleod, A. J. (1965) *Vet. Rec.*, **77**, 335.
Madin, S. H. (1958) *Diseases of swine*. Ed. Dunne. Ames, Iowa: Iowa State Univ. Press. London: Baillière, Tindall and Cox.
Madin, S. H., & Traum, J. (1953) *Vet. Med.*, **48**, 395 and 443.
Madin, S. H., & Traum, J. (1955) *Bact. Rev.*, **19**, 6.
Maitland, M. C., & Maitland, H. B. (1931) *J. comp. Path.*, **44**, 106.
Maizel, J. V., Jr., & Summers, D. F. (1968) *Virology*, **36**, 48.
Mandel, B. (1951) *Bull. N.Y. Acad. Med.*, **27**, 583.
Margalith, M., Margalith, E., Rannon, L., Goldblum, T., Leventon-Kriss, S., & Goldblum, N. (1967) *Arch. ges. Virusforsch.*, **21**, 403.
Martin, S. J., Johnston, M. D., & Clements, J. B. (1970) *J. gen. Virol.*, **7**, 103.
Martin, W. B., & Chapman, W. G. (1961) *Res. vet. Sci.*, **2**, 53.
Martin, W. B., & Edwards, L. T. (1965) *Res. vet. Sci.*, **6**, 196.
Mattern, C. F. T. (1962) *Virology*, **17**, 520.
Mattern, C. F. T., & Chi, L. L. (1962) *Virology*, **18**, 257.
Mattern, C. F. T., & Daniel, W. A. (1965) *Virology*, **26**, 646.
Mattern, C. F. T., & Dubuy, H. G. (1956) *Science*, **123**, 1037.
Mayor, H. D. (1964) *Virology*, **22**, 156.
Mayr, A. (1962) *Ann. N.Y. Acad. Sci.*, **101**, 423.

Mayr, A., & Correns, H. (1959) *Zentbl. VetMed.*, **6**, 416.
Medappa, K. C., McLean, C., & Rueckert, R. R. (1971) *Virology*, **44**, 259.
Melnick, J. L. (1962) *Am. J. publ. Hlth*, **52**, 472.
Melnick, J. L. (1965) *Viral and rickettsial diseases of man.* Eds.: Horsfall & Tamm. 4th Ed. p. 513. Philadelphia: Lippincott.
Melnick, J. L., & Hampil, B. (1970) *Bull. Wld Hlth Org.*, **42**, 847.
Melnick, J. L., & Riordan, J. T. (1947) *J. Immunol.*, **57**, 331.
Meyer, R. C., Greider, M. H., & Bohl, E. H. (1964) *Virology*, **22**, 163.
Meyer, R. C., Woods, G. T., & Simon, J. (1966) *J. comp. Path.*, **76**, 397.
Moll, T., & Davis, A. D. (1959) *Am. J. vet. Res.*, **20**, 27.
Moll, T., & Finlayson, A. V. (1957) *Science*, **126**, 401.
Montagnier, L., & Sanders, F. K. (1963) *Nature, Lond.*, **199**, 664.
Morgan, C., Howe, C., & Rose, H. M. (1959) *Virology*, **9**, 145.
Morimoto, T., Dunne, H. W., & Wang, J. T. (1968) *Am. J. vet. Res.*, **29**, 2275.
Morris, M. C. (1952) *J. Immunol.*, **68**, 97.
Moscovici, C., Gineveri, A., & Mazzaracchio, V. (1959) *Am. J. vet. Res.*, **20**, 625.
Moscovici, C., & Maisel, J. (1958) *Virology*, **6**, 769.
Mott, L. O., Patterson, W. C., Songer, J. R., & Hopkins, S. R. (1953) *N. Am. Vet.*, **34**, 782.
Mowat, G. N., Barr, D. A., Bennett, J. H. (1969) *Arch. ges. Virusforsch.*, **26**, 341.
Mowat, G. N., & Chapman, W. G. (1962) *Nature, Lond.*, **194**, 253.
Mowat, G. N., & Prydie, J. (1962) *Res. vet. Sci.*, **3**, 368.
Müller, F. (1961) *Ergebn. Mikrobiol. Immunitätsforsch. Exp. Therap.* (1961) **34**, 275. Heidelberg: Springer.
Murnane, T. C., Craighead, J. E., Mondragon, H., & Shelokov, A. (1960) *Science*, **131**, 498.
Murphy, F. A., Scherer, W. F., Harrison, A. K., Dunne, H. W., & Gary, G. W. (1970) *Virology*, **40**, 1008.
Mussgay, M .(1958) *Zentbl. Bakt. ParasitKde, I. Abt. Orig.*, **171**, 413.
Nardelli, L., Lodetti, E., Gualandi, G. L., Burrows, R., Goodridge, D., Brown, F., & Cartwright, B. (1968) *Nature, Lond.*, **219**, 1275.
Nath, N., Balaye, S., & Mohapatra, L. N. (1970) *Indian vet. J.*, **47**, 1038.
Nath, N., Balaye, S., & Mohapatra, L. N. (1971) *Nature, Lond.*, **229**, 342.
Neva, F. A., & Enders, J. F. (1954) *J. Immunol.*, **72**, 307.
Oglesby, A. S., Schaffer, F. L., & Madin, S. H. (1971) *Virology*, **44**, 329.
Oglesby, A. S. (1965) *Diss. abstract*, **26**, 632, quoted in *Vet. Bull.*, **36**, 151.
Ohlbaum, A., Figueroa, F., Grado, C., & Contreras, G. (1970) *J. gen. Virol.*, **6**, 429.
Olitsky, P. K. (1945) *Proc. Soc. exp. Biol. Med., U.S.A.*, **58**, 77.
Parker, R. C., & Hollender, A. J. (1945) *Proc. Soc. exp. Biol. Med., U.S.A.*, **60**, 88.
Parsons, R., Bynoe, M. L., Pereira, M. S., & Tyrrell, D. A. J. (1960) *Br. med. J.*, **1**, 1776.
Parsons, R., & Tyrrell, D. A. J. (1961) *Nature, Lond.*, **189**, 640.
Pay, T. W. F. (1957) *Proc. r. Soc. Med.*, **50**, 919.
Peckham, M. C. (1957) *Avian Dis.*, **1**, 247.
Perkins, J. C., Tucker, D. N., Knopf, H. L. S., Wenzel, R. P., Kapikian, A. Z., & Chanock, R. M. (1969) *Am. J. Epidemiol.*, **90**, 519.
Pette, J. (1959) *Zentbl. Bakt ParasitKde, I. Abt. Orig.*, **175**, 212.
Petrov, P. A. (1970) *Am. J. trop. Med. Hyg.*, **19**, 146.
Philipson, L., & Wesslén, T. (1958) *Arch. ges. Virusforsch.*, **8**, 77.
Phillips, B. A. (1969) *Virology*, **39**, 811.
Pindak, F. F., & Clapper, W. E. (1964) *Am. J. vet. Res.*, **25**, 52.
Pinto, C. A., & Haff, R. F. (1969) *Nature, Lond.*, **224**, 1310.
Platt, H. (1956) *J. Path. Bact.*, **72**, 299.
Platt, H. (1958) *J. Path. Bact.*, **76**, 119.
Plotkin, S. A., Cohen, B. J., & Koprowski, H. (1961) *Virology*, **15**, 473.
Plummer, G. (1962) *Nature, Lond.*, **195**, 519.
Plummer, G. (1963) *Arch. ges. Virusforsch*, **12**, 694.
Plummer, G. (1965) *Progr. med. Virol.*, **7**, 326.
Plummer, G., & Kerry, J. B. (1962) *Vet. Rec.*, **74**, 967.

Pollard, M., & Starr, T. J. (1959) *Proc. Soc. exp. Biol. Med., U.S.A.*, **101**, 521.
Polson, A., & Kipps, S. (1965) *Arch. ges. Virusforsch.*, **17**, 488.
Pons, M. (1964) *Virology*, **24**, 467.
Porterfield, J. S. (1962) *Nature, Lond.*, **194**, 1044.
Portolani, M., Palenzona, A., Bernadini, A., & La Placa, M. (1968) *Arch. ges. Virusforsch.*, **24**, 428.
Price, W. H. (1957) *Proc. natn. Acad. Sci., U.S.A.*, **43**, 790.
Prydie, J. (1966) *Vet. Rec.*, **79**, 729.
Ramos-Alvarez, M. (1957) *Ann. N.Y. Acad. Sci.*, **67**, 326.
Ramos-Alvarez, M., & Sabin, A. B. (1956) *Am. J. Publ. Hlth*, **46**, 295.
Reed, S. E., Tyrrell, D. A. J., Betts, A. D., & Watt, R. G. (1971) *J. comp. Path.*, **81**, 33.
Reich, E., Franklin, R., Shatkin, A. J., & Tatum, E. L. (1961) *Science*, **134**, 556.
Richards, W. P. C., & Savan, M. (1960) *Cornell Vet.*, **50**, 132.
Richter, W. R., Rdzok, E. J., & Moize, S. M. (1964) *Virology*, **24**, 114.
Rifkind, R. A., Godman, G. C., Howe, C., Morgan, C., & Rose, H. M. (1961) *J. exp. Med.*, **114**, 1.
Riordan, J. T., & Sá-Fleitas, M. J. (1947) *J. Immunol.*, **56**, 263.
Rispens, B. M. (1969) *Avian Dis.*, **13**, 417.
Robinson, C. R., Doane, F. W., & Rhodes, A. J. (1958) *Can. med. Ass. J.*, **79**, 615.
Robinson, L. K. (1950) *Proc. Soc. exp. Biol. Med., U.S.A.*, **75**, 570.
Roca-Garcia, M., Moyer, A. W., & Cox, H. R. (1952) *Proc. Soc. exp. Biol. Med., U.S.A.*, **81**, 519.
Rosen, L. Johnson, J. H., Huebner, R. J., & Bell, J. (1958) *Am. J. Hyg.*, **67**, 300.
Rosen, L., Melnick, J. L., Schmidt, N. J., & Wenner, H. A. (1970) *Arch. ges. Virusforsch.*, **30**, 89.
Roenquist, B. D. (1971) *Am. J. vet. Res.*, **32**, 685.
Rowlands, D. J., Cartwright, B., & Brown, F. (1969) *J. gen. Virol.*, **4**, 479.
Rowlands, D. J., Sanger, D. V., Newman, J. F. E., & Brown, F. (1971) Personal Communication.
Rueckert, R. R., & Schäfer, W. (1965) *Virology*, **26**, 333.
Rustigian, R., & Pappenheimer, A. M. (1949) *J. exp. Med.*, **89**, 69.
Sabin, A. B. (1956) *Science*, **123**, 1151.
Sabin, A. B. (1959) *B. med. J.*, **1**, 663.
Sabin, A. B. (1962) *R. Soc. Hlth J.*, **82**, 51.
Salk, J. E. (1959) *Viral and rickettsial infections of man*. Eds. Rivers and Horsfall. 3rd Ed. p. 500. London: Pitman Medical.
Sato, G., Watanabe, H., & Miura, S. (1969) *Avian Dis.*, **13**, 461.
Schaaf, K. (1959) *Avian Dis.*, **3**, 245.
Schaaf, K., & Lamoreux, W. F. (1955) *Am. J. vet. Res.*, **16**, 627.
Schaffer, F. L., Moore, H. F., & Schwerdt, C. E. (1960) *Virology*, **10**, 530.
Schaffer, F. L., & Schwerdt, C. E. (1959) *Adv. Virus Res.*, **6**, 159.
Scherer, W. F., & Hurlbut, H. S. (1967) *Am. J. Epidemiol.*, **86**, 271.
Scherer, W. F., Verna, J. C., & Richter, G. W. (1968) *Am. J. trop. Med.*, **17**, 120.
Schmidt, N. J., Dennis, J., Frommhagen, L. H., & Lennette, E. H. (1963) *J. Immunol.*, **90**, 654.
Scraba, D. G., Kay, C. M., & Colter, J. S. (1967) *J. molec. Biol.*, **26**, 67.
Schultz, E. W., & Enright, J. B. (1946) *Proc. Soc. exp. Biol., Med., U.S.A.*, **63**, 8.
Schulze, P., & Gralheer, H. (1964) *Arch. exp. Vet. med.*, **18**, 1449.
Schwerdt, C. E., & Schaffer, F. L. (1955) *Ann. N.Y. Acad. Sci.*, **61**, 740.
Sellers, R. F., Burt, L. M., Cumming, A., & Stewart, D. L. (1959) *Arch. ges. Virusforsch.*, **9**, 637.
Sellers, R. F., & Parker, J. (1969) *J. Hyg., Camb.*, **67**, 671.
Selzer, G., (1962) *J. Hyg. Camb.*, **60**, 69.
Selzer, G., & Butchart, M. (1959) *J. Hyg. Camb.*, **57**, 285.
Selzer, G., & van den Ende, M. (1956) *J. Hyg. Camb.*, **54**, 1.
Selzer, G. (1969) *Israel J. med. Sci.*, **5**, 125.

Shaver, D. N., Barron, A. L., & Karzon, D. T. (1958) *Am. J. Path.*, **34**, 943.
Shaw, M. (1953) *Proc. Soc. exp. Biol. Med., U.S.A.*, **82**, 547.
Shaw, M. (1956) *Proc. Soc. exp. Biol. Med., U.S.A.*, **92**, 390.
Shean, D. B., & Schultz, E. W. (1950) *Proc. Soc. exp. Biol. Med., U.S.A.*, **73**, 622.
Sickles, G. M., Mutterer, M., & Plager, H. (1959) *Proc. Soc. exp. Biol. Med., U.S.A.*, **102**, 742.
Simon, E. H. (1961) *Virology*, **13**, 105.
Simon, M., & Domök, I. (1963) *Acta microbiol. hung.*, **10**, 293.
Singh, K. V., Greider, M. H., & Bohl, E. H. (1961) *Virology*, **14**, 372.
Skinner, H. H. (1951) *Proc. roy. Soc. Med.*, **44**, 1041.
Skinner, H. H. (1954) *Nature, Lond.*, **174**, 1052.
Skinner, H. H. (1959) *Arch. ges. Virusforsch.*, **9**, 92.
Skinner, H. H. (1960) *Bull. Off. int. Épizoot.*, **53**, 634.
Skinner, H. H., Smith, I. M., Hollom, S. E., & Knight, E. H. (1962) *Arch. ges. Virusforsch.*, **12**, 472.
Smith, W., Sheffield, F. W., Churcher, G., & Lee, L. H. (1956) *Lancet*, **2**, 163.
Smithburn, K. C. (1952) *J. Immunol.*, **68**, 441.
Snowdon, W. A. (1966) *Nature, Lond.*, **210**, 1079.
Speir, R. W., Aliminosa, K. V., & Southam, C. M. (1962) *Proc. Soc. exp. Biol. Med., U.S.A.*, **109**, 80.
Sprunt, K., Redman, W. M., & Alexander, H. E. (1959) *Proc. Soc. exp. Biol. Med., U.S.A.*, **101**, 604.
Stott, E. J., Draper, C., Stones, P. B., & Tyrrell, D. A. J. (1969) *Arch. ges. Virusforsch.*, **28**, 89.
Stott, E. J., & Tyrrell, D. A. J. (1968) *Arch. ges. Virusforsch.*, **23**, 236.
Stuart, D. C., Jr., & Fogh, J. (1961) *Virology*, **13**, 177.
Studdert, M. J., Martin, M. C., & Peterson, J. E. (1970) *Am. J. vet. Res.*, **31**, 1723.
Stulberg, C. S., Schapira, R., & Eidam, C. R. (1952) *Proc. Soc. exp. Biol. Med., U.S.A.*, **81**, 642.
Sturman, L. S., & Tamm, I. (1969) *J. Virol.*, **3**, 8.
Tannock, G. A., Gibbs, A. J., & Cooper, P. D. (1970) *Biochem. biophys. Res. Commun.*, **38**, 298.
Taylor, P. J., & Calnek, B. W. (1962) *Avian Dis.*, **6**, 51.
Taylor, P. L., & Hanson, L. F. (1967) *Avian Dis.*, **11**, 586.
Taylor-Robinson, D., Hucker, R., & Tyrrell, D. A. J. (1962) *Br. J. exp. Path.*, **43**, 189.
Thompson, R., Harrison, V. M., & Meyers, F. P. (1951) *Proc. Soc. exp. Biol. Med., U.S.A.*, **77**, 262.
Tobin, J. O'H. (1953) *Br. med. Bull.*, **9**, 201.
Trautman, R., & Breese, S. S. (1962) *J. gen. Microbiol.*, **27**, 231.
Tyrrell, D. A. J., & Bynoe, M. L. (1966) *Lancet*, **1**, 76.
Tyrrell, D. A. J., Bynoe, M. L., Buckland, F. E., & Hayflick, L. (1962) *Lancet*, **2**, 320.
Tyrrell, D. A. J., Bynoe, M. L., Hitchcock, G., Pereira, H. G., Andrewes, C. H., & Parsons, R. (1960) *Lancet*, **1**, 235.
Tyrrell, D. A. J., & Chanock, R. M. (1963) *Science*, **141**, 152.
Tyrrell, D. A. J., Clarke, S. K. R., Heath, R. B., & Curran, R. C. (1958) *Br. J. exp. Path.*, **39**, 178.
Tyrrell, D. A. J., & Ridgewell, B. (1965) *Nature, Lond.*, **206**, 115.
Tzianabos, T. (1966) *Diss. abstr.*, **26**, 4179 (Abstract in *Vet. bull.* 1967) **34**, 458.
Urasawa, S., Urasawa, T., Chiba, S., & Kanamitsu, M. (1968) *Jap. J. med. Sci. Biol.*, **21**, 155.
Van de Woude, G. F. (1967) *Virology*, **31**, 436.
Van de Woude, G. F., & Bachrach, H. L. (1968) *Arch. ges. Virusforsch.*, **23**, 353.
Vasilenko, S., & Atsev, S. (1965) *Acta virol., Prague*, **65**, 541.
Verlinde, J. D., & Ting, L. K. (1954) *Antonie van Leuwenhoek, J. Microbiol. Serol.*, **20**, 181.
Verlinde, J. D., & Versteeg, J. (1958) *T. Diergeneesk.*, **83**, 459.

Vindel, J. A. (1963) *C. r. hebd. Séanc. Acad. Sci., Paris*, **257**, 2565.
Vizoso, A. D., Vizoso, M. R., & Hay, R. (1964) *Nature, Lond.*, **201**, 849.
von Magnus, H. (1951) *Acta path. microbiol. scand.*, **28**, 234, and **29**, 243.
von Magnus, H., Gear, J. H. S., & Paul, J. R. (1955) *Virology*, **1**, 185.
von Magnus, H., & von Magnus, P. (1949) *Acta path. microbiol. scand.*, **26**, 175.
Voroshilova, M. K., & Chumakov, M. P. (1959) *Progr. med. Virology*, **2**, 106.
Wachnik Z., & Nowacki, J. (1962) *Med. Wet., Poland*, **18**, 344.
Waldmann, O., Trautwein, K., & Pyl, G. (1931) *Zentbl. Bakt. ParasitKde, I. Abt. Org.*, **121**, 19.
Wallis, C., & Melnick, J. L. (1962) *Virology*, **16**, 504.
Warren, J., Smadel, J. E., & Russ, S. B. (1949) *J. Immunol.*, **62**, 387.
Warrington, R. E. (1967) *Arch. ges. Virusforsch.*, **21**, 78.
Wawrzkiewicz, J., Smale, C. J., & Brown, F. (1968) *Arch. ges. Virusforsch.*, **25**, 337.
Wecker, E. (1960) *Virology*, **10**, 376.
Wenner, H. A., Behbehani, A. M., & Kamitsuka, P. S. (1965) *Am. J. Epidemiol.*, **82**, 27.
Wenner, H. A., & Behbehani, A. M. (1968) *Virology Monographs*, vol. 1, pp. 1–72. Eds. S. Gard, C. Hallauer, & K. F. Meyer. Vienna New York: Springer-Verlag.
Wenner, H. A., Beran, G. W., & Wender, A. A. (1960) *Am. J. vet. Res.*, **21**, 958.
Wenner, H. A., & Chin, T. D. Y. (1957) *Spec. Publ. N.Y. Acad. Sci.*, **5**, 384.
Wild, T. F., & Brown, F. (1967) *J. gen. Virol.*, **1**, 247.
Wild, T. F., & Brown, F. (1970) *J. gen. Virol.*, **7**, 1.
Wild, T. F., Burroughs, J. N., & Brown, F. (1969) *J. gen. Virol.*, **4**, 313.
Williamson, J. D., & Grist, N. R. (1965) *J. gen. Microbiol.*, **41**, 283.
Wilson, J. C., Bryans, J. T., Doll, E. R., & Tudor, L. (1965) *Cornell Vet.*, **55**, 425.
Wilson, W. W., & Matheson, R. C. (1952) *Vet. Rec.*, **64**, 541.
Wittman, G. (1967) *Zentbl. Bakt. ParasitKde, I. Abt.* Orig., **207**, 1.
Young, N. A., Hoyer, B. H., & Martin, M. A. (1968) *Proc. natn. Acad. Sci. U.S.A.*, **61**, 548.
Zee, Y. C., & Hackett, A. J. (1967) *Arch. ges. Virusforsch.*, **20**, 473.
Zee, Y. C., Hackett, A. J., & Madin, S. H. (1967) *J. infect. Dis.*, **117**, 229.
Zee, Y. C., Hackett, A. J., & Madin, S. H. (1968b) *Am. J. vet. Res.*, **29**, 1025.
Zee, Y. C., Hackett, A. J., & Talens, L. T. (1968a) *Virology*, **34**, 596.
Zwillenberg, L. O., & Bürki, F. (1966) *Arch. ges. Virusforsch.*, **19**, 373.

2

Reovirus

Reviews: Stanley (1967). Rosen (1968). Spendlove (1970). Verwoerd (1970).

Reovirus has been adopted as the generic name for unenveloped viruses with cubic symmetry, 60 to 80 nm in diameter, containing double-stranded RNA. Typical members of the genus include 3 mammalian and 5 avian serotypes. These will be referred to as 'classical reoviruses' in contradistinction to 'possible reoviruses' which include a number of animal and plant viruses with similar properties.

CLASSICAL REOVIRUSES

Synonyms: ECHO 10. Hepato-encephalomyelitis (Stanley et al., 1953).

Description of group: Sabin (1959).

Morphology and development. Diameter 60–80 nm. In thin sections (Tournier & Plissier, 1960) virions 70 nm across, enclosing nucleoids 35 nm across, are arranged in either orderly or disorderly array in a paranuclear crescentic inclusion. Negatively stained preparations reveal an inner capsid 45–48 nm across and an outer capsid built upon it. Mayor et al. (1965) think there are 92 hollow columnar units comprising the inner capsid. Loh et al. (1965) describe the outer capsid as being composed of 92 capsomeres 11·6 × 11·0 nm in size with a central hole 4·8 nm across. Vasquez & Tournier (1964) hold, on the other hand, that there are not 92 capsomeres but 92 holes. The inner shell can be clearly seen as an electron-dense structure on removal of capsomeres by trypsin or heat (Dales et al., 1965; Müller et al., 1966). Growth takes place in the cytoplasm and is closely associated with the mitotic spindle and interphase microtubules of infected cells (Dales, 1963; Spendlove et al., 1963). Virus is released by extrusion and pinching off of cytoplasmic processes.

Chemical composition. Virions contain 14 per cent RNA and 86 per cent protein (Gomatos & Tamm, 1963). The RNA is largely double-stranded with MW estimated from the percentage RNA per virion as a minimum of $10{\cdot}2 \times 10^6$ daltons (Gomatos & Tamm, 1963) and from length measured by electronmicroscopy as 17 to 22×10^6 daltons (Vasquez & Kleinschmidt, 1968). RNAs of both mammalian and avian strains have complementary base composition with G + C content of about 40 per cent (Gomatos & Tamm, 1963a; Sekiguchi et al., 1968). The reovirus RNA occurs as separate or weakly linked subgenomic pieces. Electronmicroscopic observations (Dunnebacke & Kleinschmidt, 1967) reveal a trimodal length distribution with peaks at 0·35, 0·6 and 1·1 μm. RNA fragments of 3 distinct size classes are separated by density gradient centrifugation (Bellamy et al., 1967). Polyacrylamide gel electrophoresis (Shatkin et al., 1968; Watanabe et al., 1968) resolves 10 RNA fragments distributed into 3 size classes with a total MW of 15×10^6. Reovirus RNA has a buoyant density of 1·61 g/ml in CsCl and denatures into RNAse-sensitive single strands between 78° and 85°C with indications of specific thermo-sensitive bonds (Iglewski & Franklyn, 1967). Single-stranded adenine-rich and other poly- and oligonucleotides have been reported to make up 25 per cent to 30 per cent of the RNA of reovirus isolates from mammals (Shatkin & Sipe, 1967; Bellamy & Hole, 1970). Adenine-rich polynucleotides have also been found in avian isolates (Koide et al., 1968). Messenger RNA molecules corresponding in size and nucleotide sequence to the distinct classes of viral RNA are synthesized *in vivo* (Bellamy & Joklik, 1967; Watanabe & Graham, 1967) and *in vitro* (Skehel & Joklik, 1969). Polyacrylamide gel electrophoresis of purified virions reveals 7 classes of polypeptides (Loh & Shatkin, 1968; Smith et al., 1969). An RNA-dependent RNA polymerase (Borsa & Graham, 1968), 4 base-specific nucleoside 5′-triphosphatases (Kapuler et al., 1970) and a nucleoside triphosphate phosphorylase (Borsa et al., 1970) have been described as component parts of purified virions.

Physico-chemical characters. Virion MW 70×10^6 daltons (Gomatos & Tamm, 1963). Buoyant density in CsCl between 1·36 and 1·38 g/ml (Mayor et al., 1965; Gomatos & Tamm, 1963; Sekiguchi et al., 1968). Virus infectivity is stable between pH 2·2 and 8 and is ether-resistant. It survived treatment for an hour at room temperature with 1 per cent H_2O_2, 1 per cent phenol, 3 per cent formalin, 20 per cent lysol, but was inactivated by 70 per cent ethanol, also by 3 per cent formalin at 56°C (Stanley et al., 1953). Heating to 50–55° in the presence of $MgCl_2$ increases the titre, but below 0° the same substance inactivates the virus (Wallis et al., 1964). Virus infectivity is also enhanced by digestion with proteolytic enzymes (Spendlove et al., 1970).

Haemagglutination. Human 0 cells are agglutinated at various temperatures by all 3 mammalian types, but not by avian strains. Type 3 also agglutinates ox cells. The cell-receptor of bovine but not human cells is sensitive to RDE, but is distinct from that involved in agglutination by myxoviruses and other viruses (Gomatos & Tamm, 1962). Haemagglutinating activity has been associated with both infective and noninfective virus particles (Leers & Rozee, 1968).

Antigenic properties. Mammalian reoviruses have been separated into 3 serological types by virus neutralization (Sabin, 1959) and by haemagglutination-inhibition (Rosen, 1960), although the 3 serotypes cross-react with each other in both tests. Type 2 is antigenically heterogeneous, and its division into subtypes has been suggested (Hartley et al., 1962). Human isolates of each of the 3 types are indistinguishable from those derived from other animals (Rosen, 1962). A bovine strain tentatively described as a fourth type (Moscovici et al., 1961) is probably a type 2 (Rosen, 1962). Five distinct serotypes of avian origin have been described (Kawamura et al., 1965). Group specific antigens have been demonstrated within both mammalian (Sabin, 1959) and avian (Kawamura & Tsubahara, 1966) strains. Antigens common to mammalian and avian serotypes have also been described (Deshmukh et al., 1969). A study of reovirus types 1, 2 and 3 by immunodiffusion revealed at least 2 group-specific and 1 type-specific antigens. These could be resolved into 4 group-specific and 2 type-specific antigens by immuno-electrophoresis (Leers et al., 1968). A report of antigenic relationship between mammalian reoviruses and wound tumour virus of clover has not been confirmed (Streissle et al., 1967; Gamez et al., 1967).

Cultivation. Reoviruses of mammalian origin can be propagated in primary or continuous cell cultures derived from a wide variety of animals (Hsiung, 1958; Lenahan & Wenner, 1960). Virus multiplication leads to CPE and formation of eosinophilic cytoplasmic inclusions which gradually accumulate into perinuclear masses containing numerous virus particles (Tournier & Plissier, 1960). Plaques are produced in a variety of cells (Rhim & Melnick, 1961; Wallis et al., 1962). Avian reoviruses multiply in chick-embryo tissue cultures with CPE consisting of syncytial formation and production of eosinophilic cytoplasmic inclusions containing virus particles in regular arrays (Kawamura et al., 1965; Petek et al., 1967).

In fertile eggs, some strains of mammalian origin grow on the CAM or amniotically, but there may be difficulty in propagating them indefinitely (Stanley et al., 1953; 1954). Van Tongeren & Tiddens (1957), however, adapted a strain to chick embryos.

Avian strains may be propagated in eggs with production of pocks on the CAM (Petek et al., 1967).

Morphological and biochemical features of reovirus replication have been reviewed by Shatkin (1969).

Host range. Reoviruses have been isolated from many species of vertebrates including man, chimpanzees (Sabin, 1959), several species of monkeys (Hull et al., 1958; Malherbe et al., 1963), pigs (McFerran & Connor, 1970), dogs (Lou & Wenner, 1963; Massie & Shaw, 1966), cats (Scott et al., 1970), marsupials (Stanley et al., 1964a) and birds (Kawamura et al., 1965). Isolation of reoviruses from culicine mosquitoes has also been reported (Parker et al., 1965; Miles et al., 1965; Miles & Stenhouse, 1969), although it is uncertain whether the virus multiplies in mosquitoes. The ubiquity of reoviruses is also indicated by the frequent presence of antibodies in a wide variety of vertebrates (Rosen, 1962; Stanley et al., 1964a).

Pathogenicity. The relation of reoviruses to human disease is uncertain. Types 1 and 3 have been isolated from mild fevers in children, sometimes associated with diarrhoea, but whether they caused the illness is uncertain (Rosen et al., 1960). Considering the widespread occurrence of reoviruses in human populations, it is not surprising that these agents have been frequently isolated from patients with a variety of illnesses, including respiratory, gastro-intestinal and nervous diseases, exanthemas and hepatitis. An aetiological association of reoviruses with these conditions is dubious, but there is some evidence that these viruses may occasionally cause fatal disease in man (Krainer & Aronson, 1959; Joske et al., 1964; Tillotson & Lerner, 1967). In primates other than man, reoviruses have been associated with respiratory infections, but the majority of isolations have been from faeces (Hull et al., 1958; Malherbe et al., 1963) or from uninoculated cultures of tissues derived from apparently normal monkeys (Hull et al., 1958; Malherbe et al., 1963). Natural infections of mice have been observed both in wild (Hartley et al., 1961) and in laboratory (Cook, 1963) colonies. Symptoms, observed mainly in new-born animals, include steatorrhoea, oily hair, jaundice and stunted growth. Asymptomatic natural infection of cattle has been revealed by virus isolations from faeces (Rosen & Abinanti, 1960; Rosen et al., 1963), and a possible association with respiratory disease has been suggested (Lamont,

1968). Serological evidence indicates that natural reovirus infections are frequent in dogs (Fairchild & Cohen, 1967). An association with canine respiratory illness has also been suggested (Lou & Wenner, 1963; Massie & Shaw, 1966). In birds, reoviruses have been found causing asymptomatic infections or in association with respiratory (Fahey & Crawley, 1954) or enteric (Dutta & Pomeroy, 1967) diseases of chickens, with 'cloacal pasting' in chicks (Deshmukh & Pomeroy, 1969) and with septicemia of goslings or 'goose influenza' (see p. 69). Reoviruses have also been tentatively associated with blue comb of turkeys (see p. 68) and with Gumboro disease (see p. 69).

Experimental infections. Mice inoculated with reovirus type 3 develop jaundice, alopecia, conjunctivitis and 'oily hair' effect (due to steatorrhoea) and death results (Stanley et al., 1954). A neurotropic strain was developed from the original hepatotropic one (Stanley et al., 1954). Reovirus type 3 has also been shown to produce new-born runt disease after neonatal inoculation (Bennette et al., 1967). Inoculation of pregnant mice with reovirus type 1 may lead, according to time of infection, to fetal resorption, intra-uterine death, malformation or neonatal death (Hassan & Cochran, 1969). Reovirus type 1 causes obstructive hydrocephalus in mice, hamsters, ferrets and rats (Kilham & Margolis, 1969; Phillips et al., 1970). Chronic reovirus type 3 infections of mice and rabbits have been reported to lead to the induction of lymphomas (Keast et al., 1968; Bell et al., 1968). Subsequent findings by Levy & Huebner (1970) raise doubts regarding the relationship of lymphomas to reovirus infection. Calves experimentally infected with reovirus types 1 or 2 may remain symptomless (Rosen & Abinanti, 1960; Lamont et al., 1968) or present mild respiratory illness (Trainer et al., 1966). Pigs are susceptible to experimental infection by reovirus type 1, but show no symptoms. Monkeys (*Macaca radiata*) inoculated with reovirus type 3 may develop pyrexia, diarrhoea and leucopenia (Masillamony & John, 1970). Experimental reovirus infection of dogs may be asymptomatic (Holzinger & Griesemer, 1966) or associated with pneumonia (Lou & Wenner, 1963; Thompson et al., 1970).

Pathological lesions. Reoviruses cause pantropic infections in new-born mice (van Tongeren, 1957; Walters et al., 1963; 1965; Kundin et al., 1966). Lesions described by Stanley et al. (1953) include haemorrhages, perivascular cuffing and pyknosis throughout the brain and spinal cord, liver necrosis with lymphocytic infiltration and sometimes occlusion of ducts. Meningitis and peritoneal exsudation may occur. Stanley et al. (1964b) consider that autoimmune phenomena may be involved in the pathology of chronic reovirus infections. Ultrastructural lesions of murine encephalitis and hepatitis have been described (Papadimitriou, 1967a, b).

Ecology. Stanley et al. (1964a) discuss the ecology in various mammals, including Quokkas, a species of West Australian marsupials. The neutralizing and haemagglutinin-inhibiting substances present in sera of many vertebrates, including trout, may not be true antibodies. The virus has been recovered from mosquitoes, which possibly transmit infection mechanically (Parker et al., 1965).

PROBABLE REOVIRUSES

A number of unenveloped viruses with cubic symmetry, containing double-stranded RNA are considered as probable reoviruses either because they differ from the classical reoviruses in certain respects or because some of their properties are studied incompletely or not at all. Comparisons of some of these viruses with each other and with classical reoviruses are discussed by Verwoerd (1970).

BLUE TONGUE

Synonyms: Sore mouth. Ovine catarrhal fever.

Review: Cox (1954).

Morphology. Virions seen in negatively stained preparations are 53 to 60 nm in diameter with icosahedral symmetry (Owen & Munz, 1966; Studdert et al., 1966; Els & Verwoerd, 1969). There are conflicting reports regarding the number of capsomeres, estimated as 92 (Owen & Munz, 1966; Studdert et al., 1966) or 32 (Els & Verwoerd, 1969), although a figure of 42 cannot be excluded by the latter workers. In contrast to classical reoviruses, the capsid is single-layered (Els & Verwoerd, 1969). An outer envelope described by Owen & Munz (1966) and Bowne & Ritchie (1970) is considered by Els & Verwoerd (1969) to be an inessential part of the virion, derived from cell membranes.

Chemical composition. Virions contain 20 per cent RNA characterized as double-stranded, with a total MW of 15×10^6, resolved by density gradient centrifugation into 5 components and by polyacrylamide gel electrophoresis into 10 components (Verwoerd et al., 1970). Double-strandedness is indicated by RNAse resistance, complementary base composition (G + C content 42·4 per cent) and thermal denaturation with temperature ranging from 75°C to 95°C according to ionic strength and pH of suspending medium. No infectious RNA has been obtained.

Physico-chemical characters. The virion has a buoyant density of 1·38 g/ml in CsCl (Verwoerd et al., 1970). Ether-, chloroform- and deoxycholate-resistant, readily inactivated below pH 6·0 (Studdert, 1965; Svehag & Gorham, 1966) and generally unstable although stabilized by serum, glycerol and other additives (Howell et al., 1967) and by lyophilization (Espinosa et al., 1959). Inactivation by several chemical compounds is described by McCrory et al. (1959).

Haemagglutination. None demonstrated.

Antigenic properties. There are 16 serotypes, revealed by neutralization and other tests (Neitz, 1948; Howell, 1970). Group-specific antigens are revealed by CF (Shone et al., 1956; Robertson et al., 1965) by agar immunodiffusion (Klonitz & Svehag, 1962; Jochim & Chow, 1969) and by immunofluorescence (Pini et al., 1968). Non-infectious soluble antigens are involved in complement fixation (Kipps, 1958) and immunodiffusion (Jochim & Chow, 1969).

Cultivation. The virus grows in 6-day-old fertile eggs inoculated into the yolk sac and held at 33·5°C, or at least below 35°C (Alexander et al., 1947); virus is harvested from the embryos and, after 45–50 egg-passages, is found to be attenuated for sheep (McKercher et al., 1957). Intravenous inoculation is up to 100 times more sensitive than the yolk-sac route (Foster & Luedke, 1968; Goldsmit & Barzilai, 1968).

In tissue cultures, virus multiplication with CPE has been described in lamb-kidney cultures (Haig et al., 1956), in bovine-, hamster-, and chick-embryo primary cultures and in hamster (BHK-21) and mouse (L) cell-lines (Matumotu et al., 1970). Growth in several human cell-lines has been reported (Fernandes, 1959; Livingston & Moore, 1962). A plaque assay in mouse L cells is described by Howell et al. (1967).

Virus multiplication takes place in the cytoplasm (Livingston & Moore, 1962; Ohder et al., 1970), although inclusions extruding from the nucleus to the cytoplasm have been described (Bowne & Jochim, 1967). Ultrastructural changes associated with virus multiplication include swelling of endoplasmic reticulum, presence of mitochondrial inclusions, appearance of fine filaments and bundles of tubular elements and irregular dense bodies containing virus particles (Lecatsas, 1968). The multiplication of blue-tongue virus in the salivary glands of *Culicoides variipennis* has been described (Bowne & Jones, 1966).

Habitat. Africa, especially in the East and South. Spread in recent years to Cyprus, Palestine, Turkey, Spain, Portugal, Pakistan, India and Japan. In southern and western United States it was first identified in 1952, but may have been unrecognised earlier.

Pathogenicity. A serious disease affecting sheep, especially lambs, with fever, erosions, crusting and cyanosis around the mouth, oedema of head and neck, sometimes pulmonary oedema and lameness due to involvement of hooves and to muscle damage. Mortality in sheep varies from 5 per cent to 30 per cent in different areas. Cattle and goats suffer much milder symptoms, but may have prolonged viraemia. The virus may produce foot-lesions in pigs. Natural infection of white-tailed deer (Stair et al., 1968) and other wild ruminants (Trainer & Jochim, 1969) has been described in the United States.

Experimentally the disease is readily transmitted to sheep or cattle. Van den Ende et al. (1954) infected young mice IC; virus persisted in brains of older ones without producing symptoms. Suckling hamsters were infected IC by Cabasso et al. (1955). CF antigens were obtained from their brains or those of suckling mice. Experimental disease of white-tailed deer has been described (Karstad & Trainer, 1967; Vosdingh et al., 1968).

Pathological lesions have been described by Thomas & Neitz (1947) and Moulton (1961); they include haemorrhages, oedema, mucosal erosions and focal muscle degeneration. No inclusions are described. Congenital abnormalities are described (Richards & Cordy, 1967; McKercher et al., 1970).

Ecology. Infection is transmitted by a nocturnal vector. Virus has been recovered from *Culicoides spp.* in South Africa and from *Culicoides variipennis* in the United States, but proof that these midges are the essential vectors is incomplete. There is, however, an extrinsic incubation period in *Culicoides* (Foster et al., 1963) and injected virus can multiply in *Culicoides* (Jochim & Jones, 1966; Bowne & Jones, 1966). Preliminary evidence of transmission by the sheep kid *Melophagus ovinns* has been presented (Gray & Bannister, 1961; Luedke et al., 1965), although it is not certain whether virus multiplication takes place in this vector. There is doubtless a reservoir in wild animals and Blesbuck (*Damaliscus*) were shown to develop viraemia after inoculation (Neitz, 1933).

Control. A vaccine partly attenuated by passage in sheep has now been superseded by virus attenuated in eggs; the temperature of incubation of the eggs was important (Alexander et al., 1947). Such a vaccine is now used on a large scale and is effective if sufficiently polyvalent to deal with all strains encountered locally. A tissue-culture attenuated vaccine has been developed by Kemeny & Drehle (1961). It is not safe for pregnant ewes; cerebral and other lesions may be present in lambs born from vaccinated ewes (Young & Cordy, 1964).

AFRICAN HORSE SICKNESS

Reviews: Alexander (1935). Mornet & Gilbert (1969). Rafyi (1961).

Morphology. Polson & Deeks (1963) suggested that the diameter is 70–80 nm and that there are 92 rod-shaped subunits radiating from a spherical body. Similar findings are reported by Breese et al. (1969), but more recently Oellermann et al. (1970) describe virions with icosahedral capsids 55 nm in diameter, very similar to those of blue-tongue virus, probably with 32 capsomeres. Ozawa et al. (1966) describe oval particles 75 × 45 nm mainly present in nuclei; clusters were seen on the surface of the nucleolus. However, the findings of Lecatsas & Erasmus (1967) indicate that virus synthesis and assembly take place in the cytoplasm. Virions are occasionally contained in a 'pseudo-envelope' derived from cell membranes and not considered to be an essential part of the virus (Lecatsas & Erasmus, 1967; Oellermann et al., 1970).

Chemical composition. Virions contain double-stranded RNA resolved by sucrose density centrifugation into 5 components and by polyacrylamide gel electrophoresis into 6 components in 4 size classes (Oellermann et al., 1970).

Physico-chemical characters. Unpurified virus suspensions contain 2 infective fractions with buoyant densities in CsCl of 1·17 and 1·35 g/ml. The lighter is considered to be bound to contaminating lipids (Russell, 1967). Survives for years in the cold in an oxalate-phenol-glycerol mixture (Alexander, 1935). Virus infectivity is ether- and trypsin-resistant but acid-sensitive, being destroyed below pH 6·0 (Ozawa, 1968). It resists up to 60′ at 60°C and is stable at −70°C in different suspending media, but unstable at −22°C or −30°C in the presence of NaCl, $CaCl_2$ or $MgCl_2$; removal of salts or addition of sugars, glycerol or polyvinylpyrolidone stabilizes the virus at −20°C (Ozawa & Bahrami, 1968).

Haemagglutination. Horse cells are agglutinated by extracts of infected mouse brains, preferably at pH 6·4 and 37° for about 2 hours (Pavri, 1961).

Antigenic properties. Virus neutralization and haemagglutination-inhibition tests are type-specific, distinguishing 9 serotypes (Howell, 1962; Pavri & Anderson, 1963). Cross-protection tests in horses reveal intratypic antigenic differences (McIntosh, 1958). Complement fixation is group-specific, the antigen responsible having a diameter of 12 nm (Polson & Madsen, 1954; McIntosh, 1956).

Cultivation. All strains multiply in embryonated eggs inoculated in the yolk sac. Virus localization in embryonic tissues varies according to neuro- or viscero-tropism of particular strains (Alexander, 1938; Goldsmit, 1967). Growth without CPE is also reported in chick-embryo fibroblasts.

The virus grows in cells from many species; those of a stable monkey line and BHK (hamster) cells are most useful. Virus so grown is greatly attenuated for horses (Ozawa et al., 1965; Mirchamsy & Taslimi, 1964). A plaque assay in a stable line of monkey-kidney cells has been described by Hopkins et al. (1966). Propagation in a mosquito cell-line is described by Mirchamsy et al. (1970). Infected cells develop large cytoplasmic inclusions containing RNA and virus antigen (Ozawa, 1967). Protein inclusions free of RNA have also been described (Breese & Ozawa, 1969). Virus replication is not inhibited by halogenated deoxyribosides, actinomycin D or mitomycin C (Ozawa, 1967).

Distribution. African continent. Spread in 1944 and subsequently to the Middle East, Turkey, Pakistan and India, causing serious losses.

Pathogenicity. Causes disease in horses, mules and (in the Middle East) donkeys. In severe cases death occurs from pulmonary oedema. In more chronic ones there is cardiac involvement with oedema, especially of head and neck. Some infections are quite mild. Viraemia is often observed.

Goats and zebras are only slightly susceptible, but ferrets and dogs can be infected. Packs of hounds have suffered fatal infections after eating meat containing the virus.

Experimentally, infection has been transmitted IC to mice, rats and guinea-pigs (Nieschulz, 1933; Alexander, 1935). The neurotropic virus obtained after repeated passage in mice or guinea-pigs has lost its pathogenicity for horses, though viraemia still occurs. Other rodents, including gerbils, are susceptible to the neurotropic virus injected IC. Hamsters can be infected by IC inoculation (Sharma & Kumar, 1968). Dogs receiving virus by oral, nasal, ocular, intracerebral or intravenous routes develop symptomless infection (Dardiri & Ozawa, 1969).

Pathological lesions in horses consist of pulmonary oedema, hydrothorax, hydropericardium. They are fully described by Maurer & McCully (1963).

Ecology. Transmission is by nocturnal biting insects. Virus has been recovered from *Culicoides* (du Toit, 1944) which are mainly suspected. Tabanids and *Stomoxys* seem to be incriminated in Turkey. Experimental transmission by Anphelines and Culicines has been reported (Ozawa & Nakata, 1965).

Control. Formalinized vaccine was formerly used. An attenuated live vaccine from mouse brain is now used and is effective. Virus attenuated in tissue culture may also prove useful. A polyvalent vaccine including 8 serotypes seems adequate to protect against all strains (McIntosh, 1958). Some attenuated virus, especially type 7, may have neurotropic qualities.

COLORADO TICK FEVER

Review: Cox (1959).

Morphology. Virions seen in thin sections are 75 nm in diameter with a dense core 50 nm across (Murphy et al., 1968; Oshiro & Emmons, 1968). By negative contrast (Murphy et al., 1968) virions are 60 to 80 nm in over-all diameter, with a double-layered capsid, the inner one having a diameter of 50 nm. Pictures are consistent with icosahedral symmetry with 92 capsomers in the outer capsid. Virions with envelope derived from cell membranes are occasionally seen. Particles are often in association with granular cytoplasmic matrix and filamentous or tubular structures. Intranuclear filaments are also present.

Chemical composition. Evidence indicating that the virus contains double-stranded RNA has been presented by Green (1970).

Physico-chemical characters. Very stable in serum, surviving $3\frac{1}{2}$ years in an ordinary refrigerator and even at room temperature, but not always so stable at lower temperatures (Pickens & Luoto, 1958). Rates of inactivation at different temperatures and pH have been determined by Trent & Scott (1966). Very sensitive to acid and resistant to ether, chloroform and sodium deoxycholate (Borden et al., 1971).

Haemagglutination of cells of day-old chicks is reported.

Antigenic properties. Distinct by neutralization, CF and other tests from other arboviruses including the other tick-borne ones. Antibodies develop slowly in convalescents, but both neutralizing and CF antibodies persist for long periods.

Cultivation. Grows in fertile eggs when inoculated into the yolk sac; virus reaches highest titre in the embryo's CNS in 4–5 days. In tissue cultures of human carcinoma KB cells it grows and produces CPE. No changes were seen in monkey-kidney cells (Pickens & Luoto, 1958). Trent & Scott (1964) describe the changes produced in various tissue-culture systems. Growth in cultures of tick (*Dermacentor andersoni*) cells has been described (Yunker & Cory, 1967). Satisfactory plaques were produced in a line of hamster cells (Deig & Watkins, 1964).

Distribution. Northwestern United States, but not reaching the Pacific coast.

Pathogenicity. Affected persons, 4–5 days after a bite by an infected tick, develop fever, usually of saddle-back type, with chills, aches in head and limbs and often vomiting. Rashes are absent. There is leucopenia. Encephalitis may occur, especially in children. Virus has been recovered from several species of wild rodent, most frequently from ground-squirrels; the disease in them is inapparent.

Experimentally, hamsters can be infected IP and after adaptation the virus often kills them (Florio et al., 1944). Rhesus monkeys develop viraemia. The virus has also been adapted to mice IC and up to 8 days of age they can also be infected IP. Other species are resistant.

Pathological lesions. Hamsters show changes mainly in the spleen the follicles being full of large pale-staining mononuclear cells (Black et al., 1947). Mice show destruction of Purkinje cells (Miller et al., 1961). Cytoplasmic inclusions occur both in mice and hamsters. Suckling mice show also myocardial necrosis (Hadlow, 1957), and specific antigen is revealed with the aid of fluorescent antibody in the cytoplasm of myocardial and brain cells (Burgdorfer & Lackman, 1960).

Ecology. Infection is transmitted by the tick *Dermacentor andersoni* and the distribution of the disease and of the tick practically coincides. Ticks once infected retain the infection. It has been maintained that there is transovarial transmission, but Eklund et al. (1962) doubt this. Possibly the cycle is maintained between immature ticks and small rodents (Burgdorfer & Eklund, 1959).

Control depends largely on protection against ticks. Koprowski et al. (1950) used a living chick-embryo adapted virus as a vaccine, but the virus was not satisfactorily attenuated. Thomas et al. (1963) have described a formalized vaccine.

KEMEROVO GROUP

A group of antigenically related tick-borne viruses (Casals & Hoogstraal, 1971) including strains Kemerovo (Chumakov et al., 1963; Libikova et al., 1964a), Tribec (Libikova et al., 1964b), Lipovnik (Libikova et al., 1965), Chenuda (Taylor et al., 1966a), Wad Medani (Taylor et al., 1966b), Mono Lake and Huacho (Casals & Hoogstraal, 1971). Several members of the group have been shown to resemble reoviruses in morphology and morphogenesis (Shestopalova et al., 1964; Gresikova et al., 1969; Borden et al., 1971) and to be acid-sensitive, markedly (Kemerovo) or moderately (Chenuda, Tribec, Wad Medani) sensitive to lipid solvents (Mayer et al., 1964; Borden et al., 1971) and not stabilized by IM $MgCl_2$. Multiplication in chick embryo (Libikova et al., 1964a), human diploid cells (Karmysheva et al., 1967a, b) or in tick tissue cultures (Rehacek et al., 1969) may be accompanied by CPE and cytoplasmic inclusions containing RNA and virus antigen. Viruses of this group have been isolated from ticks or tick-borne infections in Eastern Europe, Egypt and Sudan, some (Kemerovo) in association with benign encephalitis in man (Chumakov et al., 1963) and possibly infecting horses, cattle, birds and small mammals (Libikova et al., 1964a; Gresikova et al., 1965).

OTHER PROBABLE REOVIRUSES

The following viruses resemble reoviruses in morphology and in physico-chemical properties as far as investigated.

Corriparta virus (Doherty et al., 1963; Carley & Standfast, 1969) isolated in northern Australia from culicines and from birds (Whitehead et al., 1968).

Eubenangee virus isolated from mosquitoes in northern Australia (Doherty et al., 1968; Schnagl et al., 1969).

Changuinola and Irituia. Two antigenically related viruses (Taylor, 1967) isolated from *Phlebotomus* and from small rodents in central and South America.

Lebombo and Palyam viruses isolated from mosquitoes in India (Dandawate et al., 1969) and South Africa (Dr P. Weinbren) respectively and studied by Borden et al. (1971).

Nelson Bay virus isolated from a flying fox (Gard & Compans, 1970).

Virus X of bovine serum, found in cultures of hamster cells (BHK-21) and thought to be derived from bovine serum used as medium constituent (Verwoerd, 1970; Lecatsas, 1970).

Equine encephalosis virus isolated from horses in South Africa (Verwoerd, 1970; Lecatsas, 1970).

Simian virus S.A.11 isolated from a healthy vervet monkey (Malherbe & Strickland-Cholmley, 1967; Lecatsas, 1970).

Syncytial virus of rabbits isolated from a cotton-tail rabbit (Morris et al., 1967) and also found as contaminant of tissue cultures derived from laboratory bred rabbits (Brown et al., 1970) resembles reoviruses in morphology, site of replication and cytopathology, but differs from other members of the group in being markedly ether-sensitive.

EPIZOOTIC HAEMORRHAGIC DISEASE OF DEER

Description: Shope et al. (1960).

Morphology. Virions examined by negative contrast (Tsai & Karstad, 1970) are round or hexagonal with a mean diameter of 62·3 nm. The capsid is composed of capsomeres 10·5 to 11·5 nm in diameter with an axial hole 4·5 nm wide. The number of morphological units is tentatively estimated as 42. No double-layered capsids observed. In thin sections examined by the same workers, virions appear to be formed in granular, cytoplasmic matrices. Mature virions are 58·8 nm in diameter with an electron-dense core contained in a lucent layer considered to be the capsid. Occasional virions are surrounded by a membrane thought to be of cellular origin. Infected cells contain long, hollow tubular structures with capsomere-like elements on the surface. Similar findings are reported by Borden et al. (1971).

Chemical composition. Staining by acridine orange suggests double-stranded RNA (Tsai & Karstad, 1970).

Physico-chemical characters. Virus infectivity is ether- and deoxycholate-resistant, moderately chloroform-sensitive and totally destroyed at low pH (Borden et al., 1971). Heat-sensitivity at 50°C is partly stabilized by 1 M $MgCl_2$ (Ditchfield et al., 1964).

Antigenic properties. Two serotypes designated New Jersey and South Dakota have been distinguished by virus neutralization (Shope et al., 1960). Wilhelm & Trainer (1967), however, suggest that there is only 1 serotype.

Cultivation. The New Jersey strain produces CPE in HeLa cells; there is neutralization by homologous antisera. It was still pathogenic for deer after 3 passages (Mettler et al., 1962). Pirtle & Layton (1961) cultivated the South Dakota strain in embryonic deer-kidney cells. Growth in hamster cells (BHK21) has also been reported (Tsai & Karstad, 1970).

Habitat. The United States and Canada.

Pathogenicity. Has caused fatal epizootics in Virginian white-tailed deer (*Odocoleus virginianus*), several hundred animals being killed. Mule deer and other species are insusceptible. After an incubation of 6–8 days, animals show symptoms of shock, have multiple haemorrhages and die in coma. Virus is present in blood and most tissues.

Experimentally the New Jersey virus has been passed IC in suckling mice (Mettler et al., 1962). After the first pass it was 100 per cent lethal to the mice. After 4 passages it produced only inapparent infection in the 1 deer tested (Shope et al., 1963).

Pathological lesions. There are haemorrhages in various tissues and may be retroperitoneal oedema and fluids in serous sacs (Karstad et al., 1961). Prothrombin deficiency rather than endothelial damage may cause the haemorrhages.

Ecology. Infection does not pass by direct contact; an arthropod vector is suspected.

EPIDEMIC DIARRHOEA OF INFANT MICE

Description: Cheever & Mueller (1947).

Reviews: Pappenheimer (1958); Kraft (1966).

Morphology. In thin sections of infected cells, Adams & Kraft (1967) describe particles 80 nm in diameter with 2 sets of double membranes, the outer being continuous with cellular membranes. Virions without the outer membrane, are 65 nm in diameter. Banfield et al. (1968) describe virions 75 nm in diameter, with a central body 56 nm across contained in a single shell derived from the endoplasmic reticulum. The particles are formed in cytoplasmic viroplasm and transferred to the endoplasmic reticulum. Tubules 54 to 62 nm across are frequently seen in the nucleus and the cytoplasm of infected cells.

Physico-chemical characters. Resists ether (Kraft, 1966), chloroform and pH 3·0 treatment (J. H. Blackwell, quoted by Borden et al., 1971).

Antigenic properties. Antibodies have not been demonstrated in infected mice which do not become immune as a result of infection. No protection is transferred to the young by way of placenta or milk (Kraft, 1966). Neutralizing and complement-fixation antisera can be obtained from hyperimmunized mice or rabbits (Kraft, 1966; Blackwell & Tennant, 1966; Wilsnack et al., 1969). Viral antigen can be detected in the cytoplasm of infected cells by immunofluoroescence (Wilsnack et al., 1969). No cross-reaction with reovirus 3 by VN or CF tests (Kraft, 1966).

Cultivation. Ten serial passages are reported in a continuous line (Chang) of human liver cells (Habermann, 1959). Others have reported unsuccessful attempts to propagate the virus in tissue cultures.

Pathogenicity. Attacks chiefly mice 11–15 days old; diarrhoea causes fur to be dirty and yellow. The disease is not necessarily fatal, but may be so up to 50 per cent, and many litters fail to be successfully weaned (Cheever & Mueller, 1947). In some outbreaks cytoplasmic inclusions have been found in intestinal epithelium; their relation to the disease is uncertain (Pappenheimer & Cheever, 1948).

Ecology. The disease is commoner in first rather than later litters of breeding females. Possibly the mothers gradually become more highly immune and pass on antibody in colostrum. Severity also varies with the strain of mice. Infection may be air-borne or caused by human agency. Virus is present in faeces. Recovered mice can act as carriers for a time.

Control. A clean stock has been obtained by Caesarean section and raising the babies on disease-free animals (Cheever & Mueller, 1948).

INFECTIOUS PANCREATIC NECROSIS IN TROUT

Review: Wolf, 1966.

Morphology. In thin sections virions are hexagonal in outline, approximately 55 nm in diameter, formed in the cytoplasm, frequently in association with tubular structures 45 nm in diameter. Negatively stained virions are unenveloped icosahedra about 65 nm in diameter, probably with 92 capsomeres (Moss & Gravell, 1969).

Chemical composition. Lack of inhibition by halogenated deoxyribosides (Malsberger & Cerini, 1963) suggests RNA content. Preliminary results by M. Gravell (quoted by Moss & Gravell, 1969) indicate that the RNA is double-stranded.

Physico-chemical characters. Ether-resistant. Infectivity maintained for 5 weeks at 4°C and for 36 weeks at −70°C (Malsberger & Cerini, 1965). Some infectivity persists after 1 hour at 60°C (Wolf, 1966). Survives $2\frac{1}{2}$ years at 4°C in 50 per cent glycerol (Wolf et al., 1960). Readily inactivated at pH 3·0 (Wolf, 1966).

Haemagglutination. Extensively searched for with negative results.

Antigenic characters. Neutralizing antibody obtained from recovered trout and from immunized rabbits. It is uncertain whether all strains are antigenically identical (Malsberger & Cerini, 1963).

Cultivation. Growth with CPE in cultures of various fish cells, but not in mammalian cells (Wolf et al., 1960; Malsberger & Cerini, 1963; Wolf, 1966).

Distribution: the United States; probably France (Wolf, 1966).

Pathogenicity. Eastern brook trout (*Salvelinus fontinalis*), rainbow trout (*Salmo gairdneri*) and cut-throat trout (*Salmo clarki*) are susceptible. Three species of salmon of the genus *Oncorhyncus* were insusceptible (Parisot et al., 1963). Infected fish appear grossly normal, but swim and whirl erratically and eventually die. Only young fish are affected. Pancreatic necrosis occurs with involvement of acinar and islet tissue where cytoplasmic inclusions may be found.

Control. Strict antiseptic techniques prevented spread in a fish hatchery (Snieszko et al., 1959).

AVIAN DISEASES ASSOCIATED WITH REOVIRUSES

Blue Comb

Synonyms (probably): Pullet disease, Avian diarrhoea. Infectious enteritis.

This affects young turkeys and pullets with diarrhoea, hepatic necrosis and monocytosis. The agent is filterable, unaffected by antibiotics and cultivable in 8-day chick embryos killing them in 36–72 hours (Watanabe, 1952; Jungherr & Levine, 1941; Tumlin & Pomeroy, 1958). Strains resembling reoviruses in several respects have been described in association with this disease (Deshmukh et al., 1969; Wooley & Gratzek, 1969; Fujisaki et al., 1969), but their etiologic role remains uncertain. Viruses differing from reoviruses in some properties such as ether-sensitivity have also been isolated from animals with blue comb (Deshmukh et al., 1969; Hofstad et al., 1969).

Gumboro Disease

This disease described by Cosgrove (1962) as avian nephrosis is caused by the infectious bursal agent (IBA) (Winterfield et al., 1962). Identification of this agent as a reovirus has been suggested by Mandelli et al. (1967) who found particles resembling reoviruses in morphology and site of formation in cells of the bursa of Fabricius from infected birds. Similar particles have been described by Cheville (1967). A virus diameter of 10 to 50 nm has, however, been suggested from filtration experiments (Benton et al., 1967b; Cho & Edgar, 1969). Virus infectivity persists after heating at 60°C for 90 minutes and is resistant to ether, chloroform, trypsin and low pH (Benton et al., 1967b; Cho & Edgar, 1969; Rinaldi et al., 1969).

Virus adapted to grow in embryonated eggs (Snedeker et al., 1967; Mandelli et al., 1969) reproduces the disease, but may be attenuated and possibly used for vaccination. The virus persists for long periods in the environment, and transmission is thought to be through contaminated feed or water (Benton et al., 1967a). Virus isolation from mealworms has been reported (Snedeker et al., 1967).

Infectious Myocarditis of Goslings

This name has been suggested by Rinaldi et al. (1970) for a disease previously named 'septicaemia anserum exsudativa' or 'goose influenza'. An aetiologic association with a virus-resembling reoviruses in several respects has been suggested by Csontos & Csatari (1967) and Derzsi (1967). The virus is readily propagated in embryonated goose eggs and tissue cultures with syncytial formation and cytoplasmic inclusions (Kroms, 1965; Csontos & Csatari, 1967). Virus replication is not inhibited by IUDR and infectivity resists chloroform, ether, deoxycholate, trypsin and pH 3 and its heat sensitivity is not stabilized by molar $MgCl_2$ (Csontos & Csatari, 1967). The predominant lesions seen in infected animals are degenerative changes of smooth and striated muscles (Nazy & Derzsi, 1968) resembling those caused by reovirus type 3 in mice.

REFERENCES

Adams, W. R., & Kraft, L. M. (1967) *Am. J. Path.*, **51**, 39.

Alexander, R. A. (1935) *Onderstepoort J. vet. Sci. Anim. Ind.*, **4**, 291, 323, 349, 397.

Alexander, R. A. (1938) *Onderstepoort J. vet. Sci. Anim. Ind.*, **11**, 9.

Alexander, R. A., Haig, D. A., & Adelaar, T. F. (1947) *Onderstepoort J. vet. Sci. Anim. Ind.*, **22**, 7.

Banfield, W. G., Kasnic, G., & Blackwell, J. H. (1968) *Virology*, **36**, 411.

Bell, T. M., Manube, G. M. R., & Wright, D. H. (1968) *Lancet*, **1**, 955.

Bellamy, A. R., & Hole, L. V. (1970) *Virology*, **40**, 808.

Bellamy, A. R., & Joklik, W. (1967) *J. molec. Biol.*, **29**, 19.

Bellamy, A. R., Shapiro, L., August, J. T., & Joklik, W. (1967) *J. molec. Biol.*, **29**, 1.

Bennette, J. G., Bush, P. V., & Steele, R. D. (1967) *Br. J. exp. Path.*, **48**, 251.

Benton, W. J., Cover, M. S., & Rosenberger, J. K. (1967a) *Avian Dis.*, **11**, 430.

Benton, W. J., Cover, M. S., Rosenberger, J. K., & Lake, R. S. (1967b) *Avian Dis.*, **11**, 438.

Black, W., Florio, L., & Stewart, M. O. (1947) *Am. J. Path.*, **23**, 217.

Blackwell, J., Tennant, R., & Ward, E. (1966) *Natn. Cancer Inst. Monogr.*, **20**, 63.

Borden, E. C., Murphy, F. A., Shope, R. E., Harrison, A., & Theiler, M. (1971) Personal communication.

Borsa, J., & Graham, A. F. (1968) *Biochem. Biophys. Res. Commun.*, **33**, 895.

Borsa, J., Grover, J., & Chapman, J. D. (1970) *J. Virol.*, **6**, 295.

Bowne, J. G., & Jochim, M. M. (1967) *Am. J. vet. Res.*, **28**, 1091.

Bowne, J. G., & Jones, R. H. (1966) *Virology*, **30**, 127.

Bowne, J. G., & Ritchie, A. (1970) *Virology*, **40**, 903.

Breese, S. S., Jr., & Ozawa, Y. (1969) *J. Virol.*, **4**, 109.

Breese, S. S., Jr., Ozawa, Y., & Dardiri, A. H. (1969) *J. Am. vet. Med. Ass.*, **155**, 391.

Brown, R. C., Shaw, C. N., Bloomer, M. D., & Morris, J. A. (1970) *Proc. Soc. exp. Biol. Med., U.S.A.*, **133**, 587.

Burgdorfer, W., & Eklund, C. M. (1959) *Am. J. Hyg.*, **69**, 127.

Burgdorfer, W., & Lackman, D. (1960) *J. Bact.*, **80**, 131.

Cabasso, V. J., Roberts, G. I., Douglas, J. M., Zorzi, R., Stebbins, M. R., & Cox, H. R. (1955) *Proc. Soc. exp. Biol. Med., U.S.A.*, **88**, 678.

Carley, J. G., & Standfast, H. A. (1969) *Am. J. Epidemiol.*, **89**, 583.

Casals, J., & Hoogstraal, H. (1971, in press).

Cheever, F. S., & Mueller, J. H. (1947) *J. exp. Med.*, **85**, 405.

Cheever, F. S., & Mueller, J. H. (1948) *J. exp. Med.*, **88**, 309.

Cheville, N. F. (1967) *Am. J. Path.*, **51**, 527.

Cho, Y., & Edgar, S. A. (1969) *Poult. Sci.*, **48**, 2102.

Chumakov, M. P., Sarmanova, E. S., Bychkova, M. B., Branova, G. G., Pivanova, G. P., Karpovich, L. G., Izotov, V. K., & Rzhakhova, O. E. (1963) *Vopr. Virusol.*, **8**, 440.

Cook, I. (1963) *Aust. J. exp. Biol. med. Sci.*, **41**, 651.

Cosgrove, A. S. (1962) *Avian Dis.*, **6**, 385.

Cox, H. R. (1959) In *Viral and rickettsial diseases of man.* Eds.: Rivers & Horsfall, 3rd Ed. p. 384. London: Pitman Medical.

Cox, H. R. (1954) *Bact. Rev.*, **18**, 239.

Csontos, L., & Csatari, M. M-K. (1967) *Acta vet. hung.*, **17**, 107.

Dales, S. (1963) *Proc. natn. Acad. Sci., U.S.A.*, **50**, 268.

Dales, S., Gomatos, P. J., & Hsu, K. C. (1965) *Virology*, **25**, 193.

Dandawate, C. N., Rajagopalan, P. K., Pavri, K. M., & Work, T. H. (1969) *Ind. J. med. Res.*, **57**, 1420.

Dardiri, A. H., & Osawa, Y. (1969) *J. Am. vet. Med. Ass.*, **155**, 400.

Deig, E. F., & Watkins, H. M. S. (1964) *J. Bact.*, **88**, 42.
Derzsy, D. (1967) *Acta vet. hung.*, **17**, 443.
Deshmukh, D. R., Larsen, C. T., Dutter, S. K., & Pomeroy, B. S. (1969) *Am. J. vet. Res.*, **30**, 1019.
Deshmukh, D. R., & Pomeroy, B. S. (1969) *Avian Dis.*, **13**, 427.
Deshmukh, D. R., Sayed, H. I., & Pomeroy, B. S. (1969) *Avian Dis.*, **13**, 16.
Ditchfield, J., Debbie, J. G., & Karstad, L. H. (1964) *N. Am. Wildl. nat. Resour. Conf. Trans.*, **29**, 197.
Doherty, R. L., Carley, J. G., Mackerras, M., & Marks, E. N. (1963) *Aust. J. exp. Biol. med. Sci.*, **41**, 17.
Doherty, R. L., Standfast, H. A., Wetters, E. J., Whitehead, R. H., Barrow, G. J., & Gorman, B. M. (1968) *Trans. R. Soc. trop. Med. Hyg.*, **62**, 862.
Dunnebacke, T. H., & Kleinschmidt, A. K. (1967) *Z. Naturforsch.*, **22b**, 159.
du Toit, R. M. (1944) *Onderstepoort J. vet. Sci. Anim. Ind.*, **19**, 7.
Dutta, S. K., & Pomeroy, B. S. (1967) *Avian Dis.*, **11**, 1.
Eklund, C. M., Kohls, G. M., & Kennedy, R. C. (1962) In *Symposium of the Czechoslovak academy of sciences.* Biology of Viruses of the Tick-borne Encephalitis Complex. Ed.: Libíkovǎ. New York: Academic Press.
Els, H. J., & Verwoerd, D. W. (1969) *Virology*, **38**, 213.
Ende, M. van den., Linder, A., & Kaschula, V. R. (1954) *J. Hyg., Camb.*, **52**, 155.
Espinosa, L. F., Herran, M. V., & Sanchez, G. G. (1959) *Rev. Patronato Biol. Animal*, **5**, 275.
Fahey, J. E., & Crawley, J. F. (1954) *Can. J. comp. Med.*, **18**, 13.
Fairchild, G. A., & Cohen, D. (1967) *Am. J. vet. Res.*, **28**, 1487.
Fernandes, M. V. (1959) *Am. J. vet. Res.*, **20**, 398.
Florio, L., Stewart, M. D., & Mugrage, E. R. (1944) *J. exp. Med.*, **80**, 165.
Foster, N. M., Jones, R. H., & McCrory, B. R. (1963) *Am. J. vet. Res.*, **24**, 1195.
Foster, N. M., & Luedke, A. J. (1968) *Am. J. vet. Res.*, **29**, 749.
Fujisaki, Y., Kawamura, H., & Anderson, D. P. (1969) *Am. J. vet. Res.*, **30**, 1035.
Gamez, R., Black, L. M., & MacLeod, R. (1967) *Virology*, **32**, 163.
Gard, G., & Compans, R. W. (1970) *J. Virol.*, **6**, 100.
Goldsmit, L. (1967) *Am. J. vet. Res.*, **28**, 19.
Goldsmit, L., & Barzilai, E. (1968) *J. comp. Path.*, **78**, 477.
Gomatos, P. J., & Tamm, I. (1962) *Virology*, **17**, 455.
Gomatos, P. J., & Tamm, I. (1963) *Proc. natn. Acad. Sci., U.S.A.*, **49**, 707.
Gomatos, P. J., Tamm, I., Dales, S., & Franklin, R. M. (1962) *Virology*, **17**, 441.
Gray, D. P., & Bannister, G. L. (1961) *Can. J. comp. Med.*, **25**, 230.
Green, I. J. (1970) *Virology*, **40**, 1056.
Grěsíkovǎ, M., Mrena, E., & Vachalkova, A. (1969) *Acta virol., Prague*, **13**, 67.
Grěsíkovǎ, M., Nosek, J., Kozuch, O., Ernek, E., & Lichard, M. (1965) *Acta virol., Prague*, **9**, 83.
Habermann, R. T. (1959) *Publ. Hlth Rep., Wash.*, **74**, 165.
Hadlow, W. J. (1957) *J. infect. Dis.*, **101**, 158.
Haig, D. A., McKercher, D. G., & Alexander, R. A. (1956) *Onderstepoort J. vet. Res.*, **27**, 171.
Hartley, J. W., Rowe, W. P., & Huebner, R. J. (1961) *Proc. Soc. exp. Biol. Med., U.S.A.*, **108**, 390.
Hartley, J. W., Rowe, W. P., & Austin, J. B. (1962) *Virology*, **16**, 94.
Hassan, S. A., & Cochran, K. W. (1969) *Am. J. Path.*, **55**, 147.
Hofstad, M. S., Adams, M., & Frey, M. L. (1969) *Avian Dis.*, **13**, 386.
Holzinger, E., & Griesemer, R. A. (1966) *Am. J. Epidemiol.*, **84**, 426.
Hopkins, J. G., Hazrati, A., & Ozawa, Y. (1966) *Am. J. vet. Res.*, **27**, 96.
Howell, P. G. (1962) *Onderstepoort J. vet. Res.*, **29**, 139.
Howell, P. G. (1970) *Jl S. Afr. vet. Med. Ass.*, **41**, 215.
Howell, P. G., Verwoerd, D. W., & Oellermann, R. A. (1967) *Onderstepoort J. vet. Res.* **34**, 317.

Hsiung, G. D. (1958) *Proc. Soc. exp. Biol. Med., U.S.A.*, **99**, 387.
Hull, R. N., Minner, M. R. & Mascoli, C. C. (1958) *Am. J. Hyg.*, **68**, 31.
Iglewski, W. J., & Franklin, R. M. (1967) *J. Virol.*, **1**, 302.
Jochim, M. M., & Chow, T. L. (1969) *Am. J. vet. Res.*, **30**, 33.
Jochim, M. M., & Jones, R. H. (1966) *Am. J. epidemiol.*, **841**, 241.
Joske, R. A., Keall, D. D., Leak, P. F., Stanley, N. F., & Walters, M. N.-I. (1964) *Arch. int. Med.*, **113**, 811.
Jungherr, E., & Levine, J. M. (1941) *Am. J. vet. Res.*, **2**, 261.
Kapuler, A. M., Mendelsohn, N., Klett, H., & Acs, G. (1970) *Nature, Lond.*, **225**, 1209.
Karmysheva, V. Ya., Karpovich, L. G., Sofronova, L. N., & Chumakov, M. P. (1967a) *Acta virol., Prague*, **11**, 135.
Karmysheva, V. Ya., Shestopalova, N. M., Tikhomirova, T. I., Karpovich, L. G., & Chumakov, M. P. (1967b) *Acta virol., Prague*, **11**, 140.
Karstad, L., & Trainer, D. O. (1967) *Can. vet. J.*, **8**, 247.
Karstad, L., Winter, A., & Trainer, D. O. (1961) *Am. J. vet. Res.*, **22**, 227.
Kawamura, H., Shimizu, F., Maede, M., & Tsubahara, H. (1965) *Natn Inst. Anim. Hlth Q., Tokyo*, **5**, 115.
Kawamura, H., & Tsubahara, H. (1966) *Natn. Inst. Anim. Hlth Q., Tokyo*, **6**, 187.
Keast, D., Stanley, N. F., & Phillips, P. A. (1968) *Proc. Soc. exp. Biol. Med., U.S.A.*, **128**, 1033.
Kemeny, L., & Drehle, L. E. (1961) *Am. J. vet. Res.*, **22**, 921.
Kilham, L., & Margolis, G. (1969) *Lab. Invest.*, **21**, 183.
Kipps, A. (1958) *S. Afr. J. Lab. clin. Med.*, **4**, 158.
Klontz, G. W., Svehag, S. E., & Gorham, J. R. (1962) *Arch. ges. Virusforsch.*, **12**, 259.
Koide, F., Suzuka, I., & Sekiguchi, K. (1968) *Biochem. Biophys. Res. Comm.*, **30**, 95.
Koprowski, H., Cox, H. R., Miller, M. S., & Florio, L. (1950) *Proc. Soc. exp. Biol. Med., U.S.A.*, **74**, 126.
Kraft, L. M. (1966) *Natn. Cancer Inst. Monogr.*, No. 20, p. 55.
Krainer, L., & Aronson, B. E. (1959) *J. Neuropath., exp. Neurol.*, **18**, 339.
Kroms, H. (1965) *Berl. Munch. Tierarztl. Wschr.*, **78**, 372.
Kundin, W. D., Liu, C., & Gigstad, J. (1966) *J. Immunol.*, **97**, 393.
Lamont, P. H. (1968) *J. Am. vet. Ass.*, **152**, 807.
Lamont, P. H., Darbyshire, J. H., Dawson, P. S., Omar, A. R., & Jennings, A. R. (1968) *J. comp. Path.*, **78**, 23.
Lecatsas, G. (1968) *Onderstepoort. J. vet. Res.*, **35**, 139.
Lecatsas, G. (1970) Thesis, Faculty of Science, Univ. of Pretoria, South Africa.
Lecatsas, G., & Erasmus, B. J. (1967) *Arch. ges. Virusforsch.*, **22**, 442.
Leers, W. D., & Rozee, K. R. (1968) *Arch. ges. Virusforsch.*, **24**, 155.
Leers, W. D., Rozee, K. R., & Wardlaw, A. C. (1968) *Can. J. Microbiol.*, **14**, 161.
Lenahan, M. F., & Wenner, H. A. (1960) *J. infect. Dis.*, **107**, 203.
Levy, J. A., & Huebner, R. J. (1970) *Nature, Lond.*, **225**, 949.
Libíková, H., Mayer, V., Kožuch, O., Rehaček, J., Ernek, E., & Albrecht, P. (1964a) *Acta virol., Prague*, **8**, 289.
Libíková, H., Rehaček, J., Gresikova, M., Kožuch, O., Somogyiova, J., & Ernek, E. (1964b) *Acta virol., Prague*, **8**, 96.
Libíková, H., Rehaček, J., & Somogyiova, J. (1965) *Acta virol., Prague*, **9**, 76.
Livingston, C. W., Jr., & Moore, R. W. (1962) *Am. J. vet. Res.*, **23**, 701.
Loh, P. C., Hohl, H. R., & Soergel, M. (1965) *J. Bact.*, **89**, 1140.
Loh, P. C., & Shatkin, A. J. (1968) *J. virol.*, **2**, 1353.
Lou, T. Y., & Wenner, H. A. (1963) *Am. J. Hyg.*, **77**, 293.
Luedke, A. J., Jochim, M. M., & Bowne, J. G. (1965) *Can. J. comp. Med.*, **29**, 229.
McCrory, B. R., Foster, N. M., & Bay, R. C. (1959) *Am. J. vet. Res.*, **20**, 665.
McFerran, J. B., & Connor, T. (1970) *Res. vet. Sci.*, **11**, 388.
McIntosh, B. M. (1956) *Onderstepoort J. vet. Res.*, **27**, 165.
McIntosh, B. M. (1958) *Onderstepoort J. vet. Res.*, **27**, 465.

McKercher, D. G., McGowan, B., Cabasso, V. J., Roberts, G. I., & Saito, J. K. (1957) *Am. J. vet. Res.*, **18**, 310.
McKercher, D. G., Saito, J. K., & Singh, K. V. (1970) *J. Am. vet. Med. Ass.*, **156**, 1044.
Malherbe, H., Harwin, R., & Ulrich, M. (1963) *S. Afr. Med. J.*, **37**, 407.
Malherbe, H. H., & Strickland-Cholmley, M. (1967) *Arch. ges. Virusforsch.*, **22**, 235.
Malsberger, R. G., & Cerini, C. P. (1963) *J. Bact.*, **86**, 1283.
Malsberger, R. G., & Cerini, C. P. (1965) *Ann. N.Y. Acad. Sci.*, **126**, 320.
Mandelli, G., Rinaldi, A., Cerioli, A., & Cervio, G. (1967) *Atti. Soc. Ital. Sci. Vet.*, **21**, 1.
Mandelli, G., Rinaldi, A., Cessi, D., Cervio, G., Pasencci, S., & Valeri, A. (1969) *Atti. Soc. Ital. Sci. Vet.*, **23**, 1.
Masillamony, R., & John, T. J. (1970) *Am. J. Epidemiol.*, **91**, 446.
Massie, E. L., & Shaw, E. D. (1966) *Am. J. vet. Res.*, **27**, 783.
Matumotu, M., Inaba, Y., Tanaka, Y., Morimoto, T., & Omori, T. (1970) *Jap. J. Microbiol.*, **14**, 99.
Maurer, F. D., & McCully, R. M. (1963) *Am. J. vet. Res.*, **24**, 235.
Mayor, H. D., Jamison, R. M., Jordan, L. E., & van Mitchell, M. (1965) *J. Bact.*, **89**, 1548.
Mayer, V., Kožuch, O., Libíkova, H., & Zvàda, J. (1964) *Acta virol., Prague*, **8**, 302.
Mettler, N. E., MacNamara, L. G., & Shope, R. E. (1962) *J. exp. Med.*, **116**, 665.
Miles, J. A. R., Austin, F. J., Macnamara, F. N., & Maguire, T. (1965) *Proc. Univ. Otago Med. Sch.*, **43**, 27.
Miles, J. A. R., & Stenhouse, A. S. (1969) *Am. J. trop. Med. Hyg.*, **18**, 427.
Miller, J. K., Thompkins, V. N., & Sieracki, J. C. (1961) *Arch. Path., Chicago*, **72**, 149.
Mirchamsy, H., Hazrati, A., Bahrami, S., & Shapyi, A. (1970) *Am. J. vet. Res.*, **31**, 1755.
Mirchamsy, H., & Taslimi, H. (1964) *Bull. off. int. Épizoot.*, **62**, 911.
Mornet, P., & Gilbert, Y. (1969) *L'Expansion Scientifique Francaise, Paris.*
Morris, J. A., Saglam, M., & Bozeman, F. M. (1965) *J. Infect. Dis.*, **115**, 495.
Moscovici, C., La Placa, M., Maisel, J., & Kempe, C. H. (1961) *Am. J. vet. Res.*, **22**, 852.
Moss, L. H., III, & Gravell, M. (1969) *J. virol.*, **3**, 52.
Moulton, J. E. (1961) *Am. vet. med. Ass.*, **138**, 493.
Müller, G., Schneider, C. C., & Peters, D. (1966) *Arch. ges. Virusforsch.*, **19**, 110.
Murphy, F. A., Coleman, P. H., Harrison, A. K., & Gary, G. W., Jr. (1968) *Virology*, **35**, 28.
Nazy, Z., & Derzsy, D. (1968) *Acta vet. hung.*, **18**, 3.
Neitz, W. O. (1933) *J. S. Afr. vet. med. Ass.*, **4**, 26.
Neitz, W. O. (1948) *Onderstepoort J. vet. Sci. Anim. Ind.*, **23**, 93.
Nieschulz, O. (1933) *Zentbl. Bakt. ParasitKde, Abt. Orig.*, **128**, 465.
Oellermann, R. A., Els, H. J., & Erasmus, B. J. (1970) *Arch. ges. Virusforsch.*, **29**, 163.
Ohder, H., Lund, L. J., & Whiteland, A. P. (1970) *Arch. ges. Virusforsch.*, **29**, 127.
Oshiro, L. S., & Emmons, R. W. (1968) *J. gen. Virol.*, **3**, 279.
Owen, N. C., & Munz, E. K. (1966) *Onderstepoort J. vet. Res.*, **33**, 9.
Ozawa, Y. (1967) *Arch. ges. Virusforsch.*, **23**, 155.
Ozawa, Y. (1968) *Jap. J. med. sci. Biol.*, **21**, 27.
Ozawa, Y., & Bahrami, S. (1968) *Arch. ges. Virusforsch.*, **25**, 201.
Ozawa, Y., Hazrati, A., & Erol, N. (1965) *Am. J. vet. Res.*, **26**, 154.
Ozawa, Y., Mojtabai, A., Hopkins, I. G., Hazrati, A., & Kaveh, P. (1966) *Am. J. vet. Res.*, **27**, 558.
Ozawa, Y., & Nakata, G. (1965) *Am. J. vet. Res.*, **26**, 744.
Papadimitriou, J. M. (1967a) *Am. J. Path.*, **50**, 59.
Papadimitriou, J. M. (1967b) *Br. J. exp. Path.*, **47**, 624.
Pappenheimer, A. M. (1958) *J. natn. Cancer Inst.*, **20**, 861.
Pappenheimer, A. M., & Cheever, F. S. (1948) *J. exp. Med.*, 88, 317.
Parisot, T. J., Yasutake, W. T., & Klontz, G. W. (1965) *Ann. N.Y. Acad. Sci.*, **126**, 502.
Parker, L., Baker, E., & Stanley, N. F. (1965) *Aust. J. exp. Biol. med. Sci.*, **43**, 167.

Pavri, K. M. (1961) *Nature, Lond.*, **189**, 249.
Pavri, K. M. & Anderson, C. R. (1963) *Indian J. vet. Sci.* **33**, 113.
Petek, M., Felluza, B., Borghi, C., & Baroni, A. (1967) *Arch. ges. Virusforsch.*, **21**, 413.
Phillips, P. A., Alpers, M. P., & Stanley, N. F. (1970) *Science*, **168**, 858.
Pickens, E. G., & Luoto, L. (1958) *J. infect. Dis.*, **103**, 102.
Pini, A., Ohder, H., Whiteland, A. P., & Lund, L. J. (1968) *Arch. ges. Virusforsch.*, **25**, 129.
Pirtle, E. C., & Layton, J. M. (1961) *Am. J. vet. Res.*, **22**, 104.
Polson, A., & Deeks, D. (1963) *J. Hyg., Camb.*, **61**, 149.
Polson, A., & Madsen, T. (1954) *Biochim. biophys. Acta*, **14**, 366.
Rafyi, A. (1961) *Bull. Off. int. Épizoot.*, **56**, 216.
Rehaček, J., Rajáni, J., & Grešíková, M. (1969) *Acta virol., Prague*, **13**, 439.
Rhim, J. S., & Melnick, J. L. (1961) *Virology*, **15**, 80.
Richards, W. P. C., & Cordy, D. R. (1967) *Science*, **156**, 530.
Rinaldi, A., Mandelli, G., Cervio, G., Valeri, A., Cessi, D. & Pasuicci, S. (1969) *Proc. IV Congress of the World Veterinary Poultry Association, Belgrade*, p. 303.
Rinaldi, A., Mandelli, G. C., Cessi, D., Cervio, G., Lodetti, E., & Valeri, A. (1970) *Selezione Vet.*, **11**, 250.
Robertson, A., Appel, M., Bannister, G. L., Ruckerbauer, G. M., & Boulanger, P. (1965) *Can. J. comp. Med.*, **29**, 113.
Rosen, L. (1960) *Am. J. Hyg.*, **71**, 242.
Rosen, L. (1962) *Ann. N.Y. Acad. Sci.*, **101**, 461.
Rosen, L. (1968) *Virology monographs.* Eds. Gard, Hallauer, & Meyer. Vienna/New York: Springer-Verlag.
Rosen, L., & Abinanti, F. B. (1960) *Am. J. Hyg.*, **71**, 250.
Rosen, L., Abinanti, F. R., & Hovis, J. F. (1963) *Am. J. Hyg.*, **77**, 38.
Rosen, L., Hovis, J. F., Mastrota, F. M., Bell, J. A., & Huebner, R. J. (1960) *Am. J. Hyg.*, **71**, 258 and 266.
Russell, B. (1967) *Nature, Lond.*, **215**, 983.
Sabin, A. B. (1959) *Science*, **130**, 1387.
Schnagl, R. D., Holmes, I. H. & Doherty, R. L. (1969) *Virology*, **38**, 347.
Scott, F. W., Kahn, D. E., & Gillespie, J. H. (1970) *Am. J. vet. Res.*, **31**, 11.
Sekiguchi, K., Koide, F., & Kawamura, H. (1968) *Arch. ges. Virusforsch.*, **24**, 123.
Sharma, R. N., & Kumar, S. (1968) *Indian vet. J.*, **45**, 909.
Shatkin, A. J. (1969) *Adv. vet. Res.*, **14**, 63.
Shatkin, A. J., & Sipe, J. D. (1968) *Proc. natn. Acad. Sci., U.S.A.*, **59**, 246.
Shatkin, A. J., Sipe, J. D., & Loh, P. (1968) *J. Virol.*, **2**, 986.
Shestopalova, N. M., Reingold, V. N., Tikhomirova, T. I., Karpovich, L. G., & Chumakov, M. P. (1964) *Acta virol., Prague*, **8**, 88.
Shone, D. K., Haig, D. A., & McKercher, D. G. (1956) *Onderstepoort J. vet. Res.*, **27**, 179.
Shope, R. E., MacNamara, L. G., & Mangold, R. (1960) *J. exp. Med.*, **111**, 155.
Shope, R. E., MacNamara, L. G., & Mettler, N. E. (1963) *J. exp. Med.*, **118**, 421.
Skehel, J. J., & Joklik, W. K. (1969) *Virology*, **39**, 822.
Smith, R. E., Zweerink, H. J., & Joklik, W. K. (1969) *Virology*, **39**, 791.
Snedeker, C., Wills, F. K., & Moulthorp, I. M. (1967) *Avian Dis.*, **11**, 519.
Snieszko, S. F., Wolf, K., Camper, J. E., & Pettijohn, L. L. (1959) *Am. Fisheries Soc.*, **88**, 289.
Spendlove, R. S. (1970) *Prog. med. Virol.*, **12**, 161.
Spendlove, R. S., Lennette, E. H., Knight, C. O., & Chin, J. N. (1963) *J. Immunol.*, **90**, 548.
Spendlove, R. S., McClain, M. E., & Lennette, E. H. (1970) *J. gen. Virol.*, **8**, 83.
Stair, E. L., Robinson, R. M., & Jones, L. P. (1968) *Pathologia Vet.*, **5**, 164.
Stanley, N. F. (1967) *Br. med. Bull.*, **23**, 150.
Stanley, N. F., Dorman, D. C., & Ponsford, J. (1953) *Aust. J. exp. Biol. med. Sci.*, **31**, 147.

Stanley, N. F., Dorman, D. C., & Ponsford, J. (1954) *Aust. J. exp. Biol. med. Sci.*, **32**, 543.
Stanley, N. F., Leak, P. J., Grieve, G. M., & Perrett, D. (1964a) *Aust. J. exp. Biol. med. Sci.*, **42**, 373.
Stanley, N. F., Leak, P. J., Walters, M. N.-I., & Joske, R. A. (1964b) *Br. J. exp. Path.*, **45**, 142.
Streissle, G., Rosen, L., & Tokumitsu, T. (1967) *Arch. ges. Virusforsch.*, **22**, 409.
Studdert, M. J. (1965) *Proc. Soc. exp. Biol. Med., U.S.A.*, **118**, 1006.
Studdert, M. J., Pangborn, J., & Addison, R. B. (1966) *Virology*, **29**, 509.
Svehag, S.-E., Leendertsen, L., & Gorham, J. R. (1966) *J. Hyg., Camb.*, **64**, 339.
Taylor, R. M. (1967) *Catalogue of Arthropod borne viruses of the World.* Pub. Hlth Service Publication No. 1760. Washington, D.C.: U.S. Government Printing Office.
Taylor, R. M., Hoogstraal, H., & Hurlburt, H. S. (1966b) *Am. J. tr op. Med. Hyg.*, **15**, 75.
Taylor, R. M., Hurlburt, H. S., Work, T. H., Kingston, J. R., & Hoogstraal, H. (1966a) *Am. J. trop. Med. Hyg.*, **15**, 76.
Thomas, A. D. & Neitz, W. O. (1947) *Onderstepoort J. vet. Sci. Anim. Ind.*, **22**, 27.
Thomas, L. A., Eklund, C. M., Philip, R. N., & Casey, M. (1963) *Am. J. trop. Med. Hyg.*, **12**, 678.
Thompson, H., Wright, N. G., & Cornwell, H. J. C. (1970) *Res. vet. Sci.*, **11**, 302.
Tillotson, J. R., & Lerner, A. M. (1967) *New Engl. J. Med.*, **276**, 1060.
Tongeren, H. A. E. van (1957) *Arch. ges. Virusforsch.*, **7**, 429.
Tongeren, H. A. E. van, & Tiddens, H. A. W. M. (1957) *Ned. T. Geneesk.*, **101**, 579.
Tournier, P., & Plissier, M. (1960) *C. hebd. Séanc. Acad. Sci., Paris*, **250**, 630.
Trainer, E. T., Mohanty, S. B., & Hetrich, F. M. (1966) *Am. J. Epidemiol.*, **83**, 217.
Trainer, D. O., & Jochim, M. M. (1969) *Am. J. vet. Res.*, **30**, 2007.
Trent, D. W., & Scott, L. V. (1964) *J. Bact.*, **88**, 702.
Trent, D. W., & Scott, L. V. (1966) *J. Bact.*, **91**, 1282.
Tsai, K.-S., & Karstad, L. (1970) *Can. J. Microbiol.*, **16**, 427.
Tumlin, J. T., & Pomeroy, B. S. (1958) *Am. J. vet. Res.*, **19**, 725.
Vasquez, C., & Kleinschmidt, A. K. (1968) *J. molec. Biol.*, **34**, 137.
Vasquez, P., & Tournier, P. (1964) *Virology*, **24**, 128.
Verwoerd, D. W. (1970) *Prog. med. Virol.*, **12**, 192.
Verwoerd, D. W., Louw, H., & Oellermann, R. A. (1970) *J. Virol.*, **5**, 1.
Vosdingh, R. A., Trainer, D. O., & Easterday, B. C. (1968) *Can. J. Comp. Med.*, **32**, 382.
Wallis, C., Melnick, J. L., & Bianchi, M. (1962) *Tex. Rep. Biol. Med.*, **20**, 693.
Wallis, C., Smith, K. O., & Melnick, J. L. (1964) *Virology*, **22**, 608.
Walters, M. N.-I., Leak, P. J., Joske, R. A., Stanley, N. F., & Perret, D. H. (1965) *Br. J. exp. Path.*, **46**, 200.
Walters, M. N.-I., Joske, R. A., Leak, P. J., & Stanley, N. F. (1963) *Br. J. exp. Path.*, **44**, 427.
Watanabe, M. (1952) *J. Jap. vet. Med. Ass.*, **5**, 109.
Watanabe, Y., & Graham, A. F. (1967) *J. Virol.*, **1**, 665.
Watanabe, Y., Millward, S., & Graham, A. F. (1968) *J. molec. Biol.*, **36**, 107.
Wilhelm, A. R., & Trainer, D. O. (1966) *J. Wildl. Mgmt.*, **30**, 777.
Wilsnack, R. E., Blackwell, J. H., & Parker, J. C. (1969) *Am. J. vet. Res.*, **30**, 1195.
Winterfield, R. W., Hitchner, S. B., Appleton, G. S., & Cosgrove, A. S. (1962) *L. & M. News*, **3**, 1.
Whitehead, R. H., Doherty, R. L., Domrow, R., Standfast, H. A., & Wetters, E. J. (1968) *Trans. R. Soc. trop. Med. Hyg.*, **62**, 439.
Wolf, K. (1966) *Adv. Virus Res.*, **12**, 35.
Wolf, K., Snieszko, S. F., Dumbar, C. E., & Pyle, E. (1960) *Proc. Soc. exp. Biol. Med., U.S.A.*, **104**, 105.
Wooley, R. E., & Gratzek, J. B. (1969) *Am. J. vet. Res.*, **30**, 1027.
Young, S., & Cordy, D. R. (1964) *J. Neuropath. exp. Neurol.*, **23**, 635.
Yunker, C. E., & Cory, J. (1967) *Exp. Parasitol.*, **20**, 267.

3

Togaviridae

The name 'togavirus' has been proposed (Andrewes, 1970) to cover the A and B groups of arboviruses. The name has not yet received official recognition, but the included viruses have so many characters in common that a family name to cover them all is certain to be required. Other arboviruses are included in existing genera such as reovirus (Chapter 2) and rhabdovirus (Chapter 8) or remain unclassified (Chapter 4).

Reviews: WHO (1967) (Arboviruses and human disease). Taylor (1967)—a catalogue summarizing information for all known arboviruses, including many described in Chapters 2, 4, and 8.

The following properties characterize the family: virions are 40 to 80 nm (Alphaviruses) or 20 to 50 nm (Flaviviruses) in diameter and consist of a nucleocapsid with cubic symmetry and an envelope. Studded with projections. Sensitive to ether, chloroform and deoxycholate (Theiler, 1957). Contain single-stranded RNA of MW about 3×10^6 daltons. Multiplication in the cytoplasm and maturation by budding cytoplasmic membranes. Agglutinate goose or newly hatched chick RBC (Clarke & Casals, 1958; Porterfield, 1957). Antigenic relationships within each genus revealed most easily by HAI, less by CF and least by VN. Nearly all produce encephalitis in suckling mice.

There have been many able reviews of the ecology of arboviruses: Eklund (1953), Smith (1962), Mattingly (1960), Reeves (1958; 1961). Smith (1962) dealt especially with the role of ticks, Williams et al. (1964) with the importance of bats in Africa and Miles (1964) with the ecology of arboviruses in the Pacific region.

ALPHAVIRUSES

The name 'alphaviruses' has been given to the members of the group A arboviruses of Casals & Brown (1954). These were placed in a separate group primarily on the basis of immunological tests, but other characters also serve to delineate the group. Fundamental characters are probably the same for all. The following account will specify by names or letters in parentheses which viruses have yielded information in particular tests. Sindbis (SN) and Semliki forest (SF) viruses have been particularly studied since with them there is little danger of serious infection of laboratory workers. Less information is available for the viruses of Eastern (EEE), Western (WEE), and Venezuelan Encephalitis (VEE).

Morphology. Virions are spherical or hexagonal in outline: recent estimates of diameter vary between 40 and 80 nm. There is a capsid with icosahedral symmetry; apparently 32 ring-like morphological units are arranged in a hexamere-pentamere pattern on a symmetrical lattice (Horzinek & Mussgay, 1969 (SN)). There is an outer membrane with surface-projections 6·5 to 10 nm long. (Simpson & Hauser, 1968 (SN, SF, Middleburg)). Within is a nucleoid 25–30 nm in diameter (Chain et al., 1966; Higashi et al., 1967 (Chikungunya)): tubular structures may be seen within this (Erlandson et al., 1967 (SF)). Other accounts are consistent with the above description (Faulkner & McGee-Russell 1968 (SF), Acheson & Tamm, 1967 (SF), Osterrieth & Calberg-Bacq, 1966 (SF), Lascano, 1969 (Aura)). Smaller particles, presumably precursors, have been seen; nucleoids seem to be formed in the cytoplasmic matrix: they migrate to the plasma membrane, where complete virions are formed by budding on the cell surface or into vacuoles (Morgan et al., 1961 (WEE), Acheson & Tamm, 1967 (SF), Bykovsky et al., 1969 (VEE), Grimley et al., 1968 (SF)).

Treatment with Tween separated groups of haemagglutinating particles, mostly with projections, resembling the rosettes described for myxoviruses (Mussgay & Rott, 1964). Particles seen within mosquito salivary glands resemble those seen in vertebrate cells (Janzen et al., 1970).

Chemical composition. The virions contain single-stranded RNA, but a two-stranded RNA can be extracted from infected cells and may be the form in which it first appears during virus synthesis; a third species of RNA can also be recognized. The sedimentation constants of the three RNAs are about 22S, 26S and 42S (Sonnabend et al., 1967 (SN), Ben-Ishai et al., 1968 (SN), Friedman et al., 1966 (SF), Kaariainen & Gomatos, 1969 (SF)). Dobos & Faulkner (1970) give the MW of Sindbis RNA as $3{\cdot}89-4{\cdot}45 \times 10^6$ daltons. The RNA accounts for 6·2 per cent of the weight of the virion (Wachter & Johnson 1962 (VEE)). An infectious RNA has been described for several members. Base composition of RNA (SN): A 27·4; C24·4; G26·1; U22·2 (Sonnabend et al., 1967).

Hay et al. (1968) identified 6 viral proteins: one was the lipoprotein of the envelope and another the ribonucleoprotein (SF). A glycoprotein with a MW of 53,000 is present in the virus envelope (Strauss et al., 1970; Burge & Strauss 1970 (SN)).

Physico-chemical characters. Density of complete virions 1·20 g/ml (Horzinek & Munz, 1969 (VEE)), or 1·24 g/ml (Mussgay & Rott 1964 (SF)): that of the HA released by Tween-80 was 1·23. Infectivity is unaffected by trypsin. Some are more resistant to bile salts than other arboviruses (Hardy et al., 1965). Less stable at 5·1 to to 5·7 than on either side of that range (Finkelstein et al., 1938 (EEE)). Survival at various temperatures is reported by Lockhart & Groman, 1958 (WEE) and Sanna & Angelillo, 1957 (SF). Photosensitivity was studied by Appleyard (1967 (SF)).

Haemagglutination. Unaffected by trypsin or by phospholipids (Porterfield & Rowe, 1960). Laboratory-adapted strains of WEE may fail to show haemaglutination (Chanock & Sabin, 1954).

Antigenic characters. Every member of this important group of 20 viruses shows cross-reactions in the haemagglutinin-inhibition test with one or more, usually numerous, others in the group. For this purpose one uses sera from rabbits or other animals which have had multiple inoculations. Other available serological tests are more useful for revealing differences between individual members. Active immunity tests show considerable cross-protection between members of the group. A plaque-inhibition test in cultures of chick fibroblasts (Porterfield, 1961) is particularly useful in revealing relationships and differences.

Eastern Equine Encephalomyelitis (EEE)

Review: Schaeffer & Arnold (1954).

Antigenic properties. Minor antigenic differences are demonstrated between strains of the virus coming from North and South America respectively (Casals, 1964). J. R. Henderson et al. (1967) found that freshly isolated viruses contained subpopulations differing somewhat in antigenic and other properties. On serial passage the relative proportions of these might change so that initially different viruses came to resemble each other.

Improved methods of making diagnostic antigens have been described by Pennington & Gibbs (1967) and Palmer et al. (1968).

Cultivation. In eggs, the virus grows well after inoculation by various routes and is rapidly lethal for the embryos. It grows in tissue cultures of chick embryo, mouse, hamster, monkey, guinea-pig and other mammals, also in HeLa cells. There is a rapid destructive effect. Chronic infection may be set up in some lines of human and rat cells at 31° (Bang et al., 1957). Multiplication is even reported in fish (*Gambusia*) embryos.

Distribution. The virus's distribution covers eastern United States, with separate forms in Michigan and Wisconsin, also the Caribbean, parts of Central America, and eastern South America as far as Argentina.

Pathogenicity. This is probably normally a harmless parasite of some wild birds, possibly also small rodents, reptiles and amphibia with man and horses infected incidentally by mosquito bite.

In infected horses there is a preliminary phase of fever and viraemia, followed by invasion of the CNS. There may be mild excitement, soon followed by other abnormal behaviour, somnolence, paralysis and death. Mortality is up to 90 per cent, and mortality has been heavy in the United States; in 1938, 184,000 horses were infected by either this virus or the Western strain.

In man, mortality is also high, up to 74 per cent, but incidence has never been as high as in horses. It is likely that strains from different areas vary in virulence, since in some districts there is evidence of subclinical immunization in the absence of overt disease. As in horses, the disease tends to be diphasic; in the second phase of nervous involvement, there occur convulsions, rigidity and coma. Most survivors show paralytic or mental sequelae. Laboratory infections are recorded. Outbreaks with high mortality also occur amongst pheasants (Jungherr & Wallis, 1958).

Experimentally the disease is more invasive than the Western virus (p. 80). Mice, guinea-pigs, goats, newly hatched chicks and several other species, including snakes and turtles (Hayes et al., 1964), are readily infected by various routes; with older animals IC inoculation is usually necessary (Sabin & Olitsky, 1938). Monkeys may be infected IC or IN (Wyckoff & Tesar, 1939). Chamberlain et al. (1954) have described the usefulness of chicks, emphasizing that they should be not more than half a day old. Pheasants and some other birds not native to America develop fatal infections, but there may be only viraemia in egrets, ibises and grackles. In birds the virus infects viscera, particularly liver, rather than CNS (Kissling et al., 1954). Intra-ocular infection of rabbits produces a 'toxic' corneal reaction (Evans & Bolin, 1946).

The virus multiplies after inoculation into various mosquitoes, ticks and even insects of other orders: Orthoptera, Heteroptera and larval Coleoptera and Lepidoptera (Hurlbut & Thomas, 1960).

Pathological lesions. In horses, destruction of grey matter in the cerebrum and elsewhere is evident (Kissling & Rubin, 1951). In man the main lesions are in the brain stem; pulmonary oedema also occurs.

Ecology. The virus spreads naturally amongst birds in a 'sylvan cycle', the main vector being probably *Culiseta melanura*. Various other Culicine mosquitoes (*Aëdes*, *Culex* spp) may be involved in spread to man and horses; a number of these readily transfer infection experimentally (Hayes et al., 1962). The extrinsic incubation period is from 3 to 12 days, varying with outside temperature (Chamberlain & Sudia, 1955). There is a definite association with salt marshes probably because of the important vector role which the salt-marsh mosquito *Aëdes sollicitans* plays. Stamm (1958) suggests that the normal endemic spread may at times become explosive and that only then are there infections of man and horses.

The disease is evident in late summer and doubtless depends on spread from infected birds. Isolation of virus during the winter has been made not from birds but from small rodents, reptiles and amphibia, and it is likely that there is a basic cycle involving these species; upon this the summer mosquito-bird cycle may be superimposed (Altman et al., 1967).

Infection may spread amongst pheasants in the absence of an arthropod vector (Holden, 1955). Virus is shed from infected horses in nasal secretions, urine and milk, and contact infection may occur (Kissling et al., 1956). Cannibalism may spread infection amongst mice (Traub & Kesting, 1956).

Control. A living virus attenuated by passage through pigeons was used by Traub & Ten Broeck (1935). Formolized vaccines were proved to be effective in guinea-pigs by Cox & Olitsky (1936). Similar vaccines made from infected chick embryos have been more recently and successfully used in horses (Beard et al., 1938) and, on a small scale, to protect exposed laboratory workers (Maurer et al., 1952).

Western Equine Encephalomyelitis (WEE)

Antigenic properties. Crossing with other A arboviruses in the HAI test is described by Casals (1958). Parks & Price (1958) using cross-protection tests and Porterfield (1961) using a plaque-inhibition test find evidence that WEE is more closely related to Sindbis than to others of the group—a fact also apparent from HAI tests. Neutralization, as with many other arboviruses, is improved by addition of fresh normal serum (Whitman, 1947). It may be demonstrated by tests in mice or by various tissue-culture techniques. In infected human beings neutralizing antibodies develop quickly and persist long: CF antibodies appear more slowly and are more transient. Henderson (1964) has described phase variation in the virus's antigen. As with the Eastern virus there are differences between isolates from different areas. For improved diagnostic antigens, see EEE, p. 78.

Cultivation. As with EEE, the virus can be grown in eggs inoculated by various routes and is rapidly lethal for the embryo; it can similarly be grown in tissue cultures of cells from many species, hamster-kidney cultures being especially useful; it will produce plaques on monolayers of chick-embryo cells. A colour test depending on change of pH can be used as a visible means of titration (Brown, 1958). In the chick embryo it multiplies particularly in vascular endothelium (Nir et al., 1957). A chronic inapparent infection can be set up in a continuous line of mouse (L) cells (Chambers, 1957; Lockart, 1960).

Distribution. The virus is spread over most of the United States and southern Canada, but infections of horses and man are not reported from the eastern seaboard. It also extends to South America as far as Argentina.

Pathogenicity. Presumably an avian infection as in EEE, horses and man being attacked only incidentally. Symptoms in these species are as for EEE, but the mortality is considerably lower—usually 20–30 per cent for horses (rarely 40 per cent) and 10 per cent for man, on the average. Sequelae are uncommon. Many inapparent infections apparently occur in man. Virus has been recovered from naturally infected birds, chiefly passerines, also from squirrels, cow, deer and pig.

Experimentally, the virus, given IC, will produce meningo-encephalomyelitis in a large range of laboratory and wild rodents, monkeys, rabbits, young dogs, deer, pigs, calves, goats and some birds. Hamsters of all ages and young mice and guinea-pigs can be infected by IM, IP and SC injection. As with EEE, half-day-old chicks are highly susceptible (Chamberlain et al., 1954). Virus given IV to rabbits or young mice or intra-ocularly to rabbits produces symptoms which are perhaps referable to a 'toxin' which is not separable from the virus itself (Fastier, 1952; Evans & Bolin, 1946).

Virus multiplication in various arthropods is as described under EEE (p. 80). Thomas (1963) has described the course of infection in *Culex tarsalis.*

Pathological lesions in the brain of fatal cases in man are those of diffuse encephalitis with scattered neuronal necrosis, accumulation of glial cells with only moderate cell infiltration of meninges or around blood vessels. Lesions in other viscera are not striking. Changes in other species are similar.

Ecology. As with EEE there is a summer cycle involving birds and mosquitoes; *Culex tarsalis* is of major importance in conveying infection to horses and man. There have been several suggestions as to how virus persists through the winter. It may survive for months in hibernating mosquitoes (Reeves et al., 1958, Bellamy et al., 1967). It is likely that, as with Eastern virus, a basic cycle in small rodents, reptiles and amphibia is more important (Altman et al., 1967) (see p. 80). Garter snakes (*Thamnophis*) may be reservoirs; in them viraemia may be intermittent (Thomas et al., 1960; Burton et al., 1966). Since virus is present in respiratory secretions and kidneys of infected mice, there exists the possibility of infection other than via arthropods (Froeschle, 1964).

Control. A formolized vaccine has been used, as with EEE (p. 80). A variant of low pathogenicity has been developed (Johnson, 1963); this shows promise of being useful for immunizing horses (Binn et al., 1966).

Venezuelan Equine Encephalomyelitis (VEE)

Review: Chamberlain (1968).

Antigenic properties. Crosses with other As in the HAI test (Casals, 1958). Readily separable from the others by CF and neutralization tests. Two viruses from the Amazon (Mucambo and Pixuna) are rather closely related to WEE (Shope et al., 1964). Scherer & Pancake (1970) suggest that these 2 and VEE be considered to be serological subtypes of 1 virus. Using a kinetic HAI test Young & Johnson (1969) found differences between strains from different parts of South and Central America and Florida.

Cultivation. Grows readily in *fertile eggs*, killing embryos in less than 48 hours, and in *cultures* of several primate tissues, chick embryo, L (mouse) and guinea-pig cells. Chronic infections have been set up in L cells. In chick embryo, HeLa and guinea-pig heart cultures, the virus is attenuated so that it no longer readily infects guinea-pigs, rabbits or adult mice, when inoculated by peripheral routes (Koprowski & Lennette, 1946; and other authors). Virulence was soon restored by mouse-passage.

Distribution. Venezuela, Colombia, Equador, Panama, Trinidad, Brazil. Since 1964 it has appeared in Mexico (Scherer et al., 1964) and Florida (Chamberlain et al., 1964).

Pathogenicity. Infected horses and donkeys become ill, not necessarily with encephalitis. There may be only fever, depression and diarrhoea, but some develop nervous symptoms and most often die (Kissling et al., 1956). In man, after an incubation period of 2–5 days, there is fever with severe headache, and there may be CNS involvement (tremors, diplopia and lethargy). In an outbreak in Venezuela in 1962–64, there were more than 30,000 human infections with 1,199 cases of nervous involvement and 300 deaths (Sellers et al., 1965; Kissling & Chamberlain, 1967).

Laboratory infections occur much more readily than with other A arboviruses, and many are on record (Lennette & Koprowski, 1943; Slepushkin, 1960); infection is probably by inhalation.

Experimentally, the disease is more virulent for laboratory rodents than WEE or EEE; it readily infects when given by peripheral routes. Besides horses, dogs, cats, sheep and goats are susceptible, but not cattle. Inoculated birds show no symptoms unless after very large doses; virus in their blood is of low titre. In contrast, there is viraemia at a high level in horses and other infected mammals. It is therefore suggested that the unknown natural reservoir is probably mammalian rather than avian.

Pathological lesions. The virus is more viscerotropic than neurotropic. Lesions in the brain affect vessels rather than neurones. In horses there is severe damage to blood-forming tissues—including spleen and lymph nodes—and necrotic foci in the pancreas occur regularly (Kissling *et al.*, 1956). 'Lympho-myelopoietic necrosis' is seen also in infected rodents (Victor et al., 1956). Kundin et al. (1966) studied the distribution of virus in the organs of inoculated suckling and young adult mice.

Ecology. The natural reservoir, though suspected to be mammalian, is unknown. Antibodies have, however, been found in wild rodents (*Oryzomys*, *Peromyscus* and *Sigmodon*) (Chamberlain et al., 1969; Rossi 1970; Sidwell et al., 1967). The main vectors are probably *Aëdes taeniorhynchus* and *scapularis*, though those of other genera will transfer infection experimentally. Virus was isolated from rafts of eggs laid by *Mansonia perturbans* (Chamberlain et al., 1956).

Virus is present in nose, eye, mouth, urine and milk of infected horses and direct contact infection from horse to horse can occur (Kissling et al., 1956). Contact infection also occurs between infected mice and cotton-rats (Zarate & Scherer, 1968). Virus has been found in the pharynx of infected human beings; so the same may be true of man. Transmission between horses and man or from man to man by mosquito bite is also possible as there is a high level of viraemia in both mammalian hosts (Rossi, 1967).

Control. Formolized vaccines have been prepared from infected chick embryos for protecting horses and exposed laboratory workers. However, 14 infections occurred in laboratory workers receiving ostensibly inactive vaccine (Sutton & Brooke, 1954): possibly man is a more delicate indicator of the presence of virus than are the systems used for safety-testing the vaccine. A live attenuated virus vaccine may be practicable (Alevizatos et al., 1967).

For Near East equine encephalomyelitis see p. 130.

SINDBIS VIRUS

Description: Taylor et al. (1955). This is to be the type species of the genus *Alphavirus*.

Antigenic properties. Several immunological tests indicate a closer relationship to Western Equine Encephalomyelitis (p. 81) than to other A arboviruses. There are minor antigenic differences between strains coming from different areas (Casals, 1961). The New Zealand virus Whataroa is closely related to Sindbis (Doherty, 1967; Maguire et al., 1967).

Cultivation. Multiplies in fertile eggs and is very lethal to the embryos; grows also in cultures of chick, human and monkey cells with CPE in most of them (Frothingham, 1955). Plaques are produced on chick-embryo monolayers (Shah et al., 1960; Porterfield, 1961). Frothingham (1963) found that plaque-size was larger in the presence of mumps virus. Experimentally, the virus was found to multiply in *Drosophila* (Herreng, 1967).

Distribution. Egypt, South Africa, India, Malaya, the Philippines, Australia (the only arbovirus occurring in each of 4 zoo-geographical regions).

Isolated from *Culex* spp. in Egypt and South Africa, also from several species of birds.

Pathogenicity. May be associated with fever in man (Report 1961–62).

Experimentally the virus is lethal for suckling not adult mice; in the infant mice it causes diffuse myositis and encephalitis. Studies with fluorescent antibody detected viral antigen expecially in muscle of all types and in vascular endothelium (Johnson, 1965). In inoculated monkeys of the genera *Macaca* and *Cercopithecus* and in several species of birds, there was a low-grade viraemia, but no symptoms of disease were noted (Taylor et al., 1955, Weinbren et al., 1956).

Ecology. Probably an infection of birds transmitted by *Culex* spp., but neutralizing antibodies have been found not only in birds but in sera from man and domestic ungulates in endemic areas.

SEMLIKI FOREST VIRUS

Description: Smithburn et al. (1944).

Synonym: Kumba virus.

Antigenic properties. The virus shows the usual cross-reactivity with other A arboviruses in the HAI test; it crosses rather better with Mayaro and Chikungunya than with some others (Casals, 1958). Appleyard et al. (1970) obtained 3 antigens by treating envelopes with trypsin. Antigen 1 was associated with haemagglutinating and neutralizing-antibody-blocking properties and had a MW of >200,000 daltons; the other 2 were smaller. Livers of recovered mice contain a protective substance (Shope, 1961).

Cultivation. Lethal to inoculated chick embryos (Smithburn, 1946). Grows in tissue cultures of many species. Henderson (1961) described plaque-formation in cultures of Cebus, dog, pig, hamster, guinea-pig and chick tissues. Virus in tissue cultures could be titrated rapidly by the use of a fluorescent antibody technique (Carter, 1969). Peleg (1969) described inapparent persistent infection of continuously grown *Aëdes aegypti* cells.

Distribution. Only 3 isolations are reported—from *Aëdes* or *Eretmapodites* mosquitoes in Uganda, Mozambique and Cameroon respectively (the Kumba strain from Cameroon is best considered as a Semliki virus). Serological studies (Smithburn et al., 1954) suggest that the virus may occur in Malaya and North Borneo.

Pathogenicity. Not associated with any known illness.

Laboratory infection, presumably inapparent, can occur, as shown by development of antibodies (Clarke, 1961).

Experimentally the virus causes encephalitis when injected into adult mice by various routes, or IC into guinea-pigs, rabbits, rhesus and red-tail monkeys. Apart from lesions of encephalomyelitis, most inoculated animals have shown kidney damage (Smithburn & Haddow, 1944). Viraemia lasting several days occurred in inoculated birds of several species, also in hamsters infected by various routes (Davies et al., 1955). The course of infection was studied in mice (Seamer et al., 1967) and hamsters (D. W. Henderson *et al.*, 1967); the latter could be infected by inhaling aerosols of virus.

Mims et al. (1966) found changes in the salivary glands of infected *Aëdes aegypti*; when these were maximal the mosquitoes had difficulty in obtaining a blood-meal.

Ecology. The natural hosts and vectors are unknown. Antibodies have been found in sera of human beings and 6/12 species of wild primates in Uganda (Smithburn et al., 1944). The virus multiplies in *Aëdes aegypti* which could function as an efficient vector (Woodall & Bertram, 1959). Anopheline mosquitoes can transmit infection experimentally (Collins, 1963).

Control. An antibiotic (Helenine) was found to be of some value in experimentally infected mice (Shope, 1953).

CHIKUNGUNYA

Description: Ross (1956).

The name is a native one meaning 'that which bends up' from the contorted position of a sufferer and is not, as with most arboviruses, a place name. Most properties of the virus have not yet received much study.

Antigenic properties. By HAI tests it lies closer to Mayaro and Semliki viruses than to other A arboviruses. Rises in neutralizing antibody titres occur as a result of infection in man (Mason & Haddow, 1957). There are minor antigenic differences between African and Asian strains.

Cultivation. Multiplies with CPE in chick-embryo fibroblasts and in kidney cells of Pekin duck and rhesus, also in HeLa cells (Buckley, 1959; Henderson & Taylor, 1960).

Habitat. Tanzania, Uganda, Congo, South Africa, Thailand, India.

Pathogenicity. An acute, dengue-like, fever, often biphasic; excruciating pains occur in joints and spine of affected persons. A maculo-papular rash commonly occurs in the second phase of fever. After recovery from fever, joint pains may recur for some months (Robinson, 1955). Closely related viruses have been recovered from patients with haemorrhagic fever in Thailand and in India (Rao & Anderson, 1964).

Experimentally, suckling mice die in a few days when inoculated IC, but are less readily infected IP. Older mice are much more resistant. Guinea-pigs and rabbits show no symptoms on inoculation.

A strain from Thailand produced multiple haemorrhages in intestines and elsewhere in suckling mice, rats and hamsters (Halstead & Buescher, 1961); strains from Africa have also caused haemorrhages in mice.

Pathological lesions in mice: encephalitis affecting especially Ammon's horn, myositis, myocarditis (Weinbren et al., 1958).

Ecology. Antibodies have been found in sera of wild vervet monkeys, baboons and several species of birds in southern Rhodesia (McIntosh et al., 1964) and in Thailand in sera of numerous mammals other than rodents (Halstead & Udomsakdi, 1966). Virus was isolated in Tanzania from *Anopheles* spp, *Aëdes aegypti* and *Culex fatigans* (Lumsden, 1955) and from *Aëdes africanus* in Uganda (Weinbren et al., 1958b). Lumsden considers *A. aegypti* likely to be the most important vector. The virus was unknown until isolated from an epidemic in Tanzania in 1952; at the height of this epidemic, the incidence in many villages was 60–80 per cent, all ages being attacked.

O'NYONG-NYONG

Description: Haddow et al. (1960). Williams & Woodall (1961).

Antigenic properties. Closely related in the HAI test to Chikungunya, but separable by the plaque-inhibition test (Porterfield, 1961).

Cultivation. On chick-embryo monolayers (Williams et al., 1962).

Distribution. Uganda, Kenya, Congo.

Pathogenicity. Clinical features in man similar to those of Chikungunya (p. 86) with lymphadenitis as a differentiating feature. Similar severe joint pains and rash occur (Shore, 1961).

The virus is pathogenic for suckling mice when given IC; for primary isolation it is best to use diluted sera. Older mice are resistant. Infant mice which do not die show stunting of growth and patchy alopecia (Williams & Woodall, 1961).

Ecology. The infection began and spread through northern Uganda in 1959. Five million Africans were affected (Chamberlain, 1963). Shore (1961) discusses the epidemiology. The vector appears to be *Anopheles* (*Gambiae* and *funestus*); this is thus the only arbovirus for which epidemic spread by Anophelines is known (Corbet et al., 1961).

OTHER ALPHAVIRUSES

Information is scantier concerning 6 other viruses.

Mayaro virus caused an epidemic of fever and severe headache involving more than half a force of labourers working in quarries and forest in the state of Para in Brazil and has also been troublesome in Surinam.

Uruma virus gave rise to a similar outbreak involving about half of 400 Okinawan pioneers making a settlement in Bolivia.

Middelburg virus was isolated in Cape province in South Africa from *Aëdes* captured during an epizootic amongst sheep. The virus was pathogenic for lambs, but the relation of the virus to the epizootic is uncertain since Wesselsbron virus (p. 102) was also present.

The information available concerning these and 2 other A viruses is presented in Table 5. Eight others have been described recently.

All the viruses in the table are pathogenic for suckling mice when given IC (usually IP also), but adult mice have symptomless infections.

Epidemic Exanthema and Polyarthritis

An epidemic disease occurring in Australia and having the characters indicated by the name was associated with a rise in antibodies to the A group of arboviruses (Shope & Anderson, 1960); the rises were more especially against Ross River virus (Doherty, quoted by Scherer, 1967).

FLAVIVIRUS

Synonym: 'B' group of arboviruses.

Review: Clarke & Casals (1965); Taylor (1967, catalogue).

The flaviviruses form the largest group of arboviruses; more than 30 have been described. The name is associated with the inclusion in the genus of yellow fever virus as type species. The group was originally delineated on the basis of immunological relationships.

These mutual relations are best brought out by the HAI test, using sera obtained after several inoculations of rabbits or other species.

Table 5

MISCELLANEOUS ALPHAVIRUSES FROM OLD WORLD

	Bebaru	Getah	Middelburg	Ndumu	Ross River	Whataroa	7 62 33
Where isolated	Malaya	Malaya, Japan, Australia	Central and South Africa	South Africa	Australia	New Zealand	U.S.S.R.
From what species	Mosquitoes	Mosquitoes	Mosquitoes	Mosquitoes	Birds, mosquitoes	Mosquitoes	Mosquitoes
Association with diseases		None	Isolated during epizootic in sheep		Possibly poly-arthritis in man		
Related alpha-virus (HAI)		Semliki (1-way)	Possibly Semliki Sindbis			Sindbis	
Pathogenicity for species other than mice		Viraemia in chicks	Fever in lambs; viraemia in chicks				
Antibodies found naturally		Pigs, horses, a few men & birds	Sheep; few in man & rodents		Man		
References	Scherer et al. (1962)	Kokernot et al. (1957)			Doherty, quoted by Scherer (1967)		

TABLE 6

MISCELLANEOUS ALPHAVIRUSES FROM NEW WORLD

	Aura	Highlands J	Mayaro	Mucambo	Pixuna	Una
Where isolated	Brazil	U.S.A.	Central & South America	Central & South America	Brazil	Central & South America
From what species	Mosquitoes	Rodents, birds, mosquitoes	Man, mosquitoes	Man, rodents, birds, mosquitoes	Rodents, mosquitoes	Mosquitoes, sentinel mice
Association with diseases	None		Epidemic of fever in man		None	
Related Alphaviruses (HAI)	WEE; Sindbis		Semliki	VEE	VEE	
Pathogenicity for species other than mice						
Antibodies found naturally	Very low per cent in man; various mammals		Man, monkeys			
References	Causey et al. (1963)		Casals & Whitman (1957); Causey & Maroja (1957)	Shope et al. (1964) Scherer & Pancake (1970)	Shope et al. 1964 Scherer & Pancake (1970)	Causey (1963)

Differences are revealed by the more specific neutralization and CF tests.

Many of the features of the group are shared by other arboviruses (see p. 76). They are not very different from the alphaviruses (p. 77), but differ immunologically and in that their infectivity and haemagglutins are sensitive to inactivation by trypsin.

Morphology. Virions are spherical, enveloped, 20–50 nm in diameter. They multiply in cytoplasm; specific antigen is detected especially near the nucleus. Maturation is by budding into cytoplasmic vacuoles which then pass to the cell surface and liberate virus (Japanese B—Ota, 1965). Spiky surface projections are described. There is a nucleoid about 28 nm across, consisting of a core 25 nm across surrounded by a triple-layered membrane (Japanese B and Murray Valley viruses—Filshie & Rehaček, 1968). Attempts to visualise the structure by negative staining have been difficult because particles are very unstable, but evidence suggests that symmetry is cubical (Powassan virus—Abdelwahab et al., 1964); this virus was seen in pseudocrystalline aggregates within the cell. Slavik et al. (1970). considered that tick-borne encephalitis virus resembled alphaviruses, but suggested that there might be 92 capsomeres.

Chemical composition. Contain 7–8 per cent single-stranded RNA. GC content 47–49 per cent. Essential lipid is present in the envelope. Infectious RNA has been obtained from several members (West Nile; Japanese B; yellow fever; dengue; turkey meningoencephalitis; tick-borne encephalitis). Four structural and 3 other proteins were isolated from Kunjin virus (Westaway & Reedman, 1969).

Physico-chemical characters. Buoyant density in CsCl 1·25 g/cm^3. Sensitive to ether and desoxycholic acid. Inactivated in 15′ at 56°. Labile at pH 10·7; optimum stability at pH 8·5 (Duffy & Stanley, 1945—Japanese B). Conditions for survival of tick-borne encephalitis are described by Grešikova-Kohutova (1959a & b).

Haemagglutination. As described on p. 76.

WEST NILE

Many papers refer to the Egypt 101 strain of this virus.

Original Description: Smithburn et al. (1940).

Antigenic properties. HAI tests reveal that this virus, together with Japanese B and Murray Valley and St. Louis viruses, forms a related subgroup within the B viruses. Ilhéus is also related though less closely. West Nile is more readily inactivated by sera made against these other viruses than are they by West Nile sera (Smithburn, 1942). Clarke's (1960) studies with antibody absorption are particularly valuable. Strains from different areas show minor antigenic differences: those from India stand rather apart from those from West Pakistan, Africa, the Middle East and France (Price & O'Leary, 1967).

Cultivation. Multiplies in *fertile eggs* forming plaques on the CAM and growing also after inoculation into yolk sac. Grows also in *tissue cultures* of chick embryo and many mammalian tissues, hamster kidney being especially useful. A CPE is best seen in this or in chick embryo, monkey kidney or human cancer lines. Plaques form on chick embryo or monkey-kidney monolayers under appropriate conditions (Lavillureix & Reeb, 1958; Bhatt & Work, 1957). Chronic infection may be set up in a line of mouse cells. A temperature-resistant mutant has been described by Katz & Goldblum (1968). The virus will grow in mosquito cells (Rehaček, 1968). Koprowski & Lennette (1946) record attentuation of the virus after cultivation, so that it no longer killed mice or hamsters when inoculated by peripheral routes.

Distribution. Egypt, Uganda, South Africa, probably other parts of Africa, Israel, India, south of France.

Pathogenicity. In Egypt and elsewhere it is endemic and infects man, chiefly children, as a usually silent infection. It may however be revealed, and especially in outbreaks in Israel, as a short febrile illness simulating dengue, sand-fly fever or rubella. Headache, adenopathy, maculo-papular rash, sore throat and limb pains occur (Bernkopf et al., 1953). Laboratory infections are recorded. Inoculations into cancer patients usually produced only mild fever, but 11 per cent had symptoms of mild encephalitis. Temporary regression of some cancers occurred (Southam & Moore, 1954). Horses and donkeys in Egypt may be infected (Schmidt & El Mansoury, 1963).

Experimentally the virus causes encephalitis when inoculated IC into rodents of several species, chicks or rhesus monkeys, but only fever in *Cercopithecus* (Smithburn et al., 1940). Mice and hamsters may be infected when inoculated by various routes, including exposure to a virus aerosol; after 4 to 5 days they become excited, hunched-up, rough and may show paraplegia before death. Chicks infected by mosquito bite show viraemia for several days.

Pathological lesions in mice resemble the encephalomyelitis caused by other arboviruses.

Ecology. Birds are probably the normal hosts, virus being transmitted by Culicine mosquitoes, chiefly, in Egypt, *Culex univittatus*. Secondary infection cycles may involve man and, less certainly, domestic quadrupeds. *Culex pipiens* may play a role in carrying the infection through the winter (Taylor et al., 1956). Hurlbut (1956) infected several species of *Culex*. Ticks (*Ornithodorus moubata*) can also be infected and will transmit infection, but their role in nature is uncertain (Whitman & Aitken, 1960). A virus very similar to West Nile was recovered from *Hyalomma* ticks in the southern U.S.S.R. (Kunz, 1965). Antibodies to the virus occur to varying extents in different parts of Africa in sera of human beings, monkeys, domestic quadrupeds and birds (Kokernot et al., 1956).

JAPANESE B

Synonyms: Japanese encephalitis; Russian autumn encephalitis.

Antigenic properties. See West Nile virus (p. 91). Minor differences between strains occur, and antigenic modifications may follow passage through mice (Okuno et al., 1968). Neutralizing and HAI antibodies appear within a week in affected persons and persist for years; CF antibodies appear much later and disappear sooner (Southam, 1956).

Cultivation. Growth occurs in fertile eggs, best after inoculation into the yolk sac; resulting embryo deaths could be used as a basis for titration (Howitt, 1946). Virus passed 50 times in 1-day-old chick embryos (Okuno, 1959) lost pathogenicity for mice.

Growth occurs in tissue cultures of chick embryo and various mammalian cells, sometimes without definite CPE, but strains have been adapted to produce CPE or plaques on monolayers of chick-embryo, pig, monkey or hamster-kidney cells (Bhatt & Work, 1957; Inoue et al., 1961). Multiplication occurs in cultures of moth (Suitor, 1966) or mosquito (Rehaček, 1968) tissues. On the latter plaques may develop (Suitor, 1969).

Distribution. Particularly studied in Japan, but apparently extends all over Southeast Asia from Siberia to Malaya and Southeast India.

Pathogenicity. In Japan and elsewhere incidence may be very high especially in children, but only 0·1 to 0·2 per cent of these may show encephalitis (Southam, 1956); the virus causes merely fever or inapparent infection in the others. Cases of encephalitis have varied symptoms including pareses; sequelae are common. Encephalitis occurs also in horses, but in them also there are many infections without nervous symptoms (Hale & Witherington, 1953). It may cause abortions in pigs (Burns, 1950).

Experimentally the virus causes encephalitis, usually fatal, after IC inoculation into mice, hamsters and several species of monkey. Mice can be infected, especially young ones, by peripheral injections. Inoculated rabbits, guinea-pigs, pigs, chicks and other birds usually develop inapparent infections with viraemia. Nishimura et al. (1968) describe morphological and other differences between strains with high and low pathogenicity for mice.

Pathological lesions in the brain are like those of other arbovirus encephalitides, but destruction of Purkinje cells in the cerebellum is particularly noteworthy.

Ecology. The virus is probably a parasite primarily of wild birds, especially night herons. Nestlings of these seem to act as 'amplifiers' to increase the amount of virus in an infected locality during summer months (Buescher et al., 1959; Scherer et al., 1959a). The main vector in Japan is *Culex tritaeniorrhynchus*, but other culicines may play a role elsewhere, e.g. *C. vishnui* in India (Work & Shah, 1956) and *C. gelidus* in Sarawak (Gordon Smith, 1970). Hibernating mosquitoes (Hurlbut, 1950) or bats may permit virus to survive the winter. Domestic pigs are regularly infected and have viraemia adequate to keep virus going in a secondary cycle (Scherer et al., 1959b). Man, horses and other species are only infected incidentally.

Control. Formolized vaccines for protection of human beings have been widely used, but are of dubious value. Hammon & Rhim (1963) attenuated a strain by passage in hamster-kidney cells and considered its possible development for vaccinating man. Vaccines have also been prepared from cultures in pig kidney (Inoue & Nishibe, 1967) and chick embryo (Otani, 1966). A new line of approach is to infect with an avirulent B arbovirus, e.g. West Nile, followed by a 'booster' with inactivated Japanese B (Price et al., 1961).

MURRAY VALLEY ENCEPHALITIS (MVE)

Synonym: Australian X-disease.

Review: Anderson (1954).

Morphology. Diameter between 20 and 50 nm, probably about 25 nm.

Antigenic properties. A member, by HAI tests, of the subgroup of B arboviruses containing St. Louis, West Nile and Japanese B viruses, but more closely related to the last than to the others (McLean, 1956; Pond et al., 1958). Presence of antibody in yolk of eggs laid by immune birds may prove useful in epidemiological surveys (Warner, 1957).

Cultivation. Fertile eggs can be infected by any route and die in consequence. French (1952) suggests that inoculation of the CAM with production of pocks may be the method of choice for primary isolation of virus.

Distribution. Australia, probably endemic in Northern Territory and Queensland, reaching Victoria, New South Wales and South Australia at infrequent intervals, also in Papua (New Guinea) where it may be endemic or epidemic.

Pathogenicity. Encephalitis caused by this virus in man resembles Japanese B encephalitis, but the virus attacks predominantly children in the presumably nonendemic areas of Southeastern Australia (Robertson & McLorinan, 1952). There may be troublesome sequelae. Horses may be infected, but develop encephalitis very rarely.

Experimentally, the virus produces encephalitis when inoculated IC in mice, hamsters, monkeys, sheep and newly hatched chicks. It infects suckling mice when given by peripheral routes and also, in contrast to Japanese B virus, does so in hamsters (Hammon & Sather, 1956). Rabbits, guinea-pigs and birds, including older chicks, usually show only viraemia.

Pathological lesions resemble those resulting from Japanese B.

Ecology. Probably a bird-parasite endemic in New Guinea and northern Australia. Epidemics of encephalitis in Australia in 1917, 1918, 1922 and 1925 (Australian X disease) were almost certainly caused by Murray Valley virus. A virus was isolated by Cleland & Bradley (1917) and subsequently lost. MVE virus was recovered by Miles (1952) from the next (1951) outbreak. Australian workers suggest that after heavy spring rains in the north conditions are suitable for southward spread through the agency of migrating waterbirds. The important vector is almost certainly *Culex annulirostris* (McLean, 1953; Reeves et al., 1954).

ST. LOUIS ENCEPHALITIS

Antigenic properties. See West Nile virus (p. 91).

Cultivation in *fertile eggs* is possible, inoculation being usually into yolk sac or CAM; in the latter case there is a diffuse oedematous lesion with proliferative and necrotic elements. Embryos usually die.

Virus grows in *tissue cultures* of chick, mouse and other species. Plaques are produced on chick-embryo monolayers.

Distribution. The United States, except the East; Panama; Trinidad.

Pathogenicity. Most infections of man cause only a brief febrile illness, but encephalitis may occur in any age group. More than a thousand cases of encephalitis occurred near St. Louis and Kansas City in 1932, and a still larger outbreak (also in the United States) in 1964. Sequelae are uncommon. One laboratory infection is recorded (von Magnus, 1950). Disease in horses is not reported, though they may have viraemia. Virus has been recovered from a sick grey fox (*Urocyon*) (Emmons & Lennette, 1967).

Experimentally, mice, particularly those of certain susceptible strains, develop encephalitis after inoculation IC or IN. Inoculation IP is less certain except in sucklings. Intramuscular injection may cause local myositis. Young rats, hamsters and rhesus monkeys can be infected IC; infection in rhesus is rarely fatal. Guinea-pigs, rabbits, chicks and other birds of several species inoculated by various routes develop symptomless viraemia. Young mice and hamsters can be infected by feeding. In partly resistant mice, challenge may cause a chronic infection (Webster & Clow, 1936) or a flaccid paralysis from myelitis (Cook, 1938).

Pathological lesions in the brain resemble those caused by related viruses, but focal necroses are less commonly seen. Changes in the neurons of infected mice and hamsters during virus development have been described by Zlotnik & Harris (1970).

Ecology. Naturally an infection of birds transmitted by mosquito bite. There may be rural areas in the western United States where the virus is endemic, *Culex tarsalis* being the main vector; in central and eastern regions infection is seen rather in towns, the vectors being *Culex pipiens* and *quinquefasciatus* and, in Florida, *C. nigripalpis* (Chamberlain et al., 1959; Phillips & Melnick, 1967). Virus has been isolated from several other mosquito species, also from bats. How the virus over-winters is unknown.

Control. No effective vaccine is available for man. As with Japanese B, workers are investigating the possibility of using combinations of B arboviruses to obtain effective immunity (Hammon & Sather, 1956).

ILHÉUS VIRUS

Description: Laemmert & Hughes (1947). Koprowski & Hughes (1946).

Cultivation. Grows well when inoculated on the CAM of *chick embryo*: most virus is present in the embryo itself; also in chick embryo *tissue cultures* (Koprowski & Hughes, 1946); and, in hamster-kidney tissue culture, where cytopathic effects are produced (Diercks et al., 1961).

Distribution. Brazil, Colombia, Central America and the Caribbean.

Pathogenicity. Causes infection of man which is normally inapparent, but a few cases of encephalitis are recorded. Three of these occurred amongst 9 infections of cancer patients in whom the virus was under test for possible oncolytic effects (Southam & Moore, 1951). *Experimentally* the virus will cause encephalitis when injected into mice IC; the IP or SC routes are effective only in very young mice. Viraemia was produced after injection by various routes into several species of primates, rodents and marsupials. Of these marmosets (*Callithrix*) circulated virus for the longest period—up to 7 days. In chicks, pigeons and *Sicalis* (a Brazilian canary) no viraemia occurred more than 3 days after injection (Koprowski & Hughes, 1946).

Ecology. Virus has been isolated from several mosquito species, most of them in Trinidad from *Psorophora* spp. (Anderson et al., 1956), also from man in Trinidad (Spence et al., 1962) and from birds in Panama (Galindo & Rodaniche, 1960). Antibodies have been found in sera of many persons in Brazil and Trinidad and in several species of mammals, including horses and birds. The important natural reservoir, however, is unknown.

YELLOW FEVER

Synonyms: Fiebre amarilla.

Reviews: Strode (1951). Theiler (1959). Dick (1953)—epidemiology.

Antigenic properties. Several tests are available for serological study—the IP and IC neutralization tests in mice, CF, HAI and finally plaque inhibition in cultures: they are of varying specificity, the HAI being least specific. In this test the virus shows the usual group reactions with other flaviviruses; Zika and Uganda S viruses seem to be its nearest relatives. Antibody absorption tests (Clarke, 1960) bring out differences between African and South American strains; the former contain an antigen lacking in the latter. Ethiopian strains also differ from those isolated in West Africa. The 17D strain contains an antigen absent from the strain (Asibi) from which it was derived. CF antibodies, as usual in the group, disappear sooner than neutralizing ones; they are commonly absent after vaccination (Perlowagora & Hughes, 1947). Neutralizing antibodies may persist in human sera for 75 years in the absence of opportunity for specific reinforcement (Sawyer, 1931). A precipitin may be present in sera of acutely ill monkeys (Hughes, 1933). Hatgi et al. (1966) have described the serological responses to yellow fever vaccine of people with differing experience of other flaviviruses.

Cultivation. *Fertile eggs* can be infected, usually only after previous adaptation of virus to mice or tissue culture. Inoculation is best done on to the CAM or directly into the embryo, in which highest titres of virus are found. Embryos under 11 days old commonly die. Virus will grow in *tissue cultures* of chick embryo and mouse embryo. The attenuated 17D strain used for vaccination was obtained by serial passage in chick-embryo cultures of virus already partly attenuated by passage in mice IC and then in mouse-embryo cultures (Theiler & Smith, 1937). Plaques appear on infected chick-fibroblast monolayers (Porterfield, 1959). The virus will multiply in cultures of mosquito or moth tissues (Converse & Nagle, 1967).

Distribution. Endemic in tropical Africa south of the Sahara and south to northern Rhodesia but barely reaching the Indian Ocean. Epidemics have occurred in the Sudan and Ethiopia, formerly in Europe. The outbreak in Ethiopia in 1960–62 is estimated to have affected 200,000 people with 30,000 deaths. Endemic also in tropical South America, occasionally spreading to Central America and Trinidad and formerly to the United States.

Pathogenicity. The picture in man ranges from an inapparent infection in native Africans to a fulminating disease, with high fever, albuminuria, jaundice, black vomit and other haemorrhages, and death. In children the clinical picture is atypical and hard to diagnose. A number of laboratory infections have occurred, some fatal. Fatal epidemics may occur in wild South American monkeys, especially howlers (*Alouatta*).

Experimentally the infection was first transferred to Asiatic *Macaca* monkeys by Stokes et al. (1928). In them the virus causes a fatal disease with hepatitis, though South American strains of virus tend to be less lethal. In most African primates it produces only viraemia with perhaps a little fever; 1 species of bush baby, however, *Galago crassicaudatus*, is often killed. A number of South American monkeys including marmosets and howler monkeys may die after inoculation, though species of *Cebus* are more resistant (Davis, 1930).

The virus is fatal to hedgehogs (Findlay & Clarke, 1934). It has been adapted to cause encephalitis in mice by IC inoculation; a neurotropic variant thus derived causes encephalitis in rhesus monkeys. Infant but not adult mice can be infected IP. Guinea-pigs infected IC develop encephalitis. 'Fixed' neurotropic strains can be caused to revert to pan-tropism. Even 17D, which normally causes nonfatal encephalitis in rhesus, can have full neurovirulence restored by repeated IC passage in mice.

Pathological lesions in yellow fever in man and rhesus monkeys include a fatty liver with mid-zonal necrosis, fatty degeneration in kidneys and multiple haemorrhages. Necrotic hyaline cells in the liver are called Councilman bodies. Granular eosinophilic intranuclear inclusions, commonly surrounding nucleoli, may occur in human liver, but are more regularly found in rhesus livers and in the brain and cord of mice (Cowdry & Kitchen, 1930). Use of the viscerotome to obtain samples of liver post-mortem is a valuable diagnostic measure. Histological changes in livers of Asiatic and African monkeys are described by Bearcroft (1960, 1962).

Ecology. Jungle yellow fever exists as an infection of wild primates in forests of Africa and South America. Many believe that yellow fever was originally an African disease introduced to America with the slave trade. In African forests the natural vector is probably *Aëdes africanus*, with *Aëdes simpsoni* acting as a link between monkeys in tree tops and man in villages (Smithburn et al., 1949b). The ecology is not, however, wholly clear (Haddow, 1951). Bush babies (*Galagos*) (Haddow & Ellice, 1964) and bats (Williams et al., 1964) have been suspected of playing a role. In South America *Haemagogus* spp. are the main vectors in a sylvan cycle; men working in forests are only incidentally infected (Bugher et al., 1944). Edentates, marsupials and rodents may also play a part as reservoirs. More important for man is urban yellow fever, with a cycle involving man and *Aëdes aegypti*, both in Africa and South America. This has caused devastating epidemics. Some think that endemic prevalence of dengue and other B arboviruses may check the spread to Asia and elsewhere.

Control. Urban yellow fever is readily controlled by eliminating *Aëdes aegypti*, but jungle yellow fever cannot be easily dealt with. Yellow-fever vaccines are very efficient and give immunity lasting for years. The 17D strain causes trivial reactions; vaccines are made from juice of chick embryos infected with tissue-culture virus. Eggs free from leukosis virus are used when possible. It may be injected or given as a scratch vaccine, sometimes combined with vaccinia (Dick & Horgan, 1952). It may also be used after passage into mouse-brain, but not serially (Cannon & Dewhurst, 1955). The French neurotropic strain is considered as less safe; cases of encephalitis in man have followed its use (Stones & Macnamara, 1955).

DENGUE

Synonyms: Breakbone fever. Dandy fever.

Review: Sabin (1959).

Antigenic properties. There are 4 serotypes. Nearly all the data concern types 1 and 2 which have been studied much longer than types 3 and 4. The 4 serotypes are readily distinguished by neutralization and CF tests. They are related by HAI tests rather more closely amongst themselves than to other B arboviruses. Neutralization is considerably more effective if fresh unheated serum is added to the mixture of virus and specific antibody (Sabin, 1959). Double diffusion in agar reveals antigens specific to the various serotypes, also one common to all (Ibrahim & Hammon, 1968). Some type 1 and 2 strains from Southeast Asia may have to be classified as distinct serotypes (Hammon et al., in Symposium, 1962; Hammon & Sather, 1964).

Cultivation. Type 1 virus has been adapted to grow in *fertile eggs* after previous adaptation to mice, but high titres were not obtained (Schlesinger, 1951).

Growth in *tissue culture* is reported for types 1 and 2 in monkey kidney, hamster kidney, chick embryo and other tissues (Hotta, 1957; Hotta et al., 1961) but usually only after previous mouse-adaptation. With types 3 and 4 cytophatic effects have been less readily measured. All 4 types produced good CPE in HeLa cells which had been fed with tryptose phosphate broth (Buckley, 1961). Plaques are formed under a methyl-cellulose overlay (Schulze & Schlesinger, 1963). In some trials cultures of green monkey kidney have proved more sensitive than suckling mouse brains (Russell et al., 1966); others

(Stim & Henderson, 1966) report otherwise. Type 2 led to syncytium-formation in cultures of mosquito cells (Suitor & Paul, 1969).

Distribution. *Type 1:* Hawaii, Southeast Asia from India to Japan. Temporary spread has occurred to Greece, South Africa and Australia.

Type 2: New Guinea, Thailand, Trinidad. Probably elsewhere in Southeast Asia, Central America and the Caribbean.

Type 3: Thailand, the Philippines (Hammon et al., 1960). This type or one very like it has appeared in the Caribbean area in 1964.

Type 4: Thailand, the Philippines (Hammon et al., 1960), India.

Some, at least, of the 'dengue' in tropical Africa may be caused by Chikungunya (p. 86) or other arboviruses, but dengue types 1 and 2 have recently been isolated in Nigeria (Carey et al., 1971). It has been suggested that the lack of geographical overlap between dengue and yellow fever (see p. 99) may be due to some form of interference. Such an interference is demonstrable to a limited extent in the laboratory (Sabin, 1952).

Pathogenicity. Dengue in man is an acute fever, often lasting 5 or 6 days, with severe aches in head, back and limbs and often a scarlatiniform or maculo-papular rash. Mild or inapparent infections occur. Recently dengue viruses of types 2, 3 and 4 have been recovered from patients suffering from a serious, often fatal disease, haemorrhagic fever. Dengue-associated haemorrhagic fever is reported from the Philippines, Thailand, India and other countries of East Asia (reviews by Hammon et al., 1960; Symposium, 1962; Halstead, 1966; Halstead et al., 1970). A widely held view is that the severe symptoms occur as a result of infection by 1 dengue serotype in persons partly immune as a result of infection with another serotype. From some of the patients Chikungunya virus has been recovered (see p. 87); its relation to the haemorrhagic fever is obscure; possibly there have been dual infections with this and dengue.

Experimentally, volunteers have been infected by various routes. The incubation period is usually 5–8 days.

Types 1 and 2 have been adapted to mice by IC inoculation, sucklings being most susceptible, but adaptation has usually been difficult (Hotta, 1952; Sabin, 1959). The most prominent symptom is flaccid paralysis of limbs. Chimpanzees and monkeys of several Asiatic, African or South American species normally undergo only inappararent infections, but mouse-adapted virus may produce paralytic disease in rhesus and cynomolgus (Sabin, 1955). New-born hamsters are susceptible (Meiklejohn et al., 1952). Other animal species in the laboratory have proved resistant.

Pathological lesions. In the rare fatal cases in man degenerative changes and haemorrhages have been found in various organs.

Ecology. Infection is transmitted mainly by *Aëdes aegypti*, which has domestic habits; but *A. albopictus* apparently acts as a sylvan vector. Other species may be concerned at times. There is likely to be a reservoir amongst monkeys or other jungle animals in Malaya and elsewhere in Asia (Smith, 1956; Rudnick et al., 1967), but probably not in America (Rosen, 1958). The infection is endemic in areas of Asia and Australasia within the winter 18°C isotherm; major outbreaks mainly occur outside those areas.

Control. Where *Aëdes aegypti* is the only vector, measures directed against this species should control dengue.

Types 1 and 2 have been attenuated by serial mouse-passage so that they can be used as living vaccines for protection of man. Rashes may result but no unpleasant symptoms (Sabin & Schlesinger, 1945; Sabin, 1955). Price (1968) has suggested that protection may be achieved through a programme of sequential immunization with other flaviviruses.

TURKEY MENINGO-ENCEPHALITIS

Description: Komarov & Kalmar (1960).

Cultivation. Lethal for embryonated eggs, being best given into the yolk sac. Produced plaques on chick-embryo monolayers.

Habitat. Israel.

Pathogenicity. Caused progressive and fatal paralysis and enteritis in turkeys. There was 50 per cent morbidity in the Israeli outbreak; chicks and other birds were resistant. Adult mice developed encephalitis when inoculated IC and sucklings after injection IC or IM.

Control. Virus attenuated by passage in eggs could be used for immunizing turkeys.

WESSELSBRON DISEASE

Description: Weiss et al. (1956).

Review: Weiss (1957b).

Antigenic properties are those of a B arbovirus, but the virus is not particularly closely related to any other.

Cultivation. Readily passed in series in 8-day *fertile eggs* inoculated into the yolk sac, most virus being in the body of the embryo. Mortality low and irregular. Grows also in *tissue cultures* of lamb kidney. Intracytoplasmic inclusions are found in infected cells (Parker & Stannard, 1967). Mesh-like structures are seen in cultivated hamster cells (BHK 21) infected with virulent strains (Lecatsas & Weiss, 1969). Virus attenuated by passage in cultures showed changes in various physical properties (Parker et al., 1969).

Distribution. South Africa, Rhodesia, Mozambique.

Pathogenicity. Causes epizootics in sheep, particularly giving rise to abortions and to deaths of new-born lambs and pregnant ewes. Haemorrhages and jaundice occur, and meningo-encephalitis in foetuses. Probably causes abortions in cattle also. May infect man, giving rise to fever and muscular pains; has caused numerous laboratory infections.

Experimentally, it causes abortions in ewes and may kill lambs, the incubation period being 2 to 4 days. It causes fever in cattle, sheep and pigs. Infects suckling mice IC or IP, producing fatal encephalitis. Adult mice are only infected IC. Abortions are produced in pregnant rabbits and guinea-pigs.

Pathological lesions are found mainly in the liver. There is necrobiosis of scattered hepatic cells with fatty infiltration, but lesions are variable (Le Roux, 1959).

Ecology. Transmission is by mosquito bite, the 2 important vectors being *Aëdes caballus* and *circumluteolus* (Kokernot et al., 1960). In endemic areas, antibodies are present in sera of human beings and various domestic quadrupeds.

OTHER MOSQUITO-BORNE FLAVIVIRUSES

Table 4 gives details of 9 viruses concerning which most information is available. Five others are Banzi H 336 from South Africa, Ib. An 10061 from Nigeria, Alfuy MRM from Australia, Tembusu from Malaya and Sarawak and Usutu from South Africa and Uganda. The first and last of these have been isolated from man.

TICK-BORNE FLAVIVIRUSES

There exists a family of tick-borne flaviviruses so closely interrelated antigenically that earlier work failed to differentiate sharply between them. Clarke (1962) using the tests of HAI, agar-gel precipitation and antibody absorption reported that they could be divided into 6 entities having, with one exception, discontinuous distributions. In a later paper Clarke (1964) added 2 more viruses (7 and 8) to the list.

1. Louping ill (British Isles) (LI)
2. Central European tick-borne (Central Europe, Scandinavia, Western U.S.S.R.) (CET)
3. Omsk haemorrhagic fever (Central U.S.S.R.) (OHF)
4. Kyasanur forest disease (India) (KFD)
5. Far Eastern Russian (Eastern U.S.S.R.) (FER or RSSE)
6. Langat (Malaya)
7. Powassan (North America)
8. Negishi (Japan)

The louping-ill and Negishi viruses are rather closely related to the Russian viruses, all the rest standing further away.

This account deals first in a general section with properties known or likely to be common to the group and then with the separate viruses. The particular virus dealt with in the general section will be indicated by the initials in parentheses.

Reviews: Smorodintsev (1958). Libiková (1962). Work (1963). Blaskovič et al. (1967).

Haemagglutination—as for the other flaviviruses. Adsorption and elution from rooster RBCs were studied by Salminen (1960) (CET). He obtained satisfactory results with cells from cocks, but not hens—except as young chicks (Salminen, 1959) (CET).

Antigenic properties. Crossing with other flaviviruses is seen in HAI tests, but the members of the complex are much more closely related to each other than to other Bs. Their relations, as stated above, have been clarified by Clarke (1962) using agar-gel precipitation and antibody-absorption tests. CF tests may fail to separate members of the complex, and other tests, apart from those of Clarke, yield confusing results. Immunity tests may fail to show protection against other members of the complex. Gorev & Smorodintsev (1968) found evidence of a precipitinogen common to all members, with modifications specific for the different members.

Cultivation. The viruses grow in *fertile eggs* whether inoculated on the CAM, where discrete pocks appear (LI) (Burnet, 1936a) or into the yolk-sac or embryo (LI) (Edward, 1947).

Growth also occurs in tissue cultures of mouse-embryonic tissue

(FER) (Takemori, 1949) and in HeLa and other continuous cell lines of primate origin (CET) (von Zeipel & Svedmyr, 1958; Libiková & Vilček, 1960), pig kidney (LI) (Williams, 1958), bovine, sheep, chick and other tissues. Cytopathic effects are not invariably seen, but were satisfactory in Libiková & Vilček's (1960) HeLa cells and in sheep-embryo cultures (Gaidomovich & Obukhova, 1960).

Pathogenicity. All these viruses readily produce encephalitis when inoculated IC into mice, except for Omsk haemorrhagic fever. Most of them are also pathogenic on IC inoculation into rhesus or cynomolgus monkeys, also (Silber & Soloviev, 1946) for sheep.

Ecology. The viruses are normally transmitted by ticks, and it is likely that there is a permanent reservoir in ticks; mammals and birds are only irregularly found infected in nature. There is evidence that transovarial transmission occurs in ticks (Silber & Soloviev, 1946), and virus may survive in hibernating ticks.

LOUPING-ILL

Synonym: Ovine encephalomyelitis.

Distribution. Scotland, Ireland, North and Southwest England.

Antigenic properties. A strain isolated from grouse had properties intermediate between those of typical LI and continental strains.

Pathogenicity. Normally a disease of sheep, less often cattle. The disease has 2 phases, the first characterised by high fever and viraemia, the second, several days later, by incoordination, followed by paralysis and often death. The second phase may be absent, and probably the disease is normally almost inapparent. A few cases are reported in men having contact with sheep, larger numbers in laboratory workers (Rivers & Schwentker, 1934); these have suffered from serous meningitis with some evidence of encephalitis, but the disease has been much less serious than those from the Russian viruses. A case with mild haemorrhagic symptoms is reported (Cooper et al., 1964). Rodents, deer, shrews and red grouse may be naturally infected, but so far as is known without symptoms.

Experimentally mice develop encephalitis after IC inoculation, but much less readily when infected IP, IN or SC—at least in adult mice (Edward, 1950). Many other species, including pigs, can be infected IC, but not guinea-pigs or rabbits. Rats inoculated IN develop an inapparent infection (Burnet, 1936b).

Pathological lesions. Destruction of Purkinje cells is a feature in fatal cases in sheep. Hurst (1931) found cytoplasmic inclusions in brain cells of mice, but not of other species.

Ecology. *Ixodes ricinus* is the vector and may also act as the reservoir of infection. Concurrent infection with the rickettsiae of tick-borne fever may favour invasion of the nervous system in sheep (Macleod, 1962). The role of rodents and other aspects of ecology are discussed by Smith et al. (1964a, b).

Control. Formalinized vaccines made from sheep brain, mouse brain or chick embryo have been used to protect sheep, cattle and exposed laboratory workers. The egg vaccines were not satisfactory (Edward, 1947, 1948). Brotherston & Boyce (1970) described a noninfective protective antigen, while Mayer et al. (1969) obtained seroconversions in sheep given attenuated TBE virus.

Tick-borne Encephalitis (Central European Subtype)

Synonyms: Biundulant meningo-encephalitis, diphasic milk fever.

Distribution. Central Europe from Scandinavia to the Balkans and from Germany to western U.S.S.R.

Pathogenicity. The disease in man is biphasic. An afebrile period of 4 to 10 days intervenes between the first influenza-like fever and the second phase of meningitis or meningo-encephalitis. Mild or inapparent forms occur. In severe forms there is transient or permanent paralysis and the bulbo-spinal form is usually fatal. In severity the disease is intermediate between louping-ill and the Far East form. Virus may be present in the milk of infected goats and may thus infect man (van Tongeren, 1955).

Experimentally, the virus behaves like louping-ill, but peripheral inoculation into mice is more regularly fatal. Guinea-pigs react with fever, rarely encephalitis. Young chicks may be infected IC, also lizards. Virus may localize in the mammary glands of infected goats, cows, sheep, and also mice, and appear in the milk. It may appear in the urine.

Pathological lesions are found in all parts of the CNS; the cervical cord being affected often (Grinschgl, 1955).

Ecology. The vector is *Ixodes ricinus*, and this is likely to be also an important reservoir. Radda et al. (1963) have given a good account of the ecology with special relation to populations of ticks. Transmission through milk of goats has already been mentioned, but this is unlikely to be significant in a long-term view of the ecology. Virus has been recovered from *Aëdes* mosquitoes, though these are not considered important. It is possible that mites may play a role as virus has been recovered from brains of small mammals in an area where ticks are very uncommon (Rosicky & Bardos, 1966).

Control. Protection of exposed persons against ticks is the first line of defence. Formalinized vaccine has been widely used in the U.S.S.R. (Smorodintsev, 1940). It has been made from mouse or suckling rat brains or from tissue cultures of chick embryo or cynomolgus heart (Levkovich, 1962). There is also work in progress on use of living attenuated virus for vaccination especially of cattle, sheep and goats (Blaskovič et al., 1962; Libiková & Stanček, 1965).

Tick-borne Encephalitis (Eastern Subtype)

Synonym: Far East Russian encephalitis. Russian spring-summer encephalitis.

Review: Silber & Soloviev (1946).

Virus characters: as for others of the complex.

Distribution. Eastern U.S.S.R., but a few viruses from Leningrad and elsewhere in the western U.S.S.R. belong to this subtype.

Pathogenicity. This is the most serious form of the disease for man. Flaccid paralysis followed by atrophy is common; so are symptoms due to bulbar involvement. Mortality has been around 30 per cent. Disease may also occur in naturally infected rodents and birds.

Experimentally, the virus causes encephalitis in laboratory mice inoculated by various routes, also fever in guinea-pigs. Inoculated IC it produces encephalitis in rhesus monkeys, sheep, goats and some wild rodents but not in others, also in some finches (Silber & Soloviev, 1946).

Pathological lesions resemble those of related viruses, but, in contrast to louping-ill, Purkinje cells are not particularly attacked.

Ecology. The vector is *Ixodes persulcatus* which replaces *I. ricinus* in the East, though their ranges overlap.

Control. As for the Central European virus.

Omsk Haemorrhagic Fever

Reviews: Gajdusek (1956). Netsky (1967).

Distribution. Central U.S.S.R.

Antigenic properties. Strains of this virus fall into 2 antigenic subgroups (Clarke, 1964).

Pathogenicity. Commonly a diphasic illness in man with fever, enlargement of lymph nodes, gastro-intestinal symptoms and haemorrhages from nose, stomach or uterus, but little or no CNS involvement. Mortality is 1–2 per cent. It may cause disease in musk-rats (not a native species in the area), and in man during winter months (Federova & Sizemova, 1964).

Experimentally, the virus causes fever after IP inoculation of rhesus monkeys. It does not, on first isolation, infect mice, but sucklings may not have been used.

Ecology. The vectors are *Dermacentor pictus* and *D. marginatus.* Transovarial transmission is recorded for these ticks.

Kyasanur Forest Disease

Reviews: Work (1958). Rajagopalan et al. (1968).

Distribution. Kyasanur forest, Mysore and possibly elsewhere in India.

Pathogenicity. Unknown until 1955; in 1957 an extensive epidemic occurred, chiefly amongst forest workers; numerous dead langurs (*Presbytis entellus*) and bonnet macacques (*Macaca radiata*) were picked up in the forests. Symptoms in man include headache, fever, back and limb pains and prostration, conjunctival inflammation, diarrhoea, vomiting and often bleeding from intestines and elsewhere. Symptoms from CNS involvement are lacking. Leucopenia and albuminuria are found. Death may follow especially if the dehydration is not treated. A number of laboratory infections have occurred.

Experimentally mice develop encephalitis after inoculation by various routes. Adults are less susceptible to peripheral inoculation. Virus may persist for long periods in brains and livers of surviving mice (Price, 1966). Suckling hamsters are susceptible, but other rodents were resistant. Rhesus and bonnet monkeys, injected IC, IP or SC, developed viraemia without symptoms.

Pathological lesions were similar in man and monkeys. Haemorrhages were found in lungs or elsewhere, their histopathological basis being uncertain. There were tubular necrosis in kidneys, focal necroses in liver, more so in the monkeys, and also focal necroses in some monkey brains.

Ecology. The vector appears to be a tick, *Haemaphysalis spinigera*, possibly also other *Haemaphysalis* species (Singh et al., 1964). The disease may be spread by movements of monkeys, many of which probably have mild infections, and by birds. Antibodies have been found in small forest mammals—a squirrel, a shrew and some forest rats, also in jungle fowl and a woodpecker. Squirrels (Webb, 1965) and shrews may prove to be more important in the ecology than rats. It is possible that the disease was introduced through the agency of birds. Serological evidence suggests that virus may be present in parts of India in the absence of overt disease (Boshell-Manrique, 1969).

Control. Attempts to protect by means of a formalinized RSSE virus vaccine have met with no great success. A formalinized vaccine is under study (Mansharamani et al., 1967).

Negishi Virus

A virus of the tick-borne encephalitis complex has been isolated from two fatal cases in Japan and appears to be a new member (Okuno et al., 1961).

Langat Virus

This virus was isolated from a pool of *Ixodes granulatus* in Malaya by Smith (1956).

Cultivation. Infected cells from baby mouse brains pile up in culture, showing loss of contact inhibition (Illavia & Webb, 1970). Attenuated by cultivation in chick-embryo cells or in egg yolks; 6 markers distinguished this from the parent virus (Price et al., 1970).

Pathogenicity. Comparatively low pathogenicity for mice and monkeys suggested its possible use as an immunizing agent against other tick-borne viruses.

Ecology. Antibodies were found in sera of 6/57 forest ground-rats.

Control. The attenuated virus was effective in protecting mice against homologous and even heterologous tick-borne flaviviruses. (Thind & Price, 1966). Antibodies were produced in monkeys and human volunteers (Il'enko et al., 1968; Price et al., 1970), and there were no complications. Smith (1967), however, records that encephalitis occurred in 2 injected leukaemia patients; 26 others with malignant disease had at most a little fever.

TABLE 7

MISCELLANEOUS MOSQUITO-BORNE FLAVIVIRUSES

Virus	Uganda S (=Makonde)	Zika	Spondweni	Ntaya
Where isolated	Africa; South-east Asia	Uganda; South-east Asia	South Africa; Nigeria	Uganda
From what species	*Aëdes spp* (pool)	*Aëdes africanus* Sentinel rhesus	Culicine mosquitoes; Man	Mosquitoes (pool)
Association with disease	None	None	Probably hepatitis (lab. infection)	None
Properties of virus	2 estimates of size 15–22 nm 75–112 nm (less probable)	18–26 nm	—	?81–122 nm
Related flaviviruses (HAI)	Zika; yellow fever	Uganda S. yellow fever	—	—
Pathogenicity for animals other than man	Nil in monkeys	Fever in rhesus	sheep; cattle	Nil in rhesus or hamster
Antibodies found naturally	Man; Cercopithecus	Man; Cercopithecus	Man; sheep; cattle	—
References	Dick & Haddow (1952); Ross (1956)	Dick et al. (1952)	Kokernot et al. (1957b); Macnamara (1954)	Smithburn & Haddow (1951)

Powassan Virus

This virus was isolated from the brain of a boy with fatal encephalitis in Ontario (McLean & Donohue, 1959). It infected new-born but not weaned mice, formed a haemagglutinin and showed a serological relation to members of the tick-borne encephalitis group. It was, however, less closely related to the European and Asian strains than these are to each other (Casals, 1960). Antibodies were found in sera of squirrels and chipmunks caught locally (McLean & Larke, 1963). The virus has been recovered from *Ixodes marxi* and *I. cookei*. McLean et al.

(1964) suggest that the latter tick and the woodchuck (*Marmota monax*) may be particularly important ecologically. The virus has also been recovered in New York State. A closely related virus has been obtained from *Dermacentor andersoni* in Colorado (Thomas et al., 1960). Chernesky & McLean (1969) studied the growth of the virus in *Dermacentor andersoni*. Transstadial but not transovarial transmission was shown.

Bussuquara	Kunjin	Kokobera	Edgehill	Stratford
Brazil; Colombia	Queensland; Sarawak	Queensland	Queensland	Queensland
Man; Sentinel howler monkey; Sentinel mice	Mosquitoes; man (lab. infection)	Mosquitoes	Mosquitoes	Mosquitoes
—	Fever and rash	—	—	—
—	—	—	—	—
—	Murray Valley; Kokobera; West Nile	Murray Valley; Kunjin	Murray Valley	—
Hepatitis in howler monkey (Alouatta)	—	—	—	—
—	—	Kangaroo	—	—
Gomes & Causey (1959)	Doherty et al., 1963			

FLAVIVIRUSES WITHOUT KNOWN ARTHROPOD VECTOR

The *Modoc* virus was recovered from mammary glands of a mouse (*Peromyscus manicatus*) in California. It showed some antigenic relation to Rio bravo virus in cross-immunity tests (Casals, 1960).

Rio bravo (bat salivary) virus (Burns & Farinacci, 1956) from the salivary glands of a bat (*Tadarida brasiliensis*) caught in California is identical with others from Texas and Mexico. Related to Modoc virus.

Another one, called *MML*, was isolated from *Myotis* bats in Montana (Bell & Thomas, 1964).

Dakar bat virus (Brès & Chambon, 1964) was recovered on numerous occasions from insectivorous bats (*Scotophilus*) in West Africa.

The *Bukalasa* virus was recovered from *Tadarida* bats in Uganda (Williams et al., 1964).

Entebbe bat virus was obtained from salivary glands from a pool of bats (*Tadarida*) in Uganda (Lumsden et al., 1961).

Cow-bone ridge virus was isolated from a cotton-rat (*Sigmodon sp*) in Florida.

These 7 viruses behave in general like flaviviruses, but the Rio bravo virus did not multiply in several mosquito species and when inoculated into mice peripherally showed tropism for kidneys, mammary and salivary glands. It was also responsible for 5 laboratory infections, 2 of them with orchitis (Sulkin et al., 1962). Baer & Woodall (1966) suggest that transmission of Rio bravo virus may well be transmitted otherwise than by blood-sucking arthropods.

REFERENCES

Abdelwahab, K. S. E., Almeida, J. D., Doane, F. W., & McLean, D. M. (1964) *Can. med. Ass. J.*, **90**, 1068.

Acheson, N. H., & Tamm, I. (1967) *Virology*, 32, 128.

Alevizatos, A. C., McKinney, R. W., & Feigin, R. D. (1967) *Am. J. trop. Med. Hyg.*, **16**, 762.

Altman, R., Goldfield, M., & Sussman, O. (1967) *Med. Clins N. Am.*, **51**, 661.

Anderson, C. R., Aitken, T. H. G., & Downs, W. G. (1956) *Am. J. trop. Med. Hyg.*, **5**, 621.

Anderson, S. G. (1954) *J. Hyg., Camb.*, **52**, 447.

Andrewes, C. H. (1970) *Virology*, **40**, 1070.

Appleyard, G. (1967) *J. gen. Virol.*, **1**, 143.

Appleyard, G., Olam, J. D., & Stanley, J. L. (1970) *J. gen. Virol.*, **9**, 179.

Baer, G. M., & Woodall, D. F. (1966) *Am. J. trop. Med. Hyg.*, **15**, 769.

Bearcroft, W. G. C. (1960) *J. Path. Bact.*, **80**, 19 and 421.

Bearcroft, W. G. C. (1962) *J. Path. Bact.*, **83**, 49.

Beard, J. W., Finkelstein, H., Sealy, W. C., & Wyckoff, R. W. G. (1938) *Science*, **87**, 490.

Bell, J. F., & Thomas, L. A. (1964) *Am. J. trop. Med. Hyg.*, **13**, 607.

Bellamy, R. E., Reeves, W. C., & Scrivani, R. P. (1967) *Am. J. Epidemiol.*, **85**, 282.

Ben-Ishai, Z., Goldblum, N., & Becker, Y. (1968) *J. gen. Virol.*, **2**, 365.

Bernkopf, H., Levine, S., & Nerson, R. (1953) *J. infect. Dis.*, **93**, 207.

Bhatt, P. N., & Work, T. H. (1957) *Proc. Soc. exp. Biol. Med., U.S.A.*, **96**, 213.

Binn, L. N., Sponseller, M. L., Wooding, W. L., McConnell, S. J., Spertzel, R. O., & Yager, R. H. (1966) *Am. J. vet. Res.*, **27**, 1599.

Blaskovič, D., Libíková, H., Grešiková, M., Slonim, D., Mačička, O. (1962) in *Symposium on biology of tick-borne encephalitis complex*, p. 348. Prague: Czechoslovak Academy of Science.

Blaskovič, D. et al. (1967) *Wld Hlth Org. Bull.* **36**, suppl. 1. 5 (12 papers mostly on ecology).
Boshell-Manrique J. (1969) *Am. J. trop. Med. Hyg.*, **18**, 67.
Brès, P., & Chambon, L. (1964) *Annls Inst. Pasteur, Paris*, **107**, 34.
Brown, L. V. (1958) *Am. J. Hyg.*, **67**, 214.
Brotherston, J. G., & Boyce, J. B. (1970) *J. comp. Path.*, **80**, 377.
Buckley, S. M. (1959) *Am. N.Y. Acad. Sci.*, **81**, 172.
Buckley, S. M. (1961) *Nature, Lond.*, **192**, 778.
Buescher, E. L., Scherer, W. F., McClure, H. E., Moyer, J. T., Rosenberg, M. Z., Yoshii, M., & Okada, Y. (1959) *Am. J. trop. Med.*, **8**, 678.
Bugher, J. C., Boshell-Manrique, J., Roca-Garcia, M., & Osorno-Mesa, E. (1944) *Am. J. Hyg.*, **39**, 16.
Burge, B. W., & Strauss, J. H. (1970) *J. molec. Biol.*, **47**, 449.
Burnet, F. M. (1936a) *Br. J. exp. Path.*, **17**, 294.
Burnet, F. M. (1936b) *J. Path. Bact.*, **42**, 213.
Burns, K. F. (1950) *Proc. Soc. exp. Biol. Med., U.S.A.*, **75**, 621.
Burns, K. F., & Farinacci, C. J. (1956) *Science*, **123**, 227.
Burton, A. N., McLintock, J., & Rempel, J. G. (1966) *Science*, **154**, 1029.
Bykovsky, A. F., Yershov, F. I., & Zhdanov, V. M. (1969) *J. Virol.*, **4**, 496.
Cannon, D. A. & Dewhurst, F. (1955) *Ann. trop. Med. Parasit.*, **49**, 174.
Carey, D. E., Conway, O. R., Reddy, S., & Cook, A. R. (1970) *Lancet*, **2**, 105.
Carter, G. B. (1969) *J. gen. Virol.*, **4**, 139.
Casals, J. (1958) *Proc. 6th int. Congr. trop. Med. Malar.*, **5**, 34.
Casals, J. (1960) *Can. med. Ass. J.*, **82**, 355.
Casals, J. (1961) *Abstr. 10th Pacific Science Congress.* p. 458.
Casals, J. (1964) *J. exp. Med.*, **119**, 547.
Casals, J., & Brown, L. V. (1954) *J. exp. Med.*, **99**, 429.
Casals, J., & Whitman, L. (1957) *Am. J. trop. Med. Hyg.*, **6**, 1004.
Causey, O. R., Casals, J., Shope, R. E., & Udomsakdi, S. (1963) *Am. J. trop. Med. Hyg.*, **12**, 777.
Causey, O. R., & Maroja, O. M. (1957) *Am. J. trop. Med.*, **6**, 1017.
Chain, M. M. T., Doane, F. W., & McLean, D. M. (1966) *Can. J. Microbiol.*, **5**, 895.
Chamberlain, R. W. (1963) *Proc. 7th int. Congr. trop. Med. Malar.*, **3**, 160.
Chamberlain, R. W. (1968) *Curr. topics Microbiol. Immunol.*, **42**, 38.
Chamberlain, R. W., India, W. D., & Work, T. M. (1969) *Am. J. Epidemiol.*, **89**, 197.
Chamberlain, R. W., Sikes, R. K., & Kissling, R. E. (1954) *J. Immunol.*, **73**, 106.
Chamberlain, R. W., Sikes, R. K., & Nelson, D. B. (1956) *Proc. Soc. exp. Biol. Med., U.S.A.*, **91**, 215.
Chamberlain, R. W., & Sudia, W. D. (1955) *Am. J. Hyg.*, **62**, 295.
Chamberlain, R. W., Sudia, W. D., Coleman, P. H., & Work, T. H. (1964) *Science*, **145**, 272.
Chamberlain, R. W., Sudia, W. D., & Gillett, J. D. (1959) *Am. J. Hyg.*, **70**, 221.
Chambers, V. C. (1957) *Virology*, **3**, 62.
Chernesky, M. A., & McLean, D. M. (1969) *Can. J. Microbiol.*, **15**, 1399.
Clarke, D. H. (1960) *J. exp. Med.*, **111**, 21.
Clarke, D. H. (1961) *Am. J. trop. Med. Hyg.*, **10**, 67.
Clarke, D. H. (1962) *Symposium on the biology of viruses of the tick-borne encephalitis complex.* p. 67. New York: Academic Press.
Clarke, D. H. (1964) *Bull. Wld Hlth Org.*, **31**, 45.
Clarke, D. H., & Casals, J. (1958) *Am. J. trop. Med. Hyg.*, **7**, 561.
Clarke, D. H., & Casals, J. (1965) in *Viral and rickettsial infections of man.* Eds.: Horsfall & Tamm. 4th Ed. p. 606. Philadelphia: Lippincott.
Cleland, J. B., & Bradley, B. (1917) *Med. J. Aust.*, **1**, 499.
Collins, W. E. (1963) *Am. J. Hyg.*, **77**, 109.
Converse, J. L., & Nagle, S. C. (1967) *J. Virol.*, **1**, 1096.
Cook, E. A. (1938) *J. infect. Dis.*, **63**, 206.
Cooper, W. C., Green, I. J., & Fresh, J. W. (1964) *Br. med. J.*, **2**, 1627.
Corbet, P. S., Williams, M. C., & Gillett, J. D. (1961) *Trans. r. Soc. trop. Med.*, **55**, 463.

Cowdry, E. V., & Kitchen, S. F. (1930) *Am. J. Hyg.*, **11**, 227.
Cox, H. R., & Olitsky, P. K. (1936) *J. exp. Med.*, **63**, 745.
Davies, A. M., Fendrich, J., Yoshpe-Purer, Y., & Nir, Y. (1955) *J. trop. Med. Hyg.*, **58**, 12.
Davis, N. C. (1930) *Am. J. Hyg.*, **11**, 321.
Dick, G. W. A. (1953) *Trans. r. Soc. trop. Med.*, **47**, 13.
Dick, G. W. A., & Haddow, A. J. (1952) *Trans. r. Soc. trop. Med.*, **46**, 600.
Dick, G. W. A., & Horgan, E. S. (1952) *J. Hyg., Camb.*, **50**, 376.
Dick, G. W. A., Kitchen, S. F., & Haddow, A. J. (1952) *Trans. r. Soc. trop, Med.*, **46**, 509.
Diercks, F. H., Kundin, W. D., & Porter, T. J. (1961) *Am. J. Hyg.*, **73**, 164.
Dobos, P., & Faulkner, P. (1970) *J. Virol.*, **6**, 145.
Doherty, R. L. (1967) *Jap. J. med Sci. Biol.*, **20**, 15, suppl.
Doherty, R. L., Carley, J. G., Mackerras, M. J., & Marks, E. N. (1963) *Aust. J. exp. Biol. med. Sci.*, **41**, 17.
Duffy, C. E., & Stanley, W. M. (1945) *J. exp. Med.*, **82**, 385.
Edward, D. G. ff. (1947) *Br. J. exp. Path.*, **28**, 368.
Edward, D. G. ff. (1948) *Br. J. exp. Path.*, **28**, 237, **29**, 367.
Edward, D. G. ff. (1950) *Br. J. exp. Path.*, **31**, 515.
Eklund, C. M. (1953) *Ann. Rev. Microbiol.*, **7**, 339.
Emmons, R. W., & Lennette, E. H. (1967) *Proc. Soc. exp. Biol. Med., U.S.A.*, **125**, 443.
Erlandson, R. A., Babcock, V. I., Southam, C. M., Bailey, R. B., & Shipkey, F. H. (1967) *J. Virol.*, **1**, 996.
Evans, C. A., & Bolin, V. S. (1946) *Proc. Soc. exp. Biol. Med., U.S.A.*, **61**, 106.
Fastier, L. B. (1952) *J. Immunol.*, **68**, 531.
Faulkner, P., & McGee-Russell, S. M. (1968) *Can. J. Microbiol.*, **14**, 153.
Federova, T. N., & Sizemova, G. A. (1964) *Zh. Microbiol. Epidem. Immunol., Moscow*, **41**, 134.
Filshie, B. K., & Rehaček, J. (1968) *Virology*, **34**, 435.
Findlay, G. M., & Clarke, L. P. (1934) *Trans. r. Soc. trop. Med.*, **28**, 193 and 335.
Finkelstein, H., Marx, W., Bridgers, W. H., & Beard, J. W. (1938) *Proc. Soc. exp. Biol. Med., U.S.A.*, **39**, 103.
French, E. L. (1952) *Med. J. Aust.*, **1**, 100.
Friedman, R. M., Levy, H. B., & Carter, W. B. (1966) *Proc. natn. Acad. Sci., U.S.A.*, **56**, 440.
Foreschle, J. E. (1964) *Proc. Soc. exp. Biol. Med., U.S.A.*, **115**, 881.
Frothingham, T. E. (1955) *Am. J. trop. Med.*, **4**, 863.
Frothingham, T. E. (1963) *Virology*, **19**, 583.
Gaidomovich, S. I., & Obukhova, V. R. (1960) *Probl. Virol.*, **5**, 331.
Gajdusek, D. C. (1956) *J. Pediatr.*, **60**, 841.
Galindo, P., & Rodaniche, E. de (1961) *Am. J. trop. Med. Hyg.*, **10**, 395.
Gomes, G., & Causey, O. R. (1959) *Proc. Soc. exp. Biol. Med.*, **101**, 275.
Gorev, N. E., & Smorodintsev, A. A. (1968) *Bull. Wld Hlth Org.*, **38**, 389.
Grešíková-Kohútová, M. (1959a) *Acta virol., Prague*, **3**, 159.
Grešíková-Kohútová, M. (1959b) *Acta virol., Prague*, **3**, 215.
Grimley, P. M., Berezesky, I. K., & Friedman, R. M. (1968) *J. Virol.*, **2**, 1326.
Grinschgl, G. (1955) *Bull. Wld Hlth Org.*, **12**, 535.
Haddow, A. J. (1952) *Ann. trop. Med. Parasit*, **46**, 135.
Haddow, A. J., Davies, C. W., & Walker, A. J. (1960) *Trans. r. Soc. trop. Med. Hyg.*, **54**, 517.
Haddow, A. J., & Ellice, J. M. (1964) *Trans. r. Soc. trop. Med. Hyg.*, **58**, 521.
Hale, J. H., & Witherington, D. H. (1953) *J. comp. Path.*, **63**, 195.
Halstead, S. B. (1966) *Bull. Wld Hlth Org.*, **35**, 3.
Halstead, S. B., & Buescher, E. L. (1961) *Science*, **134**, 475.
Halstead, J. B., & Udomsakdi, S. (1966) *Bull. Wld Hlth Org.*, **35**, 89.
Halstead, S. B. et al. (1970) *Yale J. Biol. Med.*, **42**, 261 *et seq* (series of 6 papers).
Hammon, W. McD., & Rhim, J. S. (1963) *Am. J. trop. Med. Hyg.*, **12**, 616.
Hammon, W. McD., Rudnick, A., & Sather, G. E. (1960) *Science*, **131**, 1102.

Hammon, W. McD., & Sather, G. E. (1956) *Proc. Soc. exp. Biol. Med., U.S.A.*, **91**, 521.
Hammon, W. McD., & Sather, G. E. (1964) *Am. J. trop. Med. Hyg.*, **13**, 629.
Hammon, W. McD., & Sather, G. E. (1966) *Am. J. trop. Med. Hyg.*, **15**, 199.
Hardy, J. L., Scherer, W. F., & Carey, J. B. (1965) *Am. J. Epidemiol.*, **82**, 73.
Hatgi, J. H., Wisseman, C. L., Rosenzweig, E. C., Harrington, B. R., & Kitaoka, M. (1966) *Am. J. trop. Med. Hyg.*, **15**, 601.
Hay, A. J., Skehel, J. J., & Burke, D. C. (1968) *J. gen. Virol.*, **3**, 175.
Hayes, R. O., Beadle, L. D., Hess, A. D., Sussman, O., & Bonese, M. J. (1962) *Am. J. trop. Med. Hyg.*, **11**, 115.
Hayes, R. O., Daniels, J. B., Maxfield, H. K., & Wheeler, R. E. (1964) *Am. J. trop. Med. Hyg.*, **13**, 595.
Henderson, D. W., Peacock, S., & Randles, W. J. (1967) *Br. J. exp. Path.*, **48**, 228.
Henderson, J. R. (1961) *Yale J. Biol. Med.*, **33**, 350.
Henderson, J. R. (1964) *J. Immunol.*, **93**, 452.
Henderson, J. R., Levine, S. I., Kerabatsos, N., & Stim, I. B. (1967) *J. Immunol.*, **99**, 925.
Henderson, J. R., & Taylor, R. M. (1960) *J. Immunol.*, **84**, 590.
Herreng, F. (1967) *C.r. hebd. Séanc. Acad. Sci., Paris*, **264**, 2854.
Higashi, N., Matsumoto, A., Tabata, K., & Nagatomo, Y. (1967) *Virology*, **33**, 55.
Holden, P. (1955) *Proc. Soc. exp. Biol. Med., U.S.A.*, **88**, 607.
Horzinek, M., & Munz, K. (1969) *Arch. ges. Virusforsch.*, **27**, 94.
Horzinek, M., & Mussgay, M. (1969) *J. Virol.*, **4**, 514.
Hotta, S. (1952) *J. infect. Dis.*, **90**, 1.
Hotta, S. (1957) *Ann. trop. Med. Parasit.*, **51**, 249.
Hotta, S., Ohyama, A., Yamada, T., & Awai, T. (1961) *Jap. J. Microbiol.*, **5**, 77.
Howitt, B. F. (1946) *Proc. Soc. exp. Biol. Med., U.S.A.*, **62**, 105.
Hughes, J. P., & Johnson, H. N. (1967) *J. Am. vet. med. Ass.*, **150**, 167.
Hughes, T. P. (1933) *J. Immunol.*, **25**, 275.
Hurlbut, H. S. (1950) *Am. J. Hyg.*, **51**, 265.
Hurlbut, H. S. (1956) *Am. J. trop. Med. Hyg.*, **5**, 76.
Hurlbut, H. S., & Thomas, J. I. (1960) *Virology*, **12**, 391.
Hurst, E. W. (1931) *J. comp. Path.*, **44**, 231.
Ibrahim, A. N., & Hammon, W. McD. (1968) *J. Immunol.*, **100**, 86.
Il'enko, V. I., Smorodintsev, A. A., Prozorova, I. N., & Platonov, V. G. (1968) *Bull. Wld Hlth Org.*, **39**, 425.
Illavia, S. J., & Webb, H. E. (1970) *Lancet*, **2**, 284.
Inoue, Y. K., Iwasaki, T., & Kato, H. (1961) *J. Immunol.*, **87**, 337.
Inoue, Y. K., & Nishibe, Y. (1967) *Arch. ges. Virusforsch.*, **21**, 192.
Janzen, H. G., Rhodes, A. J., & Doane, F. V. (1970) *Can. J. Microbiol.*, **16**, 581.
Johnson, H. N. (1963) *Am. J. trop. Med. Hyg.*, **12**, 604.
Johnson, R. T. (1965) *Am. J. Path.*, **46**, 929.
Jungherr, E. L., & Wallis, R. C. (1958) *Am. J. Hyg.*, **67**, 1.
Kaariainen, L., & Gomatos, P. J., (1969) *J. gen. Virol.*, **5**, 251.
Katz, E., & Goldblum, N. (1968) *Arch. ges. Virusforsch.*, **25**, 69.
Kissling, R. E., Chamberlain, R. W., Eidson, M. E., Sikes, R. K., & Bruce, M. A. (1954) *Am. J. Hyg.*, **60**, 237.
Kissling, R. E., Chamberlain, R. W., Nelson, D. B., & Stamm, D. D. (1956) *Am. J. Hyg.*, **63**, 274.
Kissling, R. E., & Rubin, H. (1951) *Am. J. vet. Res.*, **12**, 100.
Kissling, R. E., & Chamberlain, R. W. (1967) *Adv. vet. Sci.*, **11**, 65.
Kitaoka, M., and Nishimura, C. (1963) *Virology*, **19**, 238.
Kokernot, R. H., de Meillon, B., Paterson, H. E., Heymann, C. S., & Smithburn, K. C. (1957a) *S. Afr. J. med. Sci.*, **22**, 145.
Kokernot, R. H., Smithburn, K. C., Muspratt, J., & Hodgson, B. (1957b) *S. Afr. J. med. Sci.*, **22**, 103.
Kokernot, R. H., Smithburn, K. C., Paterson, H. E., & de Meillon, B. (1960) *S. Afr. med. J.*, **34**, 871.

Kokernot, R. H., Smithburn, K. C., & Weinbren, M. P. (1956) *J. Immunol.*, **77**, 313.
Komarov, A., & Kalmar, E. (1960) *Vet. Rec.*, **72**, 257.
Koprowski, H., & Hughes, T. P. (1946) *J. Immunol.*, **54**, 371.
Koprowski, H., & Lennette, E. H. (1946) *J. exp. Med.*, **84**, 181 and 205.
Kundin, W. D., Liu, C., & Rodina, P. (1966) *J. Immunol.*, **96**, 39.
Kunz, C. (1965) *Arch. ges. Virusforsch.*, **17**, 673.
Lascarno, E. F., Berría, M. I., & Oro, J. G. B. (1969) *J. Virol.*, **4**, 271.
Lavilloureix, J., & Reeb, E. (1958) *Bull. Soc. Path. exot.*, **51**, 941.
Laemmert, H. W., & Hughes, T. P. (1947) *J. Immunol.*, **55**, 61.
Lecatsas, G., & Weiss, K. E. (1969) *Arch. ges. Virusforsch.*, **27**, 332.
Le Roux, J. M. W. (1959) *Onderstepoort J. vet Res.*, **28**, 237.
Levkovich, E. N. (1962) in *Symposium on the biology of the viruses of the tick-borne encephalitis complex*, p. 317. Prague: Czechoslovak Academy of Sciences.
Libiková, H. (1962) (Ed.) *Symposium on the biology of viruses of the tick-borne encephalitis complex*. Prague: Czechoslovak Academy of Sciences.
Libiková, H., & Stanček, D. (1965) *Acta virol., Prague*, **9**, 481.
Libiková, H., & Vilček, J. (1960) *Acta virol., Prague*, **4**, 165.
Lockart, R. Z. (1960) *Virology*, **10**, 198.
Lumsden, W. H. R. (1955) *Trans. r. Soc. trop. Med.*, **49**, 33.
Lumsden, W. H. R., Williams, M. C., & Mason, P. J. (1961) *Ann. trop. Med. Parasit.*, **55**, 389.
McIntosh, B. M., Paterson, H. E., McGillivray, G., & de Sousa J. (1964) *Ann. trop. Med. Parasit.*, **58**, 45.
McLean, D. M. (1953) *Aust. J. exp. Biol. med. Sci.*, **31**, 481.
McLean, D. M. (1956) *Aust. J. exp. Biol. med. Sci.*, **34**, 71.
McLean, D. M., Best, J. M., Mahalingam, S., Chernesky, M. A., & Wilson, W. (1964) *Can. med. Ass. J.*, **91**, 1360.
McLean, D. M., & Donohue, W. L. (1959) *Can. med. Ass. J.*, **80**, 708.
McLean, D. M., & Larke, R. P. B. (1963) *Can. med. Ass. J.*, **88**, 182.
McLeod, J. (1962) *J. comp. Path.*, **72**, 411.
Macnamara, F. N. (1954) *Trans. r. Soc. trop. Med.*, **48**, 139.
Magnus, H. von (1950) *Acta path. microbiol. scand.*, **27**, 276
Maguire, T., Miles, J. A. R., & Casals, J. (1967) *Am. J. trop. Med. Hyg.*, **16**, 371.
Mansharami, H. J. et al. (1967) *Indian J. Path. Bact.* **10**, 9 and 25.
Mason, P. J., & Haddow, A. J. (1957) *Trans. r. Soc. trop. Med.*, **51**, 238.
Mattingly, P. F. (1960) *Trans. r. Soc. trop. Med.*, **54**, 97.
Maurer, F. D., Kuttler, K. L., Yager, R. H., & Warner, A. (1952) *J. Immunol.*, **68**, 109.
Mayer, V., Blaskovič, D., Ernak, E., & Libikova, H. (1969) *J. Hyg., Camb.*, **67**, 731.
Meiklejohn, G., England, B., & Lennette, E. H. (1952) *Am. J. trop. Med. Hyg.*, **1**, 59.
Miles, J. A. R. (1952) *Aust. J. exp. Biol. med. Sci.*, **30**, 341.
Miles, J. A. R. (1964) *Bull. Wld Hlth Org.*, **30**, 197.
Mims, C. A., Day, M. F., & Marshall, J. D. (1966) *Am. J. trop. Med. Hyg.*, **15**, 775.
Morgan, C., Howe, C., & Rose, H. M. (1961) *J. exp. Med.*, **113**, 219.
Mussgay, M., & Rott, R. (1964) *Virology*, **23**, 573.
Netzky, G. I. (1967) *Jap. J. med. Sci. Biol., suppl.*, **20**, 141.
Nir, Y., Fendrich, J., & Goldwasser, R. (1957) *J. infect. Dis.*, **100**, 207.
Nishimure, C., Nomura, M., & Kitaoka, M. (1968) *Jap. J. med. Sci. Biol.*, **21**, 1.
Okuno, T. (1959) *Jap. J. med. Sci. Biol.*, **12**, 71.
Okuno, T., Okada, T., Kondo, A., Suzuki, M., Kobayashi, M., & Oya, A. (1968) *Bull. Wld Hlth Org.*, **38**, 547.
Okuno, T., Oya, A., & Ito, T. (1961) *Jap. J. med. Sci. Biol.*, **14**, 51.
Osterrieth, P. M., & Calberg-Bacq, C. M. (1966) *J. gen. Microbiol.*, **43**, 19.
Ota, Z. (1965) *Virology*, **25**, 372.
Otani, A. (1966) *Arch. ges. Virusforsch.*, **18**, 391.
Palmer, D. E., Bucca, M. A., Bird, B. R., & Winn, J. F. (1968) *Proc. Soc. exp. Biol. Med., U.S.A.*, **127**, 514.

Parker, J. R., & Stannard, L. M. (1967) *Arch. ges. Virusforsch.*, **20**, 469.
Parker, J. R., Wunters, A. G., & Smith, M. S. (1969) *Arch. ges. Virusforsch.*, **26**, 305.
Parks, J. J., & Price, W. H. (1958) *Am. J. Hyg.*, **67**, 187.
Peleg, J. (1968) *Am. J. trop. Med. Hyg.*, **17**, 219.
Peleg, J. (1969) *J. gen. Virol.*, **5**, 463.
Pennington, R. M., & Gibbs, C. J. (1967) *Proc. Soc. exp. Biol. Med., U.S.A.*, **125**, 787.
Perlowagora, A., & Hughes, T. P. (1947) *J. Immunol.*, **55**, 103.
Phillips, C. A., & Melnick, J. L. (1967) *Progr. med. Virol.*, **9**, 159.
Pond, W. L., Russ, S. B., Rogers, N. G., & Smadel, J. E. (1955) *J. Immunol.*, **75**, 78.
Porterfield, J. S. (1957) *Nature, Lond.*, **180**, 1201.
Porterfield, J. S. (1959) *Trans. r. Soc. trop. Med.*, **53**, 458.
Porterfield, J. S. (1961) *Bull. Wld Hlth Org.*, **24**, 735.
Porterfield, J. S., & Rowe, C. E. (1960) *Virology*, **11**, 765.
Price, W. H. (1966) *Virology*, **29**, 679.
Price, W. H. (1968) *Am. J. Epidemiol.*, 88, 392.
Price, W. H., Lee, R. W., Gunkel, W. F., & O'Leary, W. (1961) *Am. J. trop. Med. Hyg.*, **10**, 403.
Price, W. H., & O'Leary, W. (1967) *Am. J. Epidemiol.*, **85**, 83.
Price, W. H., Thind, I. S., Teasdall, R. D., & O'Leary, W. (1970) *Bull. Wld Hlth Org.*, **42**, 89.
Radda, A., Loew, J., & Pretzmann, G. (1963) *Zentbl. Bakt. ParasitKde, I Abt. Orig.*, **190**, 281.
Rajagopalan, P. K. et al. (1968) *Indian J. med. Res.*, (*suppl.*), **56**, 497, et sqq (11 papers).
Rao, T. R., & Anderson, C. R. (1964) *Indian J. med. Res.*, **52**, 727.
Reeves, W. C. (1958) *Handb. Virusforsch.*, **4**, 177. Springer. Vienna.
Reeves, W. C. (1961) *Progr. med. Virol.*, **3**, 59.
Reeves, W. C., French, E. L., Marks, E. N., & Kent, N. E. (1954) *Am. J. trop. Med. Hyg.*, **3**, 147.
Rehaček, J. (1968) *Acta virol., Prague*, **12**, 241.
Report of East African Virus Research Institute (1961–62) pp. 13 and 17.
Rivers, T. M., & Schwentker, F. L. (1934) *J. exp. Med.*, **59**, 669.
Robertson, E. G., & McLorinan, H. (1952) *Med. J. Aust.*, **1**, 10.
Robinson, M. C. (1955) *Trans. r. Soc. trop. Med.*, **49**, 28.
Rosen, L. (1958) *Am. J. trop. Med. Hyg.*, **7**, 406.
Rosicky, B., & Bardos, V. (1966) *Folia parasit., Prague*, **13**, 103.
Rossi, A. L. B. (1967) *Prog. med. Virol.*, **9**, 176.
Ross, R. W. (1956) *J. Hyg., Camb.*, **54**, 177.
Rudnick, A., Marchette, N. J., & Garcia, R. (1967) *Jap. J. med. Sci. Biol., suppl.*, **20**, 69.
Russell, P. K., Buescher, E. L., McCown, J. M., & Ordonez, J. (1966) *Am. J. trop. Med. Hyg.*, **15**, 573.
Sabin, A. B. (1952) *Am. J. trop. Med. Hyg.*, **1**, 30.
Sabin, A. B. (1955) *Am. J. trop. Med. Hyg.*, **4**, 198.
Sabin, A. B. (1959) in *Viral and rickettsial infections of Man.* Eds.: Rivers and Horsfall. 3rd Ed. p. 361. London: Pitman Medical.
Sabin, A. B., & Olitsky, P. K. (1938) *Proc. Soc. exp. Biol. Med., U.S.A.*, 38, 595.
Sabin, A. B., & Schlesinger, R. W. (1945) *Science*, **101**, 640.
Salminen, A. (1959) *Ann. Med. exp. Fenn.*, 38, 267.
Salminen, A. (1960) *Acta virol., Prague*, **4**, 17.
Sawyer, W. A. (1931) *J. prev. Med.*, **5**, 413.
Schaeffer, M., & Arnold, E. H. K. (1954) *Am. J. Hyg.*, **60**, 231.
Scherer, W. F. (1967) *Jap. J. med. Sci. Biol., suppl.*, **20**, 7.
Scherer, W. F., & Pancake, B. A. (1970) *Am. J. Epidemiol.*, **91**, 225.
Scherer, W. F., Buescher, E. L., & McClure, H. E. (1959a) *Am. J. trop. Med. Hyg.*, **8**, 689.
Scherer, W. F., Dickerman, R. W., Chia, C. W., Ventura, A., Moorhouse, A., Geiger, R., & Najera, A. D. (1964) *Science*, **145**, 274.
Scherer, W. F., Izumi, T., McCown, J., & Hardy, J. L. (1962) *Am. J. trop. Med. Hyg.*, **11**, 269.

Scherer, W. F., Moyer, J. T., Izumi, T., Gresser, I., & McCown, J. (1959b) *Am. J. trop. Med. Hyg.*, **8**, 698.
Schlesinger, R. W. (1951) *Proc. Soc. exp. Biol. Med., U.S.A.*, **76**, 817.
Schmidt, J. R., & El Mansoury, H. K. (1963) *Ann. trop. Med. Parasitol.*, **57**, 415.
Schulze, I. T., & Schlesinger, R. W. (1963) *Virology*, **19**, 40.
Seamer, J., Randles, W. J., & Fitzgeorge, R. (1967) *Br. J. exp. Path.*, **48**, 395.
Sellers, R. F., Bergold, G. H., Suarez, O. M., & Morales, A. (1965) *Am. J. trop. Med. Hyg.*, **14**, 460.
Shah, K. V., Johnson, H. N., Rao, T. R., Rajagopalan, P. K., & Lamba, B. S. (1960) *Indian J. med. Res.*, **48**, 300.
Shope, R. E. (1953) *J. exp. Med.*, **97**, 627.
Shope, R. E. (1961) *J. exp. Med.*, **113**, 511.
Shope, R. E., & Anderson, S. G. (1960) *Med. J. Aust.*, **1**, 156.
Shope, R. E., Causey, O. R., & de Andrade, A. H. (1964) *Am. J. trop. Med. Hyg.*, **13**, 723.
Shore, H. (1961) *Trans. r. Soc. trop. Med.*, **55**, 361.
Sidwell, R. W., Gebhardt, L. P., & Thorpe, B. D. (1970) *Bact. Rev.*, **31**, 65.
Silber, L. A., & Soloviev, V. D. (1946) *Am. Rev. Soviet Med.*, suppl. p. 6.
Simpson, R. W., & Hauser, R. E. (1968) *Virology*, **34**, 358 and 568.
Singh, K. R. P., Pavri, K. M., & Anderson, C. R. (1964) *Indian J. Med. Res.*, **52**, 566.
Slavik, I., Morena, E., & Mayer, V. (1970) *Acta virol., Prague*, **14**, 8.
Slepushkin, A. N. (1960) *Probl. Virol.*, **4**, 54.
Smith, C. E. Gordon (1956) *Nature, Lond.*, **178**, 581.
Smith, C. E. Gordon (1959) *Br. med. Bull.*, **15**, 235.
Smith, C. E. Gordon (1960) *Trans. r. Soc. trop. Med.*, **54**, 113.
Smith, C. E. Gordon (1962) *Symp. Zool. Soc., Lond.* N9. 6, 199.
Smith, C. E. Gordon (1967) *Jap. J. med. Sci. Biol.*, **20**, 130.
Smith, C. E. Gordon (1970) *Trans. r. Soc. trop. Med. Hyg.*, **64**, 481, 519 and 522.
Smith, C. E. Gordon, & Holt, D. (1961) *Bull. Wld Hlth Org.*, **24**, 749.
Smith, C. E. Gordon, McMahon, D. A., O'Reilly, K. J., Wilson, A. L., & Robertson, J. M. (1964a) *J. Hyg., Camb.*, **62**, 53.
Smith, C. E., Gordon, Varma, M. G. R., & McMahon, D. (1964b) *Nature, Lond.*, **203**, 992.
Smithburn, K. C. (1942) *J. Immunol.*, **44**, 25.
Smithburn, K. C. (1946) *J. Immunol.*, **52**, 309.
Smithburn, K. C., & Haddow, A. J. (1944) *J. Immunol.*, **49**, 141.
Smithburn, K. C., & Haddow, A. J. (1951) *Proc. Soc. exp. Biol. Med., U.S.A.*, **77**, 130.
Smithburn, K. C., Haddow, A. J., & Lumsden, W. H. R. (1949b) *Ann. trop. Med. Parasit.*, **43**, 74.
Smithburn, K. C., Hughes, T. P., Burke, A. W., & Paul, J. H. (1940) *Am. J. trop. Med.*, **4**, 471.
Smithburn, K. C., Kerr, J. A., & Gatna, P. B. (1954) *J. Immunol.*, **72**, 248.
Smithburn, K. C., Mahaffy, A. F., & Haddow, A. J. (1944) *J. Immunol.*, **49**, 159.
Smorodintsev, A. A. (1940) *Arch. ges. Virusforsch.*, **1**, 468.
Smorodintsev, A. A. (1958) *Progr. med. Virol.*, **1**, 210.
Sonnabend, J. A., Martin, E. M., & Mecs, E. (1967) *Nature, Lond.*, **213**, 365.
Southam, C. M. (1956) *J. infect. Dis.*, **99**, 155 and 163.
Southam, C. M., & Moore, A. E. (1951) *Am. J. trop. Med.*, **31**, 724.
Southam, C. M., & Moore, A. E. (1954) *Am. J. trop. Med. Hyg.*, **3**, 19.
Spence, L., Anderson, C. R., & Downs, W. G. (1962) *Trans. r. Soc. trop. Med. Hyg.*, **56**, 504.
Stamm, D. D. (1958) *Am. J. publ. Hlth*, **48**, 328.
Stim, T. B., & Henderson, J. R. (1966) *Proc. Soc. exp. Biol. Med., U.S.A.*, **122**, 1004.
Strauss, J. H., Burge, B. W., & Darnell, J. E. (1970) *J. molec. Biol.*, **47**, 437.
Stokes, A., Bauer, J. H., & Hudson, N. P. (1928) *Am. J. trop. Med.*, **8**, 103.
Stones, P. B., & Macnamara, F. N. (1955) *Trans. r. Soc. trop. Med.*, **49**, 176.
Strode, G. K. (1951) (Ed.) *Yellow fever.* New York: McGraw-Hill.
Suitor, E. C. (1966) *Virology*, **30**, 143.

Suitor, E. C. (1969) *J. gen. Virol.*, **5**, 545.
Suitor, E. C., & Paul, F. J. (1969) *Virology*, **38**, 482.
Sulkin, S. E., Burns, K. F., Shelton, D. F., & Wallis, C. (1962) *Tex. Rep. Biol. Med.*, **20**, 113.
Sutton, L. S., & Brooke, C. C. (1954) *J. Am. med. Ass.*, **155**, 1473.
Symposium on Thai Haemorrhagic Fevers. (1962). Bangkok. SEATO Medical Research Monograph No. 2.
Takemori, N. (1949) *Jap. med. J.*, **2**, 231.
Taylor, R. M. (1967) *Catalogue of Arthropod borne viruses of the World.* Pub. Hlth Service Publication No. 1760. Washington, D.C.: U.S. Government Printing Office.
Taylor, R. M., Hurlbut, H. S., Work, T. S., Kingston, J. R., & Frothingham, T. E. (1955) *Am. J. trop. Med. Hyg.*, **4**, 844.
Taylor, R. M., Work, T. H., Hurlbut, H. S., & Rizk, F. (1956) *Am. J. trop. Med. Hyg.*, **5**, 579.
Theiler, M. (1957) *Proc. Soc. exp. Biol. Med., U.S.A.*, **96**, 380.
Theiler, M. (1959) in *Viral and rickettsial diseases of man.* 3rd Ed. Eds.: Rivers & Horsfall. p. 343. London: Pitman Medical.
Theiler, M., & Smith, H. H. (1937) *J. exp. Med.*, **65**, 767 and 787.
Thind, I. S., & Price, W. S. (1966) *Am. J. Epidemiol.*, **84**, 193, 214, and 225.
Thomas, L. A. (1963) *Am. J. Hyg.*, **78**, 150.
Thomas, L. A., Kennedy, R. C., & Eklund, C. M. (1960) *Proc. Soc. exp. Biol. Med., U.S.A.*, **104**, 355.
Tongeren, H. A. E. van (1955) *Arch. ges. Virusforsch.*, **6**, 158.
Traub, E., & Kesting, F. (1956) *Zentbl. Bakt. ParasitKde, I. Abt. Orig.*, **166**, 462.
Traub, E., & TenBroeck, C. (1935) *Science*, **81**, 572.
Victor, J., Smith, D. G., & Pollack, A. B. (1956) *J. infect. Dis.*, **98**, 55.
Wachter, R. F., & Johnson, E. W. (1962) *Fed. Proc.*, **21**, 461.
Warner, P. (1957) *Aust. J. exp. Biol. med. Sci.*, **35**, 327.
Webb, H. E. (1965) *Trans. r. Soc. trop. Med. Hyg.*, **59**, 205.
Webster, L. T., & Clow, A. D. (1936) *J. exp. Med.*, **63**, 827.
Weinbren, M. P., Haddow, A. J., & Williams, M. C. (1958) *Trans. r. Soc. trop. Med.*, **52**, 253.
Weinbren, M. P., Kokernot, R. H., & Smithburn, K. C. (1956) *S. Afr. med. J.*, **30**, 631.
Weiss, K. E. (1957b) *Bull. epizoot. Dis. Afr.*, **5**, 459.
Weiss, K. E., Haig, D. A., & Alexander, R. A. (1956) *Onderstepoort J. vet. Res.*, **27**, 183.
Westaway, E. G., & Reedman, B. M. (1969) *J. Virol.*, **4**, 688.
Whitman, L. (1947) *J. Immunol.*, **56**, 97.
Whitman, L., & Aitken, T. H. G. (1960) *Ann. trop. Med. Parasit.*, **54**, 192.
W.H.O. (1967) Techn. report ser. no. 369. *Arboviruses and human disease.*
Williams, H. E. (1958) *Nature, Lond.*, **181**, 497.
Williams, M. C., Simpson, D. I. H., & Shepherd, R. C. (1964) *Nature, Lond.*, **203**, 670.
Williams, M. C., & Woodall, J. P. (1961) *Trans. r. Soc. trop. Med.*, **55**, 135.
Williams, M. C., Woodall, J. P., & Porterfield, J. S. (1962) *Trans. r. Soc. trop. Med.*, **56**, 166.
Woodall, J. P., & Bertram, D. S. (1959) *Trans. r. Soc. trop. Med.*, **53**, 440.
Work, T. H. (1958) *Progr. med. Virol.*, **1**, 248.
Work, T. H. (1963) *Bull. Wld Hlth Org.*, **29**, 59.
Work, T. H., & Shah, K. V. (1956) *Indian J. med. Sci.*, **10**, 582.
Wyckoff, R. W. G., & Tesar, W. C. (1939) *J. Immunol.*, **37**, 329.
Young, N. A., & Johnson, K. M. (1969) *Am. J. epidemiol.*, **89**, 286.
Zarate, M. L., & Scherer, W. F. (1968) *Am. J. trop. Med. Hyg.*, **17**, 894.
Zeipel, G. von, & Svedmyr, A. (1958) *Arch. ges. Virusforsch.*, **8**, 370.
Zlotnik, I., & Harris, W. J. (1970) *B. J. exp. Path.*, **51**, 37.

4

Unclassified Arboviruses

This chapter includes the arboviruses other than those in the Alpha-, Flavi-, Reo- or Rhabdo-virus genera. They can be divided on serological grounds into groups, though some are singletons, not shown to be related to any others. Some, as will be mentioned, differ from the Alpha- and Flavi- viruses in having nucleocapsids with helical symmetry, but there is no reason to suppose that this will be the case with all.

BUNYAMWERA SUPERGROUP

Review: Bardos et al. (1969).

Members of 8 virus groups which were previously considered as separate, have been found to show some degree of serological cross-reaction with members of other groups; this may be through the CF or the HAI and VN tests. The relationships are by no means as close as those among the alphaviruses or flaviviruses. Within the super-group, HAI tests suggest that there are 2 sets of more closely related groups: group C-Guama, -Capim and Bunyamwera-California-Bwamba-Simbu (WHO, 1967).

Morphology. Holmes (1971) described the morphology of 6 viruses in the supergroup: Cache valley (Bunyamwera group), Melao (California), Oriboca (C), Oropouche and Manzanilla (Simbu), Catu (Guama). He considered that 'all were essentially similar in morphology and mode of development'. Virions were 90–100 nm in diameter and budded into vesicles in or near the Golgi apparatus. The Golgi zone localization appeared to be a fundamental property of these viruses. Within the vesicles, virions showed an electron-lucent centre, but as they passed out of the cell the cores became electron-dense. Holmes considered that the same properties were shared by a number of viruses outside the Bunyamwera supergroup—Anopheles A, Uukuniemi and probably Rift Valley fever.

Murphy et al. (1968) recorded that the Bunyamwera, Tensaw and

Maguari viruses of the Bunyamwera group were about 98 nm in diameter with a closely adherent irregular envelope; they developed by budding from membranes into intracytoplasmic cisternae. Many virions occurred in linear array in neurons. A possible similarity to coronaviruses was suggested. Viruses of the C and California groups were said to be similar.

The Calovo strain of Batai virus from Czechoslovakia, studied by von Bondsdorff et al. (1969) was reported to have an inner coiled component 12 nm in diameter and surface projections on its envelope. Inkoo from Finland was similar, but the coils were 15 nm across.

Physico-chemical characters; Haemagglutination; Cultivation and Pathogenicity. See descriptions for arboviruses generally (p. 76).

Bunyamwera Group

Review: Casals & Whitman (1960).

The group contains, at present, 13 members: Bunyamwera, Germiston and Ilesha from Africa (these 3 have all caused fevers in man); Batai from India, Malaya and Czechoslovakia; Cache Valley, Guaroa, Kairi, Wyeomyia, Tensaw and 4 others from the Americas. The Wyeomyia virus was isolated from a man with fever. All seem to be mosquito-borne; the vector of Tensaw and Guaroa viruses (Lee & Sammartin, 1967) is probably an *Anopheles*: Buttonwillow virus has been isolated from *Culicoides* (Reeves et al., 1970).

C Group

Six viruses isolated in Belem, Brazil (Causey et al., 1961) were placed in a separate serological group by Casals & Whitman (1961). They were recovered from sentinel mice or monkeys, from mosquitoes or from man. They form haemagglutinins for goose cells, working best at pH 6·2 to 7. Ardoin & Clarke (1967) could prepare satisfactory haemagglutinins from suckling mouse livers only if they used sonication followed by calcium-phosphate chromatography. The viruses produce encephalitis in suckling mice, some of them also in older mice, but not all colonies of mice are susceptible. At least 2 of them (Caraparu & Oriboca) produce plaques on monkey-kidney monolayers. Infected persons have fever, headache and malaise. Antibodies are present in a proportion of people and forest mammals in the Amazon valley, and antibodies against Oriboca virus have been found in West and South Africa. The 6 original Brazilian strains are inter-related as shown:

Related by CF *Related by HAI and neutralization*

Oriboca }
{Itaqui }
{Caraparu }
{Apeu }
{Marituba }
{Murutucu }
{with Oriboca (Shope & Causey, 1962).

Four other C viruses (Madrid, Nepuyo, Ossa and Restan) have been isolated from Belem or the Caribbean region (Trinidad, Surinam or Panama). A tenth member, Gumbo Limbo, was isolated from *Culex* mosquitoes in Florida. The serological relations of all these viruses are described by Karabatsos & Henderson (1969). Gumbo Limbo was not closely related to any of the others.

California Group

The California virus was first isolated in 1943 and 1944. Nothing was heard of it for 20 years, when it and related viruses became active in several states of the United States (Hammon & Sather, 1966). There were numerous infections; a small proportion, all children, developed meningo-encephalitis (Thompson & Evans, 1965). Young (1966) reviewed the clinical and laboratory data. Seven, perhaps 8 strains in North America have been sufficiently distinct serologically to deserve separate names (Sather & Hammon, 1967; Murphy & Coleman, 1967). One member of the group (Melao) has been found in Trinidad and Belem (Brazil): the Lumbo virus comes from Mozambique in Africa and Tahyňa from Czechoslovakia, Germany, Yugoslavia, France and Italy. There may be a reservoir in rodents, but it is suspected that rabbits and hares are of particular importance (McKiel et al., 1966; Aspock & Kunz, 1967). *Aëdes vexans* may be a vector of Tahyňa (Simkova, 1963).

Other groups of viruses within the Bunyamwera supergroups are: Guama, 6 members; Capim, 5 members; Patois, 4 members. All these are from America. Bwamba and the related Pongola virus come from Africa, and there are 13 of the Simbu group with representatives from Asia, Africa and North and South America. The Oropouche virus of the Simbu group was first isolated from a febrile patient in

Trinidad in 1955. Then in 1960 it suddenly caused an outbreak of fever affecting 7,000 people near Belem in Brazil.

RIFT VALLEY FEVER

Original description: Daubney et al. (1931).

Review: Weiss (1957), Easterday (1965).

Morphology. Levitt et al. (1963) described spheres 60–75 nm across, their surfaces covered with hollow cylinders looking like short spikes, while McGavran & Easterday (1963) suggested a diameter of 90 nm. Lecatsas & Weiss (1968) described spheres 94 nm across with a 77 nm core and a capsid-like outer layer 9 nm wide. There was a translucent layer, 8 nm wide, between the core and the 'capsid'. Virions were associated with cisternae of the endoplasmic reticulum. Holmes (1971) considered that this description was consistent with what he had found in viruses of the Bunyamwera supergroup.

Physico-chemical characters. 40′ at 56° inactivates. Survives well when frozen or lyophilized and in some circumstances is much more stable than other arboviruses. There is evidence (Francis & Magill, 1935) that infective virus persisted in a room for 3 months. It survived 1048 days in serum kept at −4° (Smithburn et al., 1949): it also withstood 0·5 per cent phenol for 6 months in the cold. 1:1000 formalin inactivated it. Ether-sensitive. Density 1·23 (Levitt et al., 1963).

Haemagglutination. A haemagglutinin for cells of day-old chicks worked best at pH 6·5 and 25°. An inhibitor present in normal mouse serum disappeared on storage in the cold. Natural haemagglutinin was more potent than an acetone-ether extracted one and also agglutinated mouse, guinea-pig and human group A cells (Mims & Mason, 1956).

Antigenic properties. Distinct from other arboviruses in all immunological tests. A CF antigen made from livers of infected mice is useful for diagnosis. The agar-gel-diffusion test is also applicable, as well as conventional neutralization and HAI tests. An atypical strain (Lunyo) is described by Weinbren et al. (1957).

Cultivation. The virus produces thickening on inoculation on to the CAM of *fertile eggs* (Saddington, 1934) and grows also when inoculated into the yolk sac. It grows in *tissue cultures* of chick, rat, mouse, human and other cells. Weiss (1957) found lamb-kidney cells very suitable, and in these cells Coackley (1963) found eosinophilic filaments in nuclei.

Distribution. Occurs naturally only in Africa—Kenya, Uganda, South Africa and probably elsewhere in Central and southern Africa.

Pathogenicity. A disease of sheep, goats and cattle causing abortions and many deaths in pregnant and new-born animals. An epizootic in South Africa in 1951 killed about 100,000 lambs. Lambs show fever, vomiting, mucopurulent nasal discharge and bloody diarrhoea. Cattle are less seriously affected. Many infections have occurred in man during these epizootics, especially in herdsmen and veterinary officers. In man it resembles dengue and is probably biphasic. It is ordinarily mild, but cases of retinal damage have occurred in South Africa. The disease is very apt to spread to laboratory workers, though people working with neurotropic virus often have mild or inapparent infections.

It may affect buffalo and camels, and antelopes in the field were reported to have died or aborted (Gear et al., 1955).

Experimentally lambs are very susceptible and may die as early as 36 hours after infection. Mice, weaned or sucklings, can be infected by various routes. They ordinarily die with hepatitis within 3 days. A neurotropic strain was obtained by IC passage in mice (McKenzie & Findlay, 1936; Smithburn, 1949). This, given SC, can be used for immunizing mice and sheep (see below), as it has largely lost its viscerotropic properties. Inoculated monkeys have a mild fever, African species being less susceptible than rhesus. The neurotropic virus, however, may kill rhesus when injected IC (Findlay et al., 1936). Rats and other laboratory rodents, but not rabbits, are susceptible, and may die. Infected guinea-pigs may abort. Ferrets infected IN (Francis & Magill, 1935) showed fever and lung consolidation. Puppies and kittens develop fatal infections, and the disease may be transmitted by contact (Walker et al., 1970). Birds are resistant.

Pathological lesions caused by pantropic virus are chiefly those of massive hepatitis in lambs and focal hepatitis in older sheep. Eosinophilic bodies resemble the Councilman bodies of yellow fever and there are intranuclear inclusions in livers. These are more homogeneous and more closely resemble Cowdry's type A inclusions than do those of yellow fever. Blood-lakes are found in the liver. There is often damage to kidneys, and there may be haemorrhages in intestines and elsewhere. Lesions in other species are similar (Daubney et al., 1931).

Ecology. Infection is mosquito-borne. The most important vectors seem to be *Eratmopodites chrysogaster* in Uganda and *Aëdes caballus* in South Africa. Contact infection probably occurs also, and infection of laboratory workers has presumably been by the respiratory route. There is probably a reservoir in some wild African animals, other than primates. Weinbren suggests that a forest rat, *Arvicanthis*, fulfils the requirements in Uganda, as virus circulates plentifully in it during an inapparent infection (Weinbren & Mason, 1957).

Control. Control involves protection of flocks from mosquitoes. Live neurotropic virus has been used on a large scale to immunize sheep; it is not safe to give it during pregnancy (Weiss, 1957). A formolized vaccine gave apparently good results in man (Randall et al., 1963). Coakley et al. (1967) reported that neurotropic virus produced only a low titre of neutralizing antibody, but that cattle had some resistance to challenge 28 months later with pantropic virus.

OTHER MOSQUITO-BORNE VIRUSES

Taylor's (1967) catalogue lists about 20 other viruses which can be grouped serologically into 2s and 3s, also 11 singletons, unrelated to any others. About half come from the Americas, the rest from Asia, Africa or Australia.

PHLEBOTOMUS FEVER

Synonyms: Sandfly fever. Pappataci fever.

Reviews: Sabin (1951, 1959).

Morphology. Diameter estimated at 17–25 nm by filtration. In the light of recent interpretation of results of filtration, this estimated size is probably too low.

Physico-chemical characters. Survives storage in solid CO_2, or when lyophilized, for 8 or 9 years.

Haemagglutination. A haemagglutinin for chick cells was demonstrable when an alcohol-soluble inhibitor was first removed. Haemagglutination was optimal at 37° and between pH 5·5 and 6·5. The haemagglutinin, unlike those of flaviviruses was inactivated by merthiolate.

Antigenic properties. The 2 known viruses, the Sicilian and Naples strains, are antigenically distinct from other arboviruses and from each other. Convalescent persons develop antibodies which are of low titre or not demonstrable at all.

Cultivation. Reports of cultivation in *fertile eggs* are unconfirmed. Grows in *tissue cultures* of human, mouse or hamster kidney. CPE may be seen in human cells only after adaptation. Virus growth is more readily demonstrated by plaque production (Henderson & Taylor, 1960; Salim, 1966). Conditions for plaque formation, for at least the Naples strain, are rather critical (Salim, 1968a). A carrier state for the Naples virus may be set up in hamster cells (Salim, 1968b).

Distribution for both Naples and Sicilian viruses is from Italy eastwards to Egypt, Iran and Pakistan.

Pathogenicity. The viruses cause in man a short sharp fever, occasionally recurrent, with pains in the eyes, head, back and limbs and gastro-intestinal disturbances. There is leucopenia. Only man is known to be naturally affected.

Experimentally volunteers have been infected (Sabin, 1951), most readily by IV or intradermal injection. The normal incubation period of 3 or 4 days may be as short as 42 hours after IV inoculation. The viruses have been adapted to cause encephalitis in suckling mice inoculated IC; after numerous passages weaned mice also have been infected. Fully adapted Naples virus would also infect mice inoculated intranasally and produced fever after IC inoculation into rhesus monkeys. Mouse-adapted viruses inoculated intradermally into man produced immunity, but no symptoms.

Pathological lesions in mice are like those caused by other arboviruses.

Ecology. The vector is *Phlebotomus papatasi*, often called sandfly. There is an extrinsic incubation period in the fly of 7–10 days. There is conflict of evidence as to the possibility of transovarial transmission.

Control. The *Phlebotomus* because of their indoor biting habits are very easily controlled by D.D.T. Use of attenuated viruses for vaccination, though theoretically possible, has not been extensively adapted, except for a time in the U.S.S.R.

OTHER PHLEBOTOMUS-BORNE VIRUSES

About a dozen other viruses have either been isolated from *Phlebotomus* or have appeared to belong in the group because of their biological properties. Two have been isolated from rodents in Brazil:

Icoaraci (Causey & Shope, 1965) and Itaporanga (Trapp et al., 1965). They were related to the Naples virus by HAI tests. Other strains have come from Iran. Ibrahim & Sweet (1970), using immunodiffusion, found no cross-reactions among 11 of 12 viruses of the group.

NAIROBI SHEEP DISEASE

Review: Zahran (1968).

Cultivation. No definite growth in fertile eggs. Grown, with CPE, in cultures of lamb, goat testis and kidney, and in a hamster cell-line (Coackley & Pini, 1965; Howarth & Terpstra, 1965). Pleomorphic cytoplasmic inclusions appear in the cultures; some of them surround the nucleus.

Distribution. Kenya and Uganda; probably also the Congo and parts of Southeast Africa.

Pathogenicity. A haemorrhagic gastro-enteritis affecting sheep and goats with high fever and a mortality of 30–70 per cent. In the disease seen in Kenya splenic enlargement and involvement of the female genital tract were noted (Daubney & Hudson, 1931). Mugera & Chema (1967) regularly saw nephritis, myocardial degeneration and necrosis of the gall bladder. Rather different symptoms were seen in an outbreak in Uganda by Weinbren et al. (1958), who frequently encountered cardiac damage and pulmonary oedema. Virus is most abundantly present in spleen and liver. Goats may undergo sub-clinical infection.

Experimentally the disease is transmissible to sheep with blood or serum. It produces encephalitis in mice inoculated IC, and sucklings can also be infected IP. They die with characteristic tonic spasms. After 110 passages in adult mice the virus was attenuated for sheep, which had symptomless viraemia (Ansell, 1957; Weinbren et al., 1958).

Ecology. Transmitted by ticks, *Rhipicephalus appendiculatus*. There may well be a reservoir in a wild animal; Daubney & Hudson (1934) suggest that a rodent, *Arvicanthis*, may serve as such.

Control. Virus attenuated by mouse passage is under trial as a vaccine.

UUKUNIEMI VIRUS

This tick-borne virus, isolated in Finland, is mentioned because its morphology has been carefully studied. It is said to have helical symmetry with nucleocapsids 2 nm wide (von Bonsdorff et al., 1969; Saikku et al., 1970) while the surface is covered with projecting hollow cylinders 9 or 10 nm long (Saikku & von Bonsdorff, 1968). The structure is considered to resemble that of the viruses in the Bunyamwera supergroup. Like them, development occurs in association with the Golgi complex (von Bonsdorff et al., 1970). A serologically related virus (Grand Arbaud) has been isolated in France (Hannoun et al., 1970).

CONGO VIRUS

Synonym: Crimean haemorrhagic fever.

This was isolated from febrile patients in Africa (Simpson et al., 1967; Woodall et al., 1967). Similar agents were recovered from cattle. It is probably transmitted by ticks, having been recovered from species of *Ornithodorus* and *Hyalomma*. Hazara virus from Pakistan and one described as causing Crimean haemorrhagic fever (Leschinskaya, 1967) are antigenically similar (Casals, 1969). In the infections in the Crimea the haemorrhagic manifestations occurred on the third to fifth day of the disease. Mortality was 16–19 per cent. Several cases occurred in workers studying the disease.

HAEMORRHAGIC FEVERS

Reviews: Gajdusek (1956). Casals et al. (1970).

A number of serious illnesses accompanied by haemorrhage are known or suspected to be arthropod-borne. In all there is a sudden onset with fever, often diphasic, purpuric rashes and haemorrhages into lungs, gastro-intestinal tract, kidneys and elsewhere with hypertensive shock and renal insufficiency. The syndrome has been seen in infections with the following viruses: dengue (p. 100), Chikungunya (p. 86), Omsk (p. 108), Crimean (p. 128), Argentinian (p. 175) and Bolivian (p. 175) haemorrhagic fevers.

The remaining example has proved very baffling.

Epidemic Haemorrhagic Fever (Manchuria, Eastern Siberia and Korea)

Synonym: Haemorrhagic nephroso-nephritis.

Reviews: Smadel (1959). Gajdusek (1956).

In this disease, which was a serious problem during the Korean war, there is particularly severe damage to kidneys and many died with shock or oliguria. Attempted transmission to experimental animals has been unsuccessful, so were many attempts at growing in tissue culture (Gey, 1954), this despite intensive work. However, Gavrilyuk et al. (1968) obtained an agent which interfered with the growth of poliomyelitis virus in cultures of human embryo and also obtained some positive results in these cultures in fluorescent antibody tests using conjugates with convalescent sera. The pathological lesions in man include retroperitoneal oedema and intense congestion of kidneys. Oliver & Macdowell (1957) have described the histopathology. The agent has been recovered from the mite *Laelaps jettmari* which normally infests field mice (*Apodemus agrarius*), but *Trombicula* spp. are more likely vectors (Traub et al., 1954). However, Chumakov (1957) considers that infection may be transmitted otherwise than through arthropods. It may not be an arbovirus (see p. 76). What may be the same disease has been reported from Hungary and elsewhere in Europe (Trencséni & Keleti, 1960).

Simian Haemorrhagic Fever

A disease with symptoms similar to the disease in man has been reported in captive rhesus monkeys in the United States and the U.S.S.R. (Palmer et al., 1968). A virus was isolated causing CPE in cultures of embryonic rhesus kidney. It contained RNA, was less than 50 nm in diameter, was labile at pH 3 and inactivated by chloroform. Cytoplasm of infected tissue-culture cells contains broad undulating lamellae (Wood et al., 1970). It was not pathogenic for mice. There was no evidence that it was arthropod-borne, but is included here for convenience.

BORNA DISEASE

(*Named from a locality in Saxony*)

Synonym: Enzootic encephalomyelitis (of horses, sheep and cattle). Near East equine encephalomyelitis.

Reviews: Nicolau & Galloway (1928, 1930).

Daubney (1967) now considers that the Near East equine encephalomyelitis described (Daubney & Mahlau, 1957) from Egypt and Syria is almost certainly Borna disease.

Morphology. Early work estimated the diameter by filtration as 85–125 nm, by UV microscopy 110–140 nm (Elford & Galloway, 1933); recently Daubney (1967) stated that the virus passed 20 nm gradocol membranes.

Antigenic properties. Immunologically distinct from viruses causing similar infections. Neutralizing antibodies not readily detected in serum. CF antigen present in brains of infected rabbits; apparently a 'soluble antigen' 15–30 nm across, stable between pH 5 and 8 (von Sprockhoff, 1958).

Cultivation is reported on the CAM of fertile eggs incubated for 5–11 days, preferably at 35°–35·5° (Rott & Nitzschke, 1958). Grows with CPE in cultures of lamb testis and monkey kidney (Daubney, 1967).

Distribution. The main focus of the disease is in Saxony, but it is reported from other parts of Germany, Poland, Rumania, Russia, and formerly elsewhere in Europe, Syria and Egypt. A virus causing staggers in horses in Nigeria (Porterfield et al., 1958) may be the same.

Pathogenicity. Affected horses show lassitude, followed by a period of excitation with tonic spasms and later paralysis. The virus may produce similar symptoms in sheep, cattle and probably deer (sporadic bovine encephalomyelitis is caused by a different agent). *Experimentally*, the disease can be transmitted to other species, most readily to rabbits (Zwick & Seifried, 1924). Inoculation IC is most effective, but infection can be produced by various other routes. The incubation period is from 20 to 60 days, but in new-born rabbits may be only 12 to 15 days (von Sprockhoff, 1958). Symptoms consist of depression, somnolence, loss of weight, salivation, later paralysis affecting first the hind limbs and finally death in coma. Infected guinea-pigs, rats and mice develop similar symptoms, but are less susceptible than rabbits; older rats and mice are more easily infected than younger ones. Fatal disease may occur in rhesus monkeys and cats inoculated IC. Human infections are not reported. Cats, dogs and ferrets are also resistant. Virus may be found in the nervous system, adrenals and in ganglia in various organs of rabbits, probably travelling along nerves. Virus may be present in the serum of recovered horses and donkeys for 6 months after recovery or exposure (Daubney, 1967).

Pathological lesions are those of meningo-encephalomyelitis, the characteristic feature being the presence of the Joest-Degen bodies in nerve cells, particularly those of the hippocampus and olfactory lobes. These are small round eosinophilic nuclear inclusions, showing evidence of internal structure. They are not constantly present in horses or rabbits, particularly in later serial passages in rabbits; they occur more regularly in infected rats and guinea-pigs.

Ecology. Earlier workers thought arthropod transmission improbable. The virus, however, has been recovered from ticks of several genera (*Hyalomma*, *Dermacentor* and *Ornithodoros*), and has been transmitted transovarially in *Hyalomma anatolicum*, which may be the chief vector in the Near East. The virus has also been recovered from the brains of herons and other wild birds (Daubney, 1967; Daubney & Mahlau, 1967). It seems justifiable, therefore, to consider this as an arbovirus. It is possible, however, that transmission may also pass by way of oral and nasal secretions.

Control. Virus in brain tissue, inactivated with phenol or phenol-glycerol has been used, apparently with success, to control the disease. Zwick et al. (1929) had previously reported success with a lapinized virus.

OTHER TICK-BORNE VIRUSES

Review: Casals (1967).

At least 25 other viruses have been isolated from ticks or are thought to be tick-borne; 15 of them can be gathered into small serologically related groups while 10 are singletons. Mention may be made of the Hughes virus apparently associated with sea birds in the Atlantic and Pacific and apparently carried by *Ornithodoros* ticks; and of the Quaranfil and Chenuda viruses from herons in Egypt and South Africa associated with the tick *Argas porsicus*. The Quaranfil virus has been isolated from human blood.

There remain 12 or more viruses isolated from bats, rodents or other species, unrelated to other arboviruses and not associated with any known vector.

SUCKLING-MOUSE CATARACT AGENT

This ether-sensitive virus was isolated in the southern United States from the tick *Haemaphysalis leporis-palustris* (Clark, 1964). When inoculated into suckling mice it caused cataracts after 20 days, sometimes nervous symptoms and stunting of growth. The virus passed a 220 nm not a 100 nm membrane. In 1 mouse strain, cataracts continued to appear in inoculated mice up to a year after inoculation (Clark & Karzon, 1968).

REFERENCES

Ansell, R. H. (1957) *Vet. Rec.*, **69**, 410.

Ardoin, P., & Clarke, D. H. (1967) *Am. J. trop. Med. Hyg.*, **16**, 357.

Aspock, H., & Kunz, C. (1967) *Zentbl. Bakt. ParasitKde, I Abt. Orig.*, **203**, 1.

Bardos, V., & co-workers (eds) (1969) *Arboviruses of the California complex and the Bunyamwera group.* Publ. house of the Slovak Academy of sciences.

Bonsdorff, C. H. von, Saikku, P., & Oker-Blom, N. (1969) *Virology*, **39**, 342.

Bonsdorff, C. H. von, Saikku, P., & Oker-Blom, N. (1970) *Acta virol., Prague*, **14**, 109.

Casals, J. (1967) *Jap. J. med. Sci. Biol., suppl.*, **20**, 119.

Casals, J. (1969) *Proc. Soc. exp. Biol. Med., U.S.A.*, **131**, 233.

Casals, J., Henderson, B. E., Hoogstraal, H., Johnson, K. M. & Shelokov, A. (1970) *J. infect. Dis.*, **122**, 437.

Casals, J., & Whitman, L. (1960) *Can. med. Ass. J.*, **82**, 355.

Casals, J., & Whitman, L. (1961) *Abstr. 10th Pacif. Sci. Congr.*, p. 458.

Causey, O. R., Causey, C. E., Maroja, O. M., & Macedo, D. G. (1961) *Am. J. trop. Med. Hyg.*, **10**, 227.

Causey, O. R., & Shope, R. E. (1965) *Proc. Soc. exp. Biol. Med., U.S.A.*, **118**, 420.

Chamberlain, R. W. (1963) *Proc. 7th Int. Conf. trop. Med. Malaria*, **3**, 160.

Chumakov, M. P. (1957) *Pub. Hlth Monogr. 50*, p. 19. Washington, D.C.: U.S. Govt Print. Office.

Clark, H. F. (1964) *J. infect. Dis.*, **114**, 476.

Clark, H. F., & Karzon, D. T. (1968) *J. Immunol.*, **101**, 776.

Coackley, W. (1963) *J. Path. Bact.*, **86**, 530.

Coackley, W., & Pini, A. (1965) *J. Path. Bact.*, **90**, 672.

Coackley, W., Pini, A., & Gosden, D. (1967) *Res. vet. Sci.*, **8**, 399 and 406.

Daubney, R. (1967) *Res. vet. Sci.*, **8**, 419.

Daubney, R., & Hudson, J. R. (1931) *Parasitology*, **23**, 507.

Daubney, R., & Hudson, J. R. (1934) *Parasitology*, **26**, 496.

Daubney, R., Hudson, J. R., & Garnham, P. C. (1931) *J. Path. Bact.*, **34**, 545.

Daubney, R., & Mahlau, E. A. (1957) *Nature, Lond.*, **179**, 584.

Daubney, R., & Mahlau, E. A. (1967) *Res. vet. Sci.*, **8**, 375.

Easterday, B. C. (1965) *Adv. vet. Sci.*, **10**, 65.

Elford, W. J., & Galloway, I. A. (1933) *Br. J. exp. Path.*, **14**, 196.

Findlay, G. M., Mackenzie, R. D., & Stern, R. O. (1936) *Br. J. exp. Path.*, **17**, 431.

Francis, T., & Magill, T. P. (1935) *J. exp. Med.*, **62**, 433.

Gajdusek, D. C. (1956) *J. Pediat.*, **60**, 841.

Gavrilyuk, B. K., Noskov, F. S., & Smorodintsev, A. A. (1968) *Acta virol., Prague*, **12**, 381.

Gear, J. H. S., de Meillon, B., Le Roux, A. F., Rofsky, R., Rose-Innes, R., Steyn, J. J., Cliff, W. D., & Schutz, K. H. (1955) *S. Afr. med. J.*, **29**, 514.
Gey, G. O. (1954) *Bull. Johns Hopkins Hosp.*, **94**, 108.
Hammon, W. Mc. D., & Sather, G. E. (1966) *Am. J. trop. Med. Hyg.*, **15**, 199.
Hannoun, C., Corniou, B., & Rageau, J. (1970) *Acta virol., Prague*, **12**, 381.
Henderson, J. R., & Taylor, R. M. (1960) *Am. J. trop. Med. Hyg.*, **9**, 32.
Holmes, I. H. (1971) *Virology*, **43**, 708.
Howarth, J. A., & Terpstra, C. (1965) *J. comp. Path.*, **75**, 437.
Ibrahim, A. N., & Sweet, B. H. (1970) *Proc. Soc. exp. Biol. Med., U.S.A.*, **135**, 23.
Karabatsos, N., & Henderson, J. R. (1969) *Acta virol., Prague*, **13**, 544.
Lecatsas, G., & Weiss, K. E. (1968) *Arch. ges. Virusforsch.*, **25**, 58.
Lee, V. H., & Sammartin, C. (1967) *Am. J. trop. Med. Hyg.*, **16**, 778.
Leschinskaya, E. V. (1967) *Jap. med. Sci. Biol., suppl.*, **20**, 143.
Levitt, J., Naude, W. du T., & Polson, A. (1963) *Virology*, **20**, 530.
McGavran, M. H., & Easterday, B. C. (1963) *Am. J. Path.*, **42**, 587.
Mackenzie, R. D., & Findlay, G. M. (1936) *Lancet*, **1**, 140.
McKiel, J. A., Hall, R. R., & Newhouse, V. F. (1966) *Am. J. trop. Med. Hyg.*, **15**, 98.
Mims, C. A., & Mason, P. J. (1956) *Br. J. exp. Path.*, **37**, 423.
Mugera, G. M., & Chema, S. (1967) *Bull. épizoot. Dis. Afr.*, **15**, 337.
Murphy, F. A., & Coleman, P. H. (1967) *J. Immunol.*, **90**, 276.
Murphy, F. A., Harrison, A. K., & Tzianabos, T. (1968) *J. virol., Prague*, **2**, 1315.
Nicolau, S., & Galloway, I. A. (1928) *Spec. Rep. Sci. med. Res. Coun., Lond.*, no. 121.
Nicolau, S., & Galloway, I. A. (1930) *Annls. Inst. Pasteur, Paris.*, **44**, 673 and **45**, 457.
Oliver, J., & Macdowell, M. (1957) *J. clin. Invest.*, **36**, 99.
Palmer, A. E., Allen, A. M., Tauraso, N. M., & Shelokov, A. (1968) *Am. J. trop. Med. Hyg.* **17**, 404, 413 and 422.
Porterfield, J. S., Hill, D. H., & Morris, A. D. (1958) *Br. vet. J.*, **114**, 425.
Randall, R., Binn, L. N., & Harrison, V. R. (1963) *Am. J. trop. Med. Hyg.*, **12**, 611.
Reeves, W. C., Scrivani, R. P., Hardy, J. L., Roberts, D. R., & Nelson, R. L. (1970) *Am. J. trop. Med. Hyg.*, **19**, 544.
Sabin, A. B. (1951) *Arch. ges. Virusforsch.*, **4**, 367.
Sabin, A. B. (1959) in *Viral and rickettsial infections of man.* Eds.: Rivers & Horsfall. 3rd Ed. p. 374. London: Pitman Medical.
Saddington, R. S. (1934) *Proc. Soc. exp. Biol. Med., U.S.A.*, **31**, 693.
Saikku, P., & von Bonsdorff, C. H. (1968) *Virology*, **34**, 804.
Saikku, P., & von Bonsdorff, C. H. (1970) *Acta virol., Prague*, **14**, 103.
Salim, A. R. (1966) *Nature, Lond.*, **210**, 466.
Salim, A. R. (1968a) *J. gen. Virol.*, **2**, 81.
Salim, A. R. (1968b) *Arch. ges. Virusforsch.*, **23**, 89.
Sather, G. E., & Hammon, W. McD. (1967) *Am. J. trop. Med. Hyg.*, **16**, 548.
Semenov, B. F. (1965) *Acta Virol., Prague*, **9**, 560.
Shope, R. E., & Causey, O. R. (1962) *Am. J. trop. Med. Hyg.*, **11**, 283.
Simkova, A. (1963) *Acta virol., Prague*, **7**, 419.
Simkova, A. (1966) *J. Hyg. Epidem. Microbiol. Immun.*, **10**, 499.
Simpson, D. I. et al. (1967) *E. Afr. med. J.*, **44**, 87.
Smadel, J. E. (1959) in *Viral and rickettsial diseases of man.* Eds.: Rivers & Horsfall. 3rd Ed. p. 400. London: Pitman Medical.
Smithburn, K. C. (1964) *Br. J. exp. Path.*, **30**, 1.
Smithburn, K. C., Mahaffy, A. F., Haddow, A. J., Kitchen, S. F., & Smith, J. F. (1949) *J. Immunol.*, **62**, 213.
Sprockhoff, H. von (1958) *Z. Immun-forsch.*, **115**, 161.
Taylor, R. M. (1967) *Catalogue of arthropod-borne viruses of the New World.* Publ. Hlth Serv. Publ. 1760. Washington, D.C.: U.S. dept. of Hlth.
Thompson, W. H., & Evans, A. S. (1965) *Am. J. Epidemiol.*, **81**, 230.
Trapp, E. E., de Andrade, A. H. P., & Shope, R. E. (1965) *Proc. Soc. exp. Biol. Med., U.S.A.*, **118**, 421.

Traub, R., Hertig, M., Lawrence, W. H., & Harriss, T. T. (1954) *Am. J. Hyg.*, **59**, 291.
Trencséni, T., & Keleti, B. (1960) *Acta med. hung.*, **16**, 303.
Walker, J. S. et al. (1970) *J. infect. Dis.*, **121**, 9, 19 and 25.
Weinbren, M. P., Gourlay, R. N., Lumsden, W. H. R., & Weinbren, B. M. (1958) *J. comp. Path.*, **68**, 174.
Weinbren, M. P., & Mason, P. J. (1957) *S. Afr. med. J.*, **31**, 427.
Weinbren, M. P., Williams, M. C., & Haddow, A. J. (1957) *S. Afr. med. J.*, **31**, 951.
Weiss, K. E. (1957) *Bull. epizoot. Dis. Afr.*, **5**, 431.
W. H. O. (1967) *Arboviruses and human disease.* Techn. Rep. Ser. no. 389.
Wood, O., Tauraso, N., & Liebhaber, H. (1970) *J. gen. Virol.*, **7**, 129.
Woodall, J. P., Williams, M. C., & Simpson, D. I. H. (1967) *E. Afr. med. J.*, **44**, 93.
Young, D. T. (1966) *Ann. int. Med.*, **65**, 419.
Zahran, G. E. D. (1968) *Handb. der Virusinf. bei Haustieren*, **3**, 1147.
Zwick, W., & Seifried, O. (1924) *Berl. Münch. tierärztl. Wehr.*, **40**, 465.
Zwick, W., Seifried, O., & Witte, J. (1929) *Arch. Tierheilk*, **59**, 511.

5

Leukoviruses

Evidence is accumulating that there exists a natural group of actually or potentially oncogenic RNA viruses infecting birds or mammals. They may be divided into 4 groups:

1. viruses of the avian leukosis-sarcoma complex
2. avian reticulo-endotheliosis
3. viruses causing leukosis and sometimes sarcomata in mammals; those infecting rodents have been chiefly studied
4. The mammary tumour virus of mice

It is likely that fundamental properties are alike in all members of the genus, but the abbreviations F (fowl), ML (mouse leukaemia) and MTV (mammary tumour virus) will indicate that particular facts have been determined for certain viruses only.

Morphology and development. Outer membranes bear projections which may be rather indefinite (F, ML) or regularly arranged ((MTV), Almeida et al., 1967). Coiled structures have been seen within, or issuing from, nucleoids; these apparently contain RNA, but there is no evidence that they are nucleocapsids like those of myxoviruses. Multiplication is by budding from cytoplasmic membranes; some host material may be carried into virus envelopes. Particles resembling leukoviruses, especially murine ones, have been seen in tissues of many species. It has been suggested that information for the production of such particles may be present and vertically transmitted in 'many, perhaps all, vertebrates' (Huebner & Todaro, 1969); it may be unexpressed, partly expressed with resulting cancer, or fully expressed with formation of infectious particles. This view is supported by the discovery that group specific antigens for RNA tumour viruses are present at some stage or other during embryonic development in mice of all strains (Huebner et al., 1970 (ML); Bentvelzen et al., 1970 (MTV)). Many members of the group (F, ML) may exist in a 'defective' state, requiring cooperation from a related 'helper virus' for full expression.

Chemical composition. The viruses contain single-stranded RNA of MW about 10–13 × 10^6 daltons; this comprises 1·5 to 2 per cent of the weight of the virion (Bather, 1957, 1958; Allison & Burke, 1962). Base composition (F): C 24·1 ± 0·42; A 25·2 ± 0·53; G 28·2 ± 0·54; U 22·5 ± 0·58 (Harel et al., 1965). Some DNA may be detected in virions as well as RNA, but this is possibly cellular DNA incorporated during virus maturation (Levinson et al., 1970). Virus growth is suppressed by inhibitors of DNA; it appears that an early stage of virus growth is dependent on activity of cellular DNA (Bader, 1964, 1966 (F); Duesberg & Robinson, 1967 (ML)). An RNA-dependent DNA-polymerase has been discovered in virions (F) (Temin & Mizutani, 1970; Baltimore, 1970; Spiegelman et al., 1970 (F, ML, MTV)), and this may lead to the formation of a DNA replicative genome (Bader & Bader, 1970 (F)).

THE FOWL LEUKOSIS-SARCOMA COMPLEX

Review: Vogt (1965a).

Viruses of this group are closely related, as is shown by their possession of a common, internal, complement-fixing antigen. This is revealed by the COFAL (Complement Fixation Avian Leukosis) test (Sarma et al., 1964). The viruses may be found as latent infections of most stocks of domestic fowls, but may give rise to malignant diseases of various sorts—sarcoma, myeloblastosis, erythroblastosis, lymphoid or visceral leukosis—or to osteopetrosis. Those most studied in the laboratory, such as the Rous sarcoma and some of the leukosis viruses, have probably become partly fixed in their properties and may now be very different from what they were in their original 'wild' state. Though now such viruses usually breed true, they do not invariably do so and injection of a leukosis virus may produce a local tumour or osteopetrosis. Neurolymphomatosis, otherwise known as fowl paralysis, or Marek's disease, was formerly confused with leukosis, but is now known to be caused by an unrelated virus (see p. 352).

Physico-chemical characters. The half-life at 37° of various viruses has been tabulated by Vogt (1965a); most determinations fall within the range 150–360 minutes. Readily oxidized, that of Rous sarcoma was preserved better in the presence of cysteine hydrochloride or HCN. Comparatively resistant to UV radiation; Rous virus was

10 times more resistant than NDV though equally sensitive to X-rays (Rubin & Temin, 1959). Sensitive to 20 per cent ether and to desoxycholic acid.

Antigenic properties. The viruses may be divided into 2 major and 2 minor groups on the basis of the antigenic properties of their coats. The best studied viruses and their subgroups were arranged in tabular form in a paper by Ishizaki & Vogt (1969). In conformity with this arrangement, fowls can be classified as regards susceptibility. They may be resistant to viruses of A, of B, of A and B or of B and C groups, while all can be infected by D viruses. This resistance has a genetic basis (Crittenden et al., 1967).

Tests for the common (COFAL) antigen are usually performed with sera of tumour-bearing hamsters (see p. 144) (Sarma et al., 1964). Sera of pigeons with tumours can also be used (Sarma et al., 1969). Immunodiffusion tests and tests using fluorescent antibody (Berman & Sarma, 1965) gave similar results (Cook et al., 1966). The group antigen is probably an internal component of the virus (Kelloff & Vogt, 1966). Some experiments, however, suggest that the antigen is also present in tissues of normal, apparently virus-free, fowls of a particular inbred line (Dougherty & Di Stefano, 1966; Payne & Chubb, 1968). The antigen may have multiple components (Roth & Dougherty, 1968; Armstrong, 1969). Its MW is estimated at 23,000 daltons (Allen, 1968).

Distribution. The viruses have been isolated only from domestic fowls; the only exception is 1 obtained from quail embryos by de Ratuld & Werner (1967, 1968). Neutralizing antibodies to Rous virus have, however, been found in sera of several species of wild birds in East Africa (Morgan, 1965). Similar tests in North America gave negative results (Rabin & Sladen, 1969), but were only carried out against 1 serotype.

It is of interest that King & Duran-Reynals (1963) found antibodies in normal ducks active only against a Rous virus which had been adapted to grow in ducks; there may have been a helper virus present, antigenically distinct from the fowl viruses.

Avian Leukosis Viruses

Reviews: Vogt (1965a).

Presence of latent leukosis virus in normal fowls has been revealed in several ways: neutralizing antibodies for Rous virus frequently develop in fowls as they grow older; particles resembling those of Rous virus can be seen in normal chick embryos and other tissues (Benedetti, 1957).

Rubin (1961) has found that normal chick embryos may contain an agent which, at least on first recovery, produces no cytopathic effects in tissue culture, but interferes with the growth of Rous virus. He called it Resistance-inducing factor (RIF). In 1 flock, 1 of every 6 hens had a chronic RIF viraemia (Rubin et al., 1962). It was serologically related to Rous virus and resembled it in its susceptibility to ether, heat and UV radiation. On passage in culture it became capable of affecting cells so that they grew in disorganized fashion. Later it was found that when 11-day embryos were inoculated, tumours or leukosis were apt to develop after they had hatched (Baluda & Jamieson, 1961), and the virus is probably an avirulent kind of leukosis virus. Other viruses called RAV (Rous-associated virus) 1, 2 and the like have been described by Rubin & Vogt (1962).

The term RIF is used in the literature in 2 senses: (*a*) to designate the virus described by Rubin, (*b*) for any virus in the family interfering with the growth of Rous virus.

A virus in the group may be defective: that is, it may continue multiplying in a cell and its descendants, conferring malignant properties on them, but unable to infect fresh cells from without. This can, however, be achieved when a related 'helper' virus such as Rubin's (1961) RIF virus or other latent leukosis virus complements it by the addition of its own coating protein. Such viruses either interfere with or activate tumour viruses according to timing or other factors. Helper viruses are of several antigenic types and the serological behaviour of a virus depends upon its coat; it may therefore belong to one or other of what are called 'pseudotypes' (Hanafusa et al., 1963, 1964; Vogt, 1965b). Thus Rous sarcoma virus (RSV) may be of various pseudotypes described as RSV (RAV 1) or RSV (RAV 2) according to which leukosis virus has furnished the coat. Many viruses under study in laboratories consist of mixtures of 2, 3 or more pseudotypes.

Recent studies suggest that defectiveness is rarely, if ever, absolute; thus apparently wholly defective viruses may be able to infect quail embryos. Hanafusa & Hanafusa (1968) therefore suggest that leukosis-negative Rous (L–R) cells may be a better term than defective or nonproducer.

Fowls seem to be in 1 or other of 2 classes. Some, congenitally infected, are immunologically tolerant, having persistent viraemia and no detectable antibodies; these transmit infection through the egg. Others undergo an infection which is usually transient and these have antibodies; some, however, may nevertheless, carry an infection in the ovary (cf. Vogt, 1965a) and also serve as sources of infection for other birds.

Manifest lymphoid leukosis may develop later in life in birds of either group (Rubin et al., 1961; Rubin, 1962). It has proved possible to obtain leukosis-free flocks by taking eggs from immune hens and protecting the hatched chicks by rigid quarantine from contact with infected birds. It is important to be able to use eggs from such flocks for producing vaccines to be used in man.

Lymphoid Leukosis

Synonym: Visceral leukosis.

Reviews: Beard et al., (1955). Biggs & Payne (1967).

The RPL 12 strain has been particularly studied.

Cultivation. As is true for the latent leukosis viruses, this has been grown in tissue culture, its presence being detected by interference with the growth of Rous sarcoma virus.

Pathogenicity. This is the commonest form of leukosis and it is increasing. Affected birds become pale and listless and show diarrhoea in the terminal stages. Except for the RPL 12 and a few other strains it has been hard to pass experimentally; those strains, however, infect young chicks which die after 4 to 36 weeks.

Pathological lesions comprise infiltration of various viscera with lymphoblasts and lymphocytes. The spleen is enlarged, the liver enlarged and either diffusely greyish-red, marbled or nodular. There is anaemia, and lymphoblasts may be found in the blood.

Earliest changes after inoculation of chicks with the RPL 12 strain are in the bursa of Fabricius. If this is removed 12 weeks after infection, no dissemination to other organs occurs (Cooper et al., 1968). The bursa is apparently an important source of malignant cells (Dent et al., 1967).

Avian Erythroblastosis

Chemical composition. In contrast to myeloblastosis, there is no associated adenosine triphosphatase nor Forssman antigen.

Cultivation. *In fertile eggs.* Passed in series by IV inoculation of 11-, 12- or 13-day-old embryos. Forty to 70 per cent of embryos died with lesions as in hatched birds (Atanasiu et al., 1957). Embryos of turkeys, quail, pheasants, guinea-fowl and ducks have also been infected (Pollard & Hall, 1941).

In tissue culture the virus is less readily established than is myeloblastosis. There is evidence that erythroblasts in culture liberate virus by a budding process, while surrounding the central body by a double membrane. Either fibroblasts or marrow could be infected *in vitro* (Lagerhof, 1960; Heine et al., 1961).

Pathogenicity. The symptoms of the disease are those of progressive anaemia and weakness.

In chicks inoculated intravenously with a big dose, the incubation period is about 9 days; death follows after a few days more. Susceptibility does not seem to vary with age. Inoculated guinea-fowls, turkeys and pheasants may show transient anaemia or even leukaemia; some fatal infections in guinea-fowls are reported by Engelbreth-Holm & Meyer (1932).

Freshly isolated strains may give rise on inoculation either to erythro- or to myelo-blastosis. After passage in the laboratory they commonly but not invariably breed true.

Infected birds show profound anaemia and the blood contains large numbers of erythroblasts. The blood changes are described by Furth (1931). Enormous numbers of virus particles are present in the plasma—up to 10^{10}/ml. There is great enlargement of the spleen, and the bone marrow is uniformly greyish-red, fat having largely disappeared.

Ecology. Natural chick-to-chick transmission is not reported.

Myeloblastosis

Morphology and development. Virus particles have been hard to find in myeloblasts; they appear to develop within the cytoplasm in viroplasts or 'grey bodies' which may arise from mitochondria and later to turn into vesicles within which virus particles may be seen. They may be rapidly liberated from cells, but no 'budding' as with erythroblastosis is described. They may be present in larger numbers in macrophages of spleen and bone marrow, possibly as a result of phagocytosis (Bonar et al., 1959). Lacour et al. (1970) found in virions of avian myeloblastosis a coiled inner component lacking any 'regularly repeating structures which could be termed helical or compared with myxovirus nucleocapsids'.

Chemical composition. Virions contain 2·3 per cent of RNA (Allison & Burke, 1962). The RNA is single-stranded with a modal length of 8·7 μm and a mean length of 8·3 μm (Gramboulan et al., 1966). An adenosine-triphosphatase is regularly associated with this virus in contrast to that of erythroblastosis; it may be derived from the host cell, as virus is budded off; its activity can be used as a method of titrating the virus (Mommaerts et al., 1954). The virus can infect kidney cells and cause a nephroblastoma; virus from such cells is largely free from the enzyme. Forssman antigen is also associated with the virus.

Cultivation. The virus grows well in tissue cultures of myeloblasts from the blood of infected birds; the changes described above under 'Morphology' are seen when cultivation is carried out in media containing 20 per cent of serum. Cells from normal chick marrow have also been infected (Beaudreau et al., 1960), also osteoblasts and other cells (Baluda & Goetz, 1961). Moscovici (1967) describes the formation of foci of transformed cells. An unusual strain from Bulgaria (MC29) causes massive transformation of chick-embryo cells in culture, or with lower doses, foci of heaped-up cells (Langlois & Beard, 1967).

Pathogenicity. As with erythroblastosis the symptoms are those of anaemia leading to progressive weakness and death. The incubation period after intravenous inoculation is proportional to the dose. The white blood count is higher than with erythroblastosis and may reach over 2,000,000/mm^3. Circulating cells are largely myeloblasts. The plasma may also contain up to 10^{12} virus particles/mm^3 (Eckert et al., 1955). Liver and spleen are enlarged, greyer and more mottled than with erythroblastosis. There may be associated sarcomata.

The MC29 strain also causes tumours in liver and kidney. Myeloblastosis given IV to 5-day turkey poults has produced mainly erythroblastosis.

Avian Sarcomata

Synonyms: These include the Rous sarcoma (chicken tumour No. 1), (Rous, 1911), the Fujinami and other fowl tumours; also tumour MH2 (Murray-Begg endothelioma).

Reviews: Rubin (1962). Vogt (1965a).

Morphology and development. Virions, as described for the group, of about 75 nm (Bernhard et al., 1956) or rather less (Gaylord, 1955). They contain an electron-dense nucleoid 35–40 nm across, surrounded by 2 membranes. They are found extracellularly or at the surface of cells or within vacuoles; these last may, indeed, be invaginations of the cell surface. The bodies may be detected in as few as 1 in 50 tumour cells (Epstein, 1956); they are more abundant in young, rapidly growing tumours produced by injecting very potent filtrates (Haguenau et al., 1958). Rubin's (1955) work suggests that virus particles are liberated quite slowly, possibly 1 per 100 cells every hour, but later work suggests that particles can at times be released much more freely. The tumour cells are certainly in some sort of equilibrium with virus, not necessarily being destroyed; the virus is in fact a 'moderate' one.

Particles, apparently morphologically perfect, can be detected by electron microscopy in propagated tumours resulting from a helper-free 'defective' virus (Dougherty & Di Stefano, 1966).

The development of the virus is a 2-stage process. Certain early events are controlled by the genome and are thus independent of what helper determines the pseudotype. These probably include formation of the COFAL antigen and the nature of the lesions produced in tissue culture or *in vivo*. The coat-protein determines the ability to infect a cell, the behaviour of the virus in neutralization tests and in the events concerned in viral interference.

Chemical composition. The RNA was broken down by Montagnier et al. (1969) into 4 smaller units of estimated $2{\cdot}5 \times 10^6$ MW.

Physico-chemical characters. The rate of inactivation at various temperatures between 37° and 60° was determined by Dougherty (1961). Its stability in various buffers is described by Bryan (1955). There is a full summary of physico-chemical characters in Vogt's (1965a) review.

Antigenic properties. Fowls with slow-growing tumours develop neutralizing antibodies; these may also be produced by immunizing other species—geese, rabbits, goats. Viruses producing histologically distinct tumours may be closely related antigenically, yet not identical (Andrewes, 1933), or they may be antigenically remote (Carr & Campbell, 1958). These differences doubtless find their explanation, at least in part, in antigenic differences between helper viruses as described in preceding paragraphs. One can also demonstrate complement fixation and agglutination of virions.

Antibodies are present in the yolks of eggs laid by immune hens (Andrewes, 1939; Aulisio & Shelokov, 1969).

Cultivation. *In fertile eggs.* Tumour viruses, especially Rous, have been propagated on the CAM giving rise to foci of ectodermal proliferation (Keogh, 1938; Harris, 1954). Lesions may consist of proliferation of epithelium or of vascular endothelium (Coates et al., 1968). Such virus behaves as usual when injected into chicks.

In tissue culture. Lo et al. (1955) described destructive effects on chick fibroblasts followed by some regeneration of cultures with development of multinucleate giant cells; the pH of infected cells rapidly dropped. Recently, workers using monolayers of chick-embryo cells have observed foci of cell proliferation which can be counted for purposes of virus assay (Manaker & Groupé, 1956; Temin & Rubin, 1958; Rubin, 1960).

Under appropriate conditions 90 per cent of cells can be 'transformed' within 24 hours (Hanafusa, 1969).

Macrophages from buffy coat were infected with virus by Rangan & Bang (1967), but, in contrast to cultures in fibroblasts, little free virus was produced. Ephrussi & Temin (1960) infected 'pure' epithelium from the iris. Different clones of virus may produce foci of different cell types, either round or fusiform (Temin, 1960). The Schmidt-Ruppin strain produced CPE in rat, guinea-pig, and mouse cells (Bergman & Jonsson, 1962). Foci of cell proliferation were produced also in human embryonic cells (Zilber & Shelvljaghyn, 1964; Jensen et al., 1964). Current studies of tissue cultures for elucidating cell-virus relations cannot be reviewed here (cf. Prince, 1960; Temin & Rubin, 1959).

Pathogenicity. Sarcomata occur sporadically amongst fowls, more in some flocks than in others. Not all are transplantable and still fewer are filterable *ab initio*. Present knowledge is mainly based on studies on the Rous No. 1 and a few other laboratory-propagated tumours.

Experimentally the Rous virus behaves very differently in different circumstances. Bryan's strain of enhanced virulence is active in very high dilutions. At other times the virus's activity is so depressed that transplantable tumours may for several passages fail to yield active filtrates. This is seen more often after inoculation of small doses of virus into older birds. Growths are usually progressive and fatal, but they occasionally regress temporarily or permanently. Lesions in viscera or elsewhere may be due to the virus itself or may be true metastases. In very young chicks the virus causes multiple haemorrhages and rapid death without evidence of neoplasia; necrosis of endothelial cells is suggested as the cause (Duran-Reynals, 1940). Some viruses, as already indicated, may cause either localized tumour growth or generalized leukosis (Stubbs & Furth, 1935). Two viruses (Rous No. 1 and MH2) produced adeno-carcinomata when inoculated into the kidney of young chicks (Carr, 1959).

Other species of birds may be infected with Rous virus—pheasants, guinea-fowls, turkeys, quails and, less readily, birds more remote zoologically, pigeons and ducks. There is a striking account of adaptation to ducks and readaptation of the duck virus to chicks by Duran-Reynals (1942). The Fujinami sarcoma goes readily in ducks (Fujinami, 1928).

Some strains of Rous virus may infect rats, causing haemorrhagic cysts (Zilber & Kryukova, 1957; Svet-Moldavsky & Skorikova, 1960) Ahlström and his colleagues have produced progressively growing, metastasizing, transplantable sarcomata in new-born rats (Ahlström & Jonsson, 1962) new-born and adult hamsters (Ahlström & Forsby, 1962); also in suckling mice and young guinea-pigs. Only temporary growth occurred in young rabbits (Ahlström, Jonsson & Forsby, 1962). Rabotti and his colleagues (1966a,b), however, produced fatal tumours in new-born rabbits and dogs by IC inoculation of centrifuged material. Tumours have also been produced in new-born monkeys of several species (Munroe & Windle, 1963; Zilber et al., 1965; Deinhardt, 1966)

Svet-Moldavsky and colleagues (1967) obtained tumours in inoculated tortoises and snakes. Tumours in mammals usually yield no free virus, but this has proved possible with a few tumours from rats (Svec et al., 1966) and hamsters (Svoboda & Klement, 1963). Most tumours, however, yield virus once more when they are transferred back to chicks or when the cells are grown in mixed culture with chick-embryo cells (Svoboda, 1964; Huebner et al., 1964). The COFAL antigen may be present in mammalian tumours but not, as a rule, the surface antigens. There is evidence that specific transplantation antigens are formed in Rous sarcomata in mice (Jonsson & Sjögren, 1965; Bubenik & Bauer, 1967). The Schmidt-Ruppin strain has been particularly used in transplantation experiments in mammals; other strains have differed in effectiveness.

Pathological lesions. Rapidly growing Rous sarcomata are soft growths containing much mucinous material mixed with blood-pigments; slow-growing ones are firm and white. Metastases are frequent, especially in lungs, liver and heart. Other filterable growths described by Rous and others are an intracanalicular spindle-celled (rifted) sarcoma, fibrosarcomata, osteochondrosarcomata, endothelioma.

Ecology. As already indicated, the avian sarcomata are probably an uncommon manifestation of the activity of an ubiquitous virus. Evidence that they are contagious would therefore be surprising.

Osteopetrosis

Synonyms: Thick leg disease. Marble bone. Diffuse osteoperiostitis.

This is a nonmalignant condition in which there is excessive activity of osteoblasts leading to great thickening of the bones. It is apt to appear in a proportion of birds inoculated with the agents of lymphoid or other forms of leukosis, but it is doubtful whether it is very closely related to the viruses of the leukosis complex. The virus is, however, reported to be ether-sensitive, to be liberated from the cell surface by budding and to resemble the Rous and similar viruses morphologically (Campbell et al., 1964; Young, 1966). Holmes (1963) succeeded in transmitting the disease from fowls to young turkeys, though it does not occur naturally in these.

Avian Reticulo-endotheliosis

Synonym: T virus.

This virus, first recovered from a turkey, is not related to the viruses of the sarcoma-leukosis complex.

Morphology. Of about the same size as leukosis viruses, but differs in having a 'chain-like or pseudohelical configuration in the nucleoid'. (Zeigel et al., 1966).

Physico-chemical properties. Ether-sensitive. Inactivated at 56° in 30 minutes.

Antigenic properties. Neutralized by specific antisera or egg yolks from immune birds, but apparently unrelated to leukosis viruses.

Cultivation. Produces 'pocks' on CAM of hens' eggs (Theilen et al., 1966). Multiplies in cultures of chick (Bose & Levine, 1967) duck, or quail embryos; no CPE and no interference with leukosis viruses.

Pathogenicity. Large doses in day-old chicks kills them 6 or 7 days later; signs of illness are usually absent until shortly before death (Sevoian et al., 1964). With smaller doses or older birds, mortality is lower. There is massive proliferation of primitive mesenchymal cells, especially in liver and spleen and around blood-vessels; whether the change is neoplastic is discussed by Olson (1967).

Distribution. There is serological evidence that the infection is widespread in the United States (Aulisio & Shelokov, 1969).

Ecology. The disease does not readily pass by contact, though some symptomless contacts have developed antibodies. Since viraemia is an important feature, the question of arthropod transmission has been raised (Thompson et al., 1968).

MAMMARY TUMOUR VIRUS OF MICE (MTV)

Synonyms: Milk factor. Bittner agent.

Reviews: Bittner (1948). Blair (1968).

Morphology and development. Much as with other leukoviruses. The virions, however, have definite regularly arranged spikes on their outer membranes. They are a little shorter than those seen in myxoviruses (Lyons & Moore, 1962): 10 nm (Hairstone et al., 1964) or 8·5 nm long (Calafat & Hageman, 1968). Their arrangement follows the pattern of an underlying recticular structure. The B particles up to 30 nm across, described by Bernhard et al. (1955), are the mature form of the virus: they have double membranes and are seen within vacuoles, budding from the cell surface (Moore et al., 1959) or within the canaliculi of affected mammary glands. They contain one sometimes more nucleoids, usually placed eccentrically. Besides these, 'A particles', only 65–70 nm across, are found close to the nucleus of tumour cells and in close relation to the Golgi apparatus; whether they are precursors of B particles is disputed. The nucleoid apparently contains tangled threads 6 nm in diameter, but does not resemble the nucleocapsid of a myxovirus (Calafat & Hageman, 1969). According to Sarkar & Moore (1970), measurement of the lengths of nucleic acid molecules shows peaks at 1·2, 2·4 and 3·6 microns. The MW of the longest molecules was $3{\cdot}6 \times 10^6$.

There have been reports that biological activity may be carried also in smaller particles, particularly associated with red blood cells and with greater specificity than for B particles as regards host range (Nandi et al., 1966).

Chemical composition. Lyons & Moore (1965) report that the lipid content is 27 per cent and the RNA 0·8 per cent on a dry-weight basis. MW 10×10^6 (Duesberg & Cardiff, 1968).

Physico-chemical characters. Buoyant density in sucrose gradient 1·16 (Duesberg & Cardiff, 1968). Fairly stable between pH 5 and 10·2. Can be preserved by freeze-drying at 76° for years. Ether-sensitive. Inactivated in 30 minutes at 56° or in 90 hours at 37°.

Antigenic properties. Neutralizing antibodies can be prepared in rabbits, guinea-pigs and mice. Tumour-bearing mice, however, do not develop antibodies, probably because of immunological tolerance. Tumours caused by the virus can be transplanted more readily into mice carrying the virus than into others (Old & Boyse, 1965). Viruses from 3 strains of mice were immunologically alike (Blair, 1960). Specific precipitins (Imagawa et al., 1948; Blair, 1965) and complement-fixing antibodies are described. Brown & Bittner (1961), however, consider such tests unreliable and prefer to look for the power of sera to prevent virus from entering cells, as tested for with fluorescent antibody. The A particles are not specifically stained by such antibody (Tanaka & Moore, 1967). According to Blair (1969) immunodiffusion tests rarely detect antigen in milk during the first lactation, but do so increasingly in later lactations. There is an antigen shared by MTV and 1 strain of mouse leukaemia and revealed by cytotoxic tests (Stück et al., 1964). There seems to be some reciprocal interference between MTV and leukaemia viruses (Squartini et al., 1967).

Besides specific viral antigens, there is evidence for existence of transplantation antigens which may or may not be identical for all tumours caused by MTV (Morton, 1969; Vaage, 1969).

Cultivation. *In fertile eggs.* Attempts at cultivation in eggs have given mainly negative results; at any rate claims to have succeeded remain unconfirmed.

In tissue cultures. Activity has been maintained in cultures of embryonic mouse skin in fibroblasts derived from mouse mammae. One cell-line derived from a tumour has continued to yield virus over 6 years (Sykes et al., 1968). The virus has also survived in organ cultures of mammary tumours (Cardiff et al., 1968).

Habitat. Wild as well as laboratory mice may harbour the virus (Andervont & Dunn, 1956).

Pathogenicity. The virus is commonly latent and only causes mammary cancers when in a genetically susceptible strain of mice and when the appropriate hormonal influences operate (breeding in ♀♀, oestrogen treatment in ♂♂). Mice can be infected, best when very young, *per os* or by parenteral injection by any route. Virus is present in the cancers, but even more abundantly in the lactating mamma, also in liver, spleen and other tissues of cancerous mice; it is not found in urine or faeces. Production of tumours has been largely relied on in titrations of virus; results are irregular, and more cancers may develop after injection of dilute inocula (Bittner, 1945).

Presence of an inhibitor may explain some of the anomalies. Tumours develop in inoculated young mice in 6 months to 2 years.

The mamma is certainly not the only 'target organ', since virus will multiply in, and be recoverable from, the spleens of mammectomized mice.

Pathological lesions. The mammary tumours are typical adeno-carcinomata. Their appearance is preceded, according to some workers, by hyperplastic changes in the mammae.

Ecology. Infection is transmitted when baby mice ingest milk of mothers of a high cancer strain such as C3H. It may then, in susceptible mice, be passed similarly for many generations, whether the mother mice develop cancer or remain latently infected. In more resistant strains of mice or after small initial infecting doses the incidence of tumours falls in succeeding generations and the infection becomes extinguished. Fetuses are not ordinarily infected *in utero* and if removed by Caesarean section and suckled on mothers of a low cancer strain, they develop few if any tumours. There are, however, reports (e.g. Bittner, 1952) that males of a high cancer strain may transmit infection to their offspring, often but not necessarily infecting the mother.

Bentvelzen et al. (1970) offer evidence that mice, perhaps all of them, transmit the viral genome of their own strain of MTV vertically as a genetic factor. Virus of a different strain, however, is passed on through the milk.

A virus resembling MTV was recovered from chemically induced mammary tumours in rats (Bergs et al., 1970).

VIRUSES RELATED TO MTV

The classical mammary tumour virus is probably only 1 of a family of viruses of varying degrees of pathogenicity. A nodule-inducing virus has been described (Pitelka et al., 1964; Nandi & De Ome, 1965). In the mamma, this produces hyperplastic nodules as MTV does, but these do not progress to become neoplastic. When present together with MTV, this virus may interfere with it and lead to a reduction in malignancy. It is apparently not regularly transmitted in milk, but may be passed on 'vertically' at conception. It is antigenically closely related to MTV, but lacks some antigenic components present in the latter (Blair & Weiss, 1966). Viruses exist, including some in wild mice, intermediate in properties between MTV and the nodule-inducing virus (Bentvelzen et al., 1970). Old & Boyse (1965) suggest that related agents may be found in rats.

Mammary Tumour Virus of Rhesus Monkey

A virus having properties similar to those of the mouse virus has been cultivated from a mammary carcinoma in a rhesus monkey (Chopra & Mason, 1970).

Mammary Cancer in Man

Particles resembling those of MTV have also been seen in the milk of women with (Chopra & Feller, 1969) or without (Moore et al., 1969) breast cancer, and there is circumstantial evidence suggesting that they may be related to human mammary cancer (Charney & Moore, 1971; Schlom et al., 1971).

MOUSE LEUKAEMIA

Review: Boiron et al. (1967).

We have now to consider a number of viruses producing leukaemia, sometimes lymphosarcoma, in mice. First comes Gross's virus which seems to be the causative agent of a lymphoid leukaemia occurring commonly in certain inbred strains of mice. The other leukaemia viruses have mostly been obtained by injecting filtrates of propagable tumours into mice, especially new-borns. These leukaemias do not correspond to any commonly occurring natural disease. It is likely that the viruses are normally wholly latent infections which may be able to reach higher levels of activity when multiplying in a neoplasm, and after propagation become 'fixed viruses' with properties differing from those they possessed when latent. They are probably not causally related to the neoplasms from which they come.

The viruses can be placed in two groups: those of the thymic and splenic types (Sinkovics, 1962; Sinkovics & Howe, 1964). The best studied in the former group are those described by Gross, Graffi & Moloney; in the latter those of Friend & Rauscher. Other authors have grouped the viruses differently, on an antigenic basis, classifying the Friend, Moloney and Rauscher viruses together (Old et al., 1964). References concerning more than 20 different strains will be found in Sinkovic's papers. There are no certain differences between the various viruses as regards morphology, chemical composition or physico-chemical characters. These properties will be described together, with some references in brackets to indicate which virus was the subject of particular observations.

Morphologically similar particles can be seen in the tissues of non-leukaemic mice (Chapman et al., 1966; Hartley et al., 1969), in tissue culture lines derived from these (Hall et al., 1967; Faras & Erikson, 1969) and in mouse lymphomas induced by radiation (Kaplan, 1967) or chemical carcinogens. Almost certainly there are related agents in all mouse stocks capable of being activated so that they will produce leukaemia; they are present even in ostensibly germ-free mice (Kajima & Pollard, 1968).

Morphology and development. The virions—the C particles of Bernhard & Guerin (1958)—are 80 to 110 nm in diameter with a central nucleoid 40–70 nm across surrounded by a thin membrane and a thick outer envelope. Projections on the surface are ill-defined. The outer shell is formed by a hollow coiled cylinder 7–7·5 nm wide similar to structures seen in some poxviruses. The internal shell may contain nucleic acid (de-Thé & O'Connor, 1966). Other (A) particles have an electron-lucent centre; their relation to the C particles is obscure (de Tkaczevski et al., 1968). The nucleoid may be formed as a result of the collapse of two concentric shells visible in immature virions (de Thé & O'Connor, 1966). Many particles show tails. There has been dispute as to whether these are artefacts. De Harven & Friend (1964) describe methods of preparation which seem to ensure that no tails will be found (Friend virus).

Chemical composition. As for other leukoviruses. Adenosine-triphosphatase and other enzymes have been located in the viral envelope of the Rauscher and Moloney viruses (de Thé, 1966); they may be of host origin. The MW of $2{\cdot}2 \times 10^8$ is reported for the virion and 10×10^6 for its RNA.

Physico-chemical Characters. Probably all ether-sensitive, though there are conflicting reports concerning Moloney's virus. Zeigel & Rauscher (1964) report the effects of various physical and chemical agents on the Rauscher virus. Inactivated in 30 minutes at 56° (Moloney & Rauscher viruses) or at 65–68° (Gross virus). Density 1·15–1·16 for Rauscher & Moloney viruses (O'Connor et al., 1964) and 1·16 to 1·18 for Gross virus (Wahren, 1966).

Antigenic properties. The viruses share a common internal complement-fixing antigen (Hartley et al., 1965); it has been referred to as COMUL (COmplement-fixation MUrine Leukaemia). The antigen is a strongly basic protein with a MW of about 26,000 (Schäfer et al., 1969), but it may comprise more than 1 component (Gregoriades & Old, 1969); it apparently contains 2·5 to 4 per cent nucleic acid.

Though there are discrepancies in the results of different workers the Friend, Moloney and Rauscher viruses are certainly closely related antigenically; the Gross virus is distinct. Old & Boyse (1965) describe 5 surface antigens best revealed by an *in vitro* cytotoxic test.

1. The G (Gross) antigen is found in cells of leukaemias induced by Gross virus and in lymphoid tissues of AKR and other mouse strains with a high incidence of spontaneous leukaemia.
2. The FMR (Friend, Moloney and Rauscher) antigen is present in cells of these 3 leukaemias. It is revealed also by neutralization and immunofluorescence tests (Old et al., 1964; Fink & Malmgren, 1963).
3. Mention has already been made (p. 147) of an antigen (ML) shared by the mammary tumour virus and some leukaemias.
4. The E-antigen is present only in leukaemias in C57 BL mice.
5. The TL antigen (Boyse et al., 1963) is found in thymus cells of some normal mice and in leukaemia cells, even those with which no virus is known to be associated.

Besides these there are transplantation antigens (Klein & Klein, 1964); they are more difficult to demonstrate than is the case with DNA oncogenic viruses. The Friend, Moloney and Rauscher viruses may be related in this test to each other but not to Gross virus (Glynn et al., 1968).

Haemagglutination. A haemagglutinin for sheep red cells is closely associated with the virion (Friend and Rauscher viruses); it is revealed after treatment with neuraminidase or phospholipase C (Schäfer & Szanto, 1969), and even then 3×10^{10} virus particles are needed. Earlier claims (Gross, 1959) could not be confirmed by Boiron et al. (1967); Gross's material was not subjected to the pretreatment.

Some particular points concerning different leukaemia viruses require mention.

Viruses of the Thymic Group

Gross's Strain

Synonyms: AK leukaemia. Spontaneous lymphoid leukaemia.

Much of the work published before about 1958 is confused by the fact that material being studied was in fact a mixture of leukaemia and polyoma viruses.

Antigenic properties. The antigenic behaviour of this virus is best explained by the existence of immunological tolerance. A soluble antigen, specific for the virus, is present in the blood and tissues of affected mice (Aoki et al., 1968).

Cultivation. Virus will multiply in cultures of embryo or spleen cells from some strains of mice, but without cytopathic effect (Gross et al., 1961; Lemonde & Clode, 1962). Cultures from thymic lymphoma induced in rats with the virus showed evidence of 'transformation' (Ioachim, 1967). The disease is a generalized lymphomatosis rather than a leukaemia, since cells in the blood are not necessarily greatly affected. Thrombocytopenia is said to precede other evidence of leukaemia (Brodsky, 1969).

Pathogenicity. The virus has been obtained from inbred mice with a very high incidence of spontaneous leukaemia (especially sublines of AK mice). Filtrates inoculated into new-born mice (especially C3H strain) with a very low leukaemia incidence induced leukaemia, but only after 8–11 months, much longer than is seen after cell transmission. By passage of filtrates through new-born C3H mice Gross obtained his A-line of high virulence which would infect mice up to 14 days old and produce leukaemia after only 3 or 4 months (Gross, 1957); it would also now cause leukaemia in day-old rats (Gross, 1961). The possibility of repeating Gross's work has depended on using mice genetically the same as his. It appears that multiplication of the virus occurs primarily in the thymus. The virus can be transmitted to rats; in them and in thymectomized mice it can produce myeloid and other forms of leukaemia (Gross, 1963).

Ecology. Transmission of the spontaneous leukaemia in AK mice is, as Gross (1955) maintains, 'vertical', being passed from one generation to the other through the ovum. Both sexes may play a part in transmission (Law, 1966). On the other hand, the virulent passage A-line may be transmitted via the milk (Gross, 1962).

Control. Development of this form of leukaemia either naturally or in inoculated mice can be prevented or diminished by thymectomy (Levinthal et al., 1959; Miller, 1962).

Leukaemias Induced by Radiation or Chemical Carcinogens

These commonly yield virus, and those from leukaemias following X-radiation have been extensively reviewed by Kaplan (1967), who adduces evidence that they are the cause of the radiation leukaemia, not mere passengers. Such viruses are in general like Gross's, but are not readily adapted to grow in tissue culture (Boiron et al., 1967).

Moloney's Strain

This is another leukaemia arising during propagation of a mouse tumour, in this case the S37 mouse sarcoma.

Original description: Moloney (1960).

Cultivation. Leukaemia cells were successfully grown *in vitro*; also virus infected normal mouse embryo and kidney-cell cultures. No CPE observed (Ginsburg & Sachs, 1961; Sinkovics & Howe, 1964).

Pathogenicity. The virus will induce lymphoid leukaemia in adult mice of several strains. First changes appear in 8 weeks. Transplants from induced leukaemias lead to lymphocytic neoplasms of varying biological behaviour (Dunn et al., 1961) (cf. fowl leukaemias, p. 136). Leukaemias were reported to occur frequently in the offspring of mice which had been inoculated in infancy (Salaman & Harvey, 1961). The virus has recently been found to infect also rats and new-born hamsters (Moloney, 1962).

Ecology. It is apparently transmitted 'vertically' to embryos *in utero*, but even better through the mother's milk (Law, 1962).

Graffi's Chloroleukaemic Strain (Graffi, 1957)

This is a myeloleukaemia obtained by inoculating filtrates of several transplantable tumours in suckling mice of the Agnes Bluhm strain (which perhaps carries the virus). Serial transplantation was at first achieved with difficulty, but, later, filtrates were passed in series in several strains of mice, and rats were also infected.

It could be cultivated in mouse-embryo cells (Graffi et al., 1963), as could a similar virus isolated by Prigozhina & Stavroskaya (1964). Antigenic specificity was revealed by antisera made in mice, but not by immune rat or rabbit sera (Levy et al., 1969).

Viruses of the Splenic Group

Friend's Strain

Synonym: Swiss mouse leukaemia.

Review: Friend (1959).

Cultivation has proved difficult in this strain; there has been, however, some evidence of multiplication in spleens of infected mice

(Moore, 1963), in mouse-embryo culture (Osato et al., 1964) or in a mouse lung cell-line (Yoshikura et al., 1967).

Pathogenicity. Experimentally, the virus when first described would readily infect adult mice but only DBA/2 and Swiss mice. (It is unusual for an agent with such limited specificity to infect particularly a strain which, like the Swiss, is not inbred.)

Injected mice show enlargement of liver and spleen within a few days; yet the disease may run a course of 2 or 3 months. The spleens soon become palpable through the abdominal wall. Mice die either within a few weeks from haemorrhage, sometimes from a ruptured spleen, or, after several months, from leukaemia. Infection has been produced with spleen suspensions diluted 10^{-3} or 10^{-1}. 'Friend disease', if it appears, develops quickly; mice which fail to show it, either naturally or as a result of splenectomy, are likely to have lymphomata later. Mirand (1968) has described a variant which causes polycythaemia. The malignant cells lack the autonomy once believed to be ordinarily a character of malignancy; for transfer is always through infection of host cells by the virus, not through cell-transplantation (Furth, 1959). Axelrad & Steeves assayed the virus by counting the macroscopic foci which were found in the spleen 9 days after intravenous injection of mice. An 'avirulent Friend-like virus' was found in wild mice in Queensland (Pope, 1963).

Several lines of work now point to the conclusion that one is dealing with a complex containing 2 viruses. Passage through rats (Dawson et al., 1966) leads to a lymphoid leukaemia, but not the spleen enlargement which is attributed to erythroleukaemia and is sometimes called 'Friend disease'; this condition is also absent from infant mice acquiring leukaemia through their mother's milk (Mirand et al., 1966). Though it was thus possible to obtain lymphoma free from Friend disease, the latter could not be obtained in a pure state. Rowson & Parr (1970), however, obtained by limiting dilutions of Friend virus material a virus with 'minimal pathogenicity'. By itself it caused only minor, nonprogressive splenomegaly. It is suggested that the virus concerned is defective, needing lymphoma virus as a helper. The 2 viruses are apparently antigenically closely related.

Pathological lesions. According to Metcalf et al. (1959) the disease is a leukaemia of reticulum-cell type, combined with erythroblastosis. The blood contains primitive cells up to 300,000/mm^3 often showing mitosis. The liver, spleen and bone marrow are massively infiltrated with similar immature cells. There is sometimes ascites.

Chamorro et al. (1962) describe the typical proliferating cell as a

'Friend cell', one of medium size, often with 2 or 3 nuclei and a kidney-shaped nucleus; they consider that this and giant cells in bone marrow are the chief producers of virus.

Ecology. Transmission takes place through mother's milk, not the placenta. No cage contact transmission.

Control. Formalinized vaccines made from leukaemia spleens produced definite immunity in mice both against filtrates and cell-suspensions used for challenge. Thymectomy does not affect the course of the infection.

Antisera will suppress Friend disease if given 24 hours before infection and may prolong life when given later (Chirigos et al., 1967). Concentrated interferon preparations have a definite sparing effect when given repeatedly after infection (Gresser et al., 1968) while the interferon-inducer statolon protects when given prophylactically (Wheelock, 1967).

Murine Erythroblastosis

This virus, described by Kirsten et al. (1967), is similar to Friend virus in that there are early lesions involving the erythropoietic system and, later, malignant lymphomas (Kirsten & Mayer, 1969). There is proliferation of red-cell precursors, but, in the blood, severe anaemia. Suckling mice of the DBA line are most sensitive, all dying after 2 or 3 months. Suckling rats are also susceptible.

Rauscher's Strain

This is very similar to Friend's and with it, too, there is evidence for presence of 2 viruses. The splenic disease component can be eliminated by repeated passage through rats (Fieldsteel et al., 1969).

Physico-chemical properties. Lipids extracted from virus preparations contained 75–80 per cent phospholipids with lecithin the main component (Johnson & Mora, 1967). After ether extraction Shibley and his colleagues (1969) obtained a suspension of nucleoids which were still infectious, possibly because of infectivity of RNA. Levy et al. (1967) carried out extensive studies of inactivation by temperature, pH and formalin.

Cultivation. Some workers have found this, like Friend's strain, difficult to cultivate. However, Wright & Lasfargues (1965) and Tyndall et al. (1965) had success using a continuous cell-line of mouse origin. The latter authors describe a CPE. Osato et al. (1966) describe *in vitro* transformation of embryonic mouse cells, while Rhim et al. (1969) state that cells in a line of hamster-embryo cells are transformed. It may even be possible to infect human cells (Wright & Korol, 1970).

Pathogenicity. Effects are similar to those of Friend's virus. As with that virus, discrete splenic lesions can be counted for purposes of assay.

Ecology. Not readily transmitted *in utero* or in the milk.

MOUSE SARCOMA VIRUSES

Two mouse sarcoma viruses have been described; they are referred to as MSV(H) and MSV(M) after their describers Harvey (1965) and Moloney (1966). MSV(M) is described as a rhabdomyosarcoma.

Harvey's virus produced tumours at the site of inoculation, either anaplastic sarcomata or angiomata, in new-born mice, rats and hamsters, also splenomegaly like that caused by Friend's virus. The virus was filterable and ether-sensitive. It arose in mice inoculated with Moloney's virus and may be a mutant of that virus. It produces foci of altered cells in mouse tissue cultures, but consists of a mixture of focus-forming and nonfocus-forming particles The latter are in excess and are probably defective like some fowl tumour viruses. Apparently virions of leukaemic viruses can act as 'helper viruses', permitting the nonfocus formers to form foci and to multiply (Hartley & Rowe, 1966). The defectiveness of MSV, however, concerns inability to form foci in cultures, not, as with Rous virus, in failure to release infectious particles (Bather et al., 1968).

Most work has been with Moloney and Rauscher viruses. In fact, the ability of these leukaemia viruses to act as helpers is the basis of the quickest, most convenient technique for titrating them. An answer can be obtained in 6 days (Fischinger & O'Connor, 1968). There has been discussion as to whether MSV(M) is also defective; its activity is certainly enhanced if a leukaemia virus is present. According to O'Connor & Fischinger (1969), preparations which are fully competent contain virions of MSV and MLV in a state of 'interviral aggregation'. The sarcoma viruses resemble the leukaemia viruses in morphological and other fundamental properties and are serologically closely related to those of the Friend-Moloney-Rauscher group.

Tumours produced by these viruses in rats and hamsters usually release no infectious virus unless a helper is added. According to which leukaemia virus is used as helper, one obtains different 'pseudotypes' as with fowl tumour viruses. However MSV(H) grown in embryonic hamster cells does release infectious virus (Simons et al., 1967), and MSV(M) tumours in new-born rats do also (Perk et al., 1968). Solid tumours have also been obtained in rats or mice inoculated with Friend virus; some have spontaneously yielded infectious virus (Odaka & Ikawa, 1968), others only when a helper was added (Fieldsteel et al., 1969).

The erythroblastosis virus (see p. 155) may cause solid lymphomata in rats (Kirsten & Mayer, 1969) or in hamsters (Klement et al., 1969). From the latter a virus was recovered which infected only hamsters and was not neutralized by antisera to the original virus. Possibly it is a naturally occurring hamster virus which can act as a helper (Kelloff et al., 1970).

A plasma-cell leukaemia of mice apparently caused by a virus has been described (Ebbesen, 1968).

OSTEOSARCOMA OF MICE

A virus causing transmissible osteosarcomata in mice was described by Finkel et al. (1966). It was serologically related to the murine leukaemia viruses. At times it caused only fibrosarcomata. It is therefore relevant that MSV(M) sometimes causes bone tumours (Soehner & Dmochowski, 1969).

AUTO-IMMUNE DISEASE IN NEW ZEALAND BLACK MICE

C particles resembling those of leukaemia viruses are abundantly present in this condition, of which glomerulitis is a feature (Howie & Helyer, 1968; Mellors et al., 1969). Their relation to the pathological changes is obscure. Hirsch et al. (1969) found that antigen-antibody complexes might accumulate in glomeruli in murine leukaemia and sarcoma.

LEUKAEMIA IN RATS

This has been described as occurring spontaneously in rats and to be of virus origin (Svec et al., 1957).

LEUKAEMIA IN GUINEA-PIGS

Leukaemia of guinea-pigs caused by filterable agent has been described by Jungeblut & Kodza (1963) and in fuller detail by Opler (1967a,b; 1968). There was general lymphomatosis, and white cell counts in the blood were of the order of 100,000 to 350,000 mm^3. There were numerous C particles in lymphoid cells differing only in minor respects from those of murine leukaemia. The incubation period was from 12 to 32 days according to whether cells or filtrates were injected. Death usually occurred within a few days of recognition of a high white cell count in the blood. The infection could be transmitted orally (Opler, 1968).

Ma et al. (1969) describe C particles as occurring in normal guinea-pigs.

LEUKAEMIA AND TUMOURS IN HAMSTERS

Graffi et al. (1968) describe a leukaemia in golden hamsters containing C particles. It was activated by injection into new-born hamsters of material from hamster warts containing papovaviruses, also (Graffi et al., 1969) by DNA extracted from such material and by some extracts from human carcinomata. The leukaemias thus obtained were serially transmitted in hamsters with filtrates.

A pigmented melanoma of hamsters was transmitted by means of filtrates; it was sometimes pigmented, sometimes not. C-type particles were seen in cisternae and budding from membranes (Epstein et al., 1968). Relationships between 4 hamster viruses, 3 oncogenic and 1 nononcogenic, are compared by Kelloff et al. (1970b).

LEUKAEMIA IN CATS

A virus causing lymphosarcoma and leukaemia in cats was described by Jarrett et al. (1964a,b). Injected kittens died 9 to 18 months after injection of glycerolated material.

The disease exists in multicentric, alimentary and thymic forms, but there is usually no great excess of circulating leukocytes.

Cultivation was successful in cells of feline, canine and human origin (Jarrett, 1969; Jarrett et al., 1969). The virus infected not only kittens but also new-born puppies (Rickard et al., 1969).

Serological studies revealed a relationship to murine leukaemias as shown by gel-diffusion tests with the common murine-leukaemia antigen (Geering et al., 1968). Sarma et al. (1970) however, found little or no cross-reaction in complement-fixation tests. The feline leukaemia virus could activate defective mouse-sarcoma viruses (Fischinger & O'Connor, 1969). These thereafter grew and produced foci of transformed cells in cat not mouse cells. Sarma et al. (1970) produced sarcoma in kittens with such a rescued virus.

LEUKAEMIA IN CATTLE

Reviews: Bendixen (1965). Dutcher et al. (1966).

Epidemiological evidence suggests that 1 form of bovine lymphosarcoma may be the result of an infectious agent. Cell cultures from such cattle and also concentrates of milk often reveal the presence of C-type particles: the cultures produce an interferon-like substance; no agent, however, has been transmissible in series nor have the cultures produced the disease experimentally (Dutcher et al., 1963).

It seems likely that there exists an infectious disease characterized by lymphocytosis and that definite leukosis may be a late development of this condition; the neoplastic change may be localized or diffuse. There is evidence that the infection can spread from affected to unaffected herds and that this spread may be halted by quarantine and slaughter (but see Bovine Syncytial Virus, p. 163).

LEUKAEMIA IN MAN

There have been a number of reports of finding C-type and other 'virus-like particles' in cells in human leukaemia. Some were certainly the EB virus described on p. 355. There is no evidence that any represent the causative agent.

C-TYPE VIRUS IN REPTILES

Zeigel & Clark (1969) recovered a virus from cultures of the spleen of a snake (*Vipera russelli*) bearing a myxofibroma. In morphology and physical properties it resembled the viruses of murine leukaemias (Gilden et al., 1970).

POSSIBLE LEUKOVIRUSES

The following 5 viruses are placed tentatively with the leukoviruses which they apparently resemble in morphological and other characters; they possess an RNA-dependent DNA-polymerase (Lin &

Thormar, 1970; Stone et al., 1971, Visna; Parks et al., 1971, foamy virus).

Visna-Maedi

Reviews: Sigurdsson et al. (1957). Sigurdsson (1954). Thormar (1965a).

The viruses known as Visna and Maedi are so similar that it is best to regard them as neurotropic and pneumotropic races of 1 virus. Their properties have been compared by Thormar (1965a).

Morphology. Virus particles 85 μm across were seen on the surface of cells in culture. They had a dense core, less than half the diameter of the particle, and a single surrounding membrane. Thormar & Cruickshank (1965) describe surface projections 10 μm long. There was a suggestion of a concentric arrangement within the core. A few helical rods 9 μm across were seen. Ether-treatment produced rosettes such as are seen in preparations of myxoviruses. Maedi particles are similar, described (Thormar 1965a) as 60–90 μm across with a 30–40 μm core. Particles seemed to be released from the cell surface from 2-walled buds; the appearances recalled those of the Bittner (p. 146) and Friend (p. 153) viruses (Thormar, 1961).

Chemical composition. Inhibited by the DNA-inhibitors BUDR and actinomycin-D. This does not imply that they are DNA-viruses; it may only be that, as with influenza and Rous sarcoma, cellular DNA is concerned in their synthesis (Thormar, 1965b).

Physico-chemical characters. Sensitive to ether, chloroform, metaperiodate and trypsin. Ninety per cent inactivated in 10–15 minutes at 50° (Thormar, 1965a). Most stable between pH 7·2 and 9·2. Survives storage for months at −50°; inactivated by 0·04 per cent formaldehyde, 4 per cent phenol and 50 per cent ethanol (Thormar, 1960—experiments on Visna). Visna is sensitive to photo-inactivation in the presence of toluidine blue (Thormar & Petersen, 1964).

Haemagglutination. Contrary to an earlier report, neither specific haemagglutination nor haemadsorption has been observed.

Antigenic properties. Neutralization tests carried out in tissue culture showed that not only Visna sheep, but also sheep suffering from Maedi frequently had antibodies to the virus. An extensive comparison of Visna and Maedi strains showed that cross-neutralization was the rule, one Visna strain being aberrant and several Maedi strains better neutralized by homologous than heterologous sera (Thormar & Helgadottír, 1965). 'Antibodies' were present also in sera of cattle and some other species. Thormar & von Magnus (1963) found neutralizing activity in the majority of human sera of all ages; the main component was heat stable and was probably a nonspecific inhibitor. Thormar (1963) has studied the kinetics of the neutralization of Visna by sheep sera. Complement-fixation tests were found convenient for diagnosis of Visna and Maedi (Gudnadottir & Kristensdottir, 1967).

Cultivation. Visna virus has been grown in cultures of ependyma or chorioid plexus from sheep brains. Multinuclear giant cells are formed; later there is cell destruction; the CPE is seen in 2–3 weeks. On passage this time is shortened to 3–15 days, when large inocula are used. Virus may persist in cultures for as long as 4 months. Third and eleventh tissue-culture passages infected sheep (Thormar & Sigurdadottír, 1962; Thormar, 1963). Maedi grows in a similar manner, though more slowly and has also been grown in other kinds of ovine cells (Sigurdadottír & Thormar, 1964). Visna also grew in cultures of pig kidney and bovine trachea (Harter et al., 1968); it produced CPE, though without evidence of multiplication, in cerebellar cells from new-born mice (Bunge & Harter, 1969). Harter & Choppin (1967) describe a plaque-assay technique.

Distribution. Visna is only reported from Iceland where it, together with Maedi, was probably introduced in sheep imported from Germany in 1933. Either Visna or Maedi or both have since been described from Holland, Kenya, India, and Montana (U.S.A.).

Pathogenicity. *Visna.* Early symptoms are abnormal head posture and lip-trembling. This is a slow demyelinating disease. Other indefinite nervous symptoms lead gradually to paraplegia or total paralysis; the disease may last for weeks or months, but is always fatal. Pleocytosis in the CSF is characteristic. The incubation period extends for months. Pleocytosis may appear in less than a month after inoculation, but clinical symptoms may be deferred till 18 months later. The disease is transmissible by intracerebral inoculation to sheep, but produces lesions like Maedi when given intranasally. The usual laboratory rodents are resistant.

Pathological lesions are those of diffuse encephalomyelitis with demyelination; they are fully described by Sigurdsson et al. (1962). In early reports Visna was confused with Rida (Sigurdsson, 1954), a disease probably identical with scrapie.

Maedi. Sheep 2 or more years old are affected with wasting and dyspnoea; the disease is highly, possibly uniformly, fatal. It lasts for 3 to 6 months, sometimes longer. The incubation period is about 2 years, but lung lesions may be detectable in animals killed 1 month after injection of infected material into nose, lungs and IV. Infection can apparently occur also after feeding and by contact. Laboratory rodents have not been infected.

Pathological lesions. Lungs are much enlarged, weighing as much as twice the normal. There is diffuse perivascular and peribronchiolar infiltration with mononuclear cells, but little fibrosis.

Maedi has been confused with Jaagsiekte, another chronic lung disease of sheep, apparently introduced into Iceland from Germany at the same time.

Control. A slaughter policy in 1951 was apparently successful in eliminating the diseases from Iceland. Some Maedi has, however, since appeared, though by 1965 Visna had not done so (H. Thormar, personal communication).

Foamy Agent (or Virus)

Review: Plummer (1962).

Several groups of workers (Enders & Peebles, 1954; Rustigian et al., 1955; Ruckle, 1958) have described changes produced by this agent, which is the commonest contaminant of monkey-kidney cultures. Jordan et al. (1965) describe pleomorphic forms resembling myxoviruses with surface projections 10 nm long arranged in hexagonal patterns. An inner helix was 10–12 nm across.

Morphology. Several papers by Clarke et al. (1967, 1968, 1969) describe virions similar to those of the murine mammary-tumour virus. Diameter 110–120 nm: nucleoid surrounded by 2 shells 50 + 90 nm across: projections on surface 15 × 5 nm. Maturation is by budding at membrane surfaces.

Physico-chemical characters. Inactivated in 15 minutes at 50°, but survived well at −20°. Ether-sensitive.

Haemagglutination. None detected.

Antigenic properties. 7 serotypes described; rhesus monkeys mainly infected with type 1, grivets with types 2 and 3 (Swack et al., 1970).

Cultivation. It produces syncytia and lace-like degeneration of monkey-kidney cells. Rather less characteristic changes are produced in HeLa and other primate cells (Paccaud, 1957). Brown (1957) was able to propagate it in rabbit kidney, and this is the most useful medium.

Distribution. Isolated from 40–65 per cent of kidneys of rhesus and cynomolgus monkeys and *Cercopithecus pygerythrus*; also isolated from a colony of baboons (*Cynocephalus*). None was isolated from 296 cultures of *Erythrocebus patas* (Plummer, 1962) or from the Malayan *Presbytis cristatus* (Thayer, 1965).

Pathogenicity. Not known to be pathogenic.

Feline Fibrosarcoma

A virus isolated from fibrosarcomata in cats is described as having virions like leukovirus C particles (Snyder & Theilen, 1969) or as resembling other syncytial viruses. Moreover, it is said to have a diameter of 109–120 nm (Snyder & Theilen, 1969; McKissick & Lamont, 1970), to be ether-sensitive and to multiply in tissue cultures of various species, producing syncytia. Cultures gave rise to sarcomata when injected into new-born kittens (Chang et al., 1970).

Bovine Syncytial Virus

A virus resembling others in this group has been isolated from cattle with lymphosarcomatosis and also from normal cattle (Malmquist et al., 1969). It produces syncytia in cultures; its development in hamster (BHK21) cells is described by Dermott et al. (1971). It is distinct from the bovine respiratory syncytial virus referred to on p. 248.

Virus from Human Nasopharyngeal Carcinoma

A similar virus of human origin is described by Achong et al. (1971). Particles were covered by radiating spikes 13 nm long.

REFERENCES

Achong, B. G., Mansell, P. W. A., Epstein, M. A., & Clifford, P. (1971) *J. natn. Cancer Inst.*, **46**, 299.
Ahlström, C. G., & Forsby, N. (1962) *J. exp. Med.*, **115**, 839.
Ahlström, C. G., & Jonsson, N. (1962) *Acta path. microbiol. scand.*, **54**, 145.

Ahlström, C. G., Jonsson, N., & Forsby, N. (1962) *Acta path. microbiol. scand., suppl.* **1**, **54**, 127.
Allen, D. W. (1968) *Biochem. biophys. Acta*, **154**, 388.
Allison, A. C., & Burke, D. (1962) *J. gen. Microbiol.*, **27**, 181.
Almeida, J. D., Watkinson, A. B., & Drewe, J. A. (1967) *J. Hyg., Camb.*, **65**, 467.
Andervont, H. B., & Dunn, T. B. (1956) *Acta int. Unio contra Cancrum*, **12**, 530.
Andrewes, C. H. (1933) *J. Path. Bact.*, **37**, 27.
Andrewes, C. H. (1939) *J. Path. Bact.*, **68**, 225.
Aoki, T., Boyse, E. A., & Old, L. J. (1968) *J. natn. Cancer Inst.*, **41**, 89.
Armstrong, D. (1969) *J. Virol.*, **3**, 133.
Atanasiu, P., Vieuchange, J., & Strunge, B. (1957) *Acta int. Unio contra Cancrum*, **7**, 213.
Aulisio, C. G., & Shelokov, A. (1967) *Proc. Soc. exp. Biol. Med., U.S.A.*, **126**, 312.
Aulisio, C. G., & Shelokov, A. (1969) *Proc. Soc. exp. Biol. Med., U.S.A.*, **130**, 178.
Axelrad, A. A., & Steeves, R. A. (1964) *Virology*, **24**, 513.
Bader, J. P. (1964) *Virology*, **22**, 462.
Bader, J. P., & Bader, A. V. (1970) *Proc. natn. Acad. Sci., U.S.A.*, **67**, 843.
Baltimore, D. (1970) *Nature, Lond.*, **226**, 1209.
Baluda, M. A., & Goetz, I. E. (1961) *Virology*, **15**, 185.
Baluda, M. A., & Jamieson, P. P. (1961) *Virology*, **14**, 33.
Bather, R. (1957) *Br. J. Cancer*, **11**, 611.
Bather, R. (1958) *Br. J. Cancer*, **12**, 256.
Bather, R., Leonard, A., & Yang, J. (1968) *J. natn. Cancer Inst.*, **40**, 551.
Beard, J. W., Sharp, D. G., & Eckert, E. A. (1955) *Adv. Virus Res.*, **3**, 149.
Beaudreau, G. S., Becker, C., Bonar, R. A., Wallbank, A. M., Beard, D., & Beard, J. W. (1960) *J. natn. Cancer Inst.*, **24**, 395.
Benedetti, E. L. (1957) *Bull. Cancer*, **44**, 473.
Bentvelzen, P., Daams, J. H., Hageman, P., & Calafat, J. (1970) *Proc. natn. Acad. Sci., U.S.A.*, **67**, 377.
Bergman, S., & Jonsson, N. (1962) *Acta path. microbiol. scand., suppl.*, **154**, 130.
Bergs, V. V., Bergs, M., & Chopre, H. C. (1970) *J. natn. Cancer Inst.*, **44**, 913.
Berman, L. D., & Sarma, P. S. (1965) *Nature, Lond.*, **207**, 263.
Bernhard, W., Bauer, A., Guérin, M., & Oberling, C. (1955) *Bull. Cancer*, **42**, 163.
Bernhard, W., & Guérin, M. (1958) *C.r. hebd. Séanc. Acad. Sci., Paris*, **247**, 1802.
Bernhard, W., Oberling, C., & Vigier, P. (1956) *Bull. Cancer*, **43**, 407.
Biggs, P. M., & Payne, L. N. (1967) *Vet. Rec.*, **80**, suppl. 7.
Bittner, J. J. (1945) *Proc. Soc. exp. Biol. Med., U.S.A.*, **59**, 43.
Bittner, J. J. (1948) *Cancer Res.*, **8**, 625.
Bittner, J. J. (1952) *Cancer Res.*, **12**, 387.
Blair, P. B. (1960) *Proc. Soc. exp. Biol. Med., U.S.A.*, **103**, 188.
Blair, P. B. (1965) *Nature, Lond.*, **208**, 165.
Blair, P. B. (1968) *Current topics in microbiology*, **45**, 1. Berlin: Springer-Verlag.
Blair, P. B. (1969) *Cancer Res.*, **29**, 745.
Blair, P. B., & Weiss, D. W. (1966) *J. natn. Cancer Inst.*, **36**, 423.
Boiron, M., Levy, J. P., & Periés, J. (1967) *Progr. med. Virol.*, **9**, 341.
Bonar, R. A., Parsons, D. F., Beaudreau, G. S., Beeker, C., & Beard, J. W. (1959) *J. natn. Cancer Inst.*, **23**, 199.
Bose, H. R., & Levine, A. S. (1967) *J. Virol.*, **1**, 1117.
Boyse, E. A., Old, L. J., & Luell, S. (1963) *J. natn. Cancer Inst.*, **31**, 987.
Brodsky, I. (1969) *Nature, Lond.*, **223**, 198.
Brown, E. R., & Bittner, J. J. (1961) *Proc. Soc. exp. Biol. Med., U.S.A.*, **106**, 303.
Brown, L. V. (1957) *Am. J. Hyg.*, **65**, 189.
Bryan, W. R. (1955) *J. natn. Cancer Inst.*, **16**, 285.
Bubenik, J., & Bauer, H. (1967) *Virology*, **31**, 489.
Bunge, R. P., & Harter, D. H. (1969) *J. Neuropath. exp. Neurol.*, **28**, 185.
Calafat, J., & Hageman, P. (1968) *Virology*, **36**, 308.
Calafat, J., & Hageman, P. (1969) *Virology*, **38**, 364.

Campbell, J. G., Young, D. E., & Carr, J. G., (1964) *J. comp. Path.*, **74**, 263.
Cardiff, R. D., Blair, P. B., & DeOme, K. B. (1968) *Virology*, **36**, 313.
Carr, J. G. (1959) *Virology*, **8**, 269.
Carr, J. G., & Campbell, J. G. (1958) *Br. J. Cancer*, **12**, 631.
Chamorro, A., Latarjet, R., Vigier, P., & Zajdela, F. (1962) in *Tumour viruses of murine origin* (Ciba symposium), p. 176. London: Churchill.
Chang, R. S., Gilden, D. H., & Boyd, H. (1970) *J. Virol.*, **6**, 599.
Chapman, A. L., Bopp, W., Brightwell, A. S., Cohen, H., & Nielsen, A. H. (1967) *Proc. Soc. exp. Biol. Med., U.S.A.*, **123**, 742.
Charney, J., & Moore, D. H. (1971) *Nature, Lond.*, **229**, 611.
Chirigos, M. A., March, R. W., Hook, W., & Hoemann, R. (1967) *Proc. Soc. exp. Biol. Med., U.S.A.*, **125**, 35.
Chopra, H. C., & Feller, W. F. (1969) *Texas Rep. Biol. Med.*, **27**, 945.
Chopra, H. C., & Mason, M. M. (1970) *Cancer Res.*, **30**, 2081.
Clarke, J. K., Attridge, J. J., Dane, D. S., & Briggs, M. (1967) *J. gen. Virol.*, **1**, 565.
Clarke, J. K., & Attridge, J. J. (1968) *J. gen. Virol.*, **3**, 185.
Clarke, J. K., Attridge, J. J., & Gay, F. W. (1969) *J. gen. Virol.*, **4**, 183.
Coates, H., Borsos, T., Foard, M., & Bang, F. B. (1968) *Int. J. Cancer*, **3**, 424.
Cook, M. K., Grochal, A. G., & Huebner, R. J. (1966) *J. natn. Cancer Inst.*, **37**, 619.
Cooper, M. D., Payne, L. N., Dent, P. B., Burmester, B. R., & Good, R. A. (1968) *J. natn. Cancer Inst.*, **41**, 373.
Crittenden, L. B., Stone, H. A., Reamer, R. H., & Ukazaki, W. (1967) *J. Virol.*, **1**, 898.
Dawson, P. J., Tacke, R. B., & Fieldsteel, A. H. (1968) *Br. J. Cancer*, **22**, 569.
de Harven, E., & Friend, C. (1964) *Virology*, **23**, 119.
Deinhardt, F. (1966) *Nature, Lond.*, **210**, 443.
Dent, P. B., Cooper, M. D., Payne, L. N., Good, R. A., & Burmester, B. R. (1967) *Perspectives in virology*, **5**, 251. New York/London: Academic Press.
Dermott, E., Clarke, J. K., & Samuels, J. (1971) *J. gen. Virol.*, **12**, 105.
de Thé, G. (1966) *Int. J. Cancer*, **1**, 119.
de Thé, G., & O'Connor, T. E. (1966) *Virology*, **28**, 713.
Dougherty, R. M. (1961) *Virology*, **14**, 371.
Dougherty, R. M., & Di Stefano, H. W. (1966) *Virology*, **29**, 586.
Duesberg, P. H., & Cardiff, R. D. (1968) *Virology*, **36**, 696.
Duesberg, P. H., & Robinson, W. S. (1967) *Virology*, **31**, 742.
Dunn, T. B., Moloney, J. B., Green, A. W., & Arnold, B. (1961) *J. natn. Cancer Inst.*, **26**, 189.
Duran-Reynals, F. (1940) *Yale J. Biol. Med.*, **13**, 77.
Duran-Reynals, F. (1942) *Cancer Res.*, **2**, 343.
Dutcher, R. M., Szekely, I. E., Larkin, E. P., Coriell, L. L., & Marshak, R. R. (1963) *Ann. N.Y. Acad. Sci.*, **108**, 1149.
Dutcher, R. M., Larkin, E. P., Tumilowicz, J. J., Marshak, R. R., & Szekely, I. E. (1966) in *Comparative leukaemia research*, p. 37. Oxford: Pergamon Press.
Ebbesen, P. (1968) *Acta path. microbiol. scand., suppl.*, 197.
Eckert, E. A., Green, I., Sharp, D. G., Beard, D., & Beard, J. W. (1955) *J. natn. Cancer Inst.*, **16**, 153.
Enders, J. F., & Peebles, T. C. (1954) *Proc. Soc. exp. Biol. Med., U.S.A.*, **86**, 277.
Engelbreth-Holm, J., & Meyer, A. R. (1932) *Acta path. microbiol. scand.*, **9**, 293.
Ephrussi, B. G., & Temin, H. M. (1960) *Virology*, **11**, 547.
Epstein, M. A. (1956) *Br. J. Cancer*, **10**, 33.
Epstein, W. L., Fukuyama, K., & Benn, M., (1968) *Nature, Lond.*, **219**, 979.
Faras, A. J., & Erikson, R. L. (1969) *J. Virol.*, **4**, 31.
Fieldsteel, A. H., Kurahara, C., & Dawson, P. J. (1969) *Nature, Lond.*, **223**, 1274.
Fink, M. A., & Malmgren, R. A. (1963) *J. natn. Cancer Inst.*, **31**, 1111.
Finkel, M. P., Biskis, B. O., & Jinkins, P. B. (1966) *Science*, **151**, 698.
Fischinger, P. J., & O'Connor, T. E. (1968) *J. natn. Cancer Inst.*, **40**, 1199.

Fischinger, P. J., & O'Connor, T. E. (1969) *Science*, **165**, 714.
Friend, C. (1959) in *Perspectives in virology*, **1**, 231, New York: John Wiley.
Fujinami, A. (1928) *Trans. Jap. path. Soc.*, **18**, 616.
Furth, J. (1931) *Arch. Path., Chicago*, **12**, 1.
Furth, J. (1959) in *Perspectives in virology*, **1**, 262. New York: John Wiley.
Gaylord, W. H. (1955) *Cancer Res.*, **15**, 80.
Geering, G., Aoki, T., & Old, L. J. (1970) *Nature, Lond.*, **226**, 265.
Gilden, R. V., Lee, Y. K., Oroszlan, S., Walker, J. L., & Huebner, R. J. (1970) *Virology*, **41**, 187.
Ginsburg, H., & Sachs, L. (1961) *Virology*, **13**, 380.
Glynn, J. P., McCoy, J. L., & Fefer, A. (1968) *Cancer Res.*, **28**, 434.
Graffi, A. (1957) *Ann. N.Y. Acad. Sci.*, **68**, 540.
Graffi, A., Baumbach, L., Schramm, T., & Bierwolf, D. (1963) *Z. Krebsforsch.*, **65**, 385.
Graffi, A., Bender, E., Schramm, T., Kuhn, W., & Schneider, F. (1969) *Proc. natn. Acad. Sci., U.S.A.*, **64**, 1172.
Graffi, A., Schramm, T., Bender, E., Graffi, I., Horn, K. H., & Bierwolf, D. (1968) *Br. J. Cancer*, **22**, 577.
Gramboulan, N., Huppert, J., & Lacour, F. (1966) *J. molec. Biol.*, **16**, 571.
Gregoriades, A., & Old, L. J. (1969) *Virology*, **37**, 189.
Gresser, I., Berman, L., de Thé, G., Brouty-Boye, D., Coppey, J., & Falcoff, E. (1968) *J. natn. Cancer Inst.*, **41**, 505.
Gross, L. (1955) *Acta haemat.*, **13**, 13.
Gross, L. (1957) *Proc. Soc. exp. Biol. Med., U.S.A.*, **94**, 767.
Gross, L. (1959) *Proc. Soc. exp. Biol. Med., U.S.A.*, **101**, 113.
Gross, L. (1961) *Proc. Soc. exp. Biol. Med., U.S.A.*, **106**, 890.
Gross, L. (1962) in *Tumour viruses of murine origin* (Ciba symposium), p. 159. London: Churchill.
Gross, L. (1963) *Proc. Soc. exp. Biol. Med., U.S.A.*, **112**, 939.
Gross, L., Dreyfuss, Y., & Moore, L. A. (1961) *Proc. Am. Soc. Cancer Res.*, **3**, 231.
Gudnadóttir, M., & Kristensdóttir, K. (1967) *J. Immunol.*, **98**, 663.
Haguenau, F., Dalton, A. J., & Moloney, J. B. (1958) *J. natn. Cancer Inst.*, **20**, 633.
Hairstone, M. A., Lyons, M. J., & Moore, D. H. (1964) *Virology*, **23**, 294.
Hall, W. T., Andresen, W. F., Sanford, K. K., Evans, V. J., & Hartley, J. W. (1967) *Science*, **156**, 85.
Hanafusa, H. (1969) *Proc. natn. Acad. Sci., U.S.A.*, **63**, 318.
Hanafusa, H., & Hanafusa, T. (1968) *Virology*, **34**, 630.
Hanafusa, H., Hanafusa, T., & Rubin, H. (1963) *Proc. natn. Acad. Sci., U.S.A.*, **49**, 572.
Hanafusa, H., Hanafusa, T., & Rubin, H. (1964) *Proc. natn. Acad. Sci., U.S.A.*, **51**, 40.
Harel, J., Huppert, J., Lacour, F., & Harel, L. (1965) *C.r. hebd. Séanc. Acad. Sci., Paris*, **261**, 2266.
Harris, R. J. C. (1954) *Br. J. Cancer*, **8**, 731.
Harter, D. H., & Choppin, P. W. (1967) *Virology*, **31**, 176.
Harter, P. H., Hsu, K. C., & Rose, H. M. (1968) *Proc. Soc. exp. Biol. Med., U.S.A.*, **129**, 295.
Hartley, J. W., & Rowe, W. P. (1966) *Proc. Nat. Acad. Sci., U.S.A.*, **55**, 780.
Hartley, J. W., Rowe, W. P., Capps, W. I., Huebner, R. J. (1965) *Proc. Nat. Acad. Sci., U.S.A.*, **53**, 931.
Hartley, J. W., Rowe, W. P., Capps, W. I., & Huebner, R. J. (1969) *J. Virol.*, **3**, 126.
Harvey, J. J. (1965) *Nature, Lond.*, **204**, 1104.
Heine, U., Beaudreau, G. S., Becker, C., Beard, D., & Beard, J. W. (1961) *J. natn. Cancer Inst.*, **26**, 359.
Hirsch, M. S., Allison, A. C., & Harvey, J. J. (1969) *Nature, Lond.*, **223**, 739.
Holmes, J. R. (1963) *J. comp. Path.*, **73**, 136.
Howie, J. B., & Helyer, B. J. (1968) *Adv. in Immunol.*, **9**, 215.
Huebner, R. J., Armstrong, D., Okuyan, M., Sarma, P. S., & Turner, H. C. (1964) *Proc. natn. Acad. Sci., U.S.A.*, **50**, 379.

Huebner, R. J., Kelloff, G. J., Sarma, P. S., Lane, W. T., Turner, H. C., Gilden, R. V., Oroszlan, S., Meier, H., Myers, D., & Peters, R. (1971) *Proc. natn. Acad. Sci., U.S.A.*, **67**, 366.
Huebner, R. J., & Todaro, G. J. (1970) *Proc. natn. Acad. Sci., U.S.A.*, **64**, 1087.
Imagawa, D. T., Green, R. G., & Halvorson, H. O. (1948) *Proc. Soc. exp. Biol. Med., U.S.A.*, **68**, 162.
Ioachim, H. L. (1967) *Science*, **155**, 585.
Ishizaki, R., & Vogt, P. K. (1966) *Virology*, **30**, 375.
Jarrett, O., Laird, H. M., & Hay, D. (1969) *Nature, Lond.*, **224**, 1208.
Jarrett, W. F. H. (1969) *Vet. Rec.*, **85**, 553.
Jarrett, W. F. H., Crawford, E. M., Martin, W. B., & Davie, F. (1964a) *Nature, Lond.*, **202**, 567.
Jarrett, W. F. H., Martin, W. B., Crighton, G. W., Dalton, R. G., & Stewart, M. F. (1964b) *Nature, Lond.*, **202**, 566.
Jensen, F. C., Girardi, A. J., Gilden, R. V., & Koprowski, H. (1964) *Proc. natn. Acad. Sci., U.S.A.*, **52**, 53.
Johnson, M., & Mora, P. T. (1967) *Virology*, **31**, 230.
Jonsson, N., & Sjögren, H. O. (1965) *J. exp. Med.*, **122**, 403 (1966) **123**, 487.
Jungeblut, C. W., & Kodza, H. (1963) *Arch. ges. Virusforsch.*, **12**, 537.
Kajima, M., & Pollard, M. (1968) *Nature, Lond.*, **218**, 188.
Kakefuda, T., & Bader, J. P. (1970) *J. Virol.*, **4**, 460.
Kaplan, H. S. (1967) *Cancer Res.*, **27**, 1325.
Kelloff, G., Huebner, R. J., Lee, Y. K., Toni, R., & Gilden, R. (1970a) *Proc. natn. Acad. Sci., U.S.A.*, **65**, 310.
Kelloff, G., Huebner, R. J., et al. (1970b) *J. gen. Virol.*, **9**, 19 and 27.
Kelloff, G., & Vogt, P. K. (1966) *Virology*, **29**, 377.
Keogh, E. V. (1938) *Br. J. exp. Path.*, **19**, 1.
King, J. W., & Duran-Reynals, F. (1943) *Yale J. Biol. Med.*, **16**, 53.
Kirsten, W. H., Mayer, L. A., Wohlmann, R. L., & Pierce, M. I. (1967) *J. natn. Cancer Inst.*, **38**, 117.
Kirsten, W. H., & Mayer, L. A. (1967) *J. natn. Cancer Inst.*, **39**, 311.
Klein, G., & Klein, E. (1964) in *Specific tumour antigens*, p. 82. Copenhagen: Munkgaard.
Klement, V., Hartley, J. W., Rowe, W. P., & Huebner, R. J. (1969) *J. natn. Cancer Inst.*, **43**, 925.
Lacour, F., Fourcade, A., Verger, C., & Delain, E. (1970) *J. gen. Virol.*, **9**, 89.
Lagerhöf, B. (1960) *Acta path. microbiol. scand.*, **49**, 361.
Langlois, A. J., & Beard, J. W. (1967) *Proc. Soc. exp. Biol. Med., U.S.A.*, **126**, 718.
Law, L. W. (1962) *Proc. Soc. exp. Biol. Med., U.S.A.*, **111**, 615.
Law, L. W. (1966) *Natn. Cancer Inst. Monogr.*, **22**, 267.
Lemonde, P., & Clode, M. (1962) *Nature, Lond.*, **193**, 1191.
Levinson, W., Bishop, J. M., Quintrell, N., & Jackson, J. (1970) *Nature, Lond.*, **227**, 1023.
Levinthal, J. T., Buffett, R. F., & Furth, J. (1959) *Proc. Soc. exp. Biol. Med., U.S.A.*, **100**, 610.
Levy, J. P., Oppenheim, S., Chenaille, P., Silvestre, D., Tavitian, A., & Boiron, M. (1967) *J. natn. Cancer Inst.*, **38**, 553.
Levy, J. P., Varet, B., Oppenheim, E., & Leclerc, J. C. (1969) *Nature, Lond.*, **224**, 606.
Lin, F. H. & Thormar, H. (1970) *J. Virol.*, **6**, 702.
Lo, W. H. Y., Gey, G. O., & Shapras, P. (1955) *Bull. Johns Hopkins Hosp.*, **97**, 248.
Lyons, M. J., & Moore, D. H. (1962) *Nature, Lond.*, **194**, 1141.
Lyons, M. J., & Moore, D. H. (1965) *J. natn. Cancer Inst.*, **35**, 549.
Ma, B. I., Schwartzendruber, D. C., & Murphy, W. H. (1969) *Proc. Soc. exp. Biol. Med., U.S.A.*, **130**, 586.
Malmquist, W. A., van der Marten, M. J., & Boothe, A. D. (1969) *Cancer Res.*, **29**, 188.
McKissick, G. E. & Lamont, P. H. (1970) *J. Virol.*, **5**, 247.
Manaker, R. A., & Groupé, V. (1956) *Virology*, **2**, 839.
Mellors, R. C., Aoki, T., & Huebner, R. J. (1969) *J. exp. Med.*, **129**, 1045.

Metcalf, D., Furth, J., & Buffett, R. F. (1959) *Cancer Res.*, **19**, 52.
Miller, J. F. A. P. (1962) in *Tumour viruses of murine origin* (Ciba symposium), p. 262. London: Churchill.
Mirand, E. A., Buffett, R. F., & Grace, T. J. (1966) *Proc. Soc. exp. Biol. Med., U.S.A.*, **112**, 970.
Mirand, E. A. (1968) *Ann. N.Y. Acad. Sci.*, **149**, 486.
Moloney, J. B. (1960) *J. natn. Cancer Inst.*, **24**, 933.
Moloney, J. B. (1962) *Fed. Proc.*, **21**, 19.
Moloney, J. B. (1966) *Natn. Cancer Inst. Monogr.*, **22**, 129.
Mommaerts, E. B., Sharp, D. G., Eckert, E. A., Beard, D., & Beard, J. W. (1954) *J. natn. Cancer Inst.*, **14**, 1011.
Montagnier, L., Goldé, A., & Vigier, P. (1969) *J. gen. Virol.*, **4**, 449.
Moore, D. H., Lasfargues, E. Y., Murray, M., Haagensen, C. D., & Pollard, E. C. (1959) *J. biophys. biochem. Cytol.*, **5**, 85.
Moore, D. H., Sarkar, W. H., Kelly, C. E., Pillsbury, N., & Charney, J. (1969) *Texas Rep. Biol. Med.*, **27**, 1027.
Morgan, H. R. (1965) *J. natn. Cancer Inst.*, **35**, 1043.
Moscovici, C. (1967) *Proc. Soc. exp. Biol. Med., U.S.A.*, **125**, 1213.
Munroe, J. S., & Windle, W. F. (1963) *Science*, **140**, 1415.
Nandi, S., & de Ome, K. B. (1965) *J. natn. Cancer Inst.*, **35**, 299.
O'Connor, T. E., & Fischinger, P. J. (1969) *J. natn. Cancer Inst.*, **43**, 487.
O'Connor, T. E., Rauscher, F. J., & Zeigel, R. F. (1964) *Science*, **144**, 1144.
Odaka, T., & Ikawa, Y. (1968) *Inst. J. Cancer*, **3**, 211.
Old, L. J., & Boyse, E. A. (1965) *Fedn Proc. Fedn Am. Socs exp. Biol.*, **24**, 1009.
Old, L. J., Boyse, E. A., & Stockert, E. (1964) *Nature, Lond.*, **201**, 777.
Olson, L. D. (1967) *Am. J. vet. Res.*, **28**, 1501.
Opler, S. R. (1967a) *J. natn. Cancer Inst.*, **38**, 797.
Opler, S. R. (1967b) *Am. J. Path.*, **51**, 1135.
Opler, S. R. (1968) *Oncology*, **22**, 273.
Osato, T., Mirand, E. A., Grace, J. T., & Price, F. (1966) *Nature, Lond.*, **209**, 779.
Paccaud, M. (1957) *Annls Inst. Pasteur, Paris*, **92**, 481.
Parks, W. P., Todaro, G. J., Solnick, E. M., & Aaronson, S. (1971) *Nature, Lond.*, **229**, 258.
Payne, L. N., & Chubb, R. C. (1968) *J. gen. Virol.*, **3**, 379.
Perk, K., Shachat, D. A., & Moloney, J. B. (1968) *Cancer Res.*, **28**, 1197.
Pitelka, D., Bern, H. A., Nandi, S., & de Ome, K. B. (1964) *J. natn. Cancer Inst.*, **33**, 867.
Plummer, G. (1962) *J. gen. Microbiol.*, **29**, 703.
Pollard, M., & Hall, W. J. (1941) *J. Am. vet. med. Ass.*, **99**, 218.
Pope, J. H. (1963) *Aust. J. exp. Biol. med. Sci.*, **41**, 349.
Prigozhina, E. L., & Stavrovskaya, A. A. (1964) *Acta virol., Prague*, **8**, 277.
Prince, A. M. (1960) *Virology*, **11**, 371.
Rabin, H., & Sladen, W. J. L. (1969) *Am. J. Epidemiol.*, **89**, 325.
Rabotti, G. F., Grove, A. S., Sellers, R. L., & Anderson, W. R. (1966b) *Nature, Lond.*, **209**, 884.
Rabotti, G. F., Sellers, R. L., & Anderson, W. A. (1966a) *Nature, Lond.*, **209**, 524.
Rangan, S. R. S., & Bang, F. B. (1967) *Proc. Soc. exp. Biol. Med., U.S.A.*, **125**, 593.
Ratuld, Y. de, & Werner, G. H. (1967) *Am. Inst. Path.*, **113**, 749.
Ratuld, Y. de, & Werner, G. H. (1968) *Am. Inst. Path.*, **115**, 122.
Rhim, J. S., Huebner, R. J., & Ting, R. C. (1969) *J. natn. Cancer Inst.*, **42**, 1053.
Rickard, C. G., Post, J. E., Noronha, F., & Barr, L. M. (1969) *J. natn. Cancer Inst.*, **42**, 987.
Roth, F. K., & Dougherty, R. M. (1969) *Virology*, **38**, 278.
Rous, P. (1911) *J. exp. Med.*, **13**, 397.
Rowson, K. E. K., & Parr, I. B. (1970) *Int. J. Cancer*, **5**, 96.
Rubin, H. (1955) *Virology*, **1**, 445.
Rubin, H. (1960) *Virology*, **10**, 29.

Rubin, H. (1961) *Virology*, **13**, 200.
Rubin, H. (1962) *Bact. Rev.*, **26**, 1.
Rubin, H., Cornelius, A., & Fanshier, L. (1961) *Proc. natn. Acad. Sci., U.S.A.*, **47**, 1058.
Rubin, H., Fanshier, L., Coinchius, A., & Hughes, W. F. (1962) *Virology*, **17**, 143.
Rubin, H., & Temin, H. M. (1959) *Virology*, **7**, 75.
Rubin, H., & Vogt, P. K. (1962) *Virology*, **17**, 184.
Ruckle, G. (1958) *Arch. ges. Virusforsch.*, **8**, 139 and 167.
Rustigian, R., Johnston, P., & Reihart, H. (1955) *Proc. Soc. exp. Biol. Med., U.S.A.*, **88**, 8.
Salaman, M. H., & Harvey, J. J. (1961) *Nature, Lond.*, **191**, 509.
Sarkar, N. H., & Moore, D. H. (1970) *J. Virol.*, **5**, 230.
Sarma, P. S., Log, T. S., Huebner, R. J., & Turner, H. C. (1969) *Virology*, **37**, 480.
Sarma, P. S., Log, T., & Huebner, R. J. (1970) *Proc. natn. Acad. Sci., U.S.A.*, **65**, 81.
Sarma, P. S., Turner, H. C., & Huebner, R. J. (1964) *Virology*, **23**, 313.
Schäfer, W., Anderer, F. A., Bauer, H., & Pister, L. (1969) *Virology*, **38**, 387.
Schäfer, W., & Szanto, J. (1969) *Z. Naturforsch.*, **24b**, 1324.
Schlom, J., Harter, D. H., Burny, A., Spiegelman, S. (1971) *Proc. natn. Acad. Sci., U.S.A.* **68**, 182.
Sevoian, M., Larose, R. N., & Chamberlain, D. M. (1964) *Avian Dis.*, **8**, 331.
Shibley, G. P., Carleton, F. J., Wright, B. S., Schidlovsky, G., Monroe, J. H., & Mayyasi, S. A. (1969) *Cancer Res.*, **29**, 905.
Sigurdadóttir, B., & Thormar, H. (1964) *J. inf. Dis.*, **114**, 55.
Sigurdsson, B. (1954) *Br. vet. J.*, **110**, 255.
Sigurdsson, B., Pálsson, P. A., & Grimsson, H. (1957) *J. Neuropath. exp.Neurol.*, **16**, 389.
Sigurdsson, B., Pálsson, P. A., & van Bogaert, L. (1962) *Acta neuropathologica*, **1**, 343.
Simons, P. J., Dourmaskhin, R. R., Turano, A., Phillips, D. E. H., & Chesterman, F. C. (1967) *Nature, Lond.*, **214**, 897.
Sinkovics, J. G. (1962) *Ann. Rev. Microbiol.*, **16**, 75.
Sinkovics, J. G., & Howe, C. D. (1964) *J. inf. Dis.*, **114**, 359.
Snyder, S. P., & Theilen, G. H. (1969) *Nature, Lond.*, **221**, 1074.
Soehner, R. L., & Dmochowski, L. (1969) *Nature, Lond.*, **224**, 191.
Spiegelman, S., Burny, A., Das, M. R., Keydar, J., Schlom, J., Právniček, M., & Watson, K. (1970) *Nature, Lond.*, **228**, 430.
Squartini, F., Olivi, M., Bolis, G. B., Ribacchi, R., & Giraldo, G. (1967) *Nature, Lond.*, **214**, 730.
Stone, L. B., Scolnick, E., Takemoto, K. K., & Aaronson, S. A. (1971) *Nature, Lond.*, **229**, 257.
Stubbs, E. L., & Furth, J. (1935) *J. exp. Med.*, **61**, 593.
Stück, B., Boyse, E. A., Old, L. J., & Carswell, E. A. (1964) *Nature, Lond.*, **203**, 1033.
Svec, F., Altaner, C., & Hlavay, E. (1966) *J. natn. Cancer Inst.*, **36**, 389.
Svec, F., Hlavay, E., Thurzo, V., & Kossey, P. (1957) *Acta haemat.*, **17**, 34.
Svet-Moldavsky, G. J., & Skorikova, A. S. (1960) *Acta virol., Prague*, **4**, 47.
Svet-Moldavsky, G. J., Trubcheninova, L., & Ravakina, L. I. (1967) *Nature, Lond.*, **214**, 300.
Svoboda, J., & Šimkovič, D. (1963) *Acta univ. internat. contre Cancer*, **19**, 302.
Svoboda, J., & Klement, V. (1963) *Folia Biol., Prague*, **9**, 403.
Swack, N. S., Schoentag, R. A., & Hsiung, G. D. (1970) *Am. J. Epidemiol.*, **92**, 79.
Sykes, J. A., Whitescarver, J., & Briggs, L. (1968) *J. natn. Cancer Inst.*, **41**, 1315.
Tanaka, H., & Moore, D. H. (1967) *Virology*, **33**, 197.
Temin, H. M. (1960) *Virology*, **10**, 182.
Temin, H. M., & Mizutani, S. (1970) *Nature, Lond.*, **226**, 1211.
Temin, H. M., & Rubin, H. (1958) *Virology*, **6**, 669.
Temin, H. M., & Rubin, H. (1959) *Virology*, **8**, 209.
Theilen, G. H., Zeigal, R. F., & Twiehaus, M. J. (1966) *J. natn. Cancer Inst.*, **37**, 731.
Thompson, K. D., Fischer, R. G., & Luecke, D. H. (1968) *Avian Dis.*, **12**, 354.
Thormar, H. (1960) *Arch. ges. Virusforsch.*, **10**, 501.

Thormar, H. (1961) *Virology*, **14**, 463.
Thormar, H. (1963) *Virology*, **19**, 273.
Thormar, H. (1965a) *Res. vet. Sci.*, **6**, 117.
Thormar, H. (1965b) *Virology*, **26**, 36.
Thormar, H., & Cruickshank, J. G. (1965) *Virology*, **25**, 145.
Thormar, H., & Helgadottír, H. (1965) *Res. vet. Sci.*, **6**, 456.
Thormar, H., & Petersen, I. (1964) *Acta path. microbiol. scand.*, **62**, 461.
Thormar, H., & Sigurdadottír, B. (1962) *Acta path. microbiol. scand.*, **55**, 186.
Thormar, H., & von Magnus, H. (1963) *Acta path. microbiol. scand.*, **57**, 261.
Tkaczevski, L. de, de Harven, E., & Friend, C. (1968) *J. Virol.*, **2**, 365.
Tyndall, R. L., Vidrine, J. C., Teeter, E., Upton, A. C., Harris, W. W., & Fink, M. A. (1965) *Proc. Soc. exp. Biol. Med., U.S.A.*, **119**, 186.
Vaage, J. (1969) *Cancer Res.*, 28, 2477.
Vogt, P. K. (1965a) *Adv. Virus Res.*, **11**, 293.
Vogt, P. K. (1965b) *Virology*, **25**, 237.
Wahren, B. (1966) *Int. J. Cancer*, **1**, 161.
Wheelock, E. F. (1967) *Proc. Soc. exp. Biol. Med., U.S.A.*, **124**, 855.
Wright, B. S., & Korol, W. (1970) *Cancer Res.*, **29**, 1886.
Wright, B. S., & Lasfargues, J. C. (1965) *J. natn. Cancer Inst.*, **35**, 319.
Yoshikura, H., Hirokawa, Y., Yamada, M. A., & Sugano, H. (1967) *Jap. J. med. Sci. Biol.*, **20**, 225.
Young, D. (1966) *J. comp. Path.*, **76**, 45.
Zeigel, R. F., & Rauscher, F. J. (1964) *J. natn. Cancer Inst.*, **32**, 1277.
Zeigel, R. F., & Clarke, H. F. (1969) *J. natn. Cancer Inst.*, **43**, 1097.
Zeigel, R. F., Theilen, G. H., & Twiehaus, M. J. (1966) *J. natn. Cancer Inst.*, **37**, 709.
Zilber, L. A., & Krynkova, I. N. (1957) *Probl. Virol.*, **2**, 247.
Zilber, L. A. A., Lapin, B. A., & Adgighytov, F. I. (1965) *Nature, Lond.*, **205**, 1123.
Zilber, L. A. A., & Shelvljaghyn, V. (1964) *Nature, Lond.*, **203**, 194.

6

Arenavirus

This genus includes lymphocytic choriomening itis virus, haemorrhagic viruses of South America and Lassa virus. The following characters define the genus. RNA viruses about 100 nm in diameter, composed of an envelope covered with projections and an interior of undetermined structure, containing a variable number of electron-dense granules giving a sandy appearance from which the generic name is derived (Rowe et al., 1970a). Sensitive to lipid solvents. Virus synthesized in cytoplasm and assembled by budding through cellular membrane. All members cross-react antigenically. Chronic infections with carrier state are frequent in the natural hosts (rodents) and pathogenicity is dependent on host immune mechanisms.

LYMPHOCYTIC CHORIOMENINGITIS (LCM)

Original account: Armstrong & Lillie (1934).

Review: Wilsnack (1966).

Morphology. Virions seen in thin sections are pleomorphic, varying in diameter between 50 and 200 nm. They consist of a 2-layered envelope with thin projections on the outer surface and an interior containing 1 or more electron-dense granules 20 to 30 nm across, resembling ribosomes (Dalton et al., 1968). Virions containing, in addition, 30 nm glycogen granules have also been described (Kajima & Majde, 1970). Granules of both types are incorporated into virions during maturation by budding at the cell surface (Dalton et al., 1968; Kajima & Majde, 1970).

Chemical composition. Virions contain single-stranded RNA separable into 2 major (28S and 22S) and 1 minor (18S) components making up a total molecular weight of $3{\cdot}5 \times 10^6$ daltons (Pedersen, 1970) assuming 1 molecule of each component per virion.

Physico-chemical characters. Virions have sedimentation coefficient 470S and buoyant density of 1·8 g/ml in sucrose and 1·17 g/ml in tartrate (Pedersen, 1970; Camyre & Pfau, 1968). Extremely thermolabile, but can be stabilized to some extent by protein or other additives (Lehman-Grube, 1968). Strains differ in sensitivity to tartrate and in stability at 4°C (Camyre & Pfau, 1968). Unstable at acid or alkali pH.

Haemagglutinin. None reported, but virus said to be adsorbed to RBCs of susceptible species (Schwartzman, 1944).

Antigenic properties. Neutralizing antibodies in recovered man and most other species persist for long periods. They may be absent in mice, but to demonstrate them incubation of serum and virus for 24 hours at 37° (Traub, 1960) or addition of fresh serum (Ackerman et al., 1962) may be necessary. Immunofluorescence is a sensitive and specific method for detecting antibody (Cohen et al., 1966). Complement fixation and precipitation, in agar gel and otherwise, are associated with a soluble antigen which is relatively stable at 56° and between pH 4·5 and 9 (Smadel et al., 1940). At least 4 or 5 distinct precipitating antigens may be detected (Chastel, 1970; Simon, 1970). Extraction with acetone-sucrose or with fluorocarbon gives a very good CF antigen (Grešiková & Casals, 1963; Chastel & LeNoc, 1968). Complement-fixing antibodies do not persist long after infection. Cross-reactions with viruses of the Tacaribe group (p. 174) and with Lassa virus (p. 176) are demonstrable by immunofluorescence and less regularly by complement fixation (Rowe et al., 1970b; Buckley et al., 1970). Immunity of mice in infected stocks may be due to 'persistent tolerated infection' in sucklings or to active immunity in adult (Weigand & Hotchin, 1961). For a time after SC injection mice may have increased susceptibility to the virus (Seamer et al., 1963). The whole question of immunity to the virus is reviewed by Traub (1963).

Cultivation. *In fertile eggs* the virus grows on the CAM, but without production of specific lesions (Tobin, 1954). It grows *in tissue culture* of chick, mouse, cattle, monkey and other species, but CPE may only be seen after adaptation. Strains of high virulence grow in a greater variety of cells. Multiplication in suspension cultures of L cells (Pedersen & Volkert, 1967) in a monkey kidney cell-line with CPE and in sarcoma 180 ascites tumours in mice is described (Simon, 1970). Plaques can be obtained in chick-embryo monolayers, but only after 12 days' incubation (Benson & Hotchin, 1960). A plaque assay based on haemadsorption interference is described by Wainwright & Mims (1967). Growth is inhibited by actinomycin D, by 6-azauridine (Buck & Pfau, 1969) and by 2(α hydroxybenzyl) benzimidazole (Pfau & Camyre, 1968).

Pathogenicity. Probably an inapparent infection in naturally infected wild house mice (*Mus musculus*), but it has also been isolated from man, monkeys, dogs, field mouse (*Apodemus*), hamsters and guinea-pigs. When first affecting a colony of laboratory mice, it caused symptoms in young mice infected *in utero*, but soon became adapted so as to be wholly latent (Traub, 1939). Some (docile) strains induce this readily while other (aggressive) strains lead to brisk reaction and death (Hotchin et al., 1962). The influence of immune mechanisms in the pathogenicity of LCM has been reviewed by Hotchin (1962) and Volkert & Larsen (1965).

Disease in stocks of guinea-pigs was described as 'Pneumopathie des Cobayes' (Lepine & Sautter, 1945); symptoms were of generalized, often fatal, disease with patchy pneumonia and exudates into serous cavities.

The disease in man, occurring in laboratory workers or in people from houses infested with mice, may be inapparent or show itself as an influenza-like fever, as meningitis or as meningo-encephalomyelitis, the last being sometimes fatal but fortunately rare. The incubation period after deliberate infection of man (Lepine et al., 1937) lasted 36–72 hours. There were 2 or 3 waves of fever.

Experimentally, the virus will infect mice of susceptible stocks when injected by various routes, the intracerebral being the most and intranasal the next most effective. Five to 12 days after infection the hair of the mice is ruffled, they tremble and have tonic convulsions in which they may die. Infection of neonatal or immunologically suppressed mice results in 'late onset disease' with runting and renal involvement (Hotchin & Collins, 1964; Hirsch et al., 1968; Mims, 1970). Infected hamsters (Smadel & Wall, 1942) show prolonged viraemia, but usually no symptoms. Monkeys of various species, chimpanzees, rats and dogs can also be infected. The virus does not readily infect rabbits except (Blanc & Bruneau, 1951) in pregnancy, causing abortion. It may, however, give rise to a specific local skin reaction when injected intradermally into rabbits (Roger, 1962) or mice (Hotchin & Benson, 1963). Different strains of virus vary greatly in virulence for different species. Neurotropic and viscerotropic strains produce rather different symptoms in mice and virus is to be found in different organs (Wilsnack & Rowe, 1964; Lehmann-Grube, 1964). In carrier mice of infected stocks, virus meningitis may be produced by intracerebral injection of sterile broth.

Pathological lesions. There is lymphocytic infiltration around blood vessels and chorioid plexuses in all species, also in various other viscera. Mice may show pleural and peritoneal effusions. Guinea-pigs, mice and monkeys may have hepatitis. Viral antibody, presumably in combination with antigen and complement is found in glomerulonephritic lesions of chronically infected mice (Oldstone & Dixon, 1969; Hirsch et al., 1968).

Ecology. Virus is excreted in stools and urine of infected mice and has been recovered from the nasopharynx of man. Infected dust is suspected as a vehicle for infection. Several reports suggest that arthropods, mosquitoes, bed-bugs (*Cimex*), monkey lice and other species (Milzer, 1942) may be concerned in transmission. At least *Aëdes* mosquitoes and *Cimex* can transmit experimentally. Virus has been recovered from ticks and also from offspring of infected ticks (Reiss-Gutfreund et al., 1962). The virus has been the subject of important studies on ecology by Traub (1936; 1938; 1939). He followed a change in host-parasite relation extending over 4 years: finally all mice were infected *in utero*, carried virus for long periods without symptoms and without developing antibodies, and became less infectious by contact, all infection being transmitted from mothers to offspring.

Control. Risk of disease in man is lessened by eliminating wild mice. Guinea-pigs may be immunized by giving virus modified by intracerebral passage in mice (Traub, 1937), while UV-irradiated virus has given slight protection to mice.

AMERICAN HAEMORRHAGIC FEVER VIRUSES

Review: Johnson et al. (1967).

The following viruses identified as arenaviruses by antigenic (Rowe et al., 1970b) and morphological (Murphy et al., 1970) characters: Junin (Parodi et al., 1958; Pirosky et al., 1959); Tacaribe (Downs et al., 1963); Machupo (Johnson et al., 1965); Amapari (Pinheiro et al., 1966); Tamiami (Calisher et al., 1970); Pichinde (Trapido, H., & Sanmartin, C., 1971); Parana (Webb et al., 1970) and Latino (Webb, P. A., Peters, C. J., & Johnson, K. M. quoted by Rowe et al., 1970b).

Morphology. Virions are round or pleomorphic 60 to 280 nm in diameter, composed of an envelope with closely spaced projections and internal electron-dense granules 20 to 25 nm in diameter. Unbounded masses of similar granules are seen in the cytoplasm of infected cells. Virus maturation by budding from plasma membranes (Murphy et al., 1970).

Chemical composition. Chemical analysis of partially purified Junin virus revealed presence of RNA and absence of DNA (Parodi et al., 1966). Virus multiplication not inhibited by BUDR (Junin, Machupo) or by Actinomycin D (Junin) (Webb et al., 1967; Coto & Vombergar, 1969; Segovia & Grazioli, 1969).

Physico-chemical characters. Sensitive to chloroform (Machupo) and deoxycholate (Tacaribe, Tamiami), very sensitive to UV (Junin), unstable at acid or alkaline pH (Junin), readily inactivated by heat (Junin, Tamiami) and stable at −70°C (Machupo, Tamiami) or lyophilized (Machupo) (Downs et al., 1963; Johnson et al., 1965; Parodi et al., 1966; Calisher et al., 1970).

Antigenic properties. Cross-reactions between LCM and all viruses of the American haemorrhagic fever group are demonstrable by immunofluorescence (Rowe et al., 1970b). Cross-reactions between haemorrhagic fever viruses can also be detected by complement fixation and virus neutralization (Casals, 1965; Wiebenga, 1965; Pinheiro et al., 1966; Calisher et al., 1970).

Haemagglutination. Failure to demonstrate haemagglutination reported by several workers.

Cultivation. All viruses of the group have been grown in tissue cultures, the most generally susceptible being the Vero line of African green monkey-kidney cells (Simizu et al., 1967; Rhim et al., 1969). Virus multiplication leads to CPE with formation of basophilic cytoplasmic inclusions and plaque production.

Pathogenicity. Junin and Machupo viruses have been associated with human haemorrhagic fevers in Argentina (Pirosky et al., 1959; Parodi et al., 1958) and in Bolivia (Johnson et al., 1965) respectively. The other viruses of the group have been isolated from bats (Tacaribe), wild rodents or their ectoparasites. The clinical, pathological and epidemiological aspects of human haemorrhagic disease caused by these viruses are comprehensively reviewed by Johnson et al. (1967).

Experimental infections can be obtained in guinea-pigs, hamsters and mice. Junin virus causes haemorrhagic disease in guinea-pigs with depression of immune response (Parodi et al., 1967) whereas Machupo and Tacaribe viruses cause subclinical infection in this host (Webb et al., 1967; Coto et al., 1967). The Panamanian marmoset (*Sanguinus geoffroyi*) is susceptible to fatal infection by Machupo virus (Webb et al., 1967). Chronic experimental infection of hamsters and a wild South American rodent, *Calomys callosus*, by Machupo virus has been described by Johnson et al. (1965). Mice are protected by thymectomy against highly virulent doses of Junin virus (Schmunis et al., 1967).

Isologous thymus or immunocompetent cell transplants counteract this protective effect (Weissenbacher et al., 1969).

The pathology of human haemorrhagic fevers caused by Junin (Pirosky et al., 1959; Polak & Jufe, 1961) and Machupo (Child et al., 1967) consists of generalized lymphadenopathy with haemorrhagic necrosis and haemorrhages and moderate congestion in many organs. The pathogenesis of Junin virus infection in hamsters has been investigated by immunofluorescence (Bruno-Lobo et al., 1968).

Ecology. Wild rodents are the common hosts, human infections appearing only when conditions become appropriate for the virus to be transmitted from rodents to man (Johnson et al., 1967). It is generally accepted that direct transmission from man to man does not occur. Although Junin and other viruses of the group have been isolated from mites and other ectoparasites of natural hosts, there is doubt regarding the transmission of infection by arthropods.

Control. Rodent control has been effective in controlling human haemorrhagic fever in Bolivia (Kuns, 1965). An inactivated Junin virus vaccine has been used (Parodi et al., 1965). Viruses of this group should be handled with great care as laboratory infections are frequent.

LASSA VIRUS

A very severe human disease observed in West Africa (Frame et al., 1970) has been shown to be caused by a virus resembling arenaviruses in morphological (Speir et al., 1970), antigenic and other properties (Buckley & Casals, 1970; Buckley et al., 1970).

A laboratory infection by this virus has been successfully treated with plasma from a person recovered from the disease (Leifer et al., 1970).

REFERENCES

Ackerman, R., Scheid, W., & Jochheim, K. A. (1962) *Zentbl. Bakt. ParasitKde, I. Abt. Orig.*, **185**, 343.

Armstrong, C., & Lillie, R. D. (1934) *Publ. Hlth Rep., Wash.*, **49**, 1019.

Benson, L. M., & Hotchin, J. E. (1960) *Proc. Soc. exp. Biol. Med., U.S.A.*, **103**, 623.

Blanc, G., & Bruneau, J. (1951) *C. r. hebd. Séanc. Acad. Sci., Paris*, **233**, 1704.

Bruno-Lobo, G. G., Bruno-Lobo, M., Johnson, K. M., Webb, P. A., & de Paola, D. (1968) *Ann. Microbiol., Rio de Janeiro*, **15**, 11.

Buck, L. L., & Pfau, C. J. (1969) *Virology*, **37**, 698.

Buckley, S. M., & Casals, J. (1970) *Am. J. trop. Med. Hyg.*, **19**, 680.

Buckley, S., Casals, J., & Downs, W. G. (1970) *Nature, Lond.*, **227**, 174.

Calisher, C. H., Tzianabos, T., Lord, R. D., & Coleman, P. H. (1970) *Am. J. trop. Med. Hyg.*, **19**, 520.

Camyre, K. P., & Pfau, C. J. (1968) *J. Virol.*, **2**, 161.

Casals, J. (1965) *Am. J. trop. Med. Hyg.*, **14**, 794.

Chastel, C. (1970) *Acta virol., Prague*, **14**, 507.
Chastel, C., & Le Noc, P. (1968) *Annls Inst. Pasteur, Paris*, **114**, 698.
Child, P. L., Mackenzie, R. B., Valverde, L. R., & Johnson, K. M. (1967) *Archs Path.*, **83**, 434.
Cohen, S. M., Triandaphili, I. A., Barlow, J. J., & Hotchin, J. (1966) *J. Immunol.*, **96**, 777.
Coto, C. E., Roy, E., & Parodi, A. S. (1967) *Arch. ges. Virusforsch.*, **20**, 31.
Coto, C. E., & Vombergar, M. D. de (1969) *Arch. ges. Virusforsch.*, **27**, 307.
Dalton, A. J., Rowe, W. P., Smith, G. H., Wilsnack, R. E., & Pugh, W. E. (1968) *J. Virol.*, **2**, 1465.
Downs, N. G., Anderson, C. R., Spence, L., Aitken, T. H. G., & Greenhall, A. H. (1963) *Am. J. trop. Med. Hyg.*, **12**, 640.
Frame, J. D., Baldwin, J. M., Jr., Gocke, D. J., & Troup, J. M. (1970) *Am. J. trop. Med. Hyg.*, **19**, 670.
Grešiková, M., & Casals, J. (1963) *Acta virol., Prague*, **7**, 380.
Hirsch, M. S., Murphy, F. A., & Hicklin, M. D. (1968) *J. exp. Med.*, **127**, 757.
Hotchin, J. (1962) *Cold Spring Harb. Symp. quant. Biol.*, **27**, 479.
Hotchin, J., & Benson, L. (1963) *J. Immunol.*, **91**, 460.
Hotchin, J., Benson, L. M., & Seamer, J. (1962) *Virology*, **18**, 71.
Hotchin, J., & Collins, D. N. (1964) *Nature, Lond.*, **203**, 1357.
Johnson, K. M., & 8 co-authors (1965) *Proc. Soc. exp. Biol. Med., U.S.A.*, **118**, 113.
Johnson, K. M., Halstead, S. B., & Cohen, S. N. (1967) *Prog. med. Virol.*, **9**, 105.
Johnson, K. M., Mackenzie, R. B., Webb, P. A., & Kuns, M. L. (1965) *Science*, **150**, 1618.
Kajima, M., & Majde, J. (1970) *Naturwissenschaften*, **57**, 93.
Kuns, M. L. (1965) *Am. J. trop. Med. Hyg.*, **14**, 813.
Lehmann-Grube, F. (1968) *Arch. ges. Virusforsch.*, **23**, 202.
Lehmann-Grube, F. (1964) *Arch. ges. Virusforsch.*, **14**, 344 and 351.
Leifer, E., Gocke, D. J., & Bourne, H. (1970) *Am. J. trop. Med. Hyg.*, **19**, 677.
Lepine, P., Mollaret, P., & Kreis, B. (1937) *C. r. hebd. Séanc. Acad. Sci., Paris*, **204**, 1846.
Lepine, P., & Sautter, V. (1945) *Annls Inst. Pasteur, Paris*, **71**, 102.
Milzer, A. (1942) *J. infect. Dis.*, **70**, 152.
Mims, C. A. (1970) *Arch. ges. Virusforsch.*, **30**, 67.
Murphy, F. A., Webb, P. A., Johnson, K. M., Whitfield, S. G., & Chappell, A. (1970) *J. Virol.*, **6**, 507.
Oldstone, M. B. A., & Dixon, F. J. (1969) *J. exp. Med.*, **129**, 483.
Parodi, A. S., Coto, C. E., Boxaca, M., Lajmanovich, S., & Gonzalez, S. (1966) *Arch. ges. Virusforsch.*, **19**, 393.
Parodi, A. S., Greenway, D. J., Rugiero, H. R., Frigerio, M., de la Barrena, J. M., Garzon, F., Boxaca, M., Guerrero, L., & Nota, N. (1958) *El Dia Med.*, **30**, 2300.
Parodi, A. S., Guerrero, L. B. de, & Weissenbacker, M. (1965) *Cienc. Invest.*, **21**, 132.
Parodi, A. S., Nota, N. R., Guerrero, L. B. de, Frigerio, M. J., Weissenbacher, M., & Rey, E. (1967) *Acta virol., Prague*, **11**, 120.
Pedersen, I. R. (1970) *J. Virol.*, **6**, 414.
Pedersen, I. R., & Volkert, M. (1966) *Acta path. microbiol. scand.*, **67**, 523.
Pfau, C. J., & Camyre, K. P. (1968) *Virology*, **35**, 375.
Pinheiro, F. P., Shope, R. E., de Andrade, A. P., Bensabath, G., Cacios, G. V., & Casals, J. (1966) *Proc. Soc. exp. Biol. Med., U.S.A.*, **122**, 531.
Pirosky, L., Zuccarini, J., Mollinelli, E. A., di Pietro, A., Barrera Oro, J. G., Martini, P., Martos, L., & d'Empaire, M. (1959) *Orientación med.*, 8, 708.
Polak, M., & Jufe, R. (1961) *Sem. med., B. Aires*, **118**, 864.
Reiss-Gutfreund, R. L., Andral, J., & Serie, C. (1962) *Annls Inst. Pasteur, Paris*, **102**, 36.
Rhim, J. S., Simizu, B., & Wiebenga, N. H. (1969) *Proc. Soc. exp. Biol. Med., U.S.A.*, **130**, 382.
Roger, F. (1962) *Annls Inst. Pasteur, Paris*, **103**, 639.
Rowe, W. P., Morphy, F. A., Bergold, G. H., Casals, J., Hotchin, J., Johnson, K. M., Lehmann-Grube, F., Mims, C. A., Traub, E., & Webb, P. A. (1970a) *J. Virol.*, **5**, 651.
Rowe, W. P., Pugh, W. E., Webb, P. A., & Peters, C. J. (1970b) *J. Virol.*, **5**, 289.

Schmunis, G., Weissenbacher, M., & Parodi, A. S. (1967) *Arch. ges. Virusforsch.*, **21**, 200.
Seamer, J., Barlow, J. L., Gledhill, A. W., & Hotchin, J. (1963) *Virology*, **21**, 309.
Segovia, Z. M., & Grazioli, F. (1969) *Acta virol., Prague*, **13**, 264.
Shwartzman, G. (1944) *J. Immunol.*, **48**, 111.
Simizu, B., Rhim, J. S., & Wiebenga, N. H. (1967) *Proc. Soc. exp. Biol. Med., U.S.A.*, **125**, 119.
Simon, M. (1970) *Acta virol., Prague*, **14**, 369.
Smadel, J. E., & Wall, M. J. (1942) *J. exp. Med.*, **75**, 581.
Smadel, J. E., Wall, M. J., & Baird, R. D. (1940) *J. exp. Med.*, **71**, 43.
Speir, R. W., Wood, O., Liebhaber, H., & Buckley, S. M. (1970) *Am. J. trop. Med. Hyg.*, **19**, 692.
Tobin, J. O'H. (1954) *Br. J. exp. Path.*, **35**, 358.
Trapido, H., & Sanmartin, C. (1971) *Am. J. trop. Med. Hyg.*, **20**, 631.
Traub, E. (1936) *J. exp. Med.*, **64**, 183.
Traub, E. (1937) *J. exp. Med.*, **66**, 317.
Traub, E. (1938) *J. exp. Med.*, **68**, 229.
Traub, E. (1939) *J. exp. Med.*, **69**, 801.
Traub, E. (1960) *Arch. ges. Virusforsch.*, **10**, 289.
Traub, E. (1963) *Arch. ges. Virusforsch.*, **14**, 65.
Volkert, M., & Larsen, J. H. (1965) *Prog. med. Virol.*, **7**, 160.
Wainwright, S., & Mims, C. A. (1967) *J. Virol.*, **1**, 1091.
Webb, P. A., Johnson, K. M., Hibbs, J. B., & Kuns, M. L. (1970) *Arch. ges. Virusforsch.*, **32**, 379.
Webb, P. A., Johnson, K. M., Mackenzie, R. B., & Kuns, M. L. (1967). *Am. J. trop. Med. & Hyg.*, **16**, 531.
Weigand, H., & Hotchin, J. E. (1961) *J. Immunol.*, **86**, 401.
Weissenbacher, M. C., Schmunis, G. A., & Parodi, A. S. (1969) *Arch. ges. Virusforsch.*, **26**, 63.
Wiebenga, N. H. (1965) *Am. J. trop. Med. Hyg.*, **14**, 802.
Wilsnack, R. E. (1966) in *Viruses of laboratory rodents. Natn. Inst. Cancer Monogr.*, **20**, 77.
Wilsnack, R. E., & Rowe, W. P. (1964) *J. exp. Med.*, **120**, 829.

7

Coronavirus

The term 'coronavirus' has been adopted as a generic name to cover the viruses of avian infectious bronchitis, mouse hepatitis, transmissible gastro-enteritis of pigs, haemagglutinating encephalomyelitis virus of pigs and certain respiratory viruses of man. The following properties characterize the genus (Leader, 1968): Approximately spherical virions 80 to 160 nm in diameter, covered by petal-like projections 20 nm long in a characteristic 'fringe' giving the appearance of a crown from which the generic name is derived. Virus assembly takes place in cytoplasmic vesicles. Sensitivity to lipid solvents and insensitivity to DNA inhibitors indicate that virions are enveloped and contain RNA.

Morphology and development. Negatively stained preparations of coronaviruses from birds (Berry et al., 1964), mice (Mallucci & Almeida, 1966), rats (Parker et al., 1970), man (Almeida & Tyrrell, 1967; McIntosh et al., 1967b) and pigs (Tajima, 1970; Phillip et al., 1971; McFerran et al., 1971) coronaviruses reveal pleomorphic but approximately spherical virions 80 to 160 nm in diameter. An envelope covered with widely spaced 20-nm long projections has been described in all members of the genus. An internal beaded filament is described in transmissible gastro-enteritis virus of pigs (Witte et al., 1968).

A common feature seen in thin sections of cells infected with murine (David-Ferreira & Manaker, 1965), porcine (Okaniwa et al., 1966), human (Hamre et al., 1967; Becker et al., 1967) and avian (Nazerian & Cunningham, 1968) coronaviruses is the accumulation of virions in cytoplasmic vesicles.

Chemical composition. Lack of inhibition of the multiplication of murine (Mallucci, 1965), porcine (Clarke, 1968; Greig & Girard, 1969a) and man (Kapikian et al., 1969) coronaviruses by halogenated deoxyribosides and by actinomycin D indicate RNA as the viral nucleic acid. The same is suggested by cytochemical studies of murine (Armstrong & Niven, 1957; Starr et al., 1960) and avian (Akers & Cunningham, 1968) strains. Biochemical analysis of the nucleic acid extracted from purified transmissible gastro-enteritis virus of pigs revealed the presence of single-stranded RNA (Caletti et al., 1968). Infectious, RNAse-sensitive nucleic acid has been obtained from materials containing the same virus (Norman et al., 1968). A 16S RNA fraction believed to represent viral RNA has been detected in livers of mice infected with mouse hepatitis virus (Tsuji et al., 1968).

HUMAN CORONAVIRUSES

Physico-chemical characters. Density 1·19 (Leader, 1968). Labile below pH 3 (Kapikian et al., 1969).

Haemagglutination. Human and vervet RBC agglutinated at 4°C; chicken, rat and mouse cells agglutinated at room temperature or at 37°C. Neuraminic acid receptors are not involved (Kaye & Dowdle, 1969). Virus purification may be achieved by adsorption to and elution from human RBC (Kaye et al., 1970).

Antigenic properties. Comparative studies by virus neutralization, complement fixation, immuno diffusion and immunofluorescence reveal cross-reactions between human and murine strains (McIntosh et al., 1969; Bradburne, 1970).

Cultivation. On first isolation, viruses grow with difficulty in primary human kidney cultures but may be propagated in human diploid cell strains with CPE (Hamre & Procknow, 1966). Virus isolation may be easier in human tracheal organ cultures (Tyrrell & Bynoe, 1965; McIntosh et al., 1967b). Adaptation to growth in a number of human cell-lines with plaque production has been described (Bradburne & Tyrrell, 1969). Some strains have also been adapted to grow in monkey-kidney cells (Bruckova et al., 1970).

Pathogenicity. Associated with acute upper respiratory illness (Hamre & Procknow, 1966; Tyrrell & Bynoe, 1965; Kapikian et al., 1969). Common colds produced experimentally in volunteers given coronaviruses (Bradburne et al., 1967). Adaptation to suckling mouse brain has been described (McIntosh et al., 1967a).

Ecology. Infection is frequent in adults with upper respiratory disease in winter and rare in children with lower respiratory disease (Kapikian et al., 1969; McIntosh et al., 1970).

MOUSE HEPATITIS VIRUS (MHV)

Synonym: JHM (a neurotropic variant described by Cheever et al., 1949 and named after J. H. Mueller).

Reviews: Gledhill (1962). Piazza (1965).

The hepatitis originally described by Gledhill & Andrewes (1951) was found to be due to synergism between two components, MHV and *Eperythrozoon coccoides* (Niven et al., 1952).

Physico-chemical characters. Inactivated in 30 minutes at 56°. Sensitive to ether and chloroform, but moderately resistant to sodium deoxycholate (Calisher & Rowe, 1966). Survives well at −76° and after lyophilization.

Antigenic properties. Neutralizing antibodies can be prepared by hyperimmunizing mice but they are of low titre. Hartley et al. (1964) found neutralizing antibodies in many human sera. Pollard & Bussell (1957) could demonstrate CF antibodies; the best antigen was in the supernatant fluid after hard centrifugation. The various strains (MHV 1, 2, 3 etc.) described below are antigenically similar, though minor differences between them may exist.

Cultivation. No growth in fertile eggs.

The MHV 1 strain was cultivated by Gompels & Niven (1953) in tissues of mouse embryos from 10–12-day pregnant mice; no CPE was described. Cultivation in mouse liver was described by Miyazaki et al. (1957), by Starr et al. (1960), and by Vainio (1961), a CPE being seen by the last worker after 3–5 days; he used the MHV 3 strain. Bang & Warwick (1959), working with the MHV 2 strain, found that macrophages in culture were selectively destroyed. An assay technique based on plaque formation in macrophage cultures has been developed (Mallucci, 1966; Shif & Bang, 1966).

Pathogenicity. The MHV 1 strain of Gledhill & Andrewes (1951), Gledhill et al. (1952) occurred as a latent infection of a varying percentage of normal mice (10 per cent for the VS strain of mice). Injected into susceptible newly weaned mice of this strain it produced negligible hepatic damages, but in the presence of the blood parasite *Eperythrozoon coccoides*, itself harmless, it caused fatal hepatitis. In new-born mice it produced a similar effect even in the absence of the *E. coccoides*. Infection was produced by injection by various routes and by feeding.

The mechanism of enhancement of the virus's pathogenicity by *E. coccoides* was studied by Gledhill et al. (1955) and Gledhill (1956).

Latent hepatitis viruses have also been activated during serial passage of leukaemias and tumours in mice, and the virus's activity has been enhanced not only by *E. coccoides* but also by cortisone, urethane, an enterotoxin from Gram-negative bacteria and by thymectomy.

Several strains of virus have been given numbers by Dick et al. (1956). MHV 1 is the originally described strain.

MHV 2 (called MHV (Pr) by Nelson) was activated by him during passage of a leukaemia in Princeton mice. It was pathogenic for these mice in the absence of *E. coccoides*, but that parasite enhanced its much lower pathogenicity for Swiss mice (Nelson, 1953).

MHV 3 is a strain obtained by Dick et al. (1956), pathogenic for weanling mice in the absence of *E. coccoides* and, after passage, uniformly fatal for them. In older mice it frequently caused ascites.

MHV 4 is suggested as a name for the neurotropic JHM virus of Cheever et al. (1949). This caused disseminated encephalomyelitis with demyelination in mice, with some focal liver necrosis. Similar neurotropic derivatives of MH 1 and MH 3 were obtained by Dick et al. (1956). Other hepatotropic strains have been recovered particularly during studies of leukaemia by workers in several countries. The neurotropic strains have also infected cotton-rats and hamsters when injected IC. Other species are insusceptible. Virus is present in blood and viscera, especially liver and kidneys; also in urine and faeces.

Pathological lesions. In fatal infections of suckling mice with MHV 1 or in older ones infected also with *E. coccoides*, the liver appears either yellow or brown and mottled with haemorrhages; the kidneys are swollen and pale. There is a widespread endothelial and mesothelial reaction in almost all tissues. Multinucleated giant cells appear in venous channels (Niven et al., 1952). Ruebner & Miyai (1962) considered that earliest lesions were in Kupffer cells. The focal necroses present in the liver soon become confluent. The lesions caused by the neurotropic MHV 4 (JHM) are described by Bailey et al. (1949). Haematological changes are described by Piazza et al. (1965).

Ecology. Virus is present in excreta of weanlings, and the virus is highly infectious amongst young mice (Rowe et al., 1963). Attempts to demonstrate transplacental transmission gave negative results (Piccinino et al., 1966).

Control. The therapeutic action of tetracyclines in the first report by Gledhill & Andrewes in 1951 is accounted for by the effect of the drug on the Eperythrozoa.

The mouse hepatitis virus of Jordan & Mirick (1955) is thought probably to be an encephalitozoon-like protozoon.

INFECTIOUS BRONCHITIS
(of birds)

Synonym: Gasping disease. The disease is distinct from infectious laryngo-tracheitis (cf. p. 349).

Review: Hofstad (1959).

Physico-chemical characters. Buoyant density of virion in CsCl = 1·2289 g/ml (Tevethia & Cunningham, 1968). Stability at 56° varies from one strain to another; most are inactivated in 30′. The virus does not survive 9–30 hours at 37°. Survived 1 hour at room temperature at pH 2 and (in 1 per cent NaOH) pH 12 (Dubose et al., 1960). One per cent formalin and 1:10,000 KMnO4 inactivated (Cunningham & Stuart, 1946). Page & Cunningham (1962) followed the rate of inactivation at 56°, 37° and 26°.

Haemagglutination. Virus modified by trypsin—later treated with a trypsin-inhibitor—will agglutinate fowl RBCs at pH 7·2, contact being at room temperature for 45–60′. Various mammalian cells were not affected. There was no elution. Egg-adapted strains were less active (Corbo & Cunningham, 1959). Separation of haemagglutinin from virions is described by Biswal et al. (1966).

Antigenic properites. Neutralization tests using death of inoculated embryos as an end-point reveal 2 antigenic variants (Connecticut and Massachusetts), showing some cross-reactivity (Hofstad, 1958). Dawson & Gough (1971) describe antigenic variation of insufficient magnitude to enable the distinction of clear serotypes. Three antigens detected by immunodiffusion in virus preparations treated with ether and heat and fractionated by CsCl density gradient centrifugation (Tevethia & Cunningham, 1968).

Young chicks may have, for a time, antibodies transmitted from their mother by way of the yolk.

Convalescent fowl sera can be used for diagnosis in gel-diffusion tests (Woernle, 1959). Antibodies have been found in human sera, especially in individuals who work with poultry (Miller & Yates, 1968).

Cultivation. *In fertile eggs*, the virus grows on the CAM without producing definite pocks (Beaudette & Hudson, 1937); also after inoculation into the allantoic cavity; infected embryos are dwarfed or curled into balls and there may be necrotic foci in liver, pneumonia and nephritis (Loomis et al., 1950). The virus grows in cultures of chick embryonic tissues. Ackers & Cunningham (1968) describe virus replication in the cytoplasm with syncytium formation, cellular necrosis and plaque production.

Pathogenicity. Chicks 2–3 days up to 4 weeks old are affected worst. There is depression, and gasping and rales are heard; the course is from 6 to 18 days; mortality is up to 90 per cent. Avian nephrosis may be caused by infectious bronchitis viruses, either the Massachusetts strain (Working Party, 1966) or perhaps by serologically distinct strains (Winterfield & Hitchner, 1962). In laying birds infection causes big drops in egg-production and eggs are defective. Pheasants may be affected. There is an increasing belief that infection may be widespread and unrecognized; nevertheless it may lie at the basis of much chronic respiratory disease occurring as a result of secondary infection with *Mycoplasma*, *E. coli* and other bacteria. Again, the only evidence of infection may be a fall in egg-production.

Experimentally the incubation period after intratracheal inoculation of chicks is 1–2 days. The disease has been transmitted in series in suckling mice inoculated IC (Simpson & Groupé, 1959).

Pathological lesions. Mucous or caseous exudate in respiratory passage is profuse and may cause blocking. Histological study reveals oedema and epithelial hypertrophy of affected mucous membranes, going on to hyperplasia, cellular exudate and repair (Cunningham, 1960). Garside (1965), however, considers that in an uncomplicated infection, epithelial hyperplasia is absent. There are lesions in the oviduct of laying birds (Sevoian & Levine, 1957).

Ecology. The agent is highly infectious and spreads by the respiratory route, but excretion via the cloaca may also be important (Pette, 1959). Birds may be infectious as long as 35 days after recovery (Hofstad, 1945), but there is no evidence that a true carrier state may result from primary infection (Cook, 1968). Progeny from a flock showing serological but no clinical signs of infection remain free from infection (Cook & Garside, 1967). Virus can be recovered from eggs and semen of experimentally infected chickens (Cook, 1971).

Control. Virus attenuated by cultivation in eggs has been administered in drinking water or as an aerosol: some success is claimed (Hoekstra & Rispens, 1960). The vaccine may be combined with one against Newcastle disease (Markham et al., 1955). Berry (1965) used a β-propiolactone vaccine with success to protect against the depression in egg-yield which follows infection with the virus.

TRANSMISSIBLE GASTRO-ENTERITIS (TGE) (of pigs)

Review: Doyle (1958).

Physico-chemical characters. Survives drying 3 days at room temperature, or $3\frac{1}{2}$ years at $-28°$. 0·05 per cent formaldehyde inactivates in 20′ at 37°C. Heating for 30′ at 56° inactivates (Bay et al., 1952; Haeltermann & Hutchings, 1956). Ether sensitive; acid-resistant.

Antigenic properties. Convalescent serum from recovered pigs neutralizes the virus (Goodwin & Jennings, 1959). A bentonite agglutination test has been described for the detection of TGE antibody (Sibinovic et al., 1966).

Cultivation. Growth in embryonated eggs reported by Eto et al. (1962). Cultivation in pig kidney and other tissue cultures with CPE and plaque formation are described by several workers (e.g. Harada et al., 1963; Cartwright et al., 1964; Witte & Easterday, 1967). Growth in organ cultures of pig tissues is reported by Rubenstein et al. (1970). The virus has also been propagated in dog kidney cultures (Welter, 1965).

Distribution. Occurs in Europe, Asia, Africa, North America and Australia.

Pathogenicity. A very fatal disease in young pigs, causing acute diarrhoea, vomiting, dehydration and often death after 5 to 7 days. Incubation period 12 to 18 hours. May affect older pigs causing scarring; and may take a chronic course. Recovery is usual in pigs over 3 weeks old. Experimental infection of dogs is described by McClurkin et al. (1970).

In experimental infections, Hooper & Haeltermann (1966) describe virus multiplication in the small intestine, especially duodenum and jejunum but viral lesions and multiplication in organs other than the digestive tract are also described (Okaniwa & Maeda, 1965; Harada et al., 1969).

Pathological lesions. Acute enteritis with hyperaemia and some necrosis but particularly distension of the gut with fluid. May be degenerative lesions in heart, muscle and kidney.

Ecology. Spread is by direct or indirect contact—imperfectly cleaned buildings may retain infection. Derbyshire et al. (1969) observed rapid spread of infection in a herd exposed to experimentally infected pigs and failed to demonstrate persistence of virus in recovered animals. Starlings have been suggested as possibly playing a role in virus spread (Pilchard, 1965).

Control. Probable protection by colostrum: young of recovered sows usually survive. Proper management, with breaks between farrowing seasons, usually eliminates infection. An inactivated vaccine prepared in dog kidney cultures was shown to be protective when given 4–11 weeks before farrowing (Welter, 1965).

HAEMAGGLUTINATING ENCEPHALOMYELITIS VIRUS (HEV) (of pigs)

Synonym: Vomiting and wasting disease of piglets.

Physico-chemical characters. Infectivity sensitive to lipid solvents rapidly inactivated at 56°C or at 37°C, but stable at low temperatures or freeze-dried (Greig & Girard, 1963).

Haemagglutination. Chicken RBC agglutinated (Greig et al., 1962).

Antigenic properties. Close antigenic relationship between Canadian strains of HEV and British strains of vomiting and wasting disease revealed by HI (Cartwright et al., 1969). All but 1 of 20 Canadian isolates were antigenically related to each other by HI and VN (Greig & Girard, 1969b).

Cultivation. Growth with syncytium formation and haemadsorption in pig-kidney cultures (Greig & Girard, 1963). No growth in monkey, bovine, ovine, rabbit or mouse cells. Viral antigen detected in the cytoplasm of infected cells by immunofluorescence (Lucas & Napthine, 1971).

Pathogenicity. The virus has been associated with highly fatal outbreaks of encephalomyelitis in sucking piglets (Mitchell, 1963) and with vomiting and wasting disease (Cartwright & Lucas, 1970). Both conditions have been reproduced experimentally in animals given virus grown in tissue cultures.

Distribution. Canada, the United Kingdom.

Ecology. Inapparent infection is frequent in adult animals which may excrete the virus for up to 10 days after infection. Virus regularly found in the respiratory tract but not in other tissues. No evidence of infection 'in utero' (Appel et al., 1965).

OTHER CORONAVIRUSES

A virus resembling coronaviruses in morphology and morphogenesis has been described in association with feline infectious peritonitis (Ward, 1970). An equine virus isolate (Ditchfield, 1969) and a pneumotropic virus of rats (Parker et al., 1970) have also been described as probable coronaviruses.

REFERENCES

Akers, T. G., & Cunningham, C. H. (1968) *Arch. ges. Virusforsch.*, **25**, 30.
Almeida, J. D., & Tyrrell, D. A. J. (1967) *J. gen. Virol.*, **1**, 175.
Appel, M., Greig, A. S., & Corner, A. H. (1965) *Res. vet. Sci.*, **6**, 482.
Armstrong, J. A., & Niven, J. S. F. (1957) *Nature, Lond.*, **180**, 1335.
Bailey, O. T., Pappenheimer, A. M., Cheever, F. S., & Daniels, J. B. (1949) *J. exp. Med.* **90**, 195.
Bang, F. B., & Warwick, A. (1959) *Virology*, **9**, 715.
Bay, W. W., Doyle, L. P., & Hutchings, L. M. (1952) *Am. J. vet. Res.*, **13**, 318.
Beaudette, F. R., & Hudson, C. B. (1937) *J. Am. vet. med. Ass.*, **90**, 51.
Becker, W. B., McIntosh, K., Dees, J. H., & Chanock, R. M. (1967) *J. Virol.*, **1**, 1019.
Berry, D. M. (1965) *J. comp. Path.*, **75**, 409.
Berry, D. M., Cruickshank, J. D., Chu, H. P., & Wells, R. J. H. (1964) *Virology*, **23**, 403.
Biswal, N., Nazerian, K., & Cunningham, C. H. (1966) *Am. J. vet. Res.*, **27**, 1157.
Bradburne, A. F. (1970) *Arch. ges. Virusforsch.*, **31**, 352.
Bradburne, A. F., Bynoe, M. L., & Tyrrell, D. A. J. (1967) *Br. med. J.*, 767.
Bradburne, A. F., & Tyrrell, D. A. J. (1969) *Arch. ges. Virusforsch.*, **28**, 133.
Bruckova, M., McIntosh, K., Kapikian, A. Z., & Chanock, R. M. (1970) *Proc. Soc. exp. Biol. Med., U.S.A.*, **135**, 431.
Caletti, E., Ristic, M., & von Lehmden-Maslin, A. A. (1968) *Am. J. vet. Res.*, **29**, 1603.
Calisher, C. H., & Rowe, W. P. (1966) in *Symp. on viruses of laboratory rodents, Natn. Cancer Inst. Monogr.*, **20**, 67.
Cartwright, S. F., Harris, H. M., Blandford, T. B., Fincham, I., & Gitten, M. (1964) *J. comp. Path.*, **75**, 387.
Cartwright, S. F., & Lucas, M. (1970) *Vet. Rec.*, **86**, 278.
Cartwright, S. F., Lucas, M., Cavill, J. P., Gush, A. F., & Blandford, T. B. (1969) *Vet. Rec.*, **84**, 175.
Cheever, F. S., Daniels, J. B., Pappenheimer, A. M., & Bailey, O . T. (1949) *J. exp. Med.* **90**, 181.
Clarke, M. C. (1968) *J. gen. Virol.*, **3**, 267.
Cook, J. K. A. (1968) *Res. vet. Sci.*, **9**, 506.
Cook, J. K. A. (1971) *J. comp. Path.*, 81, 203.
Cook, J. K. A., & Garside, J. S. (1967) *Res. vet. Sci.*, **8**, 74.
Corbo, L. T., & Cunningham, C. H. (1959) *Am. J. vet. Res.*, **20**, 876.
Cunningham, C. H. (1960) *Am. J. vet. Res.*, **21**, 498.
Cunningham, C. H., & Stuart, H. O. (1946) *Am. J. vet. Res.*, **7**, 466.
David-Ferreira, J. F., & Manaker, R. A. (1965) *J. Cell Biol.*, **24**, 57.
Dawson, P. S., & Gough, R. E. (1971) *Arch. ges. Virusforsch.*, **34**, 32.
Derbyshire, J. B., Jessett, D. M., & Newman, G. (1969) *J. comp. Path.*, **79**, 445.
Dick, G. W. A., Niven, J. S. F., & Gledhill, A. W. (1956) *Br. J. exp. Path.*, **37**, 90.

Ditchfield, W. J. B. (1969) *J. Am. vet. med. Ass.*, **155**, 384.
Doyle, L. P. (1958) in Dunne, *Diseases of swine*. Ames: Iowa State College Press.
Dubose, R. T., Grumbles, L. C., & Flowers, A. I. (1960) *Am. J. vet. Res.*, **21**, 740.
Eto, M., Ichihara T., Tsunoda, T., & Watanabe, S. (1962) *J. Jap. vet. med. Ass.*, **15**, 16.
Garside, J. S. (1965) *Vet. Rec.*, **77**, 354.
Gledhill, A. W. (1956) *J. gen. Microbiol.*, **15**, 292.
Gledhill, A. W. (1962) in *Virus diseases of laboratory animals*. Ed.: Harris, R. J. C. London/New York: Academic Press.
Gledhill, A. W., & Andrewes, C. H. (1951) *Br. J. exp. Path.*, **32**, 559.
Gledhill, A. W., Dick, G. W. A., & Andrewes, C. H. (1952) *Lancet*, **2**, 509.
Gledhill, A. W., Dick, G. W. A., & Niven, J. S. F. (1955) *J. Path. Bact.*, **69**, 299.
Gompels, A. E. H., & Niven, J. S. F. (1953) *J. Path. Bact.*, **66**, 567.
Goodwin, R. F. W., & Jennings, A. R. (1959) *J. comp. Path.*, **69**, 313.
Greig, A. S., & Girard, A. (1963) *Res. vet. Sci.*, **4**, 511.
Greig, A. S., & Girard, A. (1969a) *Res. vet. Sci.*, **10**, 509.
Greig, A. S., & Girard, A. (1969b) *Can. J. comp. Med.*, **33**, 25.
Greig, A. S., Mitchell, D., Corner, A. H., Bannister, G. L., Meads, E. B., & Julian, R. J., (1962) *Can. J. comp. Med.*, **26**, 49.
Haelterman, E. O., & Hutchings, L. M. (1956) *Ann. N.Y. Acad. Sci.*, **66**, 186.
Hamre, D., Kindig, D. A., & Mann, J. (1967) *J. Virol.*, **1**, 810.
Hamre, D., & Procknow, J. J. (1966) *Proc. Soc. exp. Biol. Med., U.S.A.*, **121**, 190.
Harada, K., Funuchi, S., Kumagai, T., & Sasahara, J. (1969) *Nat. Inst. Anim. Hlth Q., Tokyo*, **9**, 185.
Harada, K., Kumagai, T., & Sasahara, T. (1963) *Natn. Inst. Anim. Hlth Q., Tokyo*, **3**, 166.
Hartley, J. W., Rowe, W. P., Bloom, A. H., & Turner, H. C. (1964) *Proc. Soc. exp. Biol. Med., U.S.A.*, **115**, 414.
Hoekstra, J., & Rispens, B. (1960) *T. Diergeneesk*, **85**, 398.
Hofstad, M. S. (1945) *Cornell Vet.*, **35**, 32.
Hofstad, M. S. (1958) *Am. J. vet. Res.*, **19**, 740.
Hofstad, M. S. (1959) in *Diseases of poultry*. Eds. Biester and Schwarte. Iowa State Univ. Press. p. 443.
Hooper, B. E., & Haelrtermann, E. O. (1966) *Am. J. vet. Res.*, **27**, 286.
Jordan, J., & Mirick, G. S. (1955) *J. exp. Med.*, **102**, 601 and 617.
Kapikian, A. Z., James, H. D., Jr., Kelly, S. J., Dees, J. H., Turner, H. C., McIntosh, K., Kim, H. W., Parrott, R. H., Vincent, M. M., & Chanock, R. M. (1969) *J. infect. Dis.*, **119**, 282.
Kaye, H. S., & Dowdle, W. R. (1969) *J. infect. Dis.*, **120**, 576.
Kaye, H. S., Hierholzer, J. C., & Dowdle, W. R. (1970) *Proc. Soc. exp. Biol. Med., U.S.A.*, **135**, 457.
Leader (1968) *Nature, Lond.*, **220**, 650.
Loomis, L. N., Cunningham, C. H., Gray, M. L., & Thorp, F. (1950) *Am. J. vet. Res.*, **11**, 245.
Lucas, M. H., & Napthine, P. (1971) *J. comp. Path.*, **81**, 111.
McClurkin, A. W., Stark, S. L., & Norman, J. O. (1970) *Can. J. comp. Med.*, **34**, 347.
McFerran, J. B., Clarke, J. K., & Curran, W. L. (1971) *Res. vet. Sci.*, **12**, 253.
McIntosh, K., Becker, W. B., & Chanock, R. M. (1967a) *Proc. natn. Acad. Sci. U.S.A.*, **58**, 226.
McIntosh, K., Dees, J. H., Becker, B., Kapikian, A. Z., & Chanock, R. M. (1967b) *Proc. natn. Acad. Sci. U.S.A.*, **57**, 933.
McIntosh, K., Kapikian, A. Z., Hardison, K. A., Hartley, J. W., & Chanock, R. M. (1969) *J. Immunol.*, **102**, 1109.
McIntosh, K., Kapikian, A. Z., Turner, H. C., Hartley, J. W., Parrott, R. H., & Chanock, R. M. (1970) *Am. J. Epidemiol.*, **91**, 585.
Mallucci, L. (1965) *Virology*, **25**, 30.
Mallucci, L. (1966) *Virology*, **28**, 355.
Mallucci, L., & Almeida, J. D. (1966) personal communication.

Markham, F. S., Hammar, A. H., Gingher, P., Cox, H. R., & Storie, J. (1955) *Poult. Sci.*, **34**, 442.
Mitchell, D. (1963) *Res. vet. Sci.*, **4**, 506.
Miyazaki, Y., Katsuta, H., Aoyama, K., Kawai, J., & Takaoka, T. (1957) *Jap. J. exp. Med.*, **27**, 381.
Miller, L. T., & Yates, V. J., (1968) *Am. J. Epidemiol.*, **88**, 406.
Nazerian, K., & Cunningham, C. H. (1968) *J. gen. Virol.*, **3**, 469.
Nelson, J. B. (1953) *J. exp. Med.*, **99**, 433 and 441.
Niven, J. S. F., Gledhill, A. W., Dick, G. W. A., & Andrewes, C. H. (1952) *Lancet*, **2**, 1061.
Norman, J. O., McClurkin, A. W., & Bachrach, H. L. (1968) *J. comp. Path.*, **78**, 227.
Okaniwa, A., & Maeda, M. (1965/66) *Natn. Inst. Anim. Hlth Q., Tokyo*, **5**, 190, and **6**, 24.
Okaniwa, A., Maeda, M., Harada, K., & Kaji, T. (1966) *Natn. Inst. Anim. Hlth Q., Tokyo*, **6**, 119.
Page, C. A., & Cunningham, C. H. (1962) *Am. J. vet. Res.*, **23**, 1065.
Parker, J. C., Cross, S. S., & Rowe, W. P. (1970) *Arch. ges. Virusforsch.*, **31**, 293.
Pette, J. (1959) *Mh. Tierheilk*, **11**, 296.
Phillip, J. I. H., Cartwright, S. F., & Scott, A. C. (1971) *Vet. Rec.*, **88**, 311.
Piazza, M. (1965) in *Le Epatiti Sperimentale da Virus MHV1 and MHV3.* Turin: Minerva Medica.
Piazza, M., Piccinino, F., & Matano, F. (1965) *Nature, Lond.*, **205**, 1034.
Piccinino, F., Galanti, B., & Giusti, G. (1966) *Arch. ges. Virusforsch.*, **18**, 327.
Pilchard, E. I. (1965) *Am. J. vet. Res.*, **26**, 1127.
Pollard, M., & Bussell, R. H. (1957) *Science*, **126**, 1245.
Rowe, W. P., Hartley, J. W., & Capps, W. I. (1963) *Proc. Soc. exp. Biol. Med., U.S.A.*, **112**, 161.
Rubenstein, D., Tyrrell, D. A. J., Derbyshire, J. B., & Collins, A. P. (1970) *Nature, Lond.*, **227**, 1348.
Ruebner, B., & Miyai, K. (1962) *Am. J. Path.*, **40**, 425.
Sevoian, M., & Levine, P. P. (1957) *Avian Dis.*, **1**, 136.
Shif, I., & Bang, F. B. (1966) *Proc. Soc. exp. Biol. Med., U.S.A.*, **121**, 829.
Sibinovic, K. H., Ristic, M., Sibinovic, S., Alberts, J. O. (1966) *Am. J. vet Res.*, **27**, 1339.
Simpson, A. W., & Groupé, V. (1959) *Virology*, **8**, 456.
Starr, T. J., Pollard, M., Duncan, D., & Dunaway, M. R. (1960) *Proc. Soc. exp. Biol. Med., U.S.A.*, **104**, 767.
Tajima, M. (1970) *Arch. ges. Virusforsch.*, **29**, 105.
Tevethia, S. S., & Cunningham, C. H. (1968) *J. Immunol.*, **100**, 793.
Tsuji, T., Hirschowitz, B. I., & Sachs, G. (1968) *Science*, **159**, 987.
Tyrrell, D. A. J., & Bynoe, M. L. (1965) *Br. med. J.*, **1**, 1467.
Vainio, T. (1961) *Proc. Soc. exp. Biol. Med., U.S.A.*, **107**, 326.
Ward, J. M. (1970) *Virology*, **41**, 191.
Welter, C. J. (1965) *Vet. med. Small Anim. Clin.*, **60**, 1054.
Winterfield, R. W., & Hitchner, S. B. (1962) *Am. J. vet. Res.*, **23**, 1273.
Witte, K. H., & Easterday, B. C. (1967) *Arch. ges. Virusforsch.*, **20**, 327.
Witte, K. H., Tajima, M., & Easterday, B. C. (1968) *Arch. ges. Virusforsch.*, **23**, 53.
Working Party (1966) *Vet. Rec.*, **78**, 624.
Woernle, H. (1959) *Mh. Tierheilk.*, **11**, 154.

8

Rhabdovirus

Review: Howatson (1970).

The following properties characterize the genus: virions are bullet-shaped, measuring 130 to 220 nm in length and 60 to 80 nm in diameter. Contain about 2 per cent single-stranded RNA with a molecular weight of about $3{\cdot}5 \times 10^6$ daltons in a nucleocapsid with helical symmetry. This is contained in an envelope studded with projections. Virus is synthesized in the cytoplasm and assembled at cellular membranes, frequently accumulating in cytoplasmic vesicles. Most members multiply both in vertebrates and in arthropods.

Morphology and development. Electron-microscopic studies of Indiana, New Jersey (Bradish et al., 1956; Reczko, 1961; Howatson & Whitmore, 1962), Cocal (Ditchfield & Almeida, 1964), Piry (Bergold & Munz, 1970), Flanders (Murphy et al., 1966), Hart Park (Jenson et al., 1967), Kern Canyon (Murphy & Fields, 1967), Mount Elgon bat virus (Murphy et al., 1970), rabies (Davies et al., 1963; Hummeler et al., 1967; Atanasiu & Sisman, 1967), Lagos bat virus and shrew isolate Ib An 27377 (Shope et al., 1970) reveal the following common features: Virions are bullet-shaped, i.e. cylindrical, with 1 end hemispherical and the other flat. Estimates of the over-all diameter vary from 60 to 80 nm and of the mean length, from 130 to 220 nm with abnormally long (up to 410 nm) or truncated forms sometimes seen. A bullet-shaped core 45 to 50 nm in diameter is made up of a helical nucleocapsid forming a spiral with a periodicity of 5 nm around an axial channel of variable length and about 17 nm in diameter. The core is contained in a 2-layered envelope covered with closely spaced projections 8 to 10 nm in length. Filaments trailing from the flat end of virions are frequently seen. Nucleocapsids derived from spontaneously or chemically degraded vesicular stomatitis virus (Nakai & Howatson, 1968) are described as wavy ribbons 8 to 10 nm wide and 3 to 4 nm thick, made up of 1000 subunits repeated at 3 nm intervals. This ribbon forms a single-stranded helix of about 30 coils with an external diameter of 49 nm in the viral core. The mean length of the nucleocapsid from normal virions was about 3·5 μm. Simpson & Hauser (1966) describe a second helix, 17 to 18 nm in diameter, around the axial canal. Hummeler et al. (1968) describe the nucleocapsid of rabies virus as a single-stranded helix with a diameter of 16 nm, a periodicity of 7·5 nm and a length in excess of 1 μm. Several models differing from each other in some respects have been proposed to explain the structure of vesicular stomatitis virus (Simpson & Hauser, 1966; Bradish & Kirkham, 1966; Bergold & Munz, 1967; Nakai & Howatson, 1968).

The assembly of vesicular stomatitis virus takes place in cellular membranes, either at the cell surface (Howatson & Whitmore, 1962) or in cytoplasmic vesicles (Hackett et al., 1968; David-West & Labzoffsky, 1968a).

Rabies virions are assembled in association with granular cytoplasmic matrices (Matsumoto, 1962; Davies et al., 1963; Hummeler et al., 1967) which correspond to the Negri bodies of infected cells (Miyamoto & Matsumoto, 1965). The viral envelope may be formed *de novo* within or at the edges of cytoplasmic matrices, but budding of virions at the cell surface has also been described. Intracytoplasmic maturation associated with a matrix has also been observed with Lagos bat virus and the isolate from shrews Ib An 27377, both antigenically related to rabies (Shope et al., 1970).

The virus of epizootic or ephemeral fever resembles rhabdoviruses in morphology and morphogenesis (Ito et al., 1969; Holmes & Doherty, 1970), although South African strains of this virus have been described by Lecatsas et al. (1969) as conical rather than bullet-shaped with a basal diameter of 88 nm, a height of 176 nm and a core apparently consisting of a spiral of 10 turns with a pitch of 16.6 nm and a total length of approximately 2·2 μm. Also similar to rhabdoviruses are certain viruses of salmonids including the Egtved strain causing haemorrhagic septicaemia of trout (Zwillenberg et al., 1965). Finally a virus associated with a fatal haemorrhagic disease among laboratory workers handling African green monkey tissues (Marburg virus) has been shown to resemble rhabdoviruses in morphology (Siegert et al., 1968; Kissling et al., 1968; Zlotnik et al., 1968) although highly pleomorphic and filamentous forms with average length of 1 μm and occasionally reaching 2·6 μm are frequently found.

Chemical composition. Purified VSV contains 3 per cent RNA, 64 per cent protein, 13 per cent carbohydrate and 20 per cent lipid (McSharry & Wagner, 1971a). VSV (Brown et al., 1967; Huppert et al., 1967) and rabies (Sokol et al., 1969) infectious virions reveal single-stranded RNA in a single molecule with sedimentation coefficient of 36 (VSV) to 45S (VSV and rabies) and buoyant density in caesium sulphate of 1·66 g/ml (rabies). MW estimations based on S values vary between 3×10^6 and $4{\cdot}6 \times 10^6$ daltons; whereas a value of $3{\cdot}5 \times 10^6$ is derived from the nucleocapsid length of VSV (Nakai & Howatson, 1968). The base composition of the RNA of infectious VSV (Brown et al., 1967; Huppert et al., 1967) reveals about 27 to 30 per cent A and U and 20 to 22 per cent G and C. The approximate complementarity is probably fortuitous. Differences in molecular weight and base composition were observed between the RNAs of complete and truncated particles. Three major polypeptides have been identified as structural VSV proteins and correlated with the ribonucleoprotein, an envelope glycoprotein and a component holding the nucleocapsid in position (Kang & Prevec, 1969; Wagner et al., 1970; Cartwright et al., 1970a). Sokol et al. (1971) describe in rabies virus, 2 polypeptides associated with the nucleocapsid and 3 others, 1 containing sugar, in the viral envelope. Analysis of viral lipids reveals the presence of cholesterol and phospholipids at a molar ratio of 0·6 or greater and absence of unusual lipids, or fatty acids (McSharry & Wagner, 1971a). Carbohydrates found in VSV include glucose, galactose, mannose, fucose, glucosamine, galactosamine, and neuraminic acid (McSharry & Wagner, 1971b). Two nonstructural viral proteins have been detected in VSV infected cells (Wagner et al., 1970). A host-derived component (Cartwright & Pearce, 1968) and an RNA-dependent RNA polymerase (Baltimore et al., 1970) have been found as structural components of VSV.

Physico-chemical characters. Virions of rabies (Neurath et al., 1966), VSV (Warrington, 1965) and ephemeral fever (Tanaka et al., 1969) have buoyant densities of 1·19 to 1·20 g/ml in CsCl. Cores of rabies virus (Sokol et al., 1969) and nucleocapsids of VSV (Cartwright et al., 1970b) have buoyant densities of 1·32 and 1·26 g/ml respectively. Sedimentation coefficients of 600 S and 200 S have been determined for rabies virions and cores respectively (Sokol et al., 1969).

Rhabdoviruses are sensitive to lipid solvents, moderately heat sensitive and stable at low temperatures. Rabies is stable between pH 5 and 10 (Turner & Kaplan, 1967). VSV is highly sensitive to visible light (Skinner & Bradish, 1954).

Haemagglutination. Rabies, VSV (New Jersey, Indiana, Cocal) and Kern Canyon viruses agglutinate goose RBC (Halonen et al.,

1968; Arstila et al., 1969). Haemagglutination is associated with complete virions or with truncated forms (Murphy et al., 1968; Kuwert et al., 1968). The reaction takes place best in the cold, at about pH 6·0 and is very sensitive to normal serum inhibitors. Cell receptors are not destroyed by virus or by periodate. No neuraminidase activity.

Antigenic properties. A comparison of some rhabdoviruses by virus neutralization (F. Murphy, quoted by Howatson, 1970), reveals cross-reactions among Indiana, New Jersey, Cocal, Piry and Chandipura and between Hart Park and Flanders. A 1-way cross-reaction between Chandipura and Kern Canyon is also observed. Rabies and Mount Elgon bat virus did not cross-react with any of the above. Narrower cross-reactions were observed by CF and by immunodiffusion (Murphy & Fields, 1967). Other antigenic properties will be described separately for each virus.

RABIES

Synonyms: Hydrophobia. Lyssa. Wut. Tollwut. Rage.

Reviews: Atanasiu (1970). Matsumoto (1970).

Antigenic properties. Antigens of different sizes have been described (van den Ende et al., 1957; Mead, 1962; Neurath et al., 1966). Multiple precipitation lines obtained by agar immunodiffusion (Villemot & Provost, 1959; Grasset, 1967).

Cultivation. *In fertile eggs.* Some strains of virus multiply when inoculated on the CAM (Kligler & Bernkopf, 1938), also in the allantoic cavity or yolk sacs of 7-day embryos (Koprowski & Cox, 1948). Inoculated embryos are smaller than normal, but the chicks usually hatch and die later. The virus is, however, fatal when inoculated in 1-day embryos (Yoshino et al., 1956). The Flury strain adapted to eggs has been used for immunization, the high egg pass (HEP) strain being more attenuated than the LEP (low egg pass) (Koprowski & Cox, 1948). Virus has also been grown in duck eggs and used for immunization (see below) (Peck et al., 1956).

In tissue culture. Rabies virus has been propagated in primary or continuous cultures of chick embryo, mouse, hamster, pig, rabbit, dog and human cells (for references, see Atanasiu, 1970; Matsumoto, 1970). The hamster kidney cell-line BHK 21 has been particularly useful for virus production (Atanasiu et al., 1963; Hummeler et al., 1967). Virus replication may or may not be accompanied by CPE, and chronic infection or carrier states can be established (Fernandes et al., 1964). Plaque assays have been developed in cultures of chick embryo (Yoshino et al., 1966) and in BHK 21 cells (Sedwick & Wiktor, 1967). Virus replication is not inhibited by halogenated deoxyribosides or by actinomycin D, but is inhibited by arabinosylcytosine, presumably through the action of an induced cellular protein (Campbell et al., 1968).

Adaptation of rabies virus to growth in *Drosophila* with loss of pathogenicity for mice has been described (Plus & Atanasiu, 1966).

Distribution. Occurs all over the world, but has been excluded from Britain and Australia by quarantine.

Pathogenicity. The incubation period in dogs is from 10 days to 6 months or even more, usually 20–60 days. In man it is usually from 15 days to 5 months, but may be as long as a year. Infected dogs show great excitement (furious rabies) with great salivation, changed behaviour, especially biting; this is followed by depression and paralysis (dumb rabies) and, soon after, by death. At times, only dumb rabies is seen. The same 2 stages are seen in man: the characteristic feature of the excited stage is hydrophobia and death may occur at this stage. Paralytic rabies is the only form seen after vampire-bat bites. Fixed virus may cause fatal encephalomyelitis in man; this (rage de laboratoire) may follow injection of vaccine imperfectly treated. In cattle and horses the symptoms vary greatly, and diagnosis may be difficult. The disease may be chronic and inapparent in bats and probably also in skunks and other *Mustelidae*, also at times in rats (Svet-Moldavskaya, 1958) and mice (Bell, 1964). Although in most species the infection is fatal, there is more evidence than formerly that recoveries can occur; such recovered animals could be dangerous carriers. In most species the infection is usually fatal. Species naturally affected include cats, foxes, jackals, wolves, mongooses, skunks—in fact almost all terrestrial carnivores. Not only vampires but insectivorous and fruit-eating bats (including those of the old world) may be infected.

Street virus is freely excreted in the saliva of dogs and other species including man, but fixed virus is not.

Experimentally, almost every readily available mammal has proved susceptible, also chicks. On intracerebral passage through rabbits the naturally occurring 'street' virus soon becomes exalted for that species, producing a paralytic disease with an incubation period of only 4 to 6 days; simultaneously it becomes attenuated for man. This is 'fixed' virus. It may be adapted to mice by IC inoculation and may be infective for them when diluted 10^{-7}, causing a paralytic disease.

Large doses given IN to mice and guinea-pigs will kill them. Mice inoculated IC may recover after abortive infection (Lodmell et al., 1969). Egg-adapted virus is pathogenic for rodents and other species when given IC but not IM; the HEP strain loses even that power for most species (Koprowski et al., 1954).

Pathological lesions in the brain are those of encephalitis with neuronal degeneration especially in the mid-brain and medulla. Cellular infiltration is often slight. The diagnostic Negri bodies are particularly abundant in nerve cells in Ammon's horn of the hippocampus; they consist of cytoplasmic eosinophilic bodies 2–10 μ across, often multiple, with an inner structure of basophilic granules. They probably represent a cellular reaction to presence of virus and characterize infection with street virus. With most strains of fixed virus they are absent; smaller cytoplasmic bodies in nerve cells may, however, occur. Cell-degeneration and infiltration occur in infected salivary glands.

All the evidence suggests that virus reaches the central nervous system by travelling along nerves. Viraemia, at least in infected mice, is brief (Borodina, 1959). In bats, virus may localize in the brown interscapular fat (Sulkin et al., 1957). A study of the pathogenesis of experimental rabies by means of immunofluorescence is described by Johnson (1965c).

Ecology is reviewed by Johnson (1965b), Kaplan (1969) and Irvin (1970). Infection is usually transmitted by bites of infected species, particularly dogs, cats, wolves, foxes, mongooses and skunks, the saliva of which is infective. This may infect through a superficial wound without a bite. In towns, dogs and cats are of chief importance; in other parts wolves (Iran), foxes (Central Europe, North America), vampire bats (Caribbean and South America), mongooses (South Africa), skunks (North America). Rodents, e.g. ground-squirrels in Nigeria, may also act as vectors (McMillan & Boulger, 1960). Insectivorous bats in America have been found infected; the disease in them may be fatal or they may carry the virus for long periods (cf. Enright, 1951). Rabies infection of fruit bats in Thailand has been described (Smith et al., 1967). Johnson (1965b) thinks that bat rabies is an 'aberrant cycle', the natural basic reservoir being more probably in weasels, skunks or other mustelids. The incidence of rabies in wild life is reviewed by Irvin (1970), who concludes that the disease is largely maintained by *Carnivora* and *Chiroptera* although the epidemiological role of rodents and other small mammals may be important. Bats, *Mustelidae* and *Viverridae* with silent infections may be better sources of virus than hosts suffering overt disease. There is epidemiological and experimental evidence that foxes and coyotes may be infected by the respiratory route in caves frequented by bats (Frederickson & Thomas, 1965) and that insectivorous bats may transmit the infection to other hosts by way of bites (Constantine et al., 1968). Evidence for the transmission of rabies by inhalation or ingestion of infectious materials has been demonstrable under a variety of conditions (e.g. Constantine, 1962; Soave, 1966; Hronovsky & Benda, 1969; Correa-Giron et al., 1970).

Control. Six months' quarantine of imported dogs and cats suffices to keep Britain free from rabies. Immunization of dogs and cats with live attenuated Flury strain is effective; a single dose gives immunity for 3 years (Tierkel et al., 1953). The phenol-killed or Semple vaccine used in the past is less effective. The LEP strain is not sufficiently attenuated for immunizing cattle for which the HEP strain has been extensively and successfully used (Carneiro et al., 1955).

For the treatment of human beings bitten or suspected of having been bitten, various modifications have been used of Pasteur's method of massive repeated injections of fixed virus. Virus attenuated by drying, with phenol and other substances or by simple dilution has been used, but according to Greenwood (1946) there is no statistical evidence that one method is better than another. Some have disputed (Webster, 1939) that post-infection treatment is of proved value, but this scepticism is not general. The Semple phenolized vaccine is chiefly used now, but the Flury strain is under trial (Fox et al., 1957). Virus grown in duck embryos and inactivated with β-propiolactone has also been effective in producing antibodies in man (Powell & Culbertson, 1959). Use of this Flury vaccine should avoid the neuroparalytic accidents which may follow injection into man of vaccines containing much nerve tissue. The same may be achieved by the use of vaccine derived from new-born mouse brain (Acha, 1967). Further improvements may come with the development of tissue-culture vaccines (Wiktor et al., 1964) and with the use of purified virus (Sikes & Larghi, 1967; Shokeir, 1968) or subunit (Crick & Brown, 1969) vaccines.

RHABDOVIRUSES ANTIGENICALLY RELATED TO RABIES

Two African viruses, viz. Lagos bat virus and an isolate (IbAn 27377) from shrews, have been found by Shope et al. (1970) to be antigenically related to but distinguishable from rabies and each other and to resemble rabies virus in morphology and morphogenesis. The above authors suggest that these 2 African viruses, together with rabies, should form a subgroup of rhabdoviruses.

VESICULAR STOMATITIS

Synonym: Sore mouth (cattle and horses).

Review: Hanson (1952).

Antigenic properties. At least 2 antigenic types represented by the New Jersey and Indiana strains can be distinguished by complement fixation and virus neutralization. Federer et al. (1967) suggests the Indiana type should be divided into 3 subtypes represented by:

1. Cocal and an Argentinian isolate
2. a Brazilian isolate
3. Indiana C strain

Piry (Bergold & Munz, 1970) and Chandipura (Bhatt & Rodrigues, 1967) viruses are antigenically related to each other and to the Indiana and Cocal strains of VSV (F. Murphy, quoted by Howatson, 1970). An antigen common to Indiana and New Jersey serotypes has been correlated with a viral core protein which may also be found as a 20S soluble antigen, whereas a 6S antigen corresponds to type-specific envelope components (Kang & Prevec, 1970). Fractionation of different antigens from infected cell extracts or from purified virus has been described by several workers (e.g. Bradish et al., 1956; Brown et al., 1967). At least 2 precipitation antigens can be detected by gel diffusion (Brown & Crick, 1957).

Cultivation. *In fertile eggs* the virus grows well on the CAM (Burnet & Galloway, 1934); chicks either die within 1 or 2 days or survive, showing proliferative followed by necrotic changes on the membrane. Embryos 7–8 days old give most regular results especially with the NJ strain (Skinner, 1957a). There is good growth also in the allantoic cavity.

It grows well in cultures of chick embryo and kidney or other epithelial tissues of cattle, pig, rhesus monkey, guinea-pig and doubtless other species, producing a rapid destructive effect. On monolayers of kidney cells, plaques are produced. Variants producing large and small plaques were compared by Wagner et al. (1963); only the latter would produce persistent infection of L cells. Multiplication in *Aëdes* mosquitoes (Mussgay & Suarez, 1962), in insect tissue cultures (Yang et al., 1969; Bergold & Munz, 1970) and in *Drosophila melanogaster* (Peries et al., 1966; Printz, 1970) has been described. Virus replication is not DNA-dependent and takes place in the cytoplasm (David-West & Labzoffsky, 1968b). Virus penetration by viropexis is described by Simpson et al. (1969). Different virus-specific RNAs synthesized in infected cells have been described (Newman & Brown, 1969; Schincariol & Howatson, 1970).

Distribution. A disease of the New World. The virus has been introduced into Europe and perhaps South Africa, but outbreaks have been short-lived. It was formerly known especially as a disease of horses; now it is commoner among cattle. An antigenically related virus, Chandipura, has been isolated in India (Bhat & Rodrigues, 1967).

Pathogenicity. The manifestations of the infection simulate foot-and-mouth disease, though it is much milder. Differential diagnosis is practically very important. There are small papules or vesicles in the mouth of affected cattle, horses or sheep, with drooling saliva, but the lesions only last a few days. In most outbreaks secondary lesions on feet and elsewhere are uncommon. Outbreaks in cattle have been described in which the main lesions have been on the teats (Strozzi & Ramos-Saco, 1953). In pigs, foot lesions are commoner. The disease occurs naturally in racoons and perhaps deer, and these may constitute a reservoir of infection. Infection of man is not uncommon, especially among cattle handlers (Brody et al., 1967; Fields & Hawkins, 1967) or laboratory workers. It may be inapparent or resemble influenza.

Experimentally almost all species can be infected. Cattle, horses, pigs, sheep, rabbits and guinea-pigs develop lesions when inoculated on the tongue. Cattle are not infected by intramuscular injection. The virus produces encephalitis in mice and guinea-pigs after IC inoculation, or in young mice after injection intranasally, intramuscularly or elsewhere (Sabin & Olitsky, 1937). Falke & Rowe (1965) describe the different course of infection in mice of different ages: the CNS was infected via blood-stream in the young and via nerves in older ones. Guinea-pigs inoculated into foot-pads develop vesicles like those of foot-and-mouth disease; they may also show lesions in kidneys and liver. Cotton-rats can be infected IN or IC (Skinner, 1957a), and rabbits by inoculation into the tongue (Skinner, 1957b). In ferrets the virus causes vesiculation and ulceration in the mouth and on the feet (Kowalczyk & Brandly, 1954). Hamsters and chinchillas have died after IN or IC inoculation. Young chicks have been infected IC, but continued serial passage failed (Skinner, 1957a). The tongues and foot-pads of fowls and other poultry, particularly ducks and geese, were successfully infected. Geese showed secondary lesions on feet (Skinner, 1959). Experimental infection of bats by Cocal virus leads to long-term viraemia (Donaldson, 1970). A large variety of wild mammals were shown to be susceptible to VSV with development of encephalitis in young animals (Tesh et al., 1970).

Pathological lesions. The vesicles in various species closely resemble those of foot-and-mouth disease.

The chief brain lesion in fatally infected mice is neuronal destruction.

Ecology. The disease occurs rather sporadically and only between June and October. Mechanical transmission by Tabanids and *Stomoxys* has been suspected (Hanson, 1952). Virus has been recovered from wild caught *Phlebotomus* (Shelokov & Peralta, 1967), *Aëdes* mosquitoes (Sudia et al., 1968) and mites (Jonkers et al., 1964). However, the transmission of VSV by arthropods is considered to be uncertain by Jonkers (1967), who suggests that neither livestock nor rodents are essential for the virus life cycle and that there is a passive source of infection in the soil or vegetation of pastures before the onset of outbreaks.

FLANDERS-HART PARK VIRUS

Numerous mosquito isolates, antigenically related to strains Flanders (Whitney, 1964) and Hart Park, have been shown to be morphologically identical to but antigenically distinct from other rhabdoviruses (Murphy et al., 1966; Murphy & Fields, 1967). They are readily propagated intracerebrally in new-born mice, but multiply poorly if at all in embryonated eggs or tissue cultures (Whitney, 1964). They have not been associated with any disease.

KERN CANYON VIRUS

A virus isolated from a bat (*Myotis yumanensis*) in California (Johnson, 1965a) was identified as a rhabdovirus by Murphy & Fields (1967). It is antigenically unrelated to other members of the genus.

MOUNT ELGON BAT VIRUS

A virus isolated from a *Rhinolophus* insectivorous bat in Kenya (Metselaar et al., 1969) was identified by Murphy et al. (1970) as an antigenically distinct member of the genus rhabdovirus. Virions were found to be longer (mean 226 nm) than most other rhabdoviruses and to mature in association with a cytoplasmic matrix similar to that seen in cells infected with rabies virus. The virus has been propagated intracerebrally in new-born mice, but failed to grow in tissue cultures (Metselaar et al., 1969).

EPHEMERAL FEVER (of bovines)

Synonyms: Three-day stiffsickness. Bovine epizootic fever.

Viruses isolated from bovines with ephemeral (or epizootic) fever in South Africa (van der Westhuizen, 1967), Australia (Doherty et al., 1968) and Japan (Inaba et al., 1968) have been identified as probable rhabdoviruses on the basis of morphological (Ito et al., 1969; Lecatsas et al., 1969; Holmes & Doherty, 1970) and physico-chemical (Lecatsas et al., 1969; Tanaka et al., 1969; Heuschele, 1970) properties. South African strains are unique in being conical rather than bullet-shaped, but serological studies indicate close antigenic relationship between strains from the 3 continents (Inaba et al., 1969a). The virus is usually propagated in mice, but multiplication in tissue culture and particularly in hamster kidney cell-lines (BHK21) has been described (Inaba et al., 1968). A plaque assay in Vero cells has been developed by Heuschele (1970). Propagation in new-born hamsters or mice, or in BHK 21 cells led to rapid loss of pathogenicity for calves (Inaba et al., 1969b). The natural disease is of short duration, characterized mainly by respiratory symptoms increased oropharyngeal secretions and lacrimation, joint pains, tremors and stiffness.

EGTVED VIRUS

Causes haemorrhagic septicaemia amongst European salmonids (Jensen, 1963). Rainbow trout (*Salmo gairdneri*) are the chief natural hosts, but brown trout (*Salmo trutta*) and salmon (*S. salar*) are susceptible to inoculation and brook trout (*Salvelinus fontinalis*) to contact infection (Rasmussen, 1965). The virus has been propagated in cultures of trout ovarian cells (Jensen, 1963) and was shown to be strikingly similar in morphology to VSV (Zwillenberg et al., 1965; 1968).

OTHER SALMONID VIRUSES

Three viruses associated with haematopoietic necrosis in salmonid fishes, namely infectious haematopoietic necrosis, Oregon sockeye disease and Sacramento River Chinook salmon disease have been shown by Amend & Chambers (1970) to resemble rhabdoviruses in morphology. Like Egtved, these viruses are sensitive to lipid solvents and to pH 3 and are readily inactivated by heat. Some have been shown to contain RNA (see Amend & Chambers, 1970). They affect primarily haematopoietic tissues (Yasutake et al., 1965; Yasutake & Rasmussen, 1968).

MARBURG VIRUS

An agent associated with severe illness in workers handling tissues of African green monkeys was shown to resemble rhabdoviruses in some respects, but its inclusion in the genus must be considered as uncertain in view of the uniqueness of some of its properties (e.g. May & Herzberg, 1969). The virus causes fatal infections in guinea-pigs and monkeys (Simpson et al., 1968) and can be propagated in a variety of tissue cultures (Hofmann & Kunz, 1968). The virus appears in negatively stained preparations (Siegert et al., 1968; Kissling et al., 1968; Zlotnik et al., 1968), as rods or filaments 90 to 100 nm in diameter and varying in length from 130 to more than 2600 nm. An apparently helical inner structure with a periodicity of 5 to 6 nm has been described. Virus infectivity is readily inactivated by ultraviolet light, moderately heat resistant and stable at low temperatures (Bowen et al., 1969). It is ether-sensitive, and its multiplication is not inhibited by BUDR (Kissling et al., 1968).

Serological evidence indicates that natural infection of nonhuman primates may be common in Africa (Kalter et al., 1969), and it has been suggested that the virus may be transmitted by mosquitoes (Kunz et al., 1968).

REFERENCES

Acha, P. N. (1967) *Bull. Off. int. Épizoot.*, **67**, 439.
Amend, D. F., & Chambers, V. C. (1970) *J. Fish. Res. Board Can.*, **27**, 1285.
Arstila, P., Halonen, P., & Salmi, A. (1969) *Arch. ges. Virusforsch.*, **27**, 198.
Atanasiu, P. (1970) *Bull. Inst. Pasteur, Paris*, **68**, 2047.
Atanasiu, P., Lepine, P., & Dighe, P. (1963) *C. r. hebd. Séanc Acad. Sci., Paris*, **256**, 1415.
Atanasiu, P., & Sisman, J. (1967) *Bull. Off. int. Épizoot.*, **67**, 521.
Bhatt, P. N., & Rodrigues, F. M. (1967) *Indian J. med. Res.*, **55**, 1295.
Baltimore, D., Huang, A. S., & Stampfer, M. (1970) *Proc. natn. Acad. Sci., U.S.A.*, **66**, 572.
Bell, J. F. (1964) *J. infect. Dis.*, **114**, 249.
Bergold, G. H., & Munz, K. (1967) *J. Ultrastruct. Res.*, **17**, 233.
Bergold, G. H., & Munz, K. (1970) *Arch. ges. Virusforsch.*, **31**, 152.
Borodina, P. A. (1959) *Probl. Virol.*, **4**, 96.
Bowen, E. T. W., Simpson, D. I. H., Bright, W. F., Zlotnik, I., & Howard, D. M. R. (1969) *Br. J. exp. Path.*, **50**, 400.
Bradish, C. J., Brooksby, J. B., & Dillon, F. J. (1956) *J. gen. Microbiol.*, **14**, 290.

Bradish, C. J., & Kirkham, J. B. (1966) *J. gen. Microbiol.*, **44**, 359.
Brody, J. A., Fischer, G. F., & Peralta, P. H. (1967) *Am. J. Epidemiol.*, **86**, 158.
Brown, F., Cartwright, B., & Smale, C. J. (1967) *J. Immunol.*, **99**, 171.
Brown, F., & Crick, J. (1957) *Nature, Lond.*, **179**, 316.
Brown, F., Martin, S. J., Cartwright, B., & Crick, J. (1967) *J. gen. Virol.*, **1**, 479.
Burnet, F. M., & Galloway, I. A. (1934) *Br. J. exp. Path.*, **15**, 105.
Campbell, J. B., Maes, R. F., Wiktor, T. J., & Koprowski, H. (1968) *Virology*, **34**, 701.
Carneiro, V., Black, C., & Koprovski, H. (1955) *J. Am. vet. med. Ass.*, **127**, 366.
Cartwright, B., & Pearce, C. A. (1968) *J. gen. Virol.*, **2**, 207.
Cartwright, B., Smale, C. J., & Brown, F. (1970b) *J. gen. Virol.*, **7**, 19.
Cartwright, B., Talbot, P., & Brown, F. (1970a) *J. gen. Virol.*, **7**, 267.
Constantine, D. G. (1962) *Pub. Hlth Rep.*, **77**, 287.
Constantine, D. G., Solomon, G. C., & Woodall, D. F. (1968) *Am. J. vet. Res.*, **29**, 181.
Correa-Giron, E. P., Allen, R., & Sulkin, S. E. (1970) *Am. J. Epidemiol.*, **91**, 203.
Crick, J., & Brown, F. (1969) *Nature, Lond.*, **222**, 92.
David-West, T. S., & Labzoffsky, N. A. (1968a) *Arch. ges. Virusforsch.*, **23**, 105.
David-West, T. S., & Labzoffsky, N. A. (1968b) *Arch. ges. Virusforsch.*, **24**, 30.
Davies, M. C., Englert, M. E., Sharpless, G. R., & Cabasso, J. V. (1963) *Virology*, **21**, 642.
Ditchfield, J., & Almeida, J. D. (1964) *Virology*, **13**, 367.
Doherty, R. L., Standfast, H. A., & Clark, I. A. (1968) *Aust. J. Sci.*, **31**, 365.
Donaldson, A. I. (1970) *Am. J. Epidemiol.*, **92**, 132.
Enright, J. B. (1951) *A. Rev. Microbiol.*, **10**, 369.
Falke, D., & Rowe, W. P. (1965) *Arch. ges. Virusforsch.*, **18**, 549.
Federer, K. E., Burrows, R., & Brooksby, J. B. (1967) *Res. vet. Sci.*, **8**, 103.
Fernandes, M. V., Wiktor, T. J., & Koprowski, H. (1964) *J. exp. Med.*, **120**, 1099.
Fox, J. P., Koprowski, H., Conwell, W. P., Black, J., & Gelfand, H. M. (1957) *Bull. Wld Hlth Org.*, **17**, 869.
Fields, B. N., & Hawkins, K. (1967) *New Engl. J. Med.*, **227**, 989.
Frederickson, L. E., & Thomas, L. (1965) *Publ. Hlth Reps., U.S.A.*, **80**, 495.
Grasset, N. (1967) *Bull. Off. int. Épizoot.*, **67**, 535.
Greenwood, M. (1946) *Bull. Hlth Org.*, **12**, 301.
Hackett, A. J., Zee, Y. C., Schaffer, F. L., & Talens, L. (1968) *J. Virol.*, **2**, 1154.
Halonen, P. E., Murphy, F. A., Fields, B. N., & Reese, D. R. (1968) *Proc. Soc. exp. Med. Biol., U.S.A.*, **127**, 1037.
Hanson, R. P. (1952) *Bact. Rev.*, **16**, 179.
Heuschele, W. P. (1970) *Arch. ges. Virusforsch.*, **30**, 195.
Hofmann, H., & Kunz, Ch. (1968) *Zentbl. Bakt. ParasitKde, I, Abt. Orig.*, **208**, 344.
Holmes, I. H., & Doherty, R. L. (1970) *J. Virol.*, **5**, 91.
Howatson, A. F. (1970) *Adv. Virus Res.*, **16**, 195.
Howatson, A. F., & Whitmore, G. F. (1962) *Virology*, **16**, 466.
Hronovsky, V., & Benda, R. (1969) *Acta virol., Prague*, **13**, 193 and 198.
Hummeler, K., Koprowski, H., & Wiktor, T. J. (1967) *J. Virol.*, **1**, 152.
Hummeler, K., Tomassini, N., Sokol, F., Kuwert, E., & Koprowski, H. (1968) *J. Virol.*, **2**, 1191.
Huppert, J., Rosenbergova, M., Gresland, L., & Harel, L. (1967) In *The molecular biology of viruses*. Eds.: Colter & Paranchych. p. 463. New York: Academic Press.
Inaba, Y., Tanaka, Y., Sato, K., Ito, H., Omori, T., & Matumoto, M. (1968) *Jap. J. Microbiol.*, **12**, 457.
Inaba, Y., Sato, K., Tanaka, Y., Ito, H., Omori, T., & Matumoto, M. (1969a) *Jap. J. Microbiol.*, **13**, 388.
Inaba, Y., Tanaka, Y., Sato, K., Ito, H., Omori, T., & Matumoto, M. (1969b) *Jap. J. Microbiol.*, **13**, 181.
Irvin, A. D. (1970) *Vet. Rec.*, **87**, 333.
Ito, Y., Tanaka, Y., Inaba, Y., & Omori, T. (1969) *Natn. Inst. Animal Hlth Q.*, **9**, 35.
Jensen, M. H. (1963) *Bull. Off. int. Épizoot.*, **9**, 131.

Jenson, A. B., Rabin, E. R., Wende, R., & Melnick, J. (1967) *Exp. molec. Path.*, **7**, 1.
Johnson, H. N. (1965a) *Calif. Hlth.*, **23**, 35.
Johnson, H. N. (1965b) in *Viral and rickettsial infections of man.* Eds.: Horsfall & Tamm. 4th Ed. Philadelphia: Lippincott.
Johnson, R. T. (1965c) *J. Neuropathol. exp. Neurol.*, **24**, 662.
Jonkers, A. H. (1967) *Am. J. Epidemiol.*, **86**, 286.
Jonkers, A. H., Shope, R. E., Aitken, T. H. G., & Spence, L. (1964) *Am. J. vet. Res.*, **25**, 236.
Kalter, S. S., Ratner, J. J., & Heberling, R. L. (1969) *Proc. Soc. exp. Biol. Med., U.S.A.*, **130**, 10.
Kang, C. Y., & Prevec, L. (1969) *J. Virol.*, **3**, 404.
Kang, C. Y., & Prevec, L. (1970) *J. Virol.*, **6**, 20.
Kaplan, M. M. (1969) *Nature, Lond.*, **221**, 421.
Kissling, R. E., Robinson, R. Q., Murphy, F. A., & Whitfield, S. G. (1968) *Science*, **160**, 888.
Kligler, I. J., & Bernkopf, H. (1938) *Proc. Soc. exp. Biol. Med., U.S.A.*, **39**, 212.
Koprowski, H., Black, J., & Nelsen, D. J. (1954) *J. Immunol.*, **72**, 94.
Koprowski, H., & Cox, H. R. (1948) *J. Immunol.*, **60**, 533.
Kowalczyk, T., & Brandly, C. A. (1954) *Am. J. vet. Res.*, **15**, 98.
Kunz, Ch., Hofmann, H., & Aspock, H. (1968) *Zentbl. Bakt ParasitKde, I, Abt. Orig.*, **208**, 347.
Kuwert, E., Wiktor, T. J., Sokol, E., & Koprowski, J. (1968) *J. Virol.*, **2**, 1381.
Lecatsas, G., Theodoridis, A., & Erasmus, B. J. (1969) *Arch. ges. Virusforsch.*, **28**, 390.
McMillan, B., & Boulger, L. R. (1960) *Ann. trop. Med. Parasit.*, **54**, 165.
McSharry, J. J., & Wagner, R. R. (1971a) *J. Virol.*, **7**, 59.
McSharry, J. J., & Wagner, R. R. (1971b) *J. Virol.*, **7**, 412.
Matsumoto, S. (1962) *Virology*, **17**, 198.
Matsumoto, S. (1970) *Adv. Virus Res.*, **16**, 257.
May, G., & Herzberg, K. (1969) *Zentbl. Bakt. ParasitKde, I, Abt. Orig.*, **211**, 133.
Mead, T. H. (1962) *J. gen. Microbiol.*, **27**, 397 and 415.
Metselaar, D., Williams, M. C., Simpson, D. I. H., West, R., & Mutere, F. A. (1969) *Arch. ges. Virusforsch.*, **26**, 183.
Miyamoto, K., & Matsumoto, S. (1967) *J. exp. Med.*, **125**, 447.
Murphy, F. A., Coleman, P. H., & Whitfield, S. G. (1966) *Virology*, **30**, 314.
Murphy, F. A., & Fields, B. N. (1967) *Virology*, **33**, 625.
Murphy, F. A., Halonen, P. E., Gary, G. W., Jun., & Reese, D. A. (1968) *J. gen. Virol.* **3**, 289.
Murphy, F. A., Shope, R. E., Metselaar, D., & Simpson, D. I. H. (1970) *Virology*, **40**, 288.
Mussgay, M., & Suarez, O. (1962) *Virology*, **17**, 202.
Nakai, T., & Howatson, A. F. (1968) *Virology*, **35**, 268.
Neurath, A. R., Wiktor, J. J., & Koprowski, H. (1966) *J. Bact.*, **92**, 102.
Newman, J. F. E., & Brown, F. (1969) *J. gen. Virol.*, **5**, 305.
Peck, F. B., Powell, H. M., & Culbertson, C. G. (1956) *J. Am. med. Ass.*, **162**, 1373.
Peries, J., Printz, P., Canivet, M., & Chuat, J. C. (1966) *C. r. hebd. Séanc. Acad. Sci., Paris*, **262D**, 2106.
Plus, N., & Atanasiu, P. (1966) *C. r. hebd. Séanc. Acad. Sci., Paris*, **263D**, 89.
Powell, H. M., & Culbertson, C. G. (1959) *Sth-west. Vet.*, **12**, 281.
Printz, P. (1970) *Annls Inst. Pasteur, Paris*, **119**, 520.
Rasmussen, C. J. (1965) *Ann. N.Y. Acad. Sci.*, **126**, 427.
Reczko, E. (1961) *Arch. ges. Virusforsch.*, **10**, 588.
Sabin, A. B. & Obitsky, P. K. (1937) *J. exp. Med.*, **66**, 15 and 35.
Schincariol, A. L., & Howatson, A. F. (1970) *Virology*, **42**, 732.
Sedwick, W. D., & Wiktor, T. J. (1967) *J. Virol.*, **1**, 1224.
Shelokov, A., & Peralta, P. H. (1967) *Am. J. Epidemiol.*, **86**, 149.
Shokeir, A. A. (1968) *Ann. Soc. belges Med. trop. Parasit. Mycol.*, **48**, 613.

Shope, R. E., Murphy, F. A., Harrison, A. K., Causey, O. R., Kemp, G. E., Simpson, D. I. H., & Moore, D. L. (1970) *J. Virol.*, **6**, 690.

Siegert, R., Shu, H.-L., Slenczka, W., Peters, D., & Muller, G. (1968) *Ger. med. Mon.*, **13**, 1.

Sikes, R. K., & Larghi, O. P. (1967) *J. Immunol.*, **99**, 545.

Simpson, D. I. H., Zlotnik, I., & Rutter, D. A. (1968) *Br. J. exp. Path.*, **49**, 458.

Simpson, R. W., & Hauser, R. E. (1966) *Virology*, **29**, 654.

Simpson, R. W., Hauser, R. E., & Dales, S. (1969) *Virology*, **37**, 285.

Skinner, H. H. (1957a) *J. comp. Path.*, **67**, 69.

Skinner, H. H. (1957b) *J. comp. Path.*, **67**, 87.

Skinner, H. H. (1959) *Arch. ges. Virusforsch.*, **9**, 92.

Skinner, H. H., & Bradish, C. J. (1954) *J. gen. Microbiol.*, **10**, 377.

Smith, P. C., Lawhaswasdi, K., Vick, W. E., & Stanton, J. S. (1967) *Nature, Lond.*, **216**, 384.

Soave, O. A. (1966) *Am. J. vet. Res.*, **27**, 44.

Sokol, F., Schlumberger, H. D., Wiktor, T. J., Koprowski, H., & Hummeler, K. (1969) *Virology*, **38**, 651.

Sokol, F., Stancek, D., & Koprowski, H. (1971) *J. Virol.*, **7**, 241.

Strozzi, P., & Ramos-Saco, T. (1953) *J. Am. vet. Med. Ass.*, **123**, 415.

Sudia, W. D., Fields, B. N., & Calisher, C. H. (1968) *Am. J. Epidemiol.*, **76**, 598.

Sulkin, S. E., Krutzsch, P. H., Allen, R., & Wallis, C. (1957) *J. exp. Med.*, **110**, 369.

Svet-Moldavskaya, I. A. (1958) *Acta virol., Prague*, **2**, 228.

Tanaka, Y., Inaba, Y., Sato, K., Ito, H., Omori, T., & Matumoto, M. (1969) *Jap. J. Microbiol.*, **13**, 169.

Tesh, R. B., Peralta, P. H., & Johnson, K. M. (1970) *Am. J. Epidemiol.*, **91**, 216.

Tierkel, E. S., Kissling, R. E., Edison, M., & Habel, K. (1953) *Proc. 90th ann. Meeting Am. vet. Med. Ass.*, p. 443.

Turner, G. S., & Kaplan, C. (1967) *J. gen. Virol.*, **1**, 537.

van den Ende, M., Polson, A., & Turner, G. S. (1957) *J. Hyg., Camb.*, **55**, 361.

van der Westhuizen, B. (1967) *Onderstepoort J. vet. Res.*, **34**, 29.

Villemot, J. M., & Provost, A. (1959) *Annls Inst. Pasteur, Paris*, **96**, 712.

Wagner, R. R., Levy, A. H., Snyder, R. M., Ratcliff, G. A., & Hyatt, D. F. (1963) *J. Immunol.*, **91**, 112.

Wagner, R. R., Snyder, R. M., & Yamazaki, S. (1970) *J. Virol.*, **5**, 548.

Warrington, R. E. (1965) *Arch. ges. Virusforsch.*, **17**, 594.

Webster, L. T. (1939) *Am. J. Hyg.*, **30**, 113.

Whitney, E. (1964) *Am. J. trop. Med. Hyg.*, **13**, 123.

Wiktor, T. J., Fernandes, M. V., & Koprowski, H. (1964) *J. Immunol.*, **93**, 353.

Yang, Y. J., Stoltz, D. B., & Prevec, L. (1969) *J. gen. Virol.*, **5**, 473.

Yasutake, W. T., & Rasmussen, C. (1968) *Bull. Off. int. Épizoot.*, **69**, 977.

Yasutake, W. T., Parisot, T. J., & Klontz, G. W. (1965) *Ann. N.Y. Acad. Sci.*, **126**, 520.

Yoshino, K., Kuma, N., Kondo, A., & Kitaoka, M. (1956) *Jap. J. med. Sci. Biol.*, **9**, 259.

Yoshino, K., Taniguchi, S., & Arai, K. (1966) *Arch. ges. Virusforsch.*, **18**, 370.

Zlotnik, I., Simpson, D. I. H., & Howard, D. M. R. (1968) *Lancet*, **2**, 26.

Zwillenberg, L. O., Jensen, M. H., & Zwillenberg, H. L. (1965) *Arch. ges. Virusforsch.*, **17**, 1.

Zwillenberg, L. O., Pfitzner, I., & Zwillenberg, H. I. (1968) *Zentbl. Bakt. ParasitKde, I, Abt. Orig.*, **208**, 218.

9

Orthomyxovirus

'Orthomyxovirus' has been adopted as a generic name to include influenza virus types A, B and possibly C. The following characters define the genus: RNA viruses consisting of a helical nucleocapsid 9 nm in diameter contained in a pleomorphic but generally spherical envelope 80 to 120 nm in diameter, incorporating virus-coded haemagglutinating and enzymic (neuraminidase) components and host-specified carbohydrates and lipids. The viral RNA is single-stranded and occurs in multiple (probably 6) subgenomic segments with a total MW of 2×10^6 to 4×10^6 daltons. Viral replication is dependent on host DNA synthesis. The viral ribonucleoprotein is assembled intranuclearly and migrates to the periphery of the cell where virion maturation and liberation take place by budding through the modified cell membrane. Types A, B and C are distinguished by the antigenic specificity of the ribonucleoprotein. Antigenic variation within each type is accounted for by changes in the envelope proteins and occurs frequently. Genetic interactions between members within types can be readily observed.

Reviews: Hoyle (1968). Pereira (1969). Webster & Laver (1971). McQueen et al. (1968) (Animal influenza).

Morphology. Types A and B orthomyxoviruses are morphologically indistinguishable, but differ in some respects from type C (see p. 217). Virions are pleomorphic, appearing either as approximately spherical structures 80 to 120 nm in diameter or as filaments of about this diameter and up to several microns in length (Dawson & Elford, 1949). Using special methods of virus preparation for electron microscopy, Nermut & Frank (1971) describe uniform virions, probably with icosahedral envelope. In negatively stained preparations, the viral envelope is seen to be covered with uniform projections 8 to 10 nm long, spaced about 7 to 8 nm from each other (Horne et al., 1960; Hoyle et al., 1961), rarely occurring in hexagonal arrays in infectious virions or in square patterns in incomplete (von Magnus) particles (Almeida & Waterson, 1970). Purified influenza-A haemagglutinin appear as rods approximately 14 nm long and 4 nm wide, resembling the surface projections of the viral envelope, whereas the neuraminidase is an oblong structure 8·5 by 5 nm attached to a fibre 10 nm long with a terminal knob 4 nm across (Laver & Valentine, 1969). The nucleocapsid is rarely seen in spontaneously disrupted virions as a coil 40 to 60 nm wide, composed of a thread 6 nm across. The number of turns in each coil varies between 3 and 41 with a mean length of 1 to 2 μ for the extended ribonucleoprotein (Almeida & Waterson, 1970), but it has been suggested that there is a unit length of 5 to 6 turns (Apostolov et al., 1970). Nucleocapsids obtained from virions disrupted by ether or from extracts of infected cells are 9 to 10 nm wide and of variable length, with a suggestion of helical arrangement of spherical subunits 3 to 5 nm in diameter, there being 5 to 6 subunits per turn (Hoyle et al., 1961). It has been suggested that the nucleocapsid strand may be a double helix (Hoyle et al., 1961; Sigel et al., 1968). Isolated segments of nucleocapsids appearing as tubes or U-shaped structures (Hjerten et al., 1970) or as circles coiled on themselves (Pons et al., 1969) have been described. In thin sections, the viral envelope consists of an outer layer with hollow cylindrical projections, an electron-transparent layer and an inner micellar layer 4 nm thick. Compans et al. (1970) also describe an inner membrane distinct from the unit membrane which is continuous with that of the cell during the budding process which leads to viral liberation. Hollow thread-like structures about 6 nm in diameter are seen inside the membrane (Apostolov et al., 1970). Studies on the site of formation of viral components by immunofluorescence have revealed that the ribonucleoprotein is assembled in the nucleus and the envelope components in the cytoplasm (Breitenfeld & Schäfer, 1957; Franklin, 1958). Tubular inclusions 30 to 50 nm across with a crystalloid fine structure showing a periodicity of 6·5 to 7 nm have been described in the nucleus cells infected with type A or B strains (Kopp et al., 1968; Saito et al., 1970; Archetti et al., 1970). Cytoplasmic inclusions with paracrystalline structure and containing RNA have also been described (Ter Meulen & Love, 1967; Kopp et al., 1968). These inclusions occur at late stages of infection and their significance is not established. Nucleolar changes occurring at early stages of infection have been described (Compans et al., 1970).

Chemical composition. *Nucleic acid.* Virions contain approximately 1 per cent single-stranded RNA separable by sucrose density centrifugation into several pieces with sedimentation constants ranging from 7S to 42S (Agrawal & Bruening, 1966; Duesberg & Robinson, 1967; Nyak & Baluda, 1967; Barry & Davies, 1968). Polyacrylamide gel electrophoresis resolves up to 6 RNA pieces with molecular weights ranging from $1{\cdot}15 \times 10^5$ to 7×10^5 daltons (Duesberg, 1968; Pons & Hirst, 1968; Skehel, 1971a). This leads to total MW estimations as high as 4×10^6 daltons which is about twice the values obtained earlier by other methods. Separate fragments have base compositions similar to that of the whole viral RNA, but hybridization experiments by Content & Duesberg (quoted by Blair & Duesberg, 1970) suggest that separate fragments differ from each other in base sequence.

An investigation of base sequence homologies by hybridization of viral RNA with complementary RNA synthesized *in vitro* reveals varying degrees of cross-hybridization between different strains of influenza A, but not between influenza A and B (Scholtissek, 1970).

Proteins make up about 70 per cent of the dry weight of the virion. Polyacrylamide gel electrophoresis reveals up to 7 polypeptides in purified virions (Joss et al., 1969; Haslam et al., 1970a, b; Skehel & Schild, 1971). It has been possible to correlate some of the polypeptides with structural and functional virus components. An RNA-dependent RNA polymerase has been detected in influenza A and B virions (Chow & Simpson, 1971; Skehel, 1971b). Three virus-coded polypeptides distinct from those incorporated into the virion, have been described by Dimmock & Watson (1969).

Carbohydrates make up 5 per cent of the virion. Three of the structural proteins are covalently linked to sugars. The carbohydrate part of these glycoproteins probably correspond to the host-derived antigen described in influenza A and B viruses by Knight (1947), Haukenes et al. (1966) and Laver & Webster (1966).

Lipids. Estimates of the lipid content of influenza viruses vary from 18 per cent to 36 per cent (Hoyle, 1968). Chemical analysis of lipid constituents of influenza A and B viruses have been reported by Frommagen et al. (1959), Kates et al. (1961) and Blough (1971). Results by these workers and by Hoyle & Frisch-Niggemeyer (1955) and Wecker (1957) indicate that virus lipids are pre-formed constituents of the host cell.

Physico-chemical characters. Density of virion 1·19 to 1·25 g/ml in different media (Barry, 1960). Usually inactivated in 30 minutes at 56°C; 90 minutes' heating necessary for some strains. Sensitive to lipid solvents. Virus retains activity when carried down on the precipitate on treatment with protamine, alum, or calcium phosphate, or with 25 to 35 per cent methanol at −5°. Maximum stability between pH 7 and 8. Formaldehyde destroys activity (1:5000 is adequate against fairly pure preparations), so do soaps, detergents and oxidizing agents such as iodine. Inactivation by various chemicals is described by Dunham & Macneal (1943). The virus will survive for years at −70°C or after lyophilization. Infected tissues retain activity for months in 50 per cent glycerol-saline at about 0°C. Activity was demonstrated in dried dust for 14 days (Edward, 1941). Infectivity of human strains in aerosols decays more rapidly than that of avian strains (Mitchell et al., 1968). The rate of inactivation of different viral functions by UV has been determined by Henle & Henle (1947). The sequential inactivation by an ethylene-iminoquinone of the capacity to induce synthesis of different virus components is described by Scholtissek & Rott (1964).

Haemagglutination. All influenza viruses agglutinate fowl and some mammalian red cells although differences in capacity to agglutinate cells from particular animals are observed among some members of the subgroup. Agglutination results from adsorption of virus to the red cells and may be followed under appropriate conditions by elution due to the action of viral neuraminidase on red cell receptors. The rates of adsorption and elution and the sensitivity of the haemagglutinin to nonspecific inhibitors are variable with different influenza viruses (see review by Buzzell & Hanig, 1958).

Antigenic properties. Virions contain 3 virus-specific antigens: the ribonucleoprotein (RNP), the haemagglutinin (HA) and the neuraminidase (NA). The RNP is type-specific and is usually tested by complement fixation or by immunodiffusion; it is found either as part of the viral nucleocapsid or as soluble antigen. The HA and the NA are components of the viral envelope with subtype- or strain-specificity. The HA is the antigen involved in virus neutralization and haemagglutination inhibition and the NA is tested by enzyme-inhibition. Both envelope antigens may also be tested by immunodiffusion. Nonstructural antigens revealed by immunodiffusion and by immunofluorescence have been described (Dimmock, 1969). Virions also contain a host-specific antigen consisting of a polysaccharide, possibly in combination with a minor peptide. Antigenic variation is frequent, and strains with different combinations of antigenic components occur naturally or may be readily obtained by recombination (see review by Pereira, 1969, and Webster & Laver, 1971).

Cultivation. All viruses included in this subgroup can be cultivated in embryonated eggs and in cultures of certain mammalian tissues. The extensive literature on the growth cycle of influenza viruses has been comprehensively reviewed by Hoyle (1968). Virus adsorption to RDE-sensitive cell receptors leads to an irreversible virus-cell combination, followed by 'penetration' and by 'uncoating' either by fusion of virions to the host-cell wall followed by disintegration of the viral envelope and release of the nucleocapsid (Hoyle et al., 1962; Morgan & Rose, 1968) or by a process of engulfment that is referred to as viropexis (Fazekas de St Groth, 1948) and is followed by uncoating within cytoplasmic vesicles (Dales & Choppin, 1962). Virus replication can be divided into an actinomycin D-sensitive period lasting 3 to 4 hours after infection followed by an actinomycin-resistant phase (Barry, 1964; Granoff & Kingsbury, 1964). Studies on the intracellular synthesis of viral RNA have been reviewed by Kingsbury (1970) and by Blair & Duesberg (1970). Virus polypeptides are synthesized in the cytoplasm, but internal proteins migrate to the nucleus whereas envelope proteins migrate to cell membranes (White et al., 1970). Assembly of viral components into mature virions takes place at the cell membrane where host proteins are replaced by virus-specific proteins (Duc-Nguyen et al., 1966; Holland & Kiehn, 1970). Maturation is rapidly followed by virus release by budding from the cell membrane (Hoyle, 1954; Morgan et al., 1956). Viral neuraminidase appears to play a role in release (Brown & Laver, 1968; Seto & Chang, 1969).

Pathogenicity. Cause epidemic (influenza A and B) or sporadic (influenza C) respiratory infections in man. Influenza A viruses are also associated with respiratory and other infections of swine, horses and birds. Can be adapted to produce transmissible pneumonia in mice and hamsters. Viruses not so adapated will often produce in mice a non-transmissible pneumonia when inoculated intranasally in very large doses. Similarly, encephalitis may at times be produced in rodents, after adaptation; this also may or may not be transmissible in series.

ORTHOMYXOVIRUS TYPE A

These are divided into different subtypes of human, porcine, equine and avian origins. Since it is possible that an influenza virus affecting one species may pass over to and become epidemic amongst members of another species, some of these distinctions may come to be considered artificial. Members of the influenza A group share a common, ribonucleoprotein antigen. Genetic interaction has been demonstrated between subtypes originating from hosts of different species (Tumová & Pereira, 1965; Kilbourne, 1968a).

Human Orthomyxovirus Type A

Synonym: *Myxovirus influenzae-A hominis.*

Reviews: Francis (1959). Robinson (1964). Buzzell & Hanig (1958). (Haemagglutination).

Haemagglutination. Recently isolated (O phase) agglutinate human and guinea-pig better than fowl cells; after adaptation to growth in allantoic cavity (D phase) fowl cells are equally affected (Burnet & Bull, 1943). A2 (Asian) strains agglutinate cells of a wider range of species than do earlier strains (Wang & Liu, 1958) and do not show O–D variation. Most strains elute readily at 37°C, having destroyed receptors on the red cell surface or rendered them incapable of reacting with more of the same sort of virus. Mucoproteins occurring in egg white, in many mammalian sera and elsewhere, react with the virus, inhibiting haemagglutination and being themselves destroyed by an enzyme-like action. The most important of the inhibitors in sera are the α or Francis inhibitor (Francis, 1947), especially active against heated viruses and the β or Chu inhibitor which is effective only against viruses which have not been adapted to mice (Chu, 1951). A third type of inhibitor (γ inhibitor) is active against A2 (Takátsy & Barb, 1959; Cohen & Belyavin, 1969).

Antigenic properties. There are 3 subtypes designated A0, A1 and A2 (WHO, 1969) characterized on the basis of haemagglutination-inhibition and virus-neutralization tests. There are, however, minor cross-reactions detectable by the same tests between certain strains belonging to different subtypes and also between human and animal strains (see review by Pereira, 1969). Immunodiffusion tests reveal partial identity of haemagglutinins present in human subtypes A0 and A1 (Schild, 1970). Paniker (1968) describes 2 groups of human influenza A viruses with antigenically related neuraminidases: the first represented by subtypes A0 and A1 and the second by subtype A2. Antigenic relationships between neuraminidases of human and avian type A strains have been described (Webster & Pereira, 1968; Schild et al., 1969). Both the haemagglutinin and the neuraminidase show considerable antigenic variation which may occur either as a gradual antigenic drift or as sudden major shifts resulting in the emergence of new subtypes. New variants or subtypes replace pre-existing ones: the A0 subtype has been prevalent in human populations from 1933 or earlier to 1946; the A1 subtype from 1947 to 1956; and the A2 subtype from 1957 to the present, although strains appearing in Hong Kong in 1968 (Coleman et al., 1968) show a major antigenic change considered by some to justify their designation as A3 strains. Human influenza A viruses vary also in affinity for specific antibodies: Q-strains are nonavid, reacting poorly with homologous antisera, in contrast to avid or P-phase strains (van der Veen & Mulder, 1950).

Cultivation. Grow in the amniotic and allantoic cavities of fertile hens' eggs. Some strains recently isolated from human sources grow best amniotically (O phase). After adaptation (D phase) growth is equally good in the allantoic cavity (Burnet & Bull, 1943). This phase variation has not been observed with A2 strains. The virus also grows in cultures of embryonic chick and various mammalian tissues, but multiplication and cytopathogenicity are very different with different strains. Neurotropic variants are apt to show better CPE. Evidence of virus multiplication is most readily obtained by haemadsorption. Plaque formation has been described (Ledinko, 1955; Lehmann-Grube, 1963; Sugiura & Kilbourne, 1965). Monkey-kidney and human embryonic tissues can be used for primary isolations. Most strains grow best at about 35°C, but adaptation to growth at lower temperatures and a correlation of this property with attenuation and reduction of plaque-forming capacity have been described (Maasaab, 1967; Medvedeva et al., 1968). Recombination in culture between strains having various 'marker' characters was first demonstrated by Burnet & Lind (1951) and intensively studied by other workers (see review by Kilbourne, 1963). Isolation of temperature-sensitive mutants and their use in attempts to map the viral genome are reported (Simpson & Hirst, 1968; Mackenzie, 1970).

Pathogenicity. Cause influenza and sometimes pneumonia in man; encephalitis and other complications are rare. The suggestion that human strains may infect pigs under natural conditions has recently been confirmed (Kundin, 1970). An epizootic of A2/Hong Kong/68 influenza in gibbons has been reported by Johnson et al. (1971). Evidence of natural infection of several other animals rests mainly on serological evidence and requires substantiation.

Experimentally, ferrets are readily infected (Smith et al., 1933), symptoms are mainly caused by inflammation of the nasal passages; after adaptation, pneumonia also may be caused. The virus can be adapted to cause pneumonia in mice, hamsters and other mammals particularly rodents. In monkeys, horses, dogs, sheep, guinea-pigs and rats, infection is usually inapparent.

Pathological lesions. Viraemia is sometimes demonstrable in man and mice. The characteristic lesion in lungs of infected man, ferret and mouse is necrosis of epithelium of finer bronchi (Straub, 1937) and in cells of alveolar walls (Hers et al., 1962). Fatal pneumonia in man is usually caused by secondary bacterial infections; in recent years staphylococci have been the commonest. Virus pneumonia alone can cause some deaths. Winternitz et al. (1920) described the pathology of fatal cases of the 1918–19 pandemic. Strains neurotropic for mice have been obtained in the laboratory (Stuart-Harris, 1939). Transplacental passage of virus in mice with induction of fetal abnormalities has been described (Takeyama, 1966) Some strains are lethal for chick embryos, producing necrosis in developing lungs after amniotic inoculation. Apart from the effects of virus multiplication, virus particles have a toxic action demonstrable in the mouse lung (Sugg, 1949), in rabbits' eyes (Evans & Rickard, 1945) and by a pyrogenic effect in rabbits (Wagner et al., 1949).

Ecology. Transmission is from infected respiratory secretions. Factors influencing airborne transmission within a mouse colony have been described (Kilbourne, 1968b; Schulman, 1970). The frequently voiced suggestion that human influenza may have animal reservoirs remains unproved.

Control. Formolized vaccines have been proved to give some protection over at least some months; incidence in some trials has been about one-third of that in unvaccinated groups. Disrupted virus and subunit vaccines have recently been developed (Webster & Laver, 1966; Hennessy & Davenport, 1966). Efficacy of attenuated virus given intranasally has been extensively tested especially in the U.S.S.R. (Smorodinstev et al., 1961; Slepushkin, 1967). The prophylactic use of 1-adamantanamine hydrochloride against human influenza A2 has been reported (Galbraith et al., 1969; Nafta et al., 1970; Oker-Blom et al., 1970).

Porcine Orthomyxovirus Type A

Synonym: *Myxovirus influenzae-A suis.*

Antigenic properties. All porcine strains have antigenically related envelope antigens although isolates separated in time and space can be distinguished by haemagglutination-inhibition (Pereira, 1969) and by neuraminidase-inhibition (Meier-Ewert et al., 1970). The 2 antigens vary independently of each other, and both show evidence of antigenic drift. Antibodies reacting with swine virus hemagglutinin are present in many human sera, particularly from older people. The significance of the common antigen has been the subject of many studies. It is widely believed that the pandemic of 1918–19 in man was caused by a virus related to swine influenza and that this virus may have first infected pigs at that time (Laidlaw, 1935; Shope, 1935).

Distribution. North America, especially the Middle West. Formerly Europe, including the British Isles. The disease was, however, doubtfully present outside North America for many years: some reported isolations are attributable to laboratory 'pick-ups'. More recent isolations from Europe are probably genuine.

Pathogenicity. Causes influenza and pneumonia in domestic pigs, particularly when associated with *Haemophilus influenza suis* (Shope, 1931). Virus may persist for up to 3 months in experimentally infected pigs, as indicated by transmission of infection to contacts (Blăskovič et al., 1970). Causes pneumonia in ferrets and mice without the necessity for adaptation, also in lambs (Barb et al., 1962).

Ecology. Transmission is from infected respiratory secretions; also (Shope, 1941), by ingestion of earthworms containing virus-infected swine lung works: this causes no disease until the virus is activated by one or other of various stimuli. This finding was confirmed by Sen et al. (1961) and by Kammer & Hanson (1962) and supported by experimental evidence in rats (Shotts et al., 1968). Peterson et al. (1961), however, found no evidence of virus survival in earthworms.

Control. Formolized vaccines have been shown to give rise to some immunity, but are of doubtful practical use.

Equine Orthomyxovirus Type A

Synonym: *Myxovirus influenzae-A equi.*

Morphology, chemical composition and replication. Similar to other members of the genus in all features investigated (Sigel et al., 1968; Nayak & Baluda, 1968a, b).

Haemagglutination. Agglutinates horse, pig, calf, rhesus, fowl, guinea-pig and human red blood cells (Tumová & Piserova-Sovinová, 1959).

Antigenic characters. Two serotypes (A/Eq. 1 and A/Eq. 2) distinguishable from each other and from other influenza A viruses by haemagglutination-inhibition and virus neutralization. Slight antigenic variation observed within subtype A2 (Pereira et al., in press). There are minor cross-reactions referable to envelope antigens, between

A/Eq. 1 and avian strains (Tumová & Pereira, 1968) and between A/Eq. 2 and human A2/Hong Kong/68 (Coleman et al., 1968; Kasel et al., 1969). Antibodies to both equi subtypes are found in human sera, mainly in old-age groups (Masurel & Mulder, 1966; Davenport et al., 1967).

Cultivation succeeds in fertile eggs and in tissue cultures of bovine kidney, chick-embryo kidney fibroblasts, rhesus kidney and human embryo kidney (Andrewes & Worthington, 1959).

Distribution. (See review by McQueen et al., 1968). Subtype A/Equi/1 was first isolated in Czechoslovakia (Sovinová et al., 1958) and subsequently in other Central European countries. It was later recorded in North America and in several Western European countries.

Subtype A/Eq./2 was first detected in 1963 in the United States (Waddell et al., 1963) and later in several South American and European countries, sometimes causing extensive outbreaks. There are records of its occurrence in Asia.

Pathogenicity. Causes respiratory illness in horses (Tumová & Sovinová, 1959; Somma et al., 1964). Infection of foals unprotected by maternal antibody may be fatal (Miller, 1966). Horses are readily infected under experimental conditions by either subtype (Blăškovič et al., 1966, 1969; Cameron et al., 1967). Has been adapted to produce pneumonia in mice inoculated intranasally and encephalitis after intracerebral injection of suckling mice. Causes inapparent infection of ferrets (Andrewes & Worthington, 1959). Human subjects have been infected experimentally with A/Eq./2 (Kasel et al., 1965).

Control. Good antibody responses against both equine subtypes have been obtained following vaccination with inactivated virus (Brander et al., 1965; Lief & Cohen, 1965).

Avian Orthomyxovirus Type A

Synonyms: *Myxovirus pestis-galli*, classical fowl plague, virus N. *Myxovirus influenzae-A anatis*, Duck influenza virus, Tern virus.

Haemagglutination. Most strains agglutinate fowl, rhesus, guinea-pig, horse, ox, pig and monkey red cells. Duck influenza virus also agglutinates frog erythrocytes (Blăškovič et al., 1959).

Antigenic properties. All avian strains have envelope antigens directly or indirectly related to each other although the majority of isolates can be distributed into 6 antigenic groupings based on haemagglutination-inhibition tests performed with post-infection sera (Pereira et al., 1966, 1967). Further antigenic groupings will probably be required to accommodate new isolates. Minor cross-reactions between envelope antigens of avian and mammalian strains have been described (Tumová & Pereira, 1968), and some avian strains have been found to contain neuraminidase antigens closely related to those of human subtypes A0 and A2 (Webster & Pereira, 1968; Schild et al., 1969). Antigenic relationships between haemagglutinin and neuraminidase of avian strains are described by Madeley et al. (1971).

Cultivation. All strains grow readily in developing hens' eggs and in cultures of fowl and monkey-kidney tissues (Pereira et al., 1965). Some strains have also been grown in cultures of rabbit, ox and human cells. Growth in tissue cultures leads to haemadsorption and usually CPE. The majority of strains produce plaques in chick-embryo fibroblast cultures (Pereira et al., 1965). Attenuation of some strains for fowls is reported to follow cultivation in chick, pigeon or human cells (Hallaeuer & Kronauer, 1959).

Distribution. Fowl plague was formerly world-wide, but has been rarely recorded in recent years. Duck and turkey influenza is frequently observed in North America, Europe and Asia. Strains related to the tern/South Africa/61 antigenic group have been observed in Scotland and in Canada. Infection of quail, turkey, pheasant and partridge and guinea-fowl is frequently observed in Italy. Serological evidence of infection of wild birds has been reported in North America (Easterday et al., 1968) and in Australia (Dasen & Laver, 1970).

Pathogenicity. Symptoms vary from mild respiratory involvement to rapidly fatal diseases with dyspnoea, oedema of head and neck, cyanosis and diarrhoea, often with evidence of CNS involvement. Duck influenza is often associated with sinusitis (Koppel et al., 1956; Rinaldi et al., 1966). Pathological lesions vary according to virus strain and host (Jungherr et al., 1946; Uys & Becker, 1967; Rouse et al., 1968). In severe infections there is multiple focal necrosis in most organs, but some strains are particularly associated with pancreatic necrosis.

Birds of many species are susceptible, including gallinaceous birds, passerines, psittacines, pigeons, terns, ducks and geese. Epizootics among blackbirds (Maggiora & Valenti, 1903) and sparrows (Gerlach & Michalka, 1926) may have been caused by related viruses. A case of human infection by fowl plague virus has been described (DeLay et al., 1968). *Experimentally*, infection has been transmitted to mice, ferrets and other mammals, mainly by intracerebral inoculation (Findlay & Mackenzie, 1937).

Ecology. Method of transmission obscure, but probably respiratory. Factors affecting transmission under experimental conditions have been investigated by Homme et al. (1970), who suggest that stress by chilling may exacerbate persistent asymptomatic infections.

Control. Virus inactivated by formaldehyde and by phenol has been successful as a vaccine in the hands of some workers. Virus can be attenuated in several ways and used for immunization (Moses et al., 1948).

ORTHOMYXOVIRUS TYPE B

Synonym: *Myxovirus influenzae-B.*

Haemagglutination. Virus is adsorbed to and agglutinates various mammalian and avian red blood cells. No O and D phases demonstrable as with influenza A, but behaviour as to adsorption and elution very similar. Virus heated to 56°C has its haemagglutinin very easily inhibited by mucoproteins in many normal mammalian sera (Francis, 1947).

Antigenic properties. A soluble nucleoprotein antigen is shared by all strains. By the haemagglutination-inhibition test there is some antigenic diversity amongst strains, but the change from year to year is more gradual than with influenza A (Lee & Tauraso, 1968; Pereira, 1969). The neuraminidase also shows antigenic variation (Paniker, 1968). Although some strains differ considerably from others, there is a continuous antigenic spectrum and no subtypes comparable to those of type A can be distinguished.

Cultivation. Growth occurs in amniotic and allantoic cavities of hens' eggs, 13-day eggs being best for primary isolation. Cultures of monkey kidney may also be used for primary isolation, CPE being produced (Mogabgab et al., 1955) with development of basophilic cytoplasmic inclusions and sometimes eosinophilic nuclear inclusions (Berkaloff et al., 1962; Harwin et al., 1966). Laboratory-adapted strains are reported as growing, with CPE, in cultures of human embryo, chick embryo, ferret, mouse, calf and pig kidney (Haas & Wulff, 1957), but not in a number of other human cell cultures (Green et al., 1957). Growth in tissue cultures is demonstrable by haemadsorption. Plaque assays have been described (e.g. Ledinko, 1955; Lehmann-Grube, 1963).

Pathogenicity. Causes influenza and sometimes pneumonia in man. Epidemics are less widespread than those of influenza A, and there are no pandemics. Multiplies in respiratory tract of ferrets, but only irregularly produces fever or symptoms. Some strains have been adapted to produce pneumonia in mice and other rodents. Horses (Kasel et al., 1968) and pigs (Takátsy et al., 1969) can be infected experimentally.

Control. Formolized and live attenuated vaccines have been used as for influenza A.

INFLUENZA C VIRUS

Synonym: *Myxovirus influenzae-C.*

Review: Taylor (1951).

Information on the basic characters of these viruses is too scanty to allow their classification. They are here considered as a probable orthomyxovirus mainly because it has become common practice to refer to them as a type of influenza virus.

Morphology. The fine structure seen by negative contrast resembles that of orthomyxovirus types A and B (Waterson et al., 1963), but it has been repeatedly observed that the surface projections occur in regular arrays forming hexagonal and pentagonal patterns (Archetti et al., 1967; Apostolov & Flewett, 1969). The nucleocapsid has not been observed by negative contrast, but appears in sections as tubular structures 9 nm in diameter as opposed to those of types A and B which are 6 nm across when examined by the same technique (Apostolov & Flewett, 1969).

Chemical composition. No information available.

Physico-chemical properties. The virion has a buoyant density in sucrose of 1·170 to 1·188 as opposed to 1·200 to 1·210 for types A and B (Apostolov et al., 1970).

Haemagglutination. Agglutinates fowl and some mammalian red cells, but tests are better carried out at 4°C, since elution is rapid at ±20°C, although it is often incomplete. Agglutination of guinea-pig red blood cells varies between strains. Unlike other group members, its haemagglutination is feebly inhibited by egg white and other mucoprotein inhibitors, but normal rat serum inhibits. When tests are made with human O cells it appears that the same receptors are involved as with other myxoviruses (White, 1953), but if fowl cells are used, those treated with the receptor-destroying enzyme of *V. cholerae* are still agglutinable (Hirst, 1950); possibly an additional receptor is concerned. The neuraminidase activity is lower than that of types A and B (Apostolov et al., 1970).

Antigenic properties. Distinct from influenza A and B. Slight antigenic variation has been observed (Pereira, 1969).

Cultivation. In amniotic cavity of hens' eggs; growth in allantoic cavity doubtful. Some strains multiply in monkey-kidney tissue cultures (Mogabgab, 1962).

Pathogenicity. A probable cause of influenza-like infections in man (Francis et al., 1950). Though antibody rises to the virus have been detected in people with mild respiratory infections, it is only in a few instances that influenza-C virus has been definitely implicated as a cause of disease. It probably causes inapparent more often than overt infection, since antibodies are present in most adult sera, at least in Britain and the United States. Experimental disease resembling common colds has been obtained in volunteers (Joosting et al., 1968). Ready antibody formation suggests that it infects ferrets and hamsters, but it is not known to cause disease in them or in mice. Attempted serial passage in mice unsuccessful; 2 or 3 passages achieved in adult hamsters (Morozenko, 1957).

REFERENCES

Agrawal, H. O., & Bruening, G. (1966) *Proc. natn. Acad. Sci., U.S.A.*, **55**, 818.

Almeida, J. D., & Waterson, A. P. (1970). In *The biology of large RNA viruses.* Eds.: Barry & Mahy. London/New York: Academic Press.

Andrewes, C. H., & Worthington, G. (1959) *Bull. Wld Hlth Org.*, **20**, 435.

Apostolov, K., & Flewett, T. H. (1969) *J. gen. Virol.*, **4**, 365.

Apostolov, K., Flewett, T. H., & Kendal, A. P. (1970). In *The biology of large RNA viruses.* Eds.: Barry & Mahy. London/New York: Academic Press.

Archetti, I., Bereczky, E., Rosati-Valente, F., & Steve-Bocciarelli, D. (1970) *Arch. ges. Virusforsch.*, **29**, 275.

Archetti, I., Jemolo, A., Steve-Bocciarelli, D., Arangio-Ruiz, G., & Tangucci, F. (1967) *Arch. ges. Virusforsch.*, **20**, 133.

Barb, K., Farkas, E., Romváry, J., & Takátsy, G. (1962) *Acta virol., Prague*, **6**, 207.

Barry, R. D. (1960) *Aust. J. exp. Biol. med. Sci.*, 38, 499.

Barry, R. D. (1964) *Virology*, **24**, 563.

Barry, R. D., & Davies, P. (1968) *J. gen. Virol.*, **2**, 59.

Berkaloff, A., Stussi, T., & Colobert, L. (1962) *J. Microscopie*, **1**, 351.

Blair, C. D., & Duesberg, P. H. (1970) *Ann. Rev. Microbiol.*, **24**, 539.

Blåskovič, D., Jamrichova, O., Rathova, V., Kociskova, D., & Kaplan, M. M. (1970) *Bull. Wld Hlth Org.*, **42**, 767.
Blåskovič, D., Kapitancik, B., Sabo, A., Styk, B., Vrtiak, O., & Kaplan, M. (1969) *Acta virol., Prague*, **13**, 499.
Blåskovič, D., Rathova, L., & Borscky, L. (1959) *Acta virol., Prague*, **3**, 17.
Blåskovič, D., Szanto, J., Kapitancik, B., Lesso, J., Lackovic, V., & Skarda, R. (1966) *Acta virol., Prague*, **10**, 513.
Blough, H. A. (1971) *J. gen. Virol.*, **12**, 317.
Brander, G. C., Street, B. K., & Mann, G. (1965) *Vet. Rec.*, **77**, 548.
Breitenfeld, P. M., & Schäfer, W. (1957) *Virology*, **4**, 328.
Brown, J., & Laver, W. G. (1968) *J. gen. Virol.*, **2**, 291.
Burnet, F. M., & Bull, D. R. (1943) *Aust. J. exp. Biol. med. Sci.*, **21**, 55.
Burnet, F. M., & Lind, P. E. (1951) *J. gen. Microbiol.*, **5**, 59 and 67.
Buzzell, A., & Hanig, M. (1958) *Adv. Virus Res.*, **5**, 289.
Cameron, T. P., Alford, R. H., Kasel, J. A., Harvey, E. W., Byrne, R. J., & Knight, V. (1967) *Proc. Soc. exp. Biol. Med., U.S.A.*, **124**, 510.
Chow, N.-L., & Simpson, R. W. (1971) *Proc. natn. Acad. Sci., U.S.A.*, **68**, 752.
Chu, C. M. (1951) *J. gen. Microbiol.*, **5**, 739.
Cohen, A., & Belyavin, G. (1959) *Virology*, **7**, 59.
Coleman, M. T., Dowdle, W. R., Pereira, H. G., Schild, G. C., & Chang, W. K. (1968) *Lancet*, **2**, 1384.
Compans, R. W., Dimmock, N. J., & Meier-Ewert, H. (1970). In *The biology of large RNA viruses*. Eds.: Barry & Mahy. London/New York: Academic Press.
Dales, S., & Choppin, P. W. (1962) *Virology*, **18**, 489.
Dasen, C. A., & Laver, W. G. (1970) *Bull. Wld Hlth Org.*, **42**, 885.
Davenport, F. M., Hennessy, A. V., Minuse, E. (1967) *J. exp. Med.*, **126**, 1049.
Dawson, I. M., & Elford, W. J. (1949) *J. gen. Microbiol.*, **3**, 298.
DeLay, P. D., Casey, H. L., & Tubiash, H. J. (1968) *Publ. Hlth Rep., Wash.*, **82**, 615.
Dimmock, N. J. (1969) *Virology*, **39**, 224.
Dimmock, N. J., & Watson, D. H. (1969) *J. gen. Virol.*, **5**, 499.
Duc-Nguyen, H., Rose, H. M., & Morgan, C. (1966) *Virology*, **28**, 404.
Duesberg, P. H. (1968) *Proc. natn. Acad. Sci., U.S.A.*, **59**, 930.
Duesberg, P. H., & Robinson, W. S. (1967) *J. molec. Biol.*, **25**, 383.
Dunham, C. G., & Macneal, W. J. (1943) *J. Lab. clin. Med.*, **28**, 947.
Easterday, B. C., Trainer, D. O., Tumová, B., & Pereira, H. G. (1968) *Nature, Lond.*, **219**, 523.
Edward, D. G. ff. (1941) *Lancet*, **2**, 664.
Evans, C. A., & Rickard, E. R. (1945) *Proc. Soc. exp. Biol. Med., U.S.A.*, **58**, 73.
Fazekas de St Groth, S. (1948) *Nature, Lond.*, **162**, 294.
Findlay, G. M., & Mackenzie, R. D. (1937) *Br. J. exp. Path.*, **18**, 258.
Francis, T. (1947) *J. exp. Med.*, **85**, 1.
Francis, T. (1959). In *Viral and rickettsial infections of man*. 3rd Ed. p. 633. Eds.: Rivers & Horsfall. London: Pitman Medical.
Francis, T., Quilligan, J. J., & Minuse, E. (1950). *Science*, **112**, 495.
Franklin, R. M. (1958) *Virology*, **6**, 525.
Frommhagen, L. H., Knight, C. A., & Freeman, N. K. (1959) *Virology*, **8**, 176.
Galbraith, A. W., Oxford, J. S., Schild, G. C., & Watson, G. I. (1969) *Lancet*, **2**, 1026.
Gerlach, F., & Michalka, J. (1926) *Dtch. tier. Wschr.*, **34**, 897.
Granoff, A., & Kingsbury, D. W. (1964). In *Cellular biology of myxovirus infections*. Eds.: Wolstenholme & Knight. London: Churchill.
Green, I. J., Lieberman, M., & Mogabgab, W. J. (1957) *J. Immunol.*, **78**, 233.
Haas, R., & Wulff, H. (1957) *Z. Hyg. Infekt-Kr.*, **143**, 568.
Hallauer, C., & Kronauer, G. (1959) *Arch. ges. Virusforsch.*, **9**, 232.
Harwin, R. M., Joosting, A., & Gear, J. H. S. (1966) *Lancet*, **1**, 1218.
Haslam, E. A., Hampson, A. W., Egan, J. A., & White, D. O. (1970a) *Virology*, **42**, 555.
Haslam, E. A., Hampson, A. W., Radiskevics, I., & White, D. O. (1970b) *Virology*, **42**, 566.

Haukenes, G., Harboe, A., & Mortensson-Egnund, K. (1966) *Acta path. microbiol. scand.*, **66**, 510.
Henle, W., & Henle, G. (1947) *J. exp. Med.*, **85**, 347.
Hennessy, A. V., & Davenport, F. M. (1966) *J. Immunol.*, **97**, 235.
Hers, J. F. P., Mulder, J., Masurel, N., & v.d. Kuip, L. (1962) *J. Path. Bact.*, 83, 207.
Hirst, G. K. (1950) *J. exp. Med.*, **91**, 177.
Hjerten, S., Hoglund, S., & Ruttkay-Nedecky, G. (1970) *Acta virol., Prague*, **14**, 89.
Holland, J. J., & Kiehn, E. D. (1970) *Science*, **167**, 202.
Homme, P. J., Easterday, B. C., & Anderson, D. P. (1970) *Avian Dis.*, **14**, 240.
Hoyle, L. (1954) *J. Hyg., Camb.*, **52**, 180.
Hoyle, L. (1968) *The influenza viruses.* Virology Monographs. Eds.: Gard, Hallauer, & Meyer. Vienna/New York: Springer-Verlag.
Hoyle, L., Horne, R. W., & Waterson, A. P. (1962) *Virology*, **17**, 533.
Hoyle, L., & Frisch-Niggemeyer, W. (1955) *J. Hyg., Camb.*, **53**, 474.
Hoyle, L., Horne, R. W., & Waterson, A. P. (1961) *Virology*, **13**, 448.
Horne, R. W., Waterson, A. P., Wildy, P., & Farnham, A. E. (1960) *Virology*, **11**, 79.
Johnson, D. O., Wooding, W. L., Tanticharoenyos, P., & Karnjanaprakorn, C. (1971) *J. infect. Dis.*, **123**, 365.
Joss, A., Gandhi, S. S., Hay, A. J., & Burke, D. C. (1969) *J. Virol.*, **4**, 816.
Joosting, A. C. C., Head, B., Bynoe, M. L., Tyrrell, D. A. J. (1968) *Br. med. J.*, **2**, 153.
Jungherr, E. L., Tyzzer, E. E., Brandly, C. A., & Moses, H. E. (1946) *Am. J. vet. Res.*, **7**, 250.
Kammer, H., & Hanson, R. P. (1962) *J. infect. Dis.*, **110**, 99.
Kasel, J. A., Alford, R. H., Knight, V., Waddell, G. H., & Sigel, M. M. (1965) *Nature, Lond.*, **206**, 41.
Kasel, J. A., Byrne, R. J., Harvey, E. W., & Shillinger, R. (1968) *Nature, Lond.*, **219**, 968.
Kasel, J. A., Fulk, R. V., & Couch, R. B. (1969) *J. Immunol.*, **102**, 530.
Kates, M., Allison, A. C., Tyrrell, D. A. J., & James, A. T. (1961) *Biochim. biophys. Acta, Amst.*, **52**, 455.
Kilbourne, E. D. (1963) *Progr. med. Virol.*, **5**, 81.
Kilbourne, E. (1968a) *Science*, **160**, 74.
Kilbourne, E. D. (1968b) *Arch. environ. Hlth*, **14**, 768.
Kingsbury, D. W. (1970) *Progr. med. Virol.*, **12**, 49.
Knight, C. A. (1947) *J. exp. Med.*, **85**, 99.
Kopp, J. V., Kumpf, J. E., & Kroeger, A. V. (1968) *Virology*, **36**, 681.
Koppel, Z., Vrtiak, J., Vasil, M., & Spierz, S. (1956) *Veterinarstvi*, **6**, 267.
Kundin, W. D. (1970) *Nature, Lond.*, **228**, 857.
Laidlaw, P. P. (1935) *Lancet*, **1**, 1118.
Laver, W. G., & Valentine, R. C. (1969) *Virology*, **38**, 105.
Laver, W. G., & Webster, R. G. (1966) *Virology*, **30**, 104.
Ledinko, N. (1955) *Nature, Lond.*, **175**, 999.
Lee, A. M. & Tauraso, N. M. (1968) *Bull. Wld Hlth Org.*, **39**, 261.
Lehmann-Grube, F. (1963) *Virology*, **21**, 520.
Maassab, H. K. (1967) *Nature, Lond.*, **213**, 612.
Mackenzie, J. S. (1970) *J. gen. Virol.*, **6**, 63.
McQueen, J. L., Steele, J. H., & Robinson, R. Q. (1968) *Adv. vet. Sci.*, **12**, 285.
Madeley, C. R., Allan, W. H., & Kendall, A. P. (1971) *J. gen. Virol.*, **12**, 69.
Maggiora, A., & Valenti, G. L. (1903) *Zentbl. Bakt. ParasitKde, I. Abt. Orig.*, **34**, 326.
Masurel, N., & Mulder, J. (1966) *Bull. Wld Hlth Org.*, **34**, 885.
Medvedeva, T. E., Alexandrova, G. I., & Smorodintsev, A. A. (1968) *J. Virol.*, **2**, 456.
Meier-Ewert, H., Gibbs, A. J., & Dimmock, A. J. (1970) *J. gen. Virol.*, **6**, 409.
Miller, W. C. (1966) *Proc. r. Soc. Med.*, **59**, 52.
Mitchell, C. A., Guerin, L. F., & Robillard, J. (1968) *Can. J. comp. Med.*, **32**, 544.
Mogabgab, W. J. (1962) *J. Bact.*, 83, 209.
Mogabgab, W. J., Green, I. J., Dierkhising, O. C., & Philips, I. A. (1955) *Proc. Soc. exp. Biol. Med., U.S.A.*, **89**, 654.

Morgan, C., & Rose, H. M. (1968) *J. Virol.*, **2**, 925.
Morgan, C., Rose, H. M., & Moore, D. H. (1956) *J. exp. Med.*, **104**, 171.
Morozenko, M. A. (1957) *Probl. Virol.* (Eng. Ed.), **2**, 369.
Moses, H. E., Brandly, C. A., Jones, E. E., & Jungherr, E. L. (1948) *Am. J. vet. Res.*, **9**, 314.
Nafta, I., Turcann, A. G., Braun, I., Companetz, W., Simionescu, A., Brit, E., & Florea, V. (1970) *Bull. Wld Hlth Org.*, **42**, 423.
Nayak, D. P., & Baluda, M. A. (1967) *J. Virol.*, **1**, 1217.
Nayak, D. P., & Baluda, M. A. (1968b) *J. Virol.*, **2**, 99.
Nayak, D. P., & Baluda, M. A. (1968a) *Proc. natn. Acad. Sci., U.S.A.*, **59**, 184.
Nermut, M. V., & Frank, H. (1971) *J. gen. Virol.*, **10**, 37.
Oker-Blom, N., Hovi, T., Leinikki, P., Palouso, T., Pettersson, R., & Suni, J. (1970) *Br. med. J.*, **3**, 676.
Paniker, C. K. J. (1968) *J. gen. Virol.*, **2**, 385.
Pereira, H. G. (1969) *Progr. med. Virol.*, **11**, 46.
Pereira, H. G., Lang, G., Olesiuk, O. M., Snoeyenbos, G. H., Roberts, D. H., & Easterday, B. C. (1966) *Bull. Wld Hlth Org.*, **35**, 799.
Pereira, H. G., Rinaldi, A., & Nardelli, L. (1967) *Bull. Wld Hlth Org.*, **37**, 553.
Pereira, H. G., Takimoto, S., Piegas, N. S., & Ribeiro do Valle, L. A. (in press) *Bull. Wld Hlth Org.*
Pereira, H. G., Tumová, B., & Law, V. G. (1965) *Bull. Wld Hlth Org.*, **32**, 855.
Peterson, W. D., Davenport, F. M., & Francis, T. (1961) *J. exp. Med.*, **114**, 1023.
Pons, M. W., & Hirst, G. K. (1968) *Virology*, **34**, 385.
Pons, M. W., Schulze, I. T., Hirst, G. K., & Hauser, R. (1969) *Virology*, **39**, 250.
Rinaldi, A., Cervio, G., & Mandelli, G. (1966) *Boll. Inst. Sieroter. Milan*, **45**, 225.
Robinson, R. Q. (1964) *Progr. med. Virol.*, **6**, 82.
Rouse, B. T., Lang, G., & Narayan, O. (1968) *J. comp. Path.*, **78**, 525.
Saito, Y., Yoshioka, I., Igarashi, Y., & Nakagawa, S. (1970) *Virology*, **40**, 408.
Schild, G. C. (1970) *J. gen. Virol.*, **9**, 191.
Schild, G. C., Pereira, H. G., & Schettler, C. H. (1969) *Nature, Lond.*, **222**, 1299.
Scholtissek, C. (1970). In *The biology of large RNA viruses.* Eds.: Barry & Mahy. London/New York: Academic Press.
Scholtissek, C., & Rott, R. (1964) *Virology*, **22**, 169.
Schulman, J. L. (1970) *Progr. med. Virol.*, **12**, 128.
Sen, H. G., Kelly, G. W., Underdahl, N. R., & Young, G. A. (1961) *J. exp. Med.*, **113**, 517.
Seto, J. T., Chang, F. S. (1969) *J. Viriol.*, **4**, 58.
Shope, R. E. (1931) *J. exp. Med.*, **54**, 373.
Shope, R. E. (1935) Harvey lectures.
Shope, R. E. (1941) *J. exp. Med.*, **74**, 41 and 49.
Shotts, E. B., Foster, J. W., Brugh, M., Jordan, H. E., & McQueen, J. L. (1968) *J. exp. Med.*, **127**, 359.
Siegl, G., Pette, J., & Macknel, H. (1968) *Arch. ges. Virusforsch.*, **24**, 299.
Simpson, R. W., & Hirst, G. K. (1968) *Virology*, **35**, 41.
Skehel, J. J. (1971a) *J. gen. Virol.*, **11**, 103.
Skehel, J. J. (1971b) *Virology*, **45**, 793.
Skehel, J. J., & Schild, G. C. (1971) *Virology*, **44**, 396.
Slepuskin, A. N., Bobyleva, T. K., Russina, A. E., Vitkina, B. S., Ellengorm, N. S., & Zhdanov, V. M. (1967) *Bull. Wld Hlth Org.*, **36**, 385.
Smith, W., Andrewes, C. H., & Laidlaw, P. P. (1933) *Lancet*, **2**, 66.
Smorodintsev, A. A., Chalkina, O. M., Burov, S. A., & Ilyin, N. A. (1961) *J. Hyg. Epidemiol. Microbiol. Immunol., Moscow*, **5**, 60.
Somma, R. E., Vallone, E. F., Tosi, H., Giambruno, E., Infantozzi, J. M., & Salsamendi, R. (1964) *Publ. centro. Invest. vet. Migule C. Rubino*, **2**, 5.
Sovinová, O., Tumová, B., Pouska, F., & Němec, J. (1958) *Acta virol., Prague*, **2**, 52.
Straub, M. (1937) *J. Path. Bact.*, **45**, 75.
Stuart-Harris, C. H. (1939) *Lancet*, **1**, 497.

Sugg, J. Y. (1949) *J. Bact.*, **57**, 399.
Sugiura, A., & Kilbourne, E. D. (1965), *Virology*, **26**, 478.
Takátsy, G., & Barb, K. (1959) *Nature, Lond.*, **183**, 52.
Takátsy, G., Farkas, E., & Romvary, J. (1969) *Nature, Lond.*, **222**, 184.
Takeyama, T. (1966) *Tokohn J. exp. Med.*, **89**, 321.
Taylor, R. M. (1951) *Arch. ges. Virusforch.*, **4**, 485.
Ter-Meuleu, V., & Love, R. (1967) *J. Virol.*, **1**, 626.
Tumová, B., & Fiserova-Sovinová, O. (1959) *Bull. Wld Hlth Org.*, **20**, 445.
Tumová, B., & Pereira, H. G. (1965) *Virology*, **27**, 253.
Tumová, B., & Pereira, H. G. (1968) *Bull. Wld Hlth Org.*, **38**, 415.
Uys, C. J., & Becker, W. B. (1967) *J. comp. Path.*, **77**, 167.
van der Veen, J., & Mulder, J. (1950) *Studies on the antigenic composition of human influenza A virus strains.* Thesis, Leiden.
Waddell, G. H., Teigland, M. B., & Sigel, M. M. (1963) *J. Am. vet. med. Ass.*, **143**, 587.
Wagner, R. R., Bennett, I. L., & LeQuire, V. S. (1949) *J. exp. Med.*, **90**, 321.
Wang, S. P., & Liu, T. Y. (1958) *J. infect. Dis.*, **103**, 178.
Waterson, A. P., Hurrell, J. M., & Jensen, K. E. (1963), *Arch. ges. Virusforsch.*, **12**, 487.
Webster, R. G., & Laver, W. G. (1966) *J. Immunol.*, **96**, 596.
Webster, R. G., & Laver, W. G. (1971) *Progr. med. Virol.*, **13**, 271.
Webster, R. G., & Pereira, H. G. (1968) *J. gen. Virol.*, **3**, 201.
Wecker, E. (1957) *Z. Naturforsch.*, **12b**, 208.
White, D. O., Taylor, J. M., Haslam, F. A., & Hampson, A. W. (1970). In *The biology of large RNA viruses.* Eds.: Barry & Mahy. London/New York: Academic Press.
White, J. (1953) *Br. J. exp. Path.*, **34**, 668.
WHO Scientific Group Report on Respiratory Viruses (1969) *Wld Hlth Org. techn. Rep. Series No. 408.* Geneva.
Winternitz, M. C., Watson, I. M., & McNamara, F. P. (1920) *The pathology of influenza.* New Haven: Yale Univ. Press.

10

Paramyxovirus

The following characters define the genus: Approximately spherical virions consisting of (a) a helical nucleocapsid 18 nm in diameter and about 1 μm long containing a continuous single-stranded RNA genome with molecular weight of 4×10^6 to 8×10^6 daltons and (b) a lipid-containing envelope covered with projections and incorporating virus-specific haemagglutinin and neuraminidase. Virus replication is actinomycin-resistant and takes place in the cytoplasm. Viral components are assembled at the cell membrane, and virions are released by budding. A number of viruses incompletely characterized or deviating in some respects from the above definition will be described as probable paramyxoviruses.

Reviews: Chanock et al. (1963). Chanock & Coates (1964).

Morphology. Negative contrast electron microscopy of mumps (Horne et al., 1960), NDV (Horne & Waterson, 1960; Rott & Schäfer, 1961), human (HA2) and murine (Sendai) strains of parainfluenza 1 (Waterson & Hurrell, 1962; Horne et al., 1960; Hosaka et al., 1961), human (CA) and simian (SV5) strains of parainfluenza 2 (Waterson & Hurrell, 1962; Bonissol et al., 1968; Choppin & Stoeckenuis, 1964) and human and bovine strains of parainfluenza 3 (Waterson et al., 1961b; Berkaloff et al., 1965) reveal the following common features: Virions consist of a helical nucleocapsid contained in an approximately spherical envelope 100 to 200 nm in diameter, but easily deformed giving rise to aberrant forms of variable size which may reach 600 nm. The envelope is covered with projections 12 to 15 nm long and 2 to 4 nm wide. Partially disrupted virions are frequently seen, revealing an internal nucleocapsid with helical symmetry, about 18 nm in diameter, with a central hole about 5 nm across. Its surface is striated, with a periodicity of 4 to 6 nm. A herring-bone appearance has been interpreted to indicate a double helix (Horne & Waterson, 1960; Hosaka et al., 1961), but Hosaka (1968) and Finch & Gibbs (1969) present evidence indicating a single strand. The number of subunits per turn has been estimated as 15 by Hosaka (1968) and as either 11 or 13 by Finch & Gibbs (1969). A modal length of 1 μm has been estimated for the nucleocapsids of NDV (Hosaka & Shimizu, 1968; Kingsbury & Darlington, 1968), Sendai (Hosaka et al., 1966; Finch & Gibbs, 1969), mumps (Hosaka & Shimizu, 1968; Finch & Gibbs, 1969) and SV5 (Compans & Choppin, 1967). Virions containing multiple nucleocapsids have been described (Hosaka et al., 1966; Dahlberg & Simon, 1969). Envelopes devoid of nucleocapsids may exist (Sokol, 1964; Dahlberg & Simon, 1969). The morphogenesis of paramyxoviruses has been studied in cells infected with Sendai virus (Berkaloff, 1963), NDV (Feller et al., 1969), mumps (Duc-Nguyen & Rosenblum, 1967), human and simian parainfluenza 2 (Compans et al., 1967; Howe et al., 1967; Bonissol et al., 1968), human and bovine parainfluenza 3 (Reczko & Bögel, 1962; Cohen et al., 1962). In all cases, filaments, 10 to 15 nm in diameter, striated with a periodicity of about 5 nm are seen in the cytoplasm. These filaments are often present in cytoplasmic inclusions containing RNA and virus-specific antigen and almost certainly represent the viral nucleocapsid. At later stages of infection they accumulate, sometimes in regular alignment, beneath the outer cell membrane in areas where the latter is thickened and formed by 2 layers, 4 to 5 nm apart, with projections on the outward face. Nucleocapsids are then incorporated into spherical or filamentous particles by a process of invagination. In released virions, the nucleocapsid is closely applied to the inside of a 10 to 13 nm thick envelope with outer projections. In filamentous particles the nucleocapsid is regularly coiled along the whole length of the particle.

Chemical composition. The dry weight of NDV and SV5 is made up of approximately 0·9 per cent RNA, 70 per cent protein, 20 to 40 per cent lipid and 6 per cent carbohydrate (Nakajima & Obara, 1967; Klenk & Choppin, 1969). The nucleocapsid represents about 20 per cent of the virion. The RNA content of the nucleocapsid has been estimated as 9·6 per cent in mumps, 5·7 per cent in NDV (Schäfer & Rott, 1959), 3·7 per cent in Sendai virus (Hosaka, 1968) and 4·1 per cent in SV5 (Compans & Choppin, 1967). The RNAs of NDV (Duesberg & Robinson, 1965; Sokol et al., 1966; Kingsbury, 1966a), Sendai virus (Iwai et al., 1966; Blair & Robinson, 1968), SV5 (Compans & Choppin, 1968) and parainfluenza 3 (Dubois et al., 1970) have been characterized as single-stranded molecules with sedimentation coefficients of 50 to 57 S in 0·1 M NaCl, but lower at lower molarities, buoyant density in CsCl of 1·680 g/ml and MW of about 6×10^6 daltons. Similar base composition, characterized by about 30 per cent uridine and lower (approximately 20 per cent) proportions of each of the other 3 bases, has been determined for the 3 viruses listed above.

Analysis of NDV proteins by polyacrylamide gel electrophoresis (Evans & Kingsbury, 1969; Haslam et al., 1969; Bikel & Duesberg, 1969) reveals 3 major and several minor polypeptides. Yamamura et al. (1970) describe 11 protein bands in Sendai virus. Two of the structural proteins are glycoproteins (Mountcastle et al., 1971). RNA-dependent RNA polymerase has been detected in NDV (Huang et al., 1971) and Sendai (Robinson, 1971) virions. The presence of glycolipids (glyco-sphingo-lipids) has been described by Blough & Lawson (1968) in NDV and Sendai virus and by Klenk & Choppin (1970) in SV5. Soule et al. (1959) detected about 7 per cent phosphatides, of which 60 per cent was sphingomyelin, in mumps virus.

Physico-chemical characters. The buoyant density of NDV in CsCl varies from 1·212 to 1·221 g/ml and from 1·236 to 1·242 g/ml in virus grown in avian and mammalian cells respectively (Stenback & Durand, 1963). Virus infectivity is sensitive to lipid solvents, heat labile and unstable at high or low pH. Studies on the heat-sensitivity of infectivity and other properties of mumps and NDV are reviewed by Cantell (1961) and Picken (1964) respectively. Sensitivity of NDV to chemical compounds is described by Cunningham (1948).

Haemagglutination. Most strains agglutinate chicken, guinea-pig and human erythrocytes. Red-cell receptors are destroyed by neuraminidase. Virus replication in tissue cultures is readily detected by haemadsorption. Virions possess neuraminidase and haemolytic activities and the capacity to fuse mammalian cells in suspension or in monolayers.

Antigenic properties. Mumps, Newcastle disease and parainfluenza types 1 to 4 are all antigenically related although distinguishable from each other on the basis of both nucleocapsid and envelope antigens (Cook et al., 1959; Johnson et al., 1960). Criteria for the distinction of types within paramyxoviruses are discussed by Chanock and Coates (1964) who stress the importance of using post-infection guinea-pig sera for type differentiation. All members of the genus are antigenically stable.

Cultivation. All members of the genus grow in tissue cultures usually causing CPE, haemadsorption, syncytial formation and eosinophilic inclusions in the cytoplasm and occasionally in the nucleus. Most strains grow in embryonated eggs. Systems used for virus propagation and assay are described separately for each member of the genus.

Studies on the biosynthesis of Sendai and NDV have been reviewed by Blair & Duesberg (1970), who mention the following features as distinctive for these viruses:

1. Virus-specific RNA synthesis and virus production are not inhibited by actinomycin D (Barry et al., 1962; Kingsbury, 1962).
2. Replication takes place entirely in the cytoplasm (Reda et al., 1964).
3. The major products of RNA replication are single-stranded molecules with base sequence complementary to viral RNA (Kingsbury, 1966b; Bratt & Robinson, 1967; Blair & Robinson, 1968).

MUMPS

Synonyms: Epidemic parotitis. Oreillons.

Reviews: Spooner (1953). Enders & Habel (1956). Cantell (1961).

Haemagglutination. Agglutinates red blood cells of fowls, man and other species (Levens & Enders, 1945). Strains of different origin differ in the range of species whose red blood cells are agglutinated, and agglutination of different kinds of RBCs is variously affected by inhibitors. The virus is less actively adsorbed to and eluted from RBCs than is influenza. Haemadsorption is often a more sensitive index of virus activity than haemagglutination. Virus in high concentration also lyses red blood cells (Morgan et al., 1948).

Antigenic properties. Antigenically homogeneous. Two complement-fixing antigens corresponding to the viral nucleocapsid (S antigen) and envelope (V antigen) respectively can be identified (Henle et al., 1947). In man antibodies to S develop sooner than to V, but fade away sooner. A mixed antigen suffices (Burnet & Bull, 1943) for diagnostic tests. Workers disagree as to whether the CF or HAI test is more practically useful. Some feel that nonspecific inhibitors make results of HAI tests difficult to interpret. A plaque-neutralizing test with greater sensitivity than CF or HAI has been described by Ennis et al. (1968). Reactions with S antigen are revealed by gel-diffusion precipitation (Cuadrado, 1966). Crossing may occur in both CF and HAI tests with the viruses of Newcastle disease and parainfluenza 1, 2, 3 and 4. The cross-reactions are as a rule only to low titre, and they may be one-way; human sera are apt to show them better than would those of immunized animals (Chanock, 1955; DeMeio & Walker, 1957; Gardner, 1957).

An antigen for skin testing has been made by inactivating infected allantoic fluid by heat or otherwise: a positive reaction strongly indicates that the subject is immune (Enders et al., 1946).

Cultivation. Amniotic or yolk-sac inoculation can be used for primary isolations (Habel, 1945). Growth is slower than with influenza. Best to use 7–8-day-old eggs, incubated at 33°C–36°C for 4–5 days. With 'wild' virus, growth may not occur in every egg inoculated, but adaptation soon occurs. The allantoic route can then be used, and high titres of virus may be obtained with some strains (e.g. Enders) though not with all. Monkey kidney and several human tissue cultures or cell-lines can be used for virus isolation and propagation. Virus growth indicated by haemadsorption and the formation of syncytia and cytoplasmic eosinophilic inclusions. With large virus doses, cell fusion may occur independently of virus multiplication (Henle et al., 1954). Growth is also reported in mouse-embryo cultures (Kilham & Murphy, 1952) and in Maitland-type cultures of chick amnion (Watson & Cheever, 1955).

Plaque assays have been described (Hotchin et al., 1960; Ennis et al., 1968). Persistent infection of cell cultures may be demonstrated (Henle et al., 1958).

Pathogenicity. Causes, in man, parotitis, sometimes meningo-encephalitis, orchitis, oöphoritis, pancreatitis and other complications. The incubation period is usually 18–21 days. Experimentally it produces similar disease in rhesus monkeys and other primates and has been adapted to multiply in suckling hamster brains. Suckling mice can also be infected intracerebrally, more readily after previous adaptation to hamsters or after cultivation in mouse-embryo tissue culture (Kilham & Murphy, 1952), so can suckling rats (Pospisil & Brychtova, 1956). In ferrets inoculated intranasally bronchiolitis and pneumonitis were produced, but not more than 3 passages in series could be achieved (Gordon et al., 1956). Serial passage by the intranasal route was also reported in adult hamsters (Burr & Nagler, 1953). Hydrocephalus has been described in hamsters inoculated intracerebrally (Johnson et al., 1967). Intraocular infection of guinea-pigs leads to virus multiplication and production of corneal opacity (Bolin et al., 1952). A virus 'toxin' is possibly involved here, so, too, in the experiments of Ogasawara et al. (1959) when large doses given intranasally to young mice caused pulmonary consolidation not serially transmissible; even inactivated virus was effective (Wacker et al., 1962).

In infected monkeys virus lesions are found only in the salivary glands and CNS. The attack in their salivary glands is on the acinar cells (Johnson & Goodpasture, 1936).

Ecology. Transmission is from salivary secretions; patients are infectious from 6 days before clinical onset till 9 days after. Virus has been found in the urine regularly during the first 5 days and as late as the fifteenth day (Utz & Szwed, 1962). It may also be excreted in the milk (Kilham, 1951). It may be present in saliva of persons with symptomless infections (Henle et al., 1948).

Control. Killed virus vaccines have been used for immunization with varying success (e.g. Habel, 1951; Henle et al., 1951). More recently, effort has been devoted to the development and use of live attenuated vaccines (see review by Deinhardt & Shramek, 1969).

PARAINFLUENZA TYPE 1

Synnoyms: Haemadsorption virus 2 (HA2). Sendai. New-born pneumonitis virus. Haemagglutinating virus of Japan (HVJ). Influenza D.

It was proposed in 1959 (Andrewes et al.) that both Sendai and HA2 viruses should be included under parainfluenza 1, on the grounds of the close serological relationship of the 2 agents. Zhdanov (1960) has since argued in favour of their separation.

Haemagglutination is readily shown after passage in eggs or tissue culture, but on first isolation the haemadsorption test is more sensitive. Haemagglutination for human O and guinea-pig cells by human strains is best done at room temperature (Dick & Mogabgab, 1961). Cells of numerous other species (but not horse) are agglutinated by murine strains (Fukumi et al., 1954). There is adsorption to and enzymatic elution from sensitive cells. Both human and murine strains will also cause haemolysis of fowl and other RBCs; this can be avoided by using formalin-treated cells. The haemagglutinin is apparently more labile—at 47°C to 55°C—than that of influenza and also trypsin-sensitive.

Antigenic properties. Various serological tests are applicable as with other myxoviruses. For demonstrating antihaemagglutinins, Dick & Mogabgab (1961) advise mixing serum and virus, holding for 30 minutes at room temperature, then adding human O RBCs. In CF tests Sendai and HA2 strains cross-react, the former having more group specificity. In haemagglutinin inhibition and neutralization tests, the relationship is less evident (Cook et al., 1959). Bukrinskaya et al. (1962) have described differences found by the CF test between 'HA2' and 'Sendai' strains of different origins. The neuraminidase of Sendai virus is antigenically distinct from that of NDV (Tozawa et al., 1967). Cross-reactions with other paramyxoviruses are more evident in human sera than in those of immunized animals (Chanock et al., 1963). Antibodies are present in most adult human sera. Reinfections may occur during childhood. Antibodies are also present in sera of guinea-pigs of some stocks.

Cultivation. *In fertile eggs.* Some but not all newly isolated human strains will grow in fertile eggs; murine strains grow better. The amniotic route is most sensitive; on passage, growth occurs after inoculation into allantoic cavity and yolk sac. Embryos are usually not killed. Growth in the allantoic cavity is best at 35·5°C (Fukai & Suzuki, 1955).

In tissue culture. Human strains grow in tissues of man, monkey, chick embryo and swine, murine strains in kidney cells from numerous other species. CPE of human strains may be difficult to detect in living cultures, but eosinophilic cytoplasmic inclusions are often seen in stained cells. Haemadsorption is a sensitive method for detecting virus. Sendai virus can be assayed by plaque formation in human (Northrop & Walker, 1965), bovine (Nagata et al., 1965) and simian cells also (Nunen & van der Veen, 1967). Fusion of HeLa cells by human para-influenza 1 has been described by Marston (1958). The ability of in-activated Sendai virus to fuse mammalian cells (Okada, 1958) has been extensively applied to cell-fusion experiments and to the study of mammalian heterokaryons (Harris et al., 1966).

Distribution. The Sendai strain occurs as a latent infection of laboratory mice in Japan, China and the U.S.S.R. More recently it has been found to be widely distributed also in the United States both in mouse (Parker & Reynolds, 1968) and in hamster colonies (Profeta et al., 1969). Serological evidence suggests that rat colonies may be naturally infected in Great Britain (Tyrrell & Coid, 1970).

Pathogenicity. The Sendai virus causes latent infection in laboratory mice (Chu et al., 1957), being activated so that it causes fatal pneumonia after mouse-to-mouse passage. Widespread occurrence of infection without signs of disease has also been described (Parker & Reynolds, 1968). It is not generally accepted that Sendai virus causes human infections. Ability to cause pneumonia in pigs (Sasahara, 1955) seems to be well established.

The HA2 strain, on the other hand, has been isolated only from man, originally by Chanock et al. (1958), from children suffering from acute laryngo-tracheitis. The virus has only infrequently been isolated from adults, but seems likely to be a common cause of childhood infections.

Experimentally. Adults have been experimentally infected with HA2 and, after an incubation of 5 or 6 days, have developed upper respiratory symptoms, sometimes with fever, sometimes resembling common colds (Reichelderfer et al., 1958; Tyrell et al., 1959).

Strains of Sendai virus vary in pathogenicity for mice. The virus will kill mice when inoculated IC, but serial passage by this route has not succeeded. It will produce inapparent infection of ferrets, monkeys and pigs and will produce fatal pneumonitis in rats (Kuroya et al., 1953) and Chinese hamsters (Chun & Chu, 1956). It may cause symptoms in ferrets (Jensen et al., 1955).

The HA2 strain is less pathogenic for mice, producing symptomless infections in them, also (Petersen, 1959) in ferrets, hamsters and rhesus monkeys.

Pathological lesions. Mice infected with Sendai virus may show extensive lung lesions without obvious illness (Robinson et al., 1968). Virus can be recovered from lungs during the first 10 days of infection with the highest titres between 4 and 6 days (Nunen & van der Veen, 1967). Virus can be recovered in small quantities from various viscera after fatal infections.

Ecology. As already mentioned Sendai strains are latent amongst some laboratory mice and guinea-pigs. Attempts to induce chronic latent infections in clean stocks of laboratory mice have failed, though Sawicki (1962) found that virus would persist for a few weeks in a masked form. Fukumi et al. (1962) described in mouse colonies periods of prevalence followed by virtual disappearance of the virus. Parker & Reynolds (1968) describe enzootic and epizootic patterns of infection, the latter lasting 7 months in a mouse colony and then disappearing. Transmission was by direct contact or through contaminated fomites.

Control. Jensen et al. (1962) report that formolized egg-adapted virus will form a potent immunizing vaccine for man. An inactivated vaccine produced in grivet monkey-kidney cultures has been tested with promising results (Potash et al., 1966).

PARAINFLUENZA TYPE 2

Synonyms: Croup-associated (CA) virus (Chanock, 1956). Acute laryngo-tracheo-bronchitis (ALTB) virus (Beale et al., 1958).

The simian viruses SV5 (Hull et al., 1956) and SV41 (Miller et al., 1964) and the related strains of the SHV series (Emery & York, 1960), SA (Schultz & Habel, 1959) and DA (Hsiung, 1959) are classified by Chanock et al. (1963) as parainfluenza type 2 viruses. The same may apply to the WB virus shown by Liebhaber et al. (1965) to be antigenically related to SV5.

Haemagglutination. Human strains agglutinate chick erythrocytes better than human O cells. The optimum pH is 8. Virus elutes from the cells when they are warmed to 37°C; the cells then disperse, only to reagglutinate on being returned to 4°C. A neuraminidase has been described (Darrell & Howe, 1964).

Antigenic properties. VN, HI and CF are the tests most commonly used. Multiple antigens, not correlated with haemagglutinin or neuraminidase, have been detected by immunoelectrophoresis (De Vaux St. Cyr & Howe, 1966). Cross-reactions between human and simian strains are described by Chanock et al. (1961a; 1963). An antigenic variant of human parainfluenza 2 is described by Numazaki et al. (1968).

Cultivation. Human strains grow poorly in eggs inoculated amniotically, not reaching high enough titres for agglutinin to be demonstrable unless after adaptation (De Meio, 1963). Good growth in tissue cultures of HeLa cells, human embryonic lung and amnion, dog kidney, monkey kidney, the last being best for primary isolation (Mogabgab, et al., 1961). CPE may be slight at first, but becomes better on passage, appearing in 3 or 4 days. Characteristic effects are formation of syncytia and cytoplasmic inclusions (Lepine et al., 1959) containing filamentous structures resembling viral nucleocapsids (Kuhn & Harford, 1963). These inclusions are less acidophilic and develop more slowly than those of related viruses (Brandt, 1961). Plaques can be produced on monolayers of *Cercopithecus* kidney (Tytell et al., 1962).

Simian strains multiply in the amniotic and allantoic cavities of fertile eggs (Chanock et al., 1961a) and in a variety of tissue cultures, those of monkey, bovine and hamster cells being most commonly used. Plaque assay described by Choppin (1964) and by Rhim & Schell (1967).

Pathogenicity. Associated particularly with acute laryngotracheitis (croup) in children aged 6 months to 3 years; tracheotomy has had to be considered in many such cases (Beale et al., 1958). There is evidence that it may occasionally cause minor upper respiratory infections in adults. Mild illnesses resembling common colds have been produced in adult volunteers (Taylor-Robinson & Bynoe, 1963). No pathological effects have been produced by inoculation into adult or suckling mice. Simian strains are frequently present in cultures of 'normal' monkey tissues, but are not known to be naturally pathogenic. Vervet monkeys and baboons can be infected by the intranasal route and readily transmit infection to contacts (Heath et al., 1966; Larin et al., 1967). Strains closely related to SV5 and DA have been isolated from dogs with upper respiratory disease (Crandell et al., 1968) and from embryonated chicken eggs (Wagner & Enders-Ruckle, 1966). A strain serologically related but distinct from SV41 has been isolated from bats (Hollinger & Pavri, 1971).

Ecology. Parainfluenza 2 antibodies are found in high proportions of human and simian sera. Kalter et al. (1968) point out that antibody to SV5 is prevalent only in monkeys with human contact and suggest that the virus may be of human origin.

Control. An inactivated vaccine produced in grivet monkey-kidney cultures has been tested in humans with promising results (Potash et al., 1966). Monkeys have also been protected by vaccination (Tribe, 1967).

PARAINFLUENZA TYPE 3

Synonyms: Haemadsorption virus 1. Shipping fever virus.

General account: Chanock et al. (1958).

Haemagglutination. Human strains agglutinate human, guinea-pig and fowl erythrocytes, but the haemadsorption test is more sensitive for newly isolated virus, the pattern being more diffuse than with some myxoviruses (Bukrinskaiya, 1960). The virus also haemolyses sensitive red cells. Bovine strains haemagglutinate and haemolyse better at 37° than at 0°C and may be divided into 2 or 3 subgroups on the basis of HA activity for avian, bovine, ovine, porcine, guinea-pig and human cells, and on heat sensitivity (Hermondson et al., 1961). A suggestion that lack of neuraminidase may distinguish bovine from other parainfluenza viruses has not been substantiated (Dawson & Patterson, 1967; Drzeniek et al., 1967).

Antigenic properties. Strains of human and bovine origin can be differentiated by VN, HI, CF and immunodiffusion tests (Ketler et al., 1961; Abinanti et al., 1961; Fischman & Bang, 1966). Cross-relations with other myxoviruses are particularly evident with human sera (see p. 225). Serological evidence supports the idea that this virus is responsible for respiratory disease in children. The CF test is useful for diagnosis.

Cultivation. Grows when inoculated into developing eggs amniotically, not allantoically; haemagglutinins are produced in the fluids. Monkey-kidney tissue cultures are best for primary isolation, but the virus can be adapted to grow in HeLa and other human cell-lines and in chick kidney, producing CPE. Bovine strains, at least, grow in calf, goat, water buffalo and camel-kidney cultures. Syncytia and cytoplasmic inclusions are produced in HeLa and KB cells, less readily in monkey kidney (Deibel & Hotchin, 1961). Bovine strains also produce eosinophilic intranuclear inclusions (Churchill, 1963a). Virus antigens are detected only in the cytoplasm (Maassab & Loh, 1960). Growth and CPE in organ cultures of human respiratory tract tissues have been described by Craighead & Brennan (1968). A substance 'syncytin' has been suggested as concerned in syncytium formation (Chany & Cook, 1960). Plaques may be demonstrated on monolayers of a line of human cells (FC) (Deibel, 1959).

Distribution. Human strains occur throughout the world. Parainfluenza 3 has also been isolated from calves (Reisinger et al., 1959), sheep (Hore, 1966), horses (Ditchfield et al., 1965) and water buffalo (Singh & Baz, 1966). Serological evidence suggests that deer (Shah et al., 1965), camel (Singh & Ata, 1967) and dogs (Fischmann, 1967) may be naturally infected.

Pathogenicity. The virus has been frequently isolated from small children with pharyngitis, bronchiolitis and pneumonia, especially in nursery schools, much less frequently from unaffected children. These facts, supported by serological evidence, suggest that it is the cause of many of these infections (Chanock et al., 1958). It seems to be uncommon as a cause of natural respiratory infections in adults, though on inoculation into volunteers it has apparently produced some common colds (Tyrell et al., 1959). An outbreak among patas monkeys in captivity has been described; there was pneumonia with 50 per cent mortality. Experimental infection with cultures produced fever but no pneumonia (Churchill, 1963b).

Human strains produce inapparent infection or small focal lesions on IN inoculation into young hamsters (Craighead et al., 1960; Buthala & Soret, 1964).

Bovine strains have been recovered particularly from cattle with shipping fever; the facts suggest that a virus which is normally harmless may produce disease in cattle under conditions of stress. Experimental inoculation into calves has been followed by fever, conjunctivitis and muco-purulent rhinitis (Bakos & Dinter, 1960). Infection of sheep has been described as asymptomatic (Fischmann, 1967) or associated with outbreaks of respiratory disease (Hore et al., 1968). Responses of sheep and calves to experimental infection have been studied by Stevenson & Hore (1970).

Control. Killed (McLelland et al., 1961b; Tribe et al., 1968) and attenuated (Woods et al., 1968; Mayr et al., 1969) vaccines against bovine parainfluenza 3 have been developed. An inactivated vaccine has also been developed against human parainfluenza 3 (Potash et al., 1966).

PARAINFLUENZA TYPE 4

Description: Johnson et al., 1960.

Haemagglutination. Agglutinates guinea-pig RBCs at 4° and room temperature, not at 37°C. Receptors destroyed by RDE. Haemagglutinin titre enhanced by treatment with tween-ether (Henry et al., 1968). Haemadsorption demonstrable with guinea-pig or rhesus RBCs, poorly with human O, not at all with chick or rat cells.

Antigenic properties. Antibody rises occurred in sera of infected persons. Adult human sera indicate a cross-relation between this virus and mumps. Two antigenic subtypes (A and B) have been identified (Canchola et al., 1964).

Cultivation. *In eggs.* Doubtful.
In tissue cultures. Growth with CPE and cytoplasmic eosinophilic inclusions in monkey-kidney cells. No growth in HeLa and other cell-lines nor in rabbit kidney.

Pathogenicity. Relation to human disease not yet established, but serological surveys suggest widespread distribution in human populations and a possible role in the aetiology of respiratory illness (Gardner, 1969; Killgore & Dowdle, 1970).
Not pathogenic for laboratory animals, but inoculated guinea-pigs developed very good antibody rises.

NEWCASTLE DISEASE

Synonyms: Ranikhet disease. Avian pneumo-encephalitis. Atypischen Geflugelpest.

In Britain the term 'fowl-pest' is used to include this disease and fowl-plague. The Latin name *Myxovirus multiforme* (Andrewes et al., 1955) is based on the multiformity both of the clinical picture in fowls and of the virus particles.

Review: Hanson (1964).

Original description: Doyle (1927).

Haemagglutination. All strains are adsorbed to and agglutinate fowl red blood cells; some strains agglutinate a variety of avian and mammalian red cells. Agglutination tests are best read at 4°C. Elution may be rapid, but is often incomplete. The virus particle itself may act as a haemagglutinin, but within infected cells a smaller haemagglutinating component is present also (Granoff et al., 1950). Human red blood cells treated with some strains of virus are agglutinated by some infectious mononucleosis sera (Burnet & Anderson, 1946). High concentrations of virus will lyse fowl red cells (Kilham, 1949).

Antigenic properties. Immunologically distinct from other members of the group, apart from a possible relationship in the haemagglutination-inhibition test with mumps virus. Sera of persons convalescent from mumps may show HAI against NDV. However, with sera from immunized guinea-pigs, no cross-reactions can be demonstrated between nucleocapsid or envelope antigens of NDV and other paramyxoviruses (Rott, 1964). Immunological differences amongst strains from different sources have been described (Upton et al., 1953).

Cultivation. Most strains grow readily in all tissues of developing chick embryos and in chick tissue culture, in which they produce cytopathic effects. Cultivation is also reported in cultures of a wide variety of mammalian cells (see review by Bankowski, 1964). Growth in tissue cultures is usually accompanied by CPE, syncytial formation, eosinophilic cytoplasmic inclusions (Oh, 1961) and haemadsorption. Cultures may remain latently infected for long periods. A plaque technique is described (Bower, 1958). The properties of a cytotoxic factor developed in culture were described by Mason & Kaufman (1961). Kohn (1965) has described a factor, like but not the same as haemolysin, causing cell fusion. Virus cultivated in calf kidney and some other tissues loses virulence for chicks and may prove useful as an immunizing agent (Gelenezcei & Bordt, 1960; Russeff, 1962).

Distribution. Occurs all over the world in the respiratory and intestinal tracts, blood and tissues of fowls and many other avian species (see review by Levine, 1964).

Pathogenicity. Causes a fatal disease in fowls, turkeys and other species of birds. Symptoms may be primarily respiratory or nervous leading to spasms and paralysis; both may occur—hence the name pneumo-encephalitis. Closed eyes, nasal discharge and watery diarrhoea may be seen. Milder strains cause low mortality, but may affect egg production. Virulent strains are known as velogenic, milder ones as lentogenic; their epidemiological behaviour may be different. Incubation period is 2–11 days. Conjunctivitis has been caused in poultry workers and laboratory workers; lymphadenitis and some generalized symptoms may occur but recovery is rapid.

Experimentally the virus will, through its toxic effects, cause non-transmissible encephalitis in hamsters and mice. Meningo-encephalitis has also been produced in intracerebrally inoculated rhesus monkeys (Wenner et al., 1950) and pigs, and virus even survives for some time in brains of reptiles and dogfish.

Strains of various origin and history behave very differently in chicks, infected eggs and tissue culture; in birds multiplication may occur mainly in turbinates (Burnstein & Bang, 1958) or in spleen and intestine (Kohn, 1959).

Pathological lesions are described by Jungherr et al. (1946). Haemorrhages are seen, though less frequently than in fowl plague; in European strains they are common in the intestine. There may be sinusitis and inflammation of air sacs.

Ecology. Transmission may be through drinking-water or air-borne possibly by inhalation of dust from faecally contaminated litter; but air-borne infection from respiratory secretions is more important, the respiratory tract being many times more effective a portal of entry than the intestinal (Kohn, 1955). The mechanism of transmission was studied by Andrewes & Allison (1961): air-borne infection was found to be effective only over short distances with the strain they used; in the field things may be very different, and there is evidence for wind-borne transmission. Virus has been isolated from eggs of hens vaccinated with attenuated living virus (Ahmed & El-Sisi, 1965). Geese have been suspected of acting as latently infected carriers (Heller, 1957). Brown (1965) has described the history of the disease in Britain, particularly as regards its epidemiology.

Control. Wholesale slaughter and vaccination with inactivated virus have been or are being replaced by the use of live attenuated virus vaccines (Robertson, 1964; Hitchner, 1964).

YUCAIPA VIRUS

The virus was isolated from sick chicks by Bankowski et al. (1960). It was identified as a separate paramyxovirus by Dinter et al. (1964). Kawamura & Tubahara (1968) describe the growth of Yucaipa virus in chick cell cultures with CPE, syncytial formation and cytoplasmic inclusions. Plaque formation in this system was inhibited by homologous, but not by NDV antisera.

PROBABLE PARAMYXOVIRUSES

Viruses resembling paramyxoviruses in several respects are described as probable members of the genus either because their characterization is incomplete or because some of their properties differ from those of typical paramyxoviruses.

THE MEASLES-DISTEMPER-RINDERPEST TRIAD

The gathering of these viruses into a clearly defined subgroup is based on their similarity as regards antigenic, morphological and pathological properties (see review by Warren, 1960; Carlstrom, 1962; Imagawa, 1968).

Morphology. Virions of measles (Waterson et al., 1961a; Finch & Gibbs, 1969), distemper (Cruickshank et al., 1962; Norrby et al., 1963) and rinderpest (Plowright et al., 1962; Tajima, et al., 1967) are 120 to 300 nm in diameter although anomalous and long filamentous forms may be observed. A helical nucleocapsid 15 to 19 nm in diameter is contained in an envelope covered with projections. The mean length of measles virus nucleocapsids is about 1 μm (Nakai et al., 1969; Finch & Gibbs, 1970). Thin sections of cells infected with measles (Matsumoto, 1966; Nakai et al., 1969) distemper (Lawn, 1970) and rinderpest (Tajima et al., 1967) reveal accumulations of tubular structures resembling nucleocapsids in the cytoplasm and nucleus. Virus assembly takes place by budding at the cell surface although virus-like particles may be seen in cytoplasmic vesicles (Breese & de Boer, 1963—rinderpest).

Chemical composition. Measles virus contains single-stranded RNA with sedimentation coefficient of 52·2S and M.W. estimated as $6{\cdot}22 \times 10^6$ daltons (Schluederberg, 1971). Lack of inhibition by halogenated deoxyribosides indicate that distemper (Kimes & Bussell, 1968) and rinderpest (Provost et al., 1965) also contain RNA.

Physico-chemical characters. Sensitive to lipid solvents, relatively heat sensitive and unstable at low pH.

Haemagglutination. Agglutination of monkey RBC by measles virus (Peries & Chany, 1960; 1961; 1962) and of rabbit, guinea-pig, mouse, rat and monkey RBC by rinderpest virus (Liess, 1964; Provost & Borredon, 1968) has been described. Haig (1956) reports 'irregular partial haemagglutination' of chick and guinea-pig RBC by distemper virus.

Antigenic properties. Antigenic relationships among measles, distemper and rinderpest may be summarised as follows:

Measles and distemper

People with a history of measles are particularly apt to have distemper-neutralizing antibodies in their sera (Carlstrom, 1957). Measles virus produces some immunity to distemper in ferrets (Adams & Imagawa, 1957). Dogs injected with measles develop poor antibodies to distemper, but nevertheless show resistance on challenge (Warren et al., 1960; Gillespie & Karzon, 1960). Dogs recovered from distemper show antibodies to measles.

Measles and rinderpest

Sera of rinderpest-immune cattle have some neutralizing power against measles (Imagawa et al., 1960). Antibodies to rinderpest may develop in the course of measles and *vice versa* (Plowright & Ferris, 1959; Yamanouchi et al., 1969).

Distemper and rinderpest

Rinderpest has immunizing power against distemper in dogs and ferrets (Goret et al., 1960; Polding & Simpson, 1959). Rinderpest sera have neutralizing properties against distemper (Imagawa et al., 1960). Distemper virus will give some immunity against rinderpest to cattle (Gilbert et al., 1960).

Measles, distemper and rinderpest

Results of comparative studies by serological and cross-protection tests (DeLay et al., 1965) revealed direct or indirect antigenic relationships among all 3 viruses, although some immune responses were difficult or impossible to reveal depending on particular virus-host combinations. Cross-reactions referable to ribonucleoprotein and haemagglutinin antigens are described by Waterson et al (1963). The immunofluorescence test also reveals cross-reactions among the 3 viruses (Yamanouchi et al., 1970).

In summary, antibody responses both to homologous and to heterologous viruses are best in the natural host to each particular virus. Thus, measles antisera from humans, distemper antisera from dogs and rinderpest antisera from cattle all neutralize heterologous as well as homologous viruses with the possible exception of distemper antisera which react poorly if at all with measles virus (Imagawa, 1968).

Measles

Synonyms: Morbilli. Masern. Rougeole. Rubeola.

Reviews: Babbott & Gordon (1954—for older literature). McCarthy (1959). Black, Reissig & Melnick (1959). Conference (1962). Waterson (1965).

Physico-chemical characters. Mature virions have a buoyant density of 1·23 to 1·24 in CsCl (Numazaki & Karzon, 1966a). The activation energy for thermal inactivation was estimated as 62·000 calories (Arita & Matumoto, 1968). Stable between pH 5·5 and 9 for 3 hours at 0°C; inactivated below pH 4·5 (Black, 1959). Formalin in 1:4,000 inactivated in 4 days at 37°C. Rates of inactivation by ultra-violet light and gamma rays are reported by Palm & Black (1961). The action of sodium deoxycholate on infectivity (marked reduction) and other biologic properties of measles virus is described by Norrby (1966). Infectivity not inactivated by 1 M hydroxylamine (Kimes & Bussell, 1968).

Haemagglutination. A number of distinct haemagglutinating fractions extracted from infected cells or derived from purified virus have been described (Peries & Chany, 1961, 1962; Norrby et al., 1964; Norrby, 1966). A salt-dependent haemagglutinating particle (SDA) which adsorbs and agglutinates only in hypertonic ion concentrations is described by Schluederberg & Nakamura (1967). Cells treated with influenza virus or receptor-destroying enzyme are still susceptible. The haemagglutinin receptor is inactivated by formaldehyde, unlike that of typical myxoviruses (Peries & Chany, 1961). Failure to elute spontaneously from red cells suggests absence of neuraminidase from virions, but association of this enzyme with measles virus has been reported (Howe et al., 1969). Haemadsorption by measles virus is described by Rosanoff (1961) and haemolytic activity by Peries & Chany (1960). Correlations between haemagglutinating, haemolytic and cytopathic activities are described by Oddo et al. (1967).

Antigenic properties. Strains are apparently uniform antigenically. Neutralizing antibody can be readily measured by plaque techniques in patas kidney monolayers (Hsiung et al., 1958). Soluble antigens can be detected by complement fixation (Numazaki & Karzon, 1966b). Correlations between structural and antigenic viral components are described by Waterson et al. (1963), who find that cross-reactions among measles, distemper and rinderpest are referable to both the ribonucleoprotein and the haemagglutinin antigens. For other antigenic relations to distemper and rinderpest cf. p. 238.

Cultivation. *In fertile hens' eggs.* The virus was adapted to growth in the chick amnion after prior cultivation in tissue cultures of human amnion (Milovanovic et al., 1957). Adaptation to growth on CAM with inflammatory and necrotic lesions resembling those produced by canine distemper virus has been reported by Cabasso & Frickey (1966).

In tissue culture. Work on cultivation has been reviewed by Matumoto (1966). Enders & Peebles (1954) cultivated measles virus in human and monkey tissue cultures. The virus has since been propagated in a number of continuous cell-lines of primate origin, malignant or otherwise. An established line of green monkey-kidney cells (Vero) is particularly useful for the study of measles, distemper and rinderpest (Shishido et al., 1967). Bovine (Schwarz & Zirbel, 1959) and dog-kidney cultures have been used with success, and Wright (1957) reported production of intranuclear inclusions in hamster, mouse and guinea-pig cultures. The egg-adapted virus can be grown in chick-embryo tissue cultures (Katz et al., 1958). Primate kidney cultures seem best for primary isolation, but higher titres may subsequently be attained in other tissues. Plaque assays have been developed (Mares et al., 1966; Hsiung et al., 1958).

Changes produced in culture consist of formation of syncytia and of vacuoles in the cytoplasm of the syncytia, formation of nuclear and cytoplasmic inclusions and finally cell degeneration. Cascardo & Karzon (1965) describe a 'fusion factor' concerned in formation of syncytia and also the effects of the virus on mitosis and other nuclear happenings. Virus replication is not dependent on cellular DNA function (Matumoto et al., 1965). Yamanouchi et al. (1970) describe the synthesis of both nucleocapsid and envelope antigens entirely in the cytoplasm, although other workers (Rapp et al., 1960; Roizman & Schluederberg, 1961) observed intranuclear as well as cytoplasmic antigen. Accumulation of nucleocapsids both in the cytoplasm and in the nucleus is seen in electron micrographs of infected cells (Matsumoto, 1966; Nakai et al., 1969; Anisimova et al., 1970).

Pathogenicity. In man: The fever, coryza and, soon after, the rash and Koplik spots, are familiar. The incubation period is about 14 days. In virgin communities, almost 100 per cent of infections occur, affecting all ages, but in towns children are chiefly attacked. Mortality is far from negligible especially where, as in parts of West Africa, there is also protein deficiency. Broncho-pneumonia and encephalitis are responsible for many of these deaths. Otitis media and other complications are also troublesome. There is evidence, from the rare cases which die early, of giant cell formation and other changes in lymphoid tissues generally, virus being present in blood and various organs (Sherman & Ruckle, 1958). A fatal giant cell pneumonia is thought to be generally the result of measles.

An association of measles virus with subacute sclerosing panencephalitis (SSPE) was suggested by electronmicroscopic (Bouteille et al., 1965; Tellez-Nagel & Herter, 1966) and immunological (Connolly et al., 1967; 1968; Dayan et al., 1967) evidence. This was confirmed by the demonstration of measles antigen and syncytial formation in cell cultures derived from the brain of a patient with SSPE (Baublis & Payne, 1968), by propagation of measles virus from such cultures (Chen et al., 1969; Horta-Barbora et al., 1969) and by production of encephalitis with antibody response to measles virus in ferrets inoculated with human brain cultures carrying the SSPE agent or with monkey-kidney cells infected with the same agent (Katz et al., 1970). Structures resembling paramyxovirus nucleocapsids have also been found in nuclear inclusions present in muscle cells from a patient with polymositis (Chou, 1968). Serological evidence of a possible association of measles with multiple sclerosis has been presented by Panelius et al. (1970).

Experimentally. Rhesus, cynomolgus and other monkeys may be susceptible and show rash, lymphopenia and fever, but many contract an undetected infection in captivity and are thereafter immune. It is probably from such inapparently infected monkeys that an agent (MINIA) (Ruckle, 1958) indistinguishable from measles has been isolated in cultures. The acquired resistance of many monkeys means that they are most unreliable experimental animals for measles.

Production of viraemia in puppies is reported by Sergiev et al. (1959).

Tissue-culture virus has been adapted to growth in suckling mice (IC) (Imagawa & Adams, 1958). A neurotropic variant fatal in mice and hamsters was obtained by Burnstein et al. (1964). Isolates from subacute sclerosing panencephalitis have been propagated in ferrets (Katz et al., 1970) and hamsters (Byington et al., 1970).

Pathological lesions. The characteristic giant cells were reported as occurring in tonsils and pharynx early in the disease by Warthin (1931). The intranuclear inclusions were first described by Torres & Teixeira (1932); these are acidophilic, often band-like and differ from typical Cowdry type A inclusions in that they may be multiple and there is no margination of basophilic chromatin with disappearance of nucleolus. Eosinophilic cytoplasmic inclusions also occur. Taniguchi et al (1954) described intranuclear inclusions in bronchial epithelium of infected monkeys. Giant cell formation in lymphoid tissues of monkeys inoculated with measles viruses has been described by Yamanouchi et al. (1970).

Ecology. The virus is transmitted from the respiratory tract especially at times before or near the onset of the rash; the conjunctiva may be an important portal of entry. Virus has been recovered from urine (Gresser & Katz, 1960). There have been extensive studies of the epidemiology in relation to the 2-year cycle seen in many urban communities.

Control. Human γ globulin has been extensively used in the past to give an attenuated disease leaving immunity behind it.

Vaccines. Inactivated and live vaccines of several kinds have been used (Conference, 1962). Formalinized vaccines made from virus grown in tissue culture have their advocates, but most attention has been paid to living virus attenuated by growth in chick embryo or in other ways. Such vaccines are undoubtedly effective, but they do not immunize without causing in many children fever and rash, albeit much milder than the symptoms seen in the natural disease. Reactions may be controlled by giving γ globulin along with vaccine, or it may prove better to give a dose of killed virus vaccine followed later by attenuated virus. The comparative value of various vaccines has been reviewed (WHO, 1964).

Dog Distemper

Synonyms: Hard-pad. Maladie de Jeune age. Hundestaupe.

Classical account: Laidlaw & Dunkin (1926). Dunkin & Laidlaw (1926 a and b).

Physico-chemical characters. Inactivated by heating 1 hour at 55° or 30 minutes at 60°. Celiker & Gillespie (1954) found that egg-cultivated virus was more easily killed; they describe resistance to various physical and chemical agents. Bussell & Karzon (1962) determined the half-life of tissue-culture virus at various temperatures. It survives for months at −10°C and indefinitely at −76°C or when lyophilized. Labile at pH 3·0 and relatively stable at pH 4·5 or higher (Kimes & Bussell, 1968). Readily inactivated by visible light (Nemo & Cutchins, 1966). Dried virus is fairly stable at room temperature, but not above 32° (Piercy, 1961). It is ether-sensitive, inactivated by 0·1 per cent formaldehyde in a few hours, also by 1 per cent lysol. Virus inactivated by formalin or the photodynamic action of methylene blue retains antigenicity. Inactivated by 1 M hydroxylamine (Kimes & Bussell, 1968).

Antigenic properties. Neutralizing antibodies can be demonstrated in the various systems available (ferrets, dogs, eggs, mice, tissue culture). A complement-fixing antigen is present especially in the spleen of affected animals and in membranes of infected eggs. Specific precipitates are formed in agar by the Ouchterlony technique (Mansi, 1957). Relations with measles and rinderpest are discussed on p. 238.

Cultivation. *In fertile eggs.* Virus has been adapted to growth on the CAM by Haig (1948) and Cabasso & Cox (1949). Lesions appear on the membranes after a number of passages. The virus is attenuated for ferrets, and dogs and can be used for immunization.

In tissue cultures. The virus has been grown in ferret and dog kidney and other tissues, but not all strains grow well. The egg-adapted strain grows also in chick-embryo tissue cultures (Cabasso et al., 1959) and will give small plaques on chick-embryo fibroblast monolayers under an agar overlay (Bussell & Karzon, 1962). The virus could be adapted to grow and produce CPE in a number of primary and continuous cell-lines of primate origin (Bussell & Karzon, 1965).

Cytopathic effects produced in cultures include formation of giant cells and cytoplasmic inclusions (Rockborn, 1958; Hopper, 1959).

Pathogenicity. The disease particularly attacks young dogs, causing in typical cases, after 4 or 5 days' incubation, diphasic fever, discharge from eyes and nose, sometimes skin eruption, and, during the secondary fever, vomiting, diarrhoea and often pneumonia (Dunkin & Laidlaw, 1926b). Symptoms in puppies are, however, very variable; they may be quite mild. Mortality is variable, on the average about 50 per cent. Nervous symptoms, particularly fits or muscular jerkings, may occur and persist for long periods. Distemper occurs naturally in wolves, foxes, raccoons and mink and may in them simulate rabies.

The disease described as 'hard-pad' (McIntyre et al., 1948) is now regarded as a form of distemper. There is tenderness, then keratinization of the skin of the feet, usually followed by nervous involvement and commonly death. Either no virus is recoverable or that of typical distemper (Koprowski et al., 1950). Canine rhinotonsillitis is thought by Florio et al. (1965) to be closely related to distemper. Distemper is transmissible to all *Canidae*, *Mustelidae*, particularly ferrets; of *Viverridae* only the Binturong and of *Procyonidae* only raccoons appear susceptible (Goss, 1948). In ferrets dog distemper is usually fatal (Dunkin & Laidlaw, 1926a). The incubation period in this animal is about 10 days and characteristic features are vesicle formation around the mouth, discharge and later crusting of eyes and nose. Neurotropic strains of virus have been described. The virus has been adapted to multiplication in suckling mice (Morse et al., 1953; Gutierrez & Gorham, 1955) and hamsters (Cabasso et al., 1955; Motohashi et al., 1964).

Pathological lesions. In fatal cases in dogs broncho-pneumonia was the rule, but is rarely seen now. There may be gastro-intestinal inflammation often with necrosis of mucosa. Histologically there occur cytoplasmic and intranuclear inclusions; the latter may be hard to find, but they are not uncommon in the bladder. Cytoplasmic inclusions are to be found in circulating lymphocytes (Cornwell et al., 1965).

Ecology. The disease is contagious and so readily air-borne that Dunkin & Laidlaw (1926b) could not safely house normal dogs in the same building as infected ones. Virus is excreted also in urine and faeces. Epidemics may occur in virgin soil, but for the most part clinical distemper in dogs is an endemic disease. Virus may persist in carrier dogs long enough to infect the next generation of susceptibles (Gorham, 1966). Virus has been found in kidney tissue cultures from apparently normal dogs (Kasza, 1968).

Control. Several methods of vaccination have been described.

1. Formalinized virus may be given and followed by live virus to give lasting immunity (Laidlaw & Dunkin, 1928).
2. Side-to-side immunization entails the giving of living virus and antisera simultaneously, but separately.
3. Subsequently virus attenuated for dogs by continual ferret-passage ('distemperiod' virus of Green, 1939) or by cultivation in eggs (Cabasso et al., 1951) has given the best results.

Now tissue culture propagated virus seems superior. (Cabasso et al., 1962; Prydie, 1968). Use of measles vaccine to protect pups against distemper has been described (McManus, 1968).

Rinderpest

Synonym: Cattle-plague. Peste des petits ruminants.

Reviews: Curasson (1942). Scott (1964). Jacotot & Mornet (1967). Plowright (1968).

Physico-chemical characters. The half-life at 56° is short, but a small proportion may survive 50 to 60 minutes at 56°C (Plowright & Ferris, 1961) or 60°C for 30 minutes (de Boer & Barber, 1964). Optimum stability at pH 7·2 to 8. Studies on the sensitivity to various chemicals and physical agencies are comprehensively reviewed by Plowright (1968).

Antigenic properties. Can be studied by complement fixation, by immunodiffusion or by neutralization tests in animals or in tissue culture. A useful test in countries to which the virus may not be introduced is based on the power of antirinderpest serum to inhibit HA by measles virus (Waterson et al., 1963). Multiple antigens have been revealed by immunodiffusion. Stone (1960) demonstrated 1 thermostable and 1 thermolabile antigen; whereas Ishii et al (1964) obtained an additional thermostable precipitating antigen. This test is useful for detecting antigen in lymph nodes in acute cases (White, 1962). Antigenic relations with dog distemper and with measles have been reviewed (p. 238). A possible antigenic relationship with NDV is suggested by Kazimi & Majeed (1966).

Cultivation. *In fertile eggs.* Some but not all strains have proved adaptable to growth in eggs—either the CAM or yolk. Growth in the yolk sac was best after preliminary adaptation to the CAM (Shope et al., 1946). Virus passed in yolk sacs was attenuated for cattle after 20 or more passages (Jenkins & Shope, 1946). The virus was not regularly lethal for chick embryos.

In tissue culture. Virus has been cultivated in primary cultures of cattle, sheep, goat, pig, dog, hamster, monkey, human and chick-embryo tissues and in continuous lines of calf, pig, hamster, dog, rabbit, human and monkey cells (for references see Plowright, 1968). Growth is usually accompanied by the formation of syncytia and cytoplasmic and nuclear inclusions. Virus antigen appears first in the nucleus and later in the cytoplasm (Liess, 1964). A plaque assay in bovine-kidney cells is described by McKercher (1964).

Distribution. Rinderpest is a serious plague in many parts of Asia and Africa, though South Africa is now free. Until the last century Europe was affected also. On one occasion it was introduced into Brazil.

Pathogenicity. Bovines (ox, zebu, water buffalo and yak) are the most important natural hosts, but infection also occurs in other domestic animals such a sheep and goats ('peste des petits ruminants'), pigs and possibly camels (see Curasson, 1942). A number of wild animals including hippopotamus, wild pig, warthog, giraffe, deer and several wild bovines are also susceptible (see Scott, 1964). In cattle, strains vary in virulence and breeds of cattle vary in susceptibility. In acute rinderpest, there is high fever; constipation is followed by severe diarrhoea. Nasal discharge and mucosal erosions occur, and animals may be overcome and die in 2 to 6 days. With milder strains in resistant cattle the disease may be overlooked and may be inadvertently spread.

Experimentally, the infection has been adapted to goats (Edwards, 1930); it becomes virulent for them, but attenuated for cattle (caprinized virus). Rinderpest has also been adapted to rabbits (lapinized virus), producing as a rule an inapparent infection or, at the most, fever (Nakamura et al., 1938; Baker, 1946). However, after many passages, it may produce a fatal disease in rabbits (Scott, 1959). Infection has also been produced in guinea-pigs (Baker et al., 1946), suckling mice (Scott & Witcomb, 1958; Imagawa, 1965) hamsters (Nakamura et al., 1957; Scott & Witcomb, 1958), rats (Curasson, 1942) and susliks (*Citellus*) Inoue et al. (1930).

Pathological lesions. Virus in the blood is contained in leucocytes. The main lesions consist of inflammation and ulceration affecting the alimentary canal from mouth to rectum. There may be false membranes in stomach and intestines. A secondary patchy pneumonia may occur. The primary lesions appear to be in lymphoid tissues (Plowright, 1964) and an attenuated strain attacked these tissues only (Taylor & Plowright, 1965).

Ecology. Transmission is by direct contact. Virus is present in nasal discharges, urine and faeces. Carriers may transmit infection. Transmission from sheep and goats to cattle and *vice versa* has been described by Zwart & Macadam (1960a, b). Infected premises are a source of danger for at most a short period. Spread of infection to areas previously free of the virus has always been associated with introduction of live, sick animals rather than with indirect contagion through infected meat, food stuffs and the like (Anon, 1966). The importance of wild animals in the ecology of rinderpest is indicated by the demonstration of widespread infection among African game animals (Plowright & McCulloch, 1967; Taylor & Watson, 1967).

Control. Slaughter may be necessary when there is introduction into unaffected areas. Hyperimmune sera have been successfully used to give transient protection.

Vaccination. Side-to-side inoculation of serum and virus has been used; so have vaccines made by inactivating virus with formalin, phenol and glycerol or chloroform. These are being superseded by living virus attenuated by passage in goats, rabbits, eggs or tissue cultures. There is an extensive literature on the results of tests of various vaccines. It seems that different degrees of attenuation may be desirable according to the susceptibility of the breed of cattle. With some but not all vaccines it appears that some febrile reaction must be produced if good immunity is to result; also that some attenuated strains may revert to virulence on passage through cattle. A mass vaccination campaign is being carried out in Africa with encouraging results (Lepissier & McFarlane, 1967).

HUMAN RESPIRATORY SYNCYTIAL VIRUS

Synonym: Chimpanzee coryza agent (CCA). RSV.

Review: Hilleman (1963).

Morphology. Pleomorphic, 80 to 120 nm in diameter although larger forms may be found (Zakstelskaya et al., 1967; Joncas et al., 1969). Virions consist of an envelope with projections 12 to 15 nm long (Zakstelskaya et al., 1967; Norrby et al., 1970) and a helical nucleocapsid with diameter variously estimated between 12 and 18 nm (Eckert et al., 1965; Joncas et al., 1969) and a pitch of 6·5 to 7 nm (Bloth & Norrby, 1967; Joncas et al., 1969). Filaments resembling nucleocapsids are seen in cytoplasmic inclusions, and virus assembly takes place at the cell surface by a process of budding through modified cell membrane (Norrby et al., 1970).

Chemical composition. Lack of inhibition by halogenated deoxyribosides suggests that the viral nucleic acid is RNA (Hamparian et al., 1963).

Physico-chemical characters. Buoyant density in CsCl between 1·22 and 1·24 (Coates et al., 1966b). Survives quick freezing to $-70°$, but some workers report that isolations are many fewer from frozen than from fresh material. Inactivated in 1 hour at 56° (half-life 2·8′) or in 170 hours at 37° (half-life 7·2 hours) (Jordan, 1962). Hambling (1964) studied the rate of inactivation at various temperatures. Sodium and magnesium salts and glucose protect the virus against heat inactivation (Rechsteiner, 1969). Labile at pH 3.

Haemagglutination. Extensive attempts to demonstrate haemagglutination or haemadsorption have failed (Richman et al., 1961).

Antigenic characters. The neutralization and CF tests may be used for diagnosis. Some workers find the latter less sensitive. Not all strains of virus are serologically alike (Coates et al., 1963; Doggett & Taylor-Robinson, 1965). Antigenic variation is limited and does not appear to be progressive (Coates et al., 1966a). CF antigens separable from infective virus have been described (Coates et al., 1966b; Bloth & Norrby, 1966).

Cultivation. The virus grows in cultures derived from human cancer cells (KB, HeLa, Hep. 2, also Chang liver cells), less readily at times in monkey-kidney cultures. Syncytia are produced—hence the name—readily in Hep. 2, less readily in some other cell-lines (Jordan, 1962). Eosinophilic cytoplasmic inclusions are frequent. Virus antigen revealed in the cytoplasm by immunofluorescence (Bennett & Hamre, 1962). A plaque assay is described by Kisch & Johnson (1963).

Pathogenicity. Causes minor respiratory infection with rhinitis and cough. In children, especially infants, the lower respiratory tract may be involved in some 50 per cent of cases and bronchitis, bronchiolitis or broncho-pneumonia, sometimes fatal, may occur (Beem et al., 1960; Chanock et al., 1961a; McLelland et al., 1961). Antibody, present in a large number of persons over 4, does not always prevent reinfection. Such reinfections may take the form of a common cold (Johnson et al., 1961). Afebrile colds have been produced after an incubation period of about 5 days by deliberate infection of adult volunteers (Kravetz et al., 1961). Immunological reactions involving humoral antibodies play an important role in the pathogenesis of RS infections (Chanock et al., 1968).

The virus has also caused outbreaks of colds amongst captive chimpanzees, from which, indeed, the virus was originally isolated (Morris et al., 1956). Coates & Chanock (1962) recovered virus after inoculating ferrets, mink and marmosets. Serial passage in ferrets was possible. There were no symptoms, but nasal lesions with giant cells and inclusion bodies were found 3 to 6 days after inoculation. Cavallaro & Maassab (1966) adapted a strain to infect mice by IC passage; it proved fatal after 38 passages.

Ecology. Infections with the virus are prevalent during winter months in apparently normal persons during outbreaks.

Control. Use of inactivated vaccine has led to unfavourable results (Kapikian et al., 1969), and the possible use of temperature-sensitive mutants as attenuated virus vaccines is being explored (Friedwald et al., 1968).

BOVINE RESPIRATORY SYNCYTIAL VIRUS

A virus antigenically related to human respiratory syncytial virus was isolated from cattle involved in an outbreak of mild respiratory disease in Switzerland (Paccaud & Jacquier, 1970). The virus grows in bovine kidney- and lung-tissue cultures, but not in human, monkey, hamster or sheep cells. Growth in bovine cells is accompanied by the formation of syncytia and cytoplasmic eosinophilic inclusions. The virus is chloroform- and ether-sensitive, acid-sensitive, readily inactivated at 56°C and stable at −70°C. It shows a reciprocal antigenic relationship with human RS virus by virus neutralization. A similar virus has been isolated from cattle in Japan (Inaba et al., 1970) and in Belgium (Wellemans et al., 1970). These viruses should not be confused with the bovine syncytial virus described by Malmquist et al. (1969) (see p. 163).

PNEUMONIA VIRUS OF MICE

Original Account: Horsfall & Hahn (1940).

Morphology and development. Virions are approximately spherical, 80 to 120 nm in diameter; filamentous forms up to 3 μm long may be seen. Virions consist of an envelope covered with projections and a helical nucleocapsid 12 to 15 nm in diameter. Cytoplasmic inclusions contain aggregates of nucleocapsids. Virions are assembled by budding at the cell surface (Compans *et al.*, 1967).

Chemical composition. Lack of inhibition by halogenated deoxyribosides (Tennant et al., 1966) and acridine orange-staining reactions (Harter & Choppin, 1967) suggest that the viral nucleic acid is single-stranded RNA.

Physico-chemical characters. Virion density in tartrate 1·15 g/ml (Harter & Choppin, 1967). Virus infectivity is readily destroyed at 56°C, labile at room temperature. Virus survives well at −76° and after freeze-drying (Horsfall & Hahn, 1940).

Haemagglutination. Agglutinates at room temperature RBCs of mouse, hamster and, irregularly, rabbit. HA of virus in mouse-lung suspensions is enhanced by heating at 75°C (Mills & Dochez, 1944; 1945), but virus grown in cultures of hamster cells (BHK21) agglutinates without previous heating (Harter & Choppin, 1967). A non-infectious HA with density 1·13 g/ml in tartrate is described by the latter workers.

Antigenic properties. Infectivity and haemagglutination are inhibited by sera from naturally infected or immunized animals.

Cultivation. There is no good evidence of multiplication in fertile eggs, but survival after 10 serial passes in chick-embryo tissue cultures has been reported (Horsfall & Hahn, 1940). Tennant & Ward (1962) grew the virus in cultures of suckling-hamster kidney, recognizing its presence by haemadsorption. Growth in continuous line of hamster-kidney cells (BHK21) with formation of inclusions and accumulation of viral antigen in the cytoplasm is described by Harter & Choppin (1967). A plaque assay in BHK21 cells has been developed by Shimonaski & Came (1970).

Distribution. Sixty-five per cent of mouse colonies tested in the United States showed serological evidence of infection by this virus. Viruses antigenically related to it have been found in Syrian hamsters (Pearson & Eaton, 1940) and in cotton-rats (*Sigmodon*) and rabbits (Eaton & van Herick, 1944).

Pathogenicity. Virus can be activated by serial intranasal passage of lung suspensions in apparently normal mice at 7 to 9 day (not at shorter) intervals. Lung lesions may appear after 1 or 2 passes and gross consolidation after 9 or 10 (Horsfall & Hahn, 1940). It is necessary, however, to pass in mice of an uninfected stock; otherwise nothing may become apparent. Only intranasal inoculation is successful. Syrian hamsters also develop lung lesions after intranasal inoculation, but in them virus is not readily transmissible in series. Davenport (1949) passed the virus in series through 11 rabbits intratracheally, but did not increase the virus' pathogenicity for the rabbit.

Pathological lesions in lungs of infected mice showed 'dense and often bulky accumulation of cells . . . largely mononuclear' round the bronchi and blood-vessels (Horsfall & Hahn, 1940).

Ecology. Virus is poorly transmitted from mouse to mouse by contact. It is not transmitted through the placenta and is perhaps maintained as a transient inapparent infection which is focal and enzootic in mouse colonies (Tennant et al., 1966).

Control. Live or heated virus given intraperitoneally, or sublethal doses intranasally, readily immunize mice against the virus.

Polysaccharides from a number of sources—streptococci, Friedlander bacilli, Shigella, blood group A specific substance and agar—all have a sparing effect on mice inoculated with the virus (Horsfall & McCarty, 1947).

NARIVA VIRUS

This virus, isolated from forest rodents in Trinidad (Tikasingh et al. 1966) has been shown to resemble paramyxoviruses in several biological properties and on morphology (Karabatsos et al., 1969; R. Walder, 1971).

REFERENCES

Abinanti, F. R., Chanock, R. M., Cook, M. K., Wong, D., & Warfield, M. (1961) *Proc. Soc. exp. Biol. Med., U.S.A.*, **106**, 466.

Adams, J. M., & Imagawa, D. T. (1957) *Proc. Soc. exp. Biol. Med., U.S.A.*, **96**, 240.

Ahmed, A. A. S., & El-Sisi, M. A. (1965) *Vet. med. J.*, **10**, 219.

Andrewes, C. H., & Allison, A. C. (1961) *J. Hyg., Camb.*, **59**, 285.

Andrewes, C. H., Bang, F. B., & Burnet, F. M. (1955) *Virology*, **1**, 176.

Andrewes, C. H., Bang, F. B., Chanock, R. M., & Zhdanov, V. M. (1959) *Virology*, **8**, 129.

Anon. (1966) *Rev. Elev.*, **19**, 365.

Arita, M., & Matumoto, M. (1968) *Jap. J. Microbiol.*, **12**, 121.

Anisimova, E., Mares, I., & Kyncl, F. (1970) *Arch. ges. Virusforsch.*, **30**, 1.

Babbot, F. L., & Gordon, J. E. (1954) *Am. J. med. Sci.*, **228**, 334.

Baker, J. A. (1946) *Am. J. vet. Res.*, **7**, 179.

Baker, J. A., Terrence, J., & Greig, A. S. (1946) *Am. J. vet. Res.*, **7**, 189.

Bakos, K., & Dinter, Z. (1960) *Zentbl. Bakt. Parasitkde, I. Abt. Orig.*, **180**, 1.

Bankowski, R. A. (1964). Cytopathogenicity of Newcastle disease virus in *Newcastle disease virus: An evolving pathogen*. Ed. Hanson. Madison: Univ. of Wisconsin Press.

Bankowski, R. A. R., Crostvet, R. E., & Clark, G. T. (1960) *Science*, **132**, 292.

Barry, R. D., Ives, D. R., & Cruickshank, J. G. (1962) *Nature, Lond.*, **194**, 1139.

Baublis, J. V., & Payne, F. E. (1968) *Proc. Soc. exp. Biol. Med., U.S.A.*, **129**, 593.

Beale, A. J., McLeod, D. L., Stackiw, W., & Rhodes, A. J. (1958) *Br. med. J.*, **1**, 302.

Beem, M., Wright, F. H., Hamre, D., Egerer, R., & Oehme, M. (1960) *New Engl. J. Med.*, **263**, 523.

Bennett, C. R., Jr., & Hamre, D. (1962) *J. Infect. Dis.*, **110**, 8.

Berkaloff, A. (1963) *J. Microsc.*, **2**, 633.

Bikel, I., & Duesberg, P. H. (1969) *J. Virol.*, **4**, 388.

Black, F. L. (1959) *Virology*, **7**, 184.

Black, F. L., Reissig, M., & Melnick, J. L. (1959) *Adv. Virus Res.*, **6**, 205.

Blair, C. D. & Duesberg, P. H. (1970) *Ann. Rev. Microbiol.*, **24**, 539.

Blair, C. D., & Robinson, W. S. (1968) *Virology*, **35**, 537.

Blair, C. D., & Robinson, W. S. (1970) *J. Virol.*, **5**, 639.

Blough, H. A., & Lawson, D. E. M. (1968) *Virology*, **36**, 286.

Bloth, B., & Norrby, E. (1966) *Arch. ges. Virusforsch.*, **19**, 385.

Bloth, B., & Norrby, E. (1967) *Arch. ges. Virusforsch.*, **21**, 71.

Bolgel, K., & Liebelt, J. (1963) *Zentbl. Bakt. Parasitkde I. Abt. Orgi.*, **190**, 322.

Bolin, V. S., Anderson, J. A., & Leymaster, G. R. (1952) *Proc. Soc. exp. Biol. Med., U.S.A.*, **79**, 7.

Bonissol, C., Sisman, J., & Lepine, P. (1968) *Annls Inst. Pasteur, Paris*, **114**, 537.

Bouteille, M., Fontaine, C., Vedrenne, Cl., & Delareu, J. (1965) *Revue neurol.*, **118**, 454.

Bower, R. K. (1958) *J. Bact.*, **75**, 496.

Brandt, C. D. (1961) *Virology*, **14**, 1.

Bratt, M., & Robinson, W. S. (1967) *J. molec. Biol.*, **23**, 1.

Breese, S. S., & de Boer, C. (1963) *Virology*, **19**, 340.

Brown, A. C. L. (1965) *Proc. r. Soc. Med.*, **58**, 801.

Bukrinskaya, A. G. (1960) *Vop. Virusol.*, **6**, 156.

Bukrinskaya, A. G., Ho, Y-D., & Gorbunova, A. S. (1962) *Acta virol., Prague*, **6**, 352.

Burnet, F. M., & Anderson, S. G. (1946) *Br. J. exp. Path.*, **27**, 236.

Burnet, F. M., & Bull, D. R. (1943) *Aust. J. exp. Biol. med. Sci.*, **21**, 55.

Burnstein, T., & Bang, F. B. (1958) *Bull. Johns Hopkins Hosp.*, **102**, 127 and 135.

Burnstein, T., Jensen, J. H., & Waksman, B. H. (1964) *J. infect. Dis.*, **114**, 265.

Burr, M. M., & Nagler, F. R. (1953) *Proc. Soc. exp. Biol. Med., U.S.A.*, **83**, 714.

Bussell, R. H., & Karzon, D. T. (1962) *Virology*, **18**, 589.
Bussell, R. H., & Karzon, D. T. (1965) *Arch. ges. Virusforsch.*, **17**, 163.
Buthala, D. A., & Soret, M. G. (1964) *J. infect. Dis.*, **114**, 226.
Byington, D. P., Castro, A. E., & Burnstein, T. (1970) *Nature, Lond.*, **225**, 554.
Cabasso, V. J., & Cox, H. R. (1949) *Proc. Soc. exp. Biol. Med. U.S.A.*, **71**, 246.
Cabasso, V. J., Burkhart, R. L., & Leaming, J. D. (1951) *Vet. Med.*, **46**, 167.
Cabasso, V. J., Douglas, J. M., Stebbins, M. R., & Cox, H. R. (1955) *Proc. Soc. exp. Biol. Med., U.S.A.*, **88**, 199.
Cabasso, V. J., & Frickey, P. H. (1966) *Nature, Lond.*, **210**, 1233.
Cabasso, V. J., Kiser, K., & Stebbins, M. R. (1959) *Proc. Soc. exp. Biol. Med., U.S.A.*, **100**, 551.
Cabasso, V. J., Kiser, K., Stebbins, M. R., & Cooper, H. K. (1962) *Am. J. vet. Res.*, **23**, 394.
Canchola, J., Vargosko, A. J., Kim, H. W., Parrott, R. H., Christmas, E., Jeffries, B., & Chanock, R. M. (1964) *Am. J. Hyg.*, **79**, 357
Cantell, K. (1961) *Adv. Virus. Res.*, **8**, 123.
Carlstrom, G. (1957) *Lancet*, **2**, 344.
Carlstrom, G. (1962) *Am. J. Dis. Child.*, **103**, 287.
Cascardo, M. R., & Karzon, D. T. (1965) *Virology*, **26**, 311.
Cavallaro, J. J., & Maassab, H. F. (1966) *Proc. Soc. exp. Biol. Med., U.S.A.*, **121**, 37.
Celiker, A., & Gillespie, J. H. (1954) *Cornell Vet.*, **44**, 276.
Chanock, R. M. (1955) *Proc. Soc. exp. Biol. Med., U.S.A.*, **89**, 379.
Chanock, R. M. (1956) *J. exp. Med.*, **104**, 555.
Chanock, R. M., & Coates, H. V. (1964) in *Newcastle disease virus: An evolving pathogen.* Ed. Hanson. Madison: Univ. of Wisconsin Press.
Chanock, R. M., Johnson, K. M., Cook, M. K., Wong, D. C., & Vargosko, A. (1961a) *Am. Rev. resp. Dis.*, **83:2**, 125.
Chanock, R. M., Kim, H. W., Vargosko, A. J., Deleva, A., Johnson, K. M., Cumming, C., & Parrott, R. H. (1961b) *J. Am. med. Ass.*, **176**, 647.
Chanock, R. M., Parrott, R. H., Cook, M. K., Andrews, B. E., Bell, J. A., Reichelderfer, T., Kapikian, A. Z., Mastrota, F. M., & Huebner, R. J. (1958) *New Engl. J. Med.*, **258**, 207.
Chanock, R. M., Parrott, R. H., Johnson, K. M., Kapikian, A. Z., & Bell, J. A. (1963) *Am. Rev. resp. Dis.*, **88**, 152.
Chanock, R. M., Parrott, R. H., Kapikian, A. Z., Kim, H. W., & Brandt, D. D. (1968) *Perspect. Virol.*, **6**, 125.
Chany, C., & Cook, M. K. (1960) *Annls Inst. Pasteur, Paris*, **98**, 920.
Chen, T. T., Watanabe, I., Zeman, W., & Mealey, J., Jr. (1969) *Science*, **163**, 1193.
Choppin, P. (1964) *Virology*, **23**, 224.
Choppin, P. W., & Stoeckenius, W. (1964) *Virology*, **23**, 195.
Chou, S-M. (1968) *Archs Path.*, **86**, 649.
Chu, C. M., Liang, J. K., & Wen, C. C. (1957) *Scientia sinica*, **6**, 1065.
Chun, H., & Chu, C. M. (1956) *Acta microbiol. sinica*, **4**, 47.
Churchill, A. E. (1963a) *Nature, Lond.*, **197**, 409.
Churchill, A. E. (1963b) *Br. J. esp. Path.*, **44**, 529.
Coates, H. V., Alling, D. W., & Chanock, R. M. (1966a) *Am. J. Epidemiol.*, **83**, 299.
Coates, H. V., & Chanock, R. M. (1962) *Am. J. Hyg.*, **76**, 302.
Coates, H. V., Forsyth, B. R., & Chanock, R. M. (1966b) *J. Bact.*, **91**, 1263.
Coates, H. V., Kendrick, L., & Chanock, R. M. (1963) *Proc. Soc. exp. Biol. Med., U.S.A.*, **112**, 958.
Cohen, S. M., Bullivant, S., & Edwards, G. A. (1962) *Arch. ges. Virusforsch.*, **11**, 493.
Compans, R. W., & Choppin, P. W. (1968) *Virology*, **35**, 289.
Compans, R. W., & Choppin, P. W. (1967) *Virology*, **33**, 344.
Compans, R. W., Harter, D. H., & Choppin, P. W. (1967) *J. exp. Med.*, **126**, 267.
Compans, R. W., Holmes, K. V., Dales, S., & Choppin, P. W. (1966) *Virology*, **30**, 411.
Conference (1962) *Am. J. Dis. Child.*, **103**, 219.
Connolly, J. H., Allen, I. V., Hurwitz, L. J., & Millar, J. H. D. (1967) *Lancet*, **1**, 542.
Connolly, J. H., Allen, I. V., Hurwitz, L. J., & Millar, J. H. D. (1968) *Q. Jl Med.*, **37**, 625.

Cook, M. K., Andrews, B. E., Fox, H. H., Turner, H. C., James, W. D., & Chanock, R. M. (1959) *Am. J. Hyg.*, **69**, 250.
Cornwell, H. J. C., Vantsis, J. T., Campbell, R. S. F., & Penny, W. (1965) *J. comp. Path.*, **75**, 19.
Craighead, J. E., & Brennan, B. J. (1968) *Am. J. Path.*, **52**, 287.
Craighead, J. E., Cook, M. K., & Chanock, R. M. (1960) *Proc. Soc. exp. Biol. Med., U.S.A.*, **104**, 301.
Crandell, R. A., Brumlow, W. B., & Davison, V. E. (1968) *Am. J. vet. Res.*, **19**, 2141.
Cruickshank, J. G., Waterson, A. P., Kanarek, A. D., & Berry, D. M. (1962) *Res. vet. Sci.*, **3**, 485.
Cuadrado, R. R. (1966) *J. Immunol.*, **96**, 892.
Cunningham, C. H. (1948) *Am. J. vet. Res.*, **9**, 195.
Curasson, G. (1942) In *Traite de pathologie exotique Veterinaire et Comparee*, **1**, 12. Paris: Vigot Frères.
Dahlberg, J. E., & Simon, E. H. (1969) *Virology*, **38**, 666.
Darrell, R. W., & Howe, C. (1964) *Proc. Soc. exp. Biol. Med., U.S.A.*, **116**, 1091.
Davenport, F. M. (1949) *J. Immunol.*, **63**, 81.
Dawson, P. S., & Patterson, D. S. P. (1967) *Nature, Lond.*, **213**, 185.
Dayan, A. D., Gostling, J. V. T., Greaves, J. L., Stevens, D. W., & Woodhouse, M. A. (1967) *Lancet*, **1**, 980.
De Boer, C. J., & Barber, T. L. (1964) *Arch. ges. Virusforsch.*, **15**, 98.
Deibel, R. (1959) *Virology*, **8**, 262.
Deibel, R., & Hotchin, J. E. (1961) *Virology*, **14**, 66.
Deinhardt, F., & Shramek, G. J. (1969) *Prog. med. Virol.*, **11**, 126.
DeLay, P. D., Stones, S. S., Karzon, D. T., Katz, S., & Enders, J. (1965) *Am. J. vet. Res.* **26**, 1359.
De Meio, J. L. (1963) *J. Bact.*, **85**, 943,
De Meio, J. L., & Walker, D. L. (1957) *J. Immunol.*, **78**, 465.
De Vaux St Cyr, C., & Howe, C. (1966) *J. Bact.*, **91**, 1911.
Dick, E. C., & Mogabgab, W. J. (1961) *Am. J. Hyg.*, **73**, 273.
Dinter, Z., Hermodsson, S., & Hermodsson, L. (1964) *Virology*, **22**, 297.
Ditchfield, J., Macpherson, L. W., Zbitnew, A. (1965) *Can. J. comp. Med.*, **29**, 18.
Doggett, J. E., & Taylor-Robinson, D. (1965) *Arch. ges. Virusforsch.*, **15**, 601.
Doyle, T. M. (1927) *J. comp. Path.*, **40**, 144.
Drzeniek, R., Bogel, K., & Rott, R. (1967) *Virology*, **31**, 725.
Dubois, M-F., Daniel, Ph., & Mehren, A. R. (1970) *Arch. ges. Virusforsch.*, **29**, 307.
Duc-Nguyen, H., & Rosenblum, E. N. (1967) *J. Virol.*, **1**, 415.
Duesberg, P. H., & Robinson, W. S. (1965) *Proc. natn. Acad. Sci., U.S.A.*, **54**, 794.
Dunkin, G. W., & Laidlaw, P. P. (1926a) *J. comp. Path.*, **39**, 201.
Dunkin, G. W., & Laidlaw, P. P. (1926b) *J. comp. Path.*, **39**, 213.
Eaton, M. D., & Van Herick, W. (1944) *Proc. Soc. exp. Biol. Med., U.S.A.*, **57**, 89.
Eckert, E. A., Maassab, H. F., & Cavallaro, J. J. (1965) *J. Bact.*, **90**, 1784.
Edwards, J. T. (1930) *Bull. agric. Res. Inst. Pusa*, No. 199.
Emery, J. B., & Yorke, C. J. (1960) *Virology*, **11**, 313.
Enders, J. F., & Habel, K. (1956) in *Diagnostic procedures for virus and rickettsial diseases*, 2nd ed. p. 281. New York: Amer. publ. Hlth. Ass.
Enders, J. F., Kane, L. W., Maris, E. P., & Stokes, J. (1946) *J. exp. Med.*, **84**, 341.
Enders, J. F., & Peebles, T. C. (1954) *Proc. Soc. exp. Biol. Med., U.S.A.*, **86**, 277.
Ennis, F. A., Douglas, R. D., Stewart, G. L., Hopps, H. E., & Meyer, H. M., Jr. (1968) *Proc. Soc. exp. Biol. Med., U.S.A.*, **129**, 896.
Evans, M. J., & Kingsbury, D. W. (1969) *Virology*, **37**, 597.
Feller, U. Dougherty, R. M., & Distefano, H. (1969) *J. Virol.*, **4**, 753.
Finch, J. T., & Gibbs, A. J. (1969) *J. gen. Virol.*, **6**, 141.
Fischmann, H. R. (1967) *Am. J. Epidemiol.*, **85**, 272.
Fischmann, H. R., & Bang, F. B. (1966) *Proc. Soc. exp. Biol. Med., U.S.A.*, **121**, 966.

Florio, R., Joubert, L., Terry, J., Lapras, J., & van Haverbecke, G. (1965) *Rev. med. Vet.*, **116**, 161.
Friedewald, W. T., Forsyth, B. R., Smith, C. B., Charpure, M. A., & Chanock, R. M. (1968) *J. Am. med. Ass.*, **204**, 690.
Fukai, K., & Suzuki, T. (1955) *Med. J. Osaka Univ.*, **6**, 1.
Fukumi, K., Mizutani, H., Takeguchi, Y., Tajima, Y., Imaizumi, K., Tanaka, T., & Kanedo, J. I. (1962) *Jap. J. med. Sci. Biol.*, **15**, 153.
Fukumi, H., Nishikawa, F., & Kitayama, T. (1954) *Jap. J. med. Sci., Biol.*, **7**, 345.
Gardner, S. D. (1969) *J. Hyg., Camb.*, **67**, 545.
Gardner, P. S. (1957) *Br. med. J.*, **1**, 1143.
Gelenczei, E., & Bordt, D. (1960) *Am. J. vet. Res.*, **21**, 987.
Gilbert, Y., Mornet, P., & Goueffon, Y. (1960) *C.r. hebd. Séanc. Acad. Sci., Paris*, **250**, 2953.
Gillespie, J. H., & Karzon, J. T. (1960) *Proc. Soc. exp. Biol. Med., U.S.A.*, **105**, 547.
Gordon, I., Pavri, K., & Cohen, S. M. (1956) *J. Immunol.*, **76**, 328.
Goret, P., Pilet, C., Girard, M., & Camara, T. (1960) *Annls Inst. Pasteur, Paris*, **98**, 610.
Gorham, J. R. (1966) *J. Am. vet. med. Ass.*, **149**, 610.
Goss, L. J. (1948) *Am. J. vet. Res.*, **9**, 65.
Granoff, A., Liu, O. C., & Henle, W. (1950) *Proc. Soc. exp. Biol. Med., U.S.A.*, **75**, 684.
Green, R. G. (1939) *J. Am. med. vet Ass.*, **95**, 465.
Gresser, I., & Katz, S. L. (1960) *New Engl. J. Med.*, **263**, 452.
Gutierrez, J. C., & Gorham, J. R. (1955) *Am. J. vet. Res.*, **16**, 325.
Habel, K. (1945) *Publ. Hlth Rep., Wash.*, **60**, 201.
Habel, K. (1951) *Am. J. Hyg.*, **54**, 312.
Haig, D. A. (1948) *Onderstepoort J. vet. Sci.*, **23**, 149.
Haig, D. A. (1956) *Onderstepoort J. vet. Res.*, **27**, 19.
Hambling, M. H. (1964) *Br. J. exp. Path.*, **45**, 647.
Hamparian, V. V., Hilleman, M. R., & Ketler, A. (1963) *Proc. Soc. exp. Biol. Med., U.S.A.*, **112**, 1040.
Hanson, R. P. (1964) (Ed.) *Newcastle disease virus: An evolving pathogen.* Madison: Univ. of Wisconsin Press.
Harris, H., Watkins, J. F., Ford, C. E., & Schoefl, G. I. (1966), *J. Cell. Sci.*, **1**, 1.
Harter, D. H., & Choppin, P. W. (1967) *J. exp. Med.*, **126**, 251.
Haslam, E. A., Cheyne, I. M., & White, D. O. (1969) *Virology*, **29**, 118.
Heath, R. B., El. Falaki, I., Stark, J. E., Herbst-Laier, R. H., & Larin, N. M. (1966) *Br. J. exp. Path.*, **47**, 93.
Heller, O. (1957) *Mh. Vet.-Med.*, **12**, 218.
Henle, G., Baske, W. J., Burgoon, J. S., Burgoon, C. F., Hunt, C. F., Hunt, C. R., & Henle, W. (1951) *J. Immunol.*, **66**, 561.
Henle, G., Deinhardt, F., Bergs, V. V., & Henle, W. (1958) *J. exp. Med.*, **108**, 537.
Henle, G., Deinhardt, F., & Girardi, A. (1954) *Proc. Soc. exp. Biol. Med., U.S.A.*, **87**, 386.
Henle, G., Henle, W., & Harris, S. (1947) *Proc. Soc. exp. Biol. Med., U.S.A.*, **64**, 290.
Henle, G., Henle, W., Wendell, K. K., & Rosenberg, P. (1948) *J. exp. Med.*, **88**, 223.
Henry, M., Jean, M. T., & Sohier, R. (1968) *Annls Inst. Pasteur, Paris*, **115**, 1107.
Hermodsson, S., Dinter, Z., & Bakos, K. (1961) *Acta path. microbiol. scand.*, **51**, 75.
Hilleman, M. R. (1963) *Am. Rev. resp. Dis., suppl.* **88**, 189.
Hitchner, S. B. (1964) Control of Newcastle Disease in the United States by Vaccination in *Newcastle disease virus: An evolving pathogen.* Ed. Hanson. Madison: Univ. of Wisconsin Press.
Hollinger, F. B., & Pavri, K. M. (1971) *Am. J. trop. Med. Hyg.*, **20**, 131.
Hopper, P. K. (1959) *J. comp. Path.*, **69**, 78.
Hore, D. E. (1966) *Vet. Rec.*, **79**, 466.
Hore, D. E., Stevenson, R. G., Gilmour, N. J. L., Vantsis, J. T., & Thompson, D. A. (1968) *J. comp. Path.*, **78**, 259.
Horne, R. W., & Waterson, A. P. (1960) *J. molec. Biol.*, **2**, 75.
Horne, R. W., Waterson, A. P., Wildy, P., & Farnham, A. E. (1960) *Virology* **11**, 79.

Horsfall, F. L., & McCarty, M. (1947) *J. exp. Med.*, **85**, 623.
Horsfall, F. L., & Hahn, R. G. (1940) *J. exp. Med.*, **71**, 391.
Horta-Barbosa, L., Fuccillo, D. A., London, W. T., Jabbour, J. T., Zaman, W., & Sever J. L. (1969) *Proc. Soc. exp. Biol. Med., U.S.A.*, **132**, 272.
Hosaka, Y. (1968) *Virology*, **35**, 445.
Hosaka, Y., Kitano, H., & Ikeguchi, S. (1966) *Virology*, **29**, 205.
Hosaka, Y., Nichi, Y., & Fukai, K. (1961) *Biken's J.*, **4**, 243.
Hosaka, Y., & Shimizu, K. (1968) *J. molec. Biol.*, **35**, 369.
Hotchin, J. E., Deibel, R., & Benson, L. M. (1960) *Virology*, **10**, 275.
Howe, C., Morgan, C., St Cyr, de V. C., Hsu, K. C., & Rose, H. M. (1967) *J. Virol.*, **1**, 215.
Howe, C., Newcombe, E. W., & Lee, L. T. (1969) *Biochem. biophys. Res Commun.*, **34**, 388.
Hsiung, G-D. (1959) *Virology*, **9**, 717.
Hsiung, G-D., Mannini, A., & Melnick, J. L. (1958) *Proc. Soc. exp. Biol. Med., U.S.A.*, **98**, 68.
Huang, A. S., Baltimore, D., & Bratt, M. A. (1971) *J. Virol.*, **7**, 389.
Hull, R. N., Minner, J. R., & Smith, J. W. (1956) *Am. J. Hyg.*, **63**, 204.
Imagawa, D. T. (1965) *Arch. ges. Virusforsch.*, **17**, 203.
Imagawa, D. T. (1968) *Progr. med. Virol.*, **10**, 160.
Imagawa, D. T., & Adams, J. M. (1958) *Proc. Soc. exp. Biol. Med., U.S.A.*, **98**, 567.
Imagawa, D. T., Goret, P., & Adams, J. M. (1960) *Proc. natn. Acad. Sci., U.S.A.*, **46**, 1119
Inaba, Y., Tanaka, Y., Omori, T., & Matumoto, M. (1970) *Jap. J. exp. Med.*, **40**, 473.
Inoue, T., Harada, S., & Shimizu, T. (1930) *Select. Contrib. Mukden Inst. Inf. Dis., Anim.*, **1**, 221.
Ishii, S., Tokuda, G., & Watanabe, M. (1964) *Natn. Inst. Anim. Hlth Q., Tokyo*, **4**, 205,
Iwai, Y., Iwai, M., Okumoto, M., Hosokawa, Y., & Asai, T. (1966) *Biken's J.*, **9**, 241.
Jacotot, H., & Mornet, P. (1967) in *Les maladies animales a virus*. Eds.: Lepine & Goret. Paris: L'Expansion.
Jenkins, C. L., & Shope, R. E. (1946) *Am. J. vet. Res.*, **7**, 174.
Jensen, K. E., Minuse, E., & Ackermann, W. W. (1955) *J. Immunol.*, **75**, 71.
Jensen, K. E., Peeler, B. E., & Dulworth, W. G. (1962), *J. Immunol.*, **89**, 216.
Johnson, C. D., & Goodpasture, E. W. (1936) *Am. J. Path.*, **12**, 495.
Johnson, K. M. Chanock, R. M., Cook, M. K., & Huebner, R. J. (1960) *Am. J. Hyg.*, **71**, 81.
Johnson, K., Chanock, R. M., Rifkind, D., Kravetz, H. M., & Knight, V. (1961) *J. Am. med. Ass.*, **176**, 663.
Johnson, R. T., Johnson, K. P., & Edmonds, C. J. (1967) *Science*, **157**, 1066.
Joncas, J., Berthiaume, L., & Pavilanis, V. (1969) *Virology*, **38**, 493.
Jordan, W. S. (1962) *J. Immunol.*, **88**, 581.
Jungherr, E. L., Tyzzer, E. E., Brandly, C. A., & Moses, H. E. (1946) *Am. J. vet. Res.*, **7**, 250.
Kalter, S. S., Ratner, J., Kalter, G. V., Rodriquez, A. R., & Kim, C. S. (1968) *Am. J. Epidemiol.* **86**, 552.
Kapikian, A. Z., Mitchell, R. H., Chanock, R. M., Shvedoff, R. A., & Stewart, C. E. (1969) *Am. J. Epidemiol.*, **89**, 405.
Karabatsos, N., Buckley, S. M., & Ardoin, P. (1969) *Proc. Soc. exp. Biol. Med., U.S.A.*, **130**, 888.
Kasza, L. (1968) *Res. vet. Sci.*, **9**, 187.
Katz, M., Rorke, L. B., Masland, W. S., Brodano, G. B., & Koprowski, H. (1970) *J. Infect. Dis.*, **121**, 188.
Katz, S. L., Milovanovic, M. V., & Enders, J. F. (1958) *Proc. Soc. exp. Biol. Med., U.S.A.*, **97**, 23.
Kawamura, H., & Tsubahara, H. (1968) *Natn. Inst. Anim. Hlth, Tokyo*, **8**, 1.
Kazimi, S. E., & Majeed, M. A. (1966) *Pakist. J. agric. Sci.*, **3**, 16.
Ketler, A., Hamparian, V. V., & Hilleman, M. R. (1961) *J. Immunol.*, **87**, 126.
Kilham, L. (1949) *Proc. Soc. exp. Biol. Med., U.S.A.*, **71**, 63.
Kilham, L. (1951) *J. Am. med. Ass.*, **146**, 1231.

Kilham, L., & Murphy, H. W. (1952) *Proc. Soc. exp. Biol. Med., U.S.A.*, **80**, 495.
Killgore, G., & Dowdle, W. R. (1970) *Am. J. Epidemiol.*, **91**, 308.
Kimes, R. C., & Bussell, R. H. (1968) *Arch. ges. Virusforsch.*, **24**, 387.
Kingsbury, D. W. (1962) *Biochem. biophys. Res. Commun.*, **9**, 156.
Kingsbury, D. W. (1966a) *J. molec. Biol.*, **18**, 195.
Kingsbury, D. W. (1966b) *J. molec. Biol.*, **18**, 204.
Kingsbury, D. W., & Darlington, R. W. (1968) *J. Virol.*, **2**, 248.
Kisch, A. L., & Johnson, K. M. (1963) *Proc. Soc. exp. Biol. Med., U.S.A.*, **112**, 583.
Klenk, H.-D., & Choppin, P. W. (1969) *Virology*, **37**, 155.
Klenk, H.-D., & Choppin, P. W. (1970) *Proc. natn. Acad. Sci., U.S.A.*, **66**, 57.
Kohn, A. (1955) *Am. J. vet. Res.* **16**, 450.
Kohn, A. (1959) *Am. J. Hyg.*, **69**, 167.
Kohn, A. (1965) *Virology*, **26**, 228.
Koprowski, H., Jervis, G. A., James, T. R., Burkhart, R. D., & Poppensiek, G. C. (1950) *Am. J. Hyg.*, **51**, 63.
Kravetz, H. M., Knight, V., Chanock, R. M., Morris, J. A., Johnson, K. M., Rifkind, D., & Utz, J. P. (1961) *J. Am. med. Ass.*, **176**, 657.
Kuhn, N. O., & Harford, C. G. (1963) *Virology*, **21**, 527.
Kuroya, M., Ishida, N., & Shiratori, T. (1953) *Yokohama med. Bull.*, **4**, 217.
Laidlaw, P. P., & Dunkin, G. W. (1926) *J. comp. Path.*, **39**, 222.
Laidlaw, P. P., & Dunkin, G. W. (1928) *J. comp. Path.*, **41**, 1.
Larin, N. M., Herbst-Laier, R. H., Copping, M. P., & Wenham, R. B. M. (1967) *Nature, Lond.*, **213**, 827.
Lawn, A. M. (1970) *J. gen. Virol.*, **8**, 157.
Lepine, P., Chany, C., Droz, B., & Robbe-Fossat, F. (1959) *Ann. N.Y. Acad. Sci.*, **81**, 62.
Lepissier, H.-E., & MacFarlane, I. M. (1967) *Bull. Off. int. Épizoot.*, **68**, 655.
Levens, J. H., & Enders, J. F. (1945) *Science*, **102**, 117.
Levine, P. P. (1964) World dissemination of Newcastle disease In *Newcastle disease virus: An evolving pathogen.* Ed. Hanson. Madison: Univ. of Wisconsin Press.
Liebhaber, H., Krugman, S., McGregor, G., & Giles, J. P. (1965) *J. exp. Med.*, **122**, 1135.
Liess, B. (1964) *Arch. exp. Vet. Med.*, **20**, 157.
Maassab, H. F., & Loh, P. C. (1960) *Proc. Soc. exp. Biol. Med., U.S.A.*, **109**, 897.
McCarthy, K. (1959) *Br. med. Bull.*, **15**, 201.
MacIntyre, A. B., Trevan, D. J., & Montgomerie, R. F. (1948) *Vet. Rec.*, **60**, 635.
McKercher, P. D. (1964) *Can. J. comp. Med.*, **28**, 113.
McLelland, L., Hilleman, M. R., Hamparian, V. V., Ketler, A., Reilly, C. M., Cornfeld, D., & Stokes, J. (1961a) *New Engl. J. Med.*, **264**, 1169.
McLelland, L., Hampil, B., Hamparian, V. V., Potash, L., Ketler, A., & Hilleman, M. R. (1961b) *J. Immunol.*, **87**, 134.
McManus, K. P. (1968) *Aust. vet. J.*, **44**, 231.
Malmquist, W. A., Van der Maaten, M. J., & Boothe, A. D. (1969) *Cancer Res.*, **29**, 188.
Mansi, W. (1957) *J. comp. Path.*, **67**, 297.
Mares, I., Casny, J., Macek, M., Drevo, M., & Regacova, D. (1966) *Acta virol., Prague*, **10**, 211.
Marston, R. Q. (1958) *Proc. Soc. exp. Biol. Med., U.S.A.*, **98**, 853.
Mason, E. J., & Kaufmann, N. (1961) *Br. J. exp. Path.*, **42**, 118.
Matsumoto, N. (1966) *Bull. Yamaguchi med. Sch.*, **13**, 167.
Matumoto, M. (1966) *Bact. Rev.*, **30**, 152.
Matumotu, M., Arita, M., & Oda, M. (1965) *Jap. J. exp. Med.*, **35**, 319.
Mayr, A., Wizigmann, G., Schels, H., & Plank, P. (1969) *Zentbl. VetMed.*, **16B**, 454.
Miller, R. H., Pursell, A. R., Mitchell, F. E., & Johnson, K. M. (1964) *Am. J. Hyg.*, **80**, 365.
Mills, K. C., & Dochez, A. R. (1940) *Proc. Soc. exp. Biol. Med., U.S.A.*, **57**, 140.
Mills, K. C., & Dochez, A. R. (1945) *Proc. Soc. exp. Biol. Med., U.S.A.*, **60**, 141.
Milovanovich, M. V., Enders, J. F., & Mitus, A. (1957) *Proc. Soc. exp. Biol. Med., U.S.A.*, **95**, 120.
Mogabgab, W. J., Dick, E. C., & Holmes, B. (1961) *Am. J. Hyg.*, **74**, 304.

Morgan, H. R., Enders, J. F., & Wagley, P. F. (1948) *J. exp. Med.*, **88**, 503.
Morris, J. A., Blount, R. E., & Savage, R. E. (1956) *Proc. Soc. exp. Biol. Med., U.S.A.*, **92**, 544.
Morse, H. G., Chow, T. L., & Brandly, C. A. (1953) *Proc. Soc. exp. Biol. Med., U.S.A.*, **84**, 10.
Motohashi, T., Kishi, S., & Nakamura, J. (1964) *Am. J. vet. Res.*, **25**, 825.
Mountcastle, W. E., Compans, R. W., & Choppin, P. W. (1971) *J. Virol.*, **7**, 47.
Nagata, I., Maeno, K., Yoshii, S., & Matsumoto, T. (1965) *Arch. ges. Virusforsch.*, **15**, 257.
Nakai, T., Shand, F. L., & Howatson, A. T. (1969) *Virology*, **38**, 50.
Nakajima, H., & Obara, J. (1967) *Arch. ges. Virusforsch.*, **20**, 287.
Nakamura, J., Kishi, S., Matsuzawa, H., Kinshi, J., & Miyamoto, T. (1957) *Bull. Nippon Inst. biol. Sci.*, **2**, 1.
Nakamura, J., Wagatsuma, S., & Fukusho, K. (1938) *J. Jap. Soc. vet. Sci.*, **17**, 25.
Nemo, G. J., & Cutchins, E. C. (1966) *J. Bact.*, **91**, 798.
Norrby, E. (1966) *Proc. Soc. exp. Biol. Med., U.S.A.*, **121**, 948.
Norrby, E., Friding, B., Rockborn, B., & Gard, S. (1963) *Arch. ges. Virusforsch.*, **13**, 335.
Norrby, E., Magnusson, P., Falksveden, L.-G., & Gronberg, M. (1964) *Arch. ges. Virusforsch.*, **14**, 462.
Norrby, E., Marusyk, H., & Orvell, C. (1970) *J. Virol.*, **6**, 237.
Northrop, R. L., & Walker, D. L. (1965) *Proc. Soc. exp. Biol. Med., U.S.A.*, **118**, 698.
Numazaki, Y., & Karzon, D. T. (1966a) *J. Immunol.*, **97**, 458.
Numazaki, Y., & Karzon, D. T. (1966b) *J. Immunol.*, **97**, 470.
Numazaki, Y., Shigeta, S., Yano, N., Takai, S., & Ishida, N. (1968) *Proc. Soc. exp. Biol. Med., U.S.A.*, **127**, 992.
Nunen, M. C. J. van & van der Veen, J. (1967) *Arch. ges. Virusforsch.*, **22**, 388.
Oddo, F. G., Chiarini, A., & Sinatra, A. (1967) *Arch. ges. Virusforsch.*, **22**, 35.
Ogasawara, K., Sugai, K., & Iijima, S. (1959) *Virology*, **9**, 714.
Oh, J. O. (1961) *Br. J. exp. Path.*, **42**, 424.
Okada, Y. (1958) *Biken's J.*, **1**, 103.
Paccaud, M. F., & Jacquier, Cl. (1970) *Arch. ges. Virusforsch.*, **30**, 327.
Palm, C. R., & Black, F. L. (1961) *Proc. Soc. exp. Biol. Med., U.S.A.*, **107**, 588.
Panelius, M., Salmi, A., & Halonen, P. (1970) *Acta path. microbiol. scand.*, **78**, 588.
Parker, J. C., & Reynolds, R. K. (1968) *Am. J. Epidemiol.*, **88**, 112.
Pearson, H. E., & Eaton, M. D. (1940) *Proc. Soc. exp. Biol. Med., U.S.A.*, **45**, 677.
Peries, J. R., & Chany, C. (1960) *C.r. hebd. Séanc. Acad. Sci., Paris*, **251**, 820.
Peries, J. R., & Chany, C. (1961) *C.r. hebd. Séanc. Acad. Sci., Paris*, **252**, 2956.
Peries, J. R., & Chany, C. (1962) *Proc. Soc. exp. Biol. Med., U.S.A.*, **110**, 477.
Petersen, K. B. (1959) *Acta path. microbiol. scand.*, **45**, 213.
Picken, J. C. (1964) In *Newcastle disease virus: An evolving pathogen*, p. 167. Madison: Univ. of Wisconsin Press.
Piercy, S. E. (1961) *Vet. Rec.*, **73**, 898.
Plowright, W. (1964) *J. Hyg., Camb.*, **62**, 257.
Plowright, W. (1968) *Virology monographs*, **3**, 25. Eds.: Gard, Hallauer, & Meyer. Vienna/New York: Springer Verlag.
Plowright, W., Cruikshank, J. G., & Waterson, A. P. (1962) *Virology*, **17**, 118.
Plowright, W., & Ferris, R. D. (1959) *J. comp. Path.*, **69**, 152.
Plowright, W., & Ferris, R. D. (1961) *Arch. ges. Virusforsch.*, **11**, 516.
Plowright, W., & McCulloch, B. (1967) *J. Hyg., Camb.*, **65**, 343.
Polding, J. B., & Simpson, R. W. (1959) *Vet. Rec.*, **69**, 582.
Pospišil, L., & Brychtova, J. (1956) *Zentbl. Bakt. ParasitKde, I. Abt. Orig.*, **165**, 1.
Potash, L., Tytell, A. A., Sweet, B. H., Machlowitz, R. A., Stokes, J., Weibel, R. E., Woodhour, A. F., & Hilleman, M. R. (1966) *Am. Rev. resp. Dis.*, **93**, 536.
Profeta, M. L., Lief, F. S., & Plotkin, S. A. (1969) *Am. J. Epidemiol.*, **89**, 316.
Provost, A., & Borredon, C. (1968) *Revue Élev. Méd. vét. Pays trop.*, **21**, 33.
Provost, A., Queval, R., & Borredon, C. (1965) *Revue Élev. Méd. vét. Pays trop.*, **18**, 371.

Prydie, J. (1968) *Res. vet. Sci.*, **9**, 443.
Rapp, F., Gordon, I., & Baker, R. F. (1960) *J. biophys. biochem. Cytol.*, **7**, 43.
Rechsteiner, J. (1969) *J. gen. Virol.*, **5**, 397.
Reczko, E., & Bögel, K. (1962) *Arch. ges. Virusforsch.*, **12**, 404.
Reda, I. M., Rott, R., & Schafer, W. (1964) *Virology*, **22**, 422.
Reichelderfer, T. E., Chanock, R. M., Craighead, J. C., Huebner, R. J., Turner, H. C., James, W., & Ward, T. G. (1958) *Science*, **128**, 779.
Reisinger, R. C., Heddleston, K. L., & Manthei, C. A. (1959) *J. Am. Vet. med. Ass.*, **135**, 147.
Rhim, J. S., & Schell, K. (1967) *Nature, Lond.*, **216**, 271.
Richman, A. V., Pedreira, F. A., Tauraso, N. M. (1971) *Appl. Microbiol.*, **21**, 1099.
Robertson, A. (1964) Methods of Control of Newcastle Disease and their limitations. In *Newcastle disease virus: An evolving pathogen.* Ed.: Hanson. Madison: Univ. of Wisconsin Press.
Robinson, T. W. E., Cureton, R. J. R., & Heath, R. B. (1968) *J. med. Microbiol.*, **1**, 89.
Robinson, W. S. (1971) *J. Virol.*, **8**, 81.
Rockborn, G. (1958) *Arch. ges. Virusforsch.*, **8**, 485.
Roizman, B., & Schluederberg, A. E. (1961) *Proc. Soc. exp. Biol. Med., U.S.A.*, **106**, 320.
Rosanoff, E. I. (1961) *Proc. Soc. exp. Biol. Med., U.S.A.*, **106**, 563.
Rott, A. (1964) In *Newcastle disease: An evolving pathogen*, p. 133. Ed.: Hanson. Madison: Univ. of Wisconsin Press.
Rott, R., & Schäfer, W. (1961) *Virology*, **14**, 298.
Ruckle, G. (1958) *Arch. ges. Virusforsch.*, **8**, 139.
Russeff, C. (1962) *Zentbl. Bakt ParasitKde, I. Abt. Orig.*, **184**, 403.
Sasahara, J. (1955) *Exp. Rep.* 3, *Natn. Inst. Anim. Hlth, Tokyo*, p. 13.
Sawicki, L. (1962) *Acta virol., Prague*, **6**, 347.
Schäfer, W., & Rott, R. (1959) *Z. Naturforsch.*, **14b**, 629.
Schluederberg, A. (1971) *Biophys. biochem. Res. Comm.*, **42**, 1012.
Schluederberg, A., & Nakamura, M. (1967) *Virology*, **33**, 297.
Schultz, E. W., & Habel, K. (1959) *J. Immunol.*, **82**, 274.
Schwartz, A. T. F., & Zirbel, L. W. (1959) *Proc. Soc. exp. Biol. Med., U.S.A.*, **102**, 711.
Scott, G. R. (1964) *Adv. vet. Sci.*, **9**, 113.
Scott, G. R. (1959) *J. comp. Path.*, **69**, 148.
Scott, G. R., & Witcomb, M. A. (1958) *A.R.E. Afr. vet. Res. Org., 1956–57*, p. 15.
Sergiev, P. G., Rizantseva, N. F., & Smirnova, E. V. (1959) *Probl. Virol.*, **4**, 50.
Shah, K. V., Schaller, G. B., Flyger, V., & Herman, C. M. (1965) *Bull. Wildl. Dis. Ass.*, **1**, 31.
Sherman, F. E., & Ruckle, G. (1957) *Archs Path.*, **65**, 587.
Shimonaski, G., & Came, P. E. (1970) *Appl. Microbiol.*, **20**, 775.
Shishido, A., Yamanouchi, K., Hikita, M., Sato, T., Fukuda, A., & Kobune, F. (1967) *Arch. ges. Virusforsch.*, **22**, 364.
Shope, R. E., Griffiths, H. J., & Jenkins, D. L. (1946) *Am. J. vet. Res.*, **7**, 135.
Singh, K. V., & Ata, F. A. (1967) *Vet. Rec.*, **81**, 84.
Singh, K. V. & Baz, T. I. (1966) *Nature, Lond.*, **210**, 656.
Sokol, F., Neurath, A. R., & Vilcek, J. (1964) *Acta virol., Prague*, **8**, 59.
Sokol, F., Skacianska, E., & Pivec, L. (1966) *Acta virol., Prague*, **10**, 291.
Soule, D. W., Marinetti, G. V., & Morgan, H. R. (1959) *J. exp. Med.*, **110**, 93.
Spooner, E. T. C. (1953) *Br. med. Bull.*, **9**, 212.
Stenback, W. A., & Durand, D. P. (1963) *Virology*, **20**, 545.
Stevenson, R. G., & Hore, D. E. (1970) *J. comp. Path.*, **80**, 613.
Stone, S. S. (1960) *Virology*, **11**, 638.
Tajima, M., Ushijima, T., Kishi, S., & Nakamura, J. (1967) *Virology*, **31**, 92.
Taniguchi, T., Kamahora, J., Kato, S., & Hagiwara, K. (1954) *Med. J. Osaka Univ.*, **5**, 367.
Taylor, W. P., & Plowright, W. (1965) *J. Hyg., Camb.*, **63**, 263.
Taylor, W. P., & Watson, R. M. (1967) *J. Hyg., Camb.*, **65**, 537.
Taylor-Robinson, D., & Bynoe, M. L. (1963) *J. Hyg., Camb.*, **61**, 407.

Tellez-Nagel, I., & Harter, D. H. (1966) *Science*, **154**, 899.
Tennant, R. W., Parker, J. C., & Ward, T. G. (1966) In *Viruses of laboratory rodents*. Natn. Cancer Inst. Monogr., **20**, 93.
Tennant, R. W., & Ward, T. G. (1962) *Proc. Soc. exp. Biol. Med., U.S.A.*, **111**, 395.
Tikasingh, E. S., Jonkers, A. H., Spence, L., & Aitken, T. H. G. (1966) *Am. J. trop. Med. Hyg.*, **15**, 235.
Torres, G. M., & Teixeira, J. de C. (1932) *C.r. Séanc. Soc. Biol.*, **109**, 138.
Tozawa, H., Homma, M., & Ishida, N. (1967) *Proc. Soc. exp. Biol. Med., U.S.A.*, **124**, 734.
Tribe, G. W. (1966) *Br. J. exp. Path.*, **47**, 472.
Tribe, G. W., Kanarek, A. D., & White, G. (1968) *Res. vet. Sci.*, **9**, 152.
Tyrrell, D. A. J., & Coid, C. R. (1970) *Vet. Rec.*, **86**, 164.
Tyrrell, D. A. J., Petersen, K. B., Sutton, R. N. P., & Pereira, M. S. (1959) *Br. med. J.*, **2**, 909.
Tytell, A. A., Torop, H. A., & McCarthy, F. J. (1962) *Proc. Soc. exp. Biol. Med., U.S.A.*, **108**, 723.
Upton, E., Hanson, R. P., & Brandly, C. A. (1953) *Proc. Soc. exp. Biol. Med., U.S.A.*, **84**, 691.
Utz, J. P., & Szwed, C. F. (1962) *Proc. Soc. exp. Biol. Med., U.S.A.*, **110**, 841.
Wacker, W. B., Lipton, M. M., & Steigman, A. J. (1962) *Proc. Soc. exp. Biol. Med., U.S.A.*, **109**, 955.
Wagner, K., & Enders-Ruckle, G. (1966) *Zentbl. VetMed.*, **13B**, 215.
Walder, R. (1971) *J. gen. Virol.*, **11**, 123.
Warren, J. (1960) *Adv. Virus Res.*, **7**, 27.
Warren, J., Nadel, M. K., Slater, E., & Millian, S. J. (1960) *Am. J. vet. Res.*, **21**, 111.
Warthin, A. S. (1931) *Archs Path.*, **11**, 864.
Waterson, A. P. (1965) *Arch. ges. Virusforsch.*, **16**, 57.
Waterson, A. P., & Hurrell, J. M. W. (1962) *Arch. ges. Virusforsch.*, **12**, 138.
Waterson, A. P., Jansen, K. E., Tyrrell, D. A. J., & Horne, R. W. (1961a) *Virology*, **14**, 374.
Waterson, A. P., Rott, R., & Ruckle-Enders, G. (1963) *Z. Naturf.*, **18b**, 377.
Waterson, A. P., Rott, R., & Schäfer, W. (1961b) *Z. Naturforsch.*, **16b**, 154.
Watson, B. K., & Cheever, F. S. (1955) *J. Immunol.*, **75**, 161.
Wellemans, G., Leunen, J., & Luchsinger, E. (1970) *Ann. med. Vet.*, **114**, 89.
Wenner, A. A., Monley, A., & Todd, R. N. (1950) *J. Immunol.*, **64**, 305.
White, G. (1962) *Vet. Rec.*, **74**, 1477.
WHO (1964) *WHO Chron.*, **18**, 81.
Woods, G. T., Mansfield, M. E., Cmarik, G., Zinzilieta, M., & Marquis, G. (1968) *Am. J. vet. Res.*, **29**, 1349.
Wright, J. (1957) *Lancet*, **1**, 669.
Yamamura, K., Shibuta, H., Hirano, K., & Yamamoto, T. (1970) *Jap. J. exp. Med.*, **40**, 123.
Yamanouchi, K., Egashira, Y., Uchida, N., Kodama, H., Kobune, F., Hayami, M., Fukuda, A., & Shishido, A. (1970) *Jap. J. med. Sci., Biol.*, **23**, 131.
Yamanouchi, K., Fukuda, A., Kobune, F., Hayami, M., & Shishido, A. (1969) *Am. J. vet. Res.*, **30**, 1831.
Yamanouchi, K., Kobune, F., Fukuda, A., Hayami, M., & Shishido, A. (1970) *Arch. ges. Virusforsch.*, **29**, 90–100.
Zakstelskaya, L. Ya., Almeida, J. D., & Bradstreet, C. M. P. (1967) *Acta virol., Prague*, **11**, 420.
Zhdanov, V. M. (1960) *Virology*, **10**, 146.
Zwart, D., & Macadam, I. (1967a) *Res. vet. Sci.*, **8**, 37.
Zwart, D., & Macadam, I. (1967b) *Res. vet. Sci.*, **8**, 53.

11

Unclassified RNA Viruses

RUBELLA

Synonym: German measles.

Reviews: Ingalls et al. (1960) (epidemiology and teratology). Neva et al. (1964); Schiff & Sever (1966).

Morphology. Electron microscopy of both thin sections and negatively stained preparations reveals virions 50 to 70 nm in diameter, consisting of a pleomorphic triple-layered envelope about 8 nm thick, surrounding an electron-lucid layer about 11 nm thick and an electron-dense core about 30 nm in diameter (Best et al., 1967; Edwards et al., 1969; Holmes et al., 1969). Virus maturation takes place by budding at cytoplasmic membranes with accumulation of virions in cisternae of the endoplasmic reticulum or, more frequently, with virus liberation when budding takes place at marginal membranes (Murphy et al., 1968; Tuchinda et al., 1969; Hamvas et al., 1969).

A resemblance between rubella virus and togaviruses (see Chapter 3) has been noted by several workers (e.g. Holmes et al., 1969).

Chemical composition. Infectious, single-stranded RNA in a single piece with MW of about 3×10^6 daltons has been obtained from virions or from infected cells (Hovi & Vaheri, 1970a; Sedwick & Sokol, 1970).

Physico-chemical characters. Buoyant density of virions in CsCl variously estimated 1·085 g/ml (Russell et al., 1967), 1·120 g/ml (Thomsen et al., 1968) and 1·32 (Amstey et al., 1968). Density in sucrose or potassium tartrate is increased from 1·16 or 1·10 to 1·18 or 1·20 by addition of EDTA (McCombs & Rawls, 1968) or tween 80 (Magnusson & Skaaret, 1967). Sedimentation coefficient in sucrose estimated as 342 S (Russell et al., 1967) and 240 S (Thomsen et al., 1968). Virus infectivity is sensitive to lipid solvents. Rather labile at pH 5 and quickly inactivated at pH 3 (Chagnon & Laflamme, 1964).

The kinetics of thermal and UV inactivation have been determined by Fabiyi et al. (1966). Inactivation by gamma radiation suggests a radio-sensitive target of 2·6 to 4·0 $\times$ 10^6 daltons (Kenny et al., 1969).

Haemagglutination of 1-day-old chick RBC is maximal at 4°C, less pronounced at 25°C and negative at 37°C. Adult goose and sheep RBC also agglutinated, but to lower titres (Stewart et al., 1967). Methods for obtaining high titre HA and for enhancement of HA activity are described (Furukawa et al., 1967; Halonen et al., 1967b; Laufs & Thompsen, 1968a; Liebhaber, 1970). Association of lanthanum-staining material present on the surface of RBC with HA activity is suggested by Dickinson et al. (1969). A lipoprotein inhibitor is present in fetal calf serum (Laufs & Thompsen, 1968b). Haemadsorption is described by Schmidt et al. (1968a).

Antigenic properties. Antibodies detected in convalescent or immune sera by virus neutralization (Weller & Neva, 1962; Parkman et al., 1964), complement fixation (Sever et al., 1965; Halonen et al., 1967a), immunodiffusion (Salmi, 1970), haemagglutination inhibition (Schmidt et al., 1968a). Fractionation of HA and CF antigen is described by Schmidt et al. (1968b). Extensive testing failed to reveal cross-reactions with arboviruses (Mettler et al., 1968). No antigenic variation detectable by kinetic haemagglutination-inhibition tests (Best & Banatvala, 1970).

Cultivation. *In eggs* growth has been demonstrated in embryonated duck eggs (Fucillo et al., 1967).

In tissue cultures, rubella virus was first propagated in primary human (Weller & Neva, 1962) and monkey (Parkman et al., 1962) tissues and subsequently in a wide variety of other cells, the most commonly used being rabbit (Beale et al., 1963; McCarthy et al., 1963) and hamster (BHK21) cell-lines (Vaheri et al., 1965; 1967). Carrier cultures derived from tissues of infants with congenital rubella have been studied by Rawls & Melnick (1966). Virus replication in tissue cultures can be demonstrated indirectly by interference with the growth of other viruses (Parkman et al., 1962) or directly by CPE (Weller & Neva, 1962; Beale et al., 1963; McCarthy et al., 1963), plaque formation (Rhim et al., 1967) or immunofluorescence (Hobbins & Smith, 1968). Plaque assays based on positive (Schmidt et al., 1969) or negative (Rawls et al., 1967a) haemadsorption have been described. Cytochemical and immunofluorescence studies reveal that virus replication is entirely cytoplasmic (Bonissol, 1970; Woods et al., 1966) and is not inhibited by actinomycin D (Maes et al., 1966).

Different virus-specific RNAs synthesized in infected cells are described (Hovi & Vaheri, 1970b; Sedwick & Sokol, 1970) with tentative identification of replicative and replicative-intermediate forms.

Pathogenicity. Rubella is a mild infection with generalized rash, enlargement of lymph nodes, especially the post-cervical. There is usually little fever or constitutional disturbance. The incubation period is 16–18 days. Meningo-encephalitis and other complications are rare. The disease is important because of the high incidence of congenital abnormalities in children whose mothers had rubella during the first 3 or 4 months of pregnancy (Gregg, 1945; Swan et al., 1946). Incidence has been different at different times and places; abnormalities include deaf-mutism, cataract, cardiac and dental malformations and microcephaly; purpura may be present at birth.

The virus has been recovered not only from fetuses in the early months of pregnancy but from infants born with the 'rubella-syndrome' and even from normal infants whose mothers had rubella later in pregnancy. Virus may be present in the ocular lens, pharyngeal secretions, CSF, blood and urine and may persist as late as 3 years after birth (Alford et al., 1964; Monif et al., 1965; Menser et al., 1967). Infants carrying virus in this way have antibodies in their sera (Weller et al., 1965). The literature on congenital rubella has been reviewed by Rawls (1968) and Sever & White (1968).

Experimentally, infection has been transmitted to rhesus monkeys (Habel, 1942); these developed leucopenia and, in some cases, a light pink rash. Other workers have infected rhesus (Parkman et al., 1965), *Cercopithecus* (Sigurdadottir et al., 1963) and *Erythrocebus* (Draper & Laurence, 1969) monkeys as shown by the presence of viraemia and development of antibodies, but most accounts say that no rashes were seen. Experimental infections of rabbits, hamsters, guinea-pigs and ferrets by the respiratory route are described by Oxford (1967). Infection of rats (Cotlier et al., 1967) and mice (Carver et al., 1967) is also reported. Vertical transmission with growth retardation of congenitally infected animals may be observed (London et al., 1969; Kono et al., 1969). The disease has also appeared in volunteers inoculated SC or IN, after an incubation period of 13–20 days (Anderson, 1949); some infections were apparently subclinical.

Ecology. Major outbreaks may be spaced by 9–10-year intervals (Krugman, 1965). Serological surveys suggest that infection occurs early in life, reaching 80 to 87 per cent by the age of 17 to 22 years (Rawls et al., 1967b). Persistence of virus in congenitally affected infants may well favour survival of virus in nature (Weller et al., 1965).

Control. Live attenuated vaccines prepared in cultures of monkey kidney (Parkman et al., 1966), rabbit kidney (Huygelen et al., 1969), human fibroblasts (Plotkin et al., 1967) or duck embryos (Weibel et al., 1968) have been extensively tested and used with success (see Symposium, 1969; Parkman & Meyer, 1969).

SWINE FEVER

Synonym: Hog cholera.

Reviews: Dunne (1958). Fuchs (1968). Wilsdon (1958). (Symptoms and Control).

Morphology. Approximately spherical virions about 40 nm in diameter, with a core about 29 nm across and an envelope 6 nm thick (Horzinek et al., 1967; Mayr et al., 1967; Cunliffe & Rebers, 1968).

Chemical composition. Lack of inhibition by halogenated deoxyribosides suggests that the viral nucleic acid is RNA (Dinter 1963; Loan, 1964).

Physico-chemical characters. Virions have buoyant density in CsCl of 1·14 to 1·20 g/ml (Horzinek, 1966; Mayr et al., 1967) and sedimentation coefficient of 108 ± 24 S (Horzinek, 1967). A very stable virus, more so in blood held between pH 4·8 and 5·1 than at neutral pH (Chapin et al., 1939). A pH $< 1{\cdot}4$ or > 13 is necessary to kill it within an hour (Slavin, 1938). When dried *in vacuo* and sealed in ampoules it survives well even at 37°, but is soon inactivated when dried in air in the field; however, it persists a long time in infected pork or garbage. In defibrinated blood the virus was only inactivated after 30 minutes at 69° or 1 hour at 66° (Torrey & Prather, 1964). Virus keeps well in the cold when mixed with glycerol or in the presence of 0·5 per cent phenol. It is sensitive to lipid solvents, moderately sensitive to trypsin and not stabilized by $MgCl_2$ (Dinter 1963; Loan, 1964).

Antigenic properties. Neutralization and CF tests are applicable; also precipitation in agar gel-diffusion tests (Molńar, 1954; Mansi, 1957), lymph nodes or pancreas from infected pigs being the best antigens for this last test. Gel-diffusion tests also reveal an antigenic component in common with mucosal disease of cattle (see p. 267) (Darbyshire, 1962). A soluble antigen seems to be concerned (Pirtle, 1964). An immunofluorescence assay has been developed (Mengeling et al., 1963). Using Gillespie's culture-adapted strain (1960b), Coggins & Baker (1964) evolved a useful neutralization test in tissue culture.

Recovered pigs are permanently immune. A variant virus is reported from the United States giving rise to 'breaks' in the course of attempted immunization. It is poorly neutralized by specific antisera, but no distinct stable antigenic type has been distinguished. Variant virus may be analogous to the Q variant of influenza virus which is poorly neutralized by homologous sera (see p. 211).

Cultivation. Adaptation of virus to duck embryos (Coronel & Albis, 1950) and to chick embryos (Fontanelli et al., 1959) has been claimed, but further studies are not yet reported. On the other hand, the virus will multiply in cultures of swine tissues, usually with no cytopathic effects or only (Gustafson & Pomerat, 1957) minimal ones. However, Gillespie et al. (1960b) adapted 1 strain to pig-kidney tissues in lamb serum and saw definite CPE. Special conditions for regular production of CPE in porcine tissue cultures have been described (Mayr & Mahnel, 1964; Segre, 1964; Crawford et al., 1968). A plaque assay is described by van Bekkum & Barteling (1970). The susceptibility of primary or continuous cultures of cells derived from 29 species of mammals was studied by Pirtle & Kniazeff (1968), who found evidence of multiplication demonstrable by immunofluorescence in a wide variety of cells. Persistent infection of porcine tissue cultures (Soekawa & Isawa, 1960) may lead to virus attenuation (Sato et al., 1964; Loan & Gustafson, 1964). Kumagai et al. (1958) report that in the presence of swine fever, the CPE of Newcastle disease virus on swine-testis monolayers is greatly accelerated. The usefulness of this test was confirmed by Korn & Nishimura (1963) and Loan (1965), though it was unreliable for weakly virulent viruses for which an assay based on interference with NDV has been developed (Sato et al., 1969).

Distribution. The disease apparently first appeared in midwestern North America about 1833 and thence in the course of the nineteenth century spread all over the world. Its possible origin is discussed by Hanson (1957).

Pathogenicity. Usually exceedingly contagious and fatal amongst herds of pigs. These show fever, apathy, vomiting, eye-discharge, diarrhoea, cutaneous haemorrhages, but mild strains have been reported (Keast et al., 1962). Symptoms referable to encephalomyelitis are frequent. Some strains of virus are unusually neurotropic. Pigs may die of the virus infection or, after secondary infection with *Salmonella cholerae-suis* or other bacteria, of pneumonia or ulcerative enteritis. Only pigs are naturally affected. Transplacental infection with attenuated (vaccine) strains may lead to stillbirths or diseased newborns (Young et al., 1955; Huck, 1964) sometimes with congenital tremor (Harding et al., 1966).

Experimentally, the disease is easily reproduced by inoculating pigs by various routes, the incubation period being from 3 to 7 days. Asymptomatic infection can be produced experimentally in peccaries, calves, goats, sheep and deer, but not in wild mice, cotton-tail rabbits, sparrows, rats, raccoons or pigeons (Loan & Storm, 1968). Baker (1946) and Koprowski et al. (1946) have succeeded by zigzagging, i.e. making alternate passages between pigs and rabbits; others have found this unnecessary. Infected rabbits show nothing but transient fever. Lapinized virus is attenuated for pigs and has been used for immunization.

Pathological lesions comprise degeneration of small blood vessels leading to small haemorrhages in kidney, bladder, skin, lymph nodes and infarction especially in the spleen, but pigs dying early of pure virus infection show very little macroscopically. Characteristic lesions, including cuffing and microgliosis, may, however, be found in the CNS (Done, 1957). With later deaths there are seen pneumonia and 'button ulcers' in intestines. Leucopenia is the rule. Studies using immunofluorescence reveal specific antigen; it appears mainly and earlier in cytoplasm; later there is some in the nucleus (Karasszon & Bodon, 1963; Stair et al., 1963; Sirbu et al., 1964). Inclusion bodies have been described but are doubtfully specific.

Ecology. Spread is by direct contact and by feeding contaminated garbage. Virus apparently disappears between outbreaks; Shope (1958) has produced evidence suggesting that it may persist in swine lung-worms. The role of 'pregnant carrier' sows in maintenance of infection is suggested by Huck & Aston (1964). Korn & Hecke (1964) think there may be very weak strains only causing deaths in very young pigs and that these may acquire virulence after repeated passages, acquiring ability to cause disease in older animals.

Control. Hyperimmune sera can be used to give temporary protection when the disease appears in a herd. Several methods of active immunization are used:

1. Simultaneous inoculation of virulent virus and antisera at separate sites; this method is being replaced by others, as it leads to persistence of infection in the locality.
2. Virus inactivated by crystal violet (McBryde & Cole, 1936) is widely and successfully used, also Boynton's (1933) eucalyptol-treated vaccine. These, however, give only temporary immunity.

3. Viruses attenuated by passage in rabbits (Baker, 1947) give much promise, also those attenuated in tissue culture (Casselberry et al., 1953; Loan & Gustafson, 1964; Sato et al., 1964; Sampson et al., 1965). They may be given with or without antisera. There is still doubt, however, as to whether the character of attenuation is stable.

Attenuated virus may cause abortions or malformations amongst litters (Bontcheff et al., 1959; Young et al., 1955) and may be dangerous when given to baby pigs (Baker & Sheffy, 1960). Korn (1964) maintains that weakly virulent nonimmunizing strains actually sensitize pigs so that their reaction to challenge with more virulent strains is more severe. The virus of mucosal disease of cattle, shown to be antigenically related to swine fever, is being tested for its ability to immunize swine (see p. 267).

Since 1963 it has been possible to introduce an eradication policy for Britain.

VIRUS DIARRHOEA
(bovine)

Synonyms: Mucosal disease. Bovine diarrhoea (New York, Oregon and Indiana strains).

Reviews: Johnston (1959). Pritchard (1963) Mills et al. (1965).

Morphology. Maes & Reczko (1970) describe approximately spherical virions 57 ± 7 nm in diameter with a 24 ± 4 nm wide core and an envelope without projections. Virions with mean diameter of 46 nm or 60 nm according to method of preparation and with a core 25 ± 7 nm across are described by Hafez et al. (1968). Ritchie & Fernelius (1969) describe 3 size classes of particles considered to be soluble antigen (15 to 20 nm), unenveloped (30 to 50 nm) and enveloped (80 to >100 nm) virions. Both helical (Ditchfield & Doane, 1964) and cubic (Maes & Reczko, 1970) symmetry have been suggested for the viral nucleocapsid.

Chemical composition. Infectious RNA, sensitive to RNAse, has been extracted by phenol from virus (Diderholm & Dinter, 1966).

Physico-chemical characters. The virion has a buoyant density of 1·15 g/ml in CsCl, sucrose or tartrate (Fernelius, 1968). It is sensitive to lipid solvents, readily inactivated at 56°C, not stabilized by $MgCl_2$ at 50°C, stable at low temperatures, unstable below pH 3 (Dinter, 1963; Burki, 1966; Tanaka et al., 1968). Similarities with swine-fever virus, with equine-arteritis virus and with flaviviruses have been pointed out by Dinter (1963) and Burki (1966).

Antigenic properties. Different reports state that the New York and Indiana strains do not cross-immunize (in cattle) or do show cross-neutralization (in tissue culture) (Gillespie et al., 1960a). There is also said to be cross-neutralization with mucosal disease (Gillespie et al., 1961; Taylor et al., 1963). Huck & Cartwright (1964) classified 53 virus strains into 7 serotypes, of which a number were associated with mucosal disease. One serotype was isolated from 75 per cent of cattle with infertility.

Gel-diffusion tests show an antigen in common between mucosal-disease virus and swine fever (Darbyshire, 1962), and Sheffy et al. (1962) have found that mucosal-disease virus will give some protection to pigs against swine fever. Gutenkunst & Malmquist (1965) and Darbyshire (1967) describe the properties of the soluble antigen.

Cultivation. Darbyshire (1963) encountered a strain which would multiply and produce pocks on chorioallantoic membranes of eggs. Cultivation of mucosal-disease virus (Noice & Schipper, 1959; Bekkum, 1959) and of bovine diarrhoea (Gillespie et al., 1959; 1960a) in cultures of bovine tissues has been reported. Bovine-diarrhoea virus has been adapted to grow in swine-kidney cultures (Malmquist et al., 1965). Growth is usually unaccompanied by CPE, but adaptation to a bovine kidney cell-line with regular CPE has been described (Marcus & Moll, 1968). Evidence of virus multiplication may also be obtained by enhancement of the CEP of NDV (Inaba et al., 1968) or by immunofluorescence (Fernelius, 1969).

Pathogenicity. The disease picture described for mucosal disease includes, besides diarrhoea, fever and oral ulcerations. There is intestinal catarrh, and necrotic lesions occur in the mucosa together with lesions in the hooves, lymph nodes and elsewhere. The disease transmitted experimentally is usually mild (Huck, 1957). Antibodies may be present in most members of a herd of cattle in the absence of disease (Bögel & Voss, 1964). This or a related virus is apparently widely distributed among sheep flocks in southern Germany (Bögel, 1964). Natural infection of pigs in Australia has been suggested (Snowdon & French, 1968). The virus has been isolated from deer in a reservation adjacent to a cattle farm (Romvary, 1965). Serial passage in rabbits leading to attenuation has been described (Baker et al., 1954; Fernelius et al., 1969).

Control. Virus attenuated by serial passages in rabbits immunized against virulent virus (Baker et al., 1954). Combined use of immune serum and live virus vaccine confers some protection (Simony et al.,

1968). Vaccines combined with other viruses (e.g. bovine rhinotracheitis, parainfluenza 3) have been used by several workers (e.g. Rosner, 1968; McKercher et al., 1968).

INFECTIOUS ARTERITIS OF HORSES

This has been confused with rhinopneumonitis or equine abortion (p. 346), since it, too, is a cause of abortion in mares.

Synonyms: Pink-eye. Epizootic cellulitis. 'Fievre typhoide du cheval'. 'Pferdestaupe'. Equine influenza.

Description: Doll et al. (1957).

Morphology. Virions are approximately spherical, enveloped, 50 to 70 nm in diameter with a core 20 to 30 nm (Maess et al., 1970; Magnusson et al., 1970). In thin sections, particles about 58 nm in diameter, with double membranes (Estes & Cheville, 1970) mature by budding into cisternae of the endoplasmic reticulum (Magnusson et al., 1970).

Chemical composition. Lack of inhibition by halogenated deoxyribosides suggests that the viral nucleic acid is RNA (Burki, 1966).

Physico-chemical characters. Buoyant density in CsCl 1·18 g/ml; in sucrose, 1·17 g/ml, in potassium tartrate, 2 peaks, 1·17 and 1·24 g/ml (Hyllseth, 1970). Survives for 6 years at −20°C, 75 days at 4°C, only 2 days at 37°C, 20′ not 30′ at 56°C (McCollum et al., 1961). Inactivation at 50°C enhanced by molar $MgCl_2$; sensitive to lipid solvents and low pH; trypsin-resistant (Burki, 1966).

Antigenic properties. Antibodies demonstrable by virus neutralization and complement fixation (McCollum et al., 1961; Matumoto et al., 1965; Burki & Gerber, 1966).

Cultivation. None reported in fertile eggs. Grows in monolayers of horse kidney, producing CPE, becoming attenuated so as to be of use for immunization (McCollum et al., 1961). Later it was adapted to grow in hamster-kidney cultures (Wilson et al., 1962). Growth with plaque production in hamster cells (BHK21) is described by Hyllseth (1969).

Pathogenicity. Causes fever, conjunctivitis, rhinitis, oedema of the legs or trunk, enteritis and colitis. Virus is present in the tissues and various body fluids, but not in the urine. Is one cause of abortions in mares, virus having invaded the fetus. Pneumonia may occur and is usually fatal.

Experimentally. The incubation period in horses is 5–10 days. Other species are not susceptible.

Pathological lesions. Broncho-pneumonia and pleural effusions are seen in fatal cases. Gelatinous swellings occur round the larynx and elsewhere. The essential lesions are those of medial necrosis of smaller arteries; when endothelial intima is involved thrombosis and infarction follow with the characteristic haemorrhages and oedema.

Ecology. It is believed to be this virus which has caused widespread epizootics in the past. It is extremely contagious, infecting mainly young animals, probably through the respiratory tract. Some maintain that recovered stallions may be carriers and transmit infection.

Control. Prospects for a vaccine made from virus attenuated in tissue culture seem good (McCollum, 1969).

INFECTIOUS ANAEMIA OF HORSES

Synonym: Swamp fever.

Reviews: Dreguss & Lombard (1954). Ishii (1963). Hyslop (1966).

Morphology. Negatively stained virions are described as approximately spherical, 90 to 140 nm in diameter, with an outer envelope possibly studded with small projections, but no well-defined internal component (Nakajima et al., 1969a). In sections, particles 80 to 150 nm in diameter, with a nucleoid 40 to 60 nm across, are formed by budding through modified plasma membrane with thin surface projections (Ito et al., 1969; Tajima et al., 1969; Kono et al., 1970).

Physico-chemical characters. Lack of inhibition by IUDR (Kono et al., 1970) and incorporation of ^3H = uridine (Nakajima et al., 1970) suggest that the viral nucleic acid is RNA. Mean buoyant density in CsCl, 1·18 g/ml (Nakajima et al., 1969b). Trypsin-resistant (Nakajima et al., 1969c). Virus in serum is inactivated in 60′ at 60°. It is resistant to drying and putrefaction, but not to sunlight. It survived 158 days in tap water at room temperature (Rodionov & Oleynik, 1946). Ether-sensitive (Nakajima & Obara, 1964). Virus in serum inactivated by 0·1 per cent formaldehyde in 1 month at 5°, but it survived 0·5 per cent phenol under these conditions (Stein & Gates, 1952). It was inactivated by 4 per cent NaOH in 15′, but oxidizing agents (hypochlorite and $KMnO_4$) were not very effective.

Haemagglutination of fowl (M. N. Dreguss, quoted by Ishii, 1963) and frog (*Rana esculenta*) cells by infective serum is reported (Fedotov, 1950). Agglutination of group O human RBC by serum of infected horses is described by Saurino et al. (1966).

Antigenic properties. Antibodies may be detected by virus neutralization (Tanaka & Sakaki, 1962), complement fixation (Kono & Kobayashi, 1966) and immunoprecipitation (Moore et al., 1966; Nakajima & Ushimi, 1971). Strains can be distinguished by virus neutralization, but possess a common CF antigen (Kono et al., 1971). Saurino et al. (1966) describe a specific antigen in the serum of infected horses, detectable by immunoadherence, passive haemagglutination, haemagglutination inhibition and complement fixation.

Cultivation. Propagation in embryonated eggs described by Dreguss & Lombard (1954). Multiplication in embryonic equine tissues without CPE described by Watanabe (1960). Interference with vesicular stomatitis multiplication has been used as a method of assay (El-Zein et al., 1968). Growth in primary (Kobayashi & Kono, 1967) and continuous (Moore et al., 1970) cultures of horse leukocytes is accompanied by CPE.

Habitat. Distribution is world-wide. The disease is more prevalent in low-lying marshy country.

Pathogenicity. Certainly pathogenic only for horses and other Equidae. Horses may have acute or chronic infections, the latter often consisting of a succession of acute episodes with remissions; these may last for years, but the disease is usually fatal. The incubation period is usually 12–15 days, but it may be much longer before symptoms appear even though viraemia occurs earlier. There occur fever, anaemia, wasting, serous nasal discharge and areas of subcutaneous oedema. Viraemia may be present for years even during remissions. Typical sublingual haemorrhages may be useful indications of carrier state (Steck, 1967). A chronic case of human infection is reported (Peters, 1924); in this the blood remained infectious for horses over a long period.

Experimentally, several workers have reported successful transmission to rabbits, rats, mice, swine, pigeons and other species, few or no symptoms being produced, but other workers (e.g. Stein & Mott, 1944) have failed to confirm these claims (see also Hyslop, 1966).

Pathological lesions. In fatal cases one finds numerous haemorrhages, areas of subcutaneous oedema, degeneration of liver and kidney, replacement of yellow by red bone marrow in long bones.

Ecology. Biting insects, especially *Stomoxys* and Tabanids, are suspected of transmitting the virus, probably mechanically, but transmission by contact or even oral infection is believed to be possible also.

Virus occurs in milk, semen, saliva and urine. Since viraemia is so prolonged, even up to 18 years, transmission by inadequately sterilized syringes is also to be guarded against.

Control. No specific prophylaxis is at present possible. Infected animals should be slaughtered. A number of vaccines have been developed, but none has proved to be more than marginally effective (see Hyslop, 1966; Johnson, 1966).

LACTIC DEHYDROGENASE VIRUS

Synonyms: Riley virus. LDH Agent.

Original description: Riley et al. (1960).

Reviews: Riley (1963). Notkins (1965).

Morphology. Viruses are elliptical or oblong 36–42 × 45–75 nm (de Thé & Notkins, 1965). Different forms, usually elongated but with varying dimensions, have been described (Crispens & Burns, 1964; Prosser & Evans, 1967). A nucleoid 25 or 33 nm in diameter, contained in a double membrane, is seen in sections (du Buy & Johnson, 1965; Prosser & Evans, 1967).

Chemical composition. Infectious RNA has been obtained (Notkins & Scheele, 1963a). A genome with a MW of 6 to 8 × 10^6 daltons is suggested by radiation target size (Rowson et al., 1968).

Physico-chemical characters. Ether-sensitive. Inactivated in 10 minutes at 100°, 30 minutes at 80° or 40 minutes at 60°. Withstands lyophilization; also pH 4 for 3 hours (Notkins, 1965).

Haemagglutination. None demonstrated.

Antigenic properties. Neutralizing antibody has been demonstrated in plasma in late stages of infection coexisting with viraemia (Rowson et al., 1966). Infectious virus-antibody complexes (Notkins et al., 1966) may be neutralized by anti γ globulin (Notkins et al., 1968). Prevention of the development of immune tolerance (Mergenhagen et al., 1967) and promotion of antibody production (Notkins et al., 1966) have been observed in infected mice.

Cultivation. Growth in mouse-embryo tissue cultures has been observed without CPE (Yaffe, 1962; Plagemann & Swim, 1966), but Frantsi & Gregory (1969) describe CPE in mouse-embryo liver cell cultures. Cultivation in mouse macrophages (Evans & Salaman,

1965) is more generally used. Virus replication does not require DNA-dependent RNA synthesis (du Buy & Johnson, 1970).

Habitat. The virus is present in a number of different transplantable mouse tumours in which it is a mere 'passenger'. Doubtless present also in apparently normal mice.

Pathogenicity. Nonpathogenic. Its presence is recognised by a greatly raised titre of lactic dehydrogenase in the blood of infected mice. There is apparently impaired clearance of this and other enzymes when these are injected (Notkins & Scheele, 1964), and it has been suggested that the raised levels are caused by blocking of the reticulo-endothelial system by the virus's action. There is evidence of association of infection with splenomegaly, and the virus is probably identical with the wild mouse virus of Pope & Rowe (1964); this caused chronic splenic hyperplasia. Synergism between LDH virus and *Eperythrozoon coccoides* has been described (Riley, 1964). The LDH agent persists in infected mice for long periods. The titre may reach 10^{10} particles/ml.

Ecology. Virus is present in saliva, urine and for longer periods in faeces. It can be transmitted from mother to infant mice (Crispens, 1964; 1965; Notkins & Scheele, 1963b) and is apparently transmitted between cage mates, but not to mice in adjacent cages.

REFERENCES

Alford, C. A., Neva, F. A., & Weller, T. H. (1964) *New Engl. J. Med.*, **271**, 1275.
Amstey, M. S., Hobbins, T. E., & Parkman, P. D. (1968) *Proc. Soc. exp. Biol. Med., U.S.A.*, **127**, 1231.
Anderson, S. G. (1949) *J. Immunol.*, **62**, 29.
Baker, J. A. (1946) *Proc. Soc. exp. Biol. Med., U.S.A.*, **63**, 183.
Baker, J. A. (1947) *J. Am. vet. med. Ass.*, **111**, 503.
Baker, J. A., & Sheffy, B. E. (1960) *Proc. Soc. exp. Biol. Med., U.S.A.*, **105**, 675.
Baker, J. A., Yorke, C. J., Gillespie, J. H., & Mitchell, G. B. (1954) *Am. J. vet. Res.*, **15**, 525.
Beale, A. J., Christofinis, G. C., & Furminger, I. G. S. (1963) *Lancet*, **2**, 640.
Bekkum, J. G. van (1959) *Proc. 16th. int. vet. Congr., Madrid*, **2**, 477.
Bekkum, J. G. van, & Barteling, S. J. (1970) *Arch. ges. Virusforsch.*, **32**, 185.
Best, J. M., & Banatvala, J. E. (1970) *J. gen. Virol.*, **9**, 215.
Best, J. M., Banatvala, J. E., Almeida, J. D., & Waterson, A. P. (1967) *Lancet*, **2**, 237.
Bögel, K. (1964) *Zentbl. VetMed.*, **11b**, 687.
Bögel, K., & Voss, H. J. (1964) *Zentbl. VetMed.*, **11b**, 181.
Bonissol, C. (1970) *Annls. Inst. Pasteur, Paris*, **118**, 102.
Bontcheff, N., Ivanoff, M., & Bojadjieff, S. (1959) *Bull. Off. int. Épizoot.*, **51**, 252.
Boynton, W. H. (1933) *J. Am. vet. med. Ass.*, **83**, 747.
Burki, F. (1966) *Arch. ges. Virusforsch.*, **19**, 123.
Burki, F., & Gerber, H. (1966) *Berl. Munch. tierarztl. Wschr.*, **79**, 391.
Carver, D. H., Seto, D. S. Y., Marcus, P. I., & Rodrigues, L. (1967) *J. Virol.*, **1**, 1089.
Casselberry, N. H., Malinquist, H. A., Houlihan, W. A., & Boynton, W. H. (1953) *Vet. Med.*, **48**, 24.

Chagnon, A., & Laflamme, P. (1964) *Can. J. Microbiol.*, **10**, 501.
Chapin, R. M., Powick, W. C., McBryde, C. N., & Cole, C. G. (1939) *J. Am. vet. med. Ass.*, **95**, 494.
Coggins, L., & Baker, J. A. (1964) *Am. J. vet. Res.*, **25**, 408.
Coronel, A. B., & Albis, F. J. (1950) *Phillipp. J. anim. Med.*, **11**, 127.
Cotlier, E., Fox, J., Bohigian, G., Beaty, C., & du Pree, A. (1968) *Nature, Lond.*, **217**, 38.
Crawford, J. G., Dayhuff, T. R., & Gallian, M. J. (1968) *Am. J. vet. Res.*, **29**, 1733.
Crispens, C. G. (1964) *J. natn. Cançer Inst.*, **32**, 497.
Crispens, C. G. (1965) *J. natn. Cancer Inst.*, **34**, 331.
Crispens, C. G., & Burns, T. A. (1964) *Nature, Lond.*, **204**, 1302.
Cunliffe, H. R., & Rebers, P. A. (1968) *Can. J. comp. Med.*, **32**, 409.
Darbyshire, J. H. (1962) *Res. vet. Sci.*, **3**, 118.
Darbyshire, J. H. (1963) *J. comp. Path.*, **73**, 309.
Darbyshire, J. H. (1967) *J. comp. Path.*, **77**, 107.
de Thé, G., & Notkins, A. L. (1965) *Virology*, **26**, 512.
Diderholm, H., & Dinter, Z. (1966) *Zentbl. Bakt. ParasitKde, I. Abt. Orig.*, **201**, 270.
Dickinson, P. C. T., Chang, T.-W., & Weinstein, L. (1969) *Proc. Soc. exp. Biol. Med., U.S.A.*, **132**, 55.
Dinter, Z. (1963) *Zentbl. Bakt. ParasitKde, I. Abt. Orig.*, **188**, 475.
Ditchfield, J., & Doane, F. W. (1964) *Can. J. comp. Med.*, **28**, 184.
Doll, E. R., Bryans, J. T., McCollum, W. H., & Crowe, M. E. (1957) *Cornell Vet.*, **47**, 3.
Done, J. T. (1957) *Vet. Rec.*, **69**, 1341.
Draper, C. C., & Lawrence, G. D. (1969) *J. med. Microbiol.*, **2**, 249.
Dreguss, M. N., & Lombard, L. S. (1954) *Experimental studies in equine infectious anaemia.* Philadelphia: Univ. of Pennsylvania Press.
du Buy, H. G., & Johnson, M. L. (1965) *J. exp. Med.*, **122**, 587.
du Buy, H. G., & Johnson, M. L. (1970) *Proc. Soc. exp. Biol. Med., U.S.A.*, **133**, 1023.
Dunne, H. W. (1958) *Diseases in swine.* Ames: Iowa State College Press.
Edwards, M. R., Cohen, S. M., Bruno, M., & Deibel, R. (1969) *J. Virol.*, **3**, 439.
El-Zein, A., Myers, W. L., & Serge, D. (1968) *J. infect. Dis.*, **118**, 473.
Estes, P. C., & Cheville, N. F. (1970) *Am. J. Path.*, **58**, 235.
Evans, R., & Salaman, M. H. (1965) *J. exp. Med.*, **122**, 993.
Fabiyi, A., Sever, J. L., Ratner, N., & Caplan, B. (1966) *Proc. Soc. exp. Biol. Med., U.S.A.*, **122**, 392.
Fedotov, A. I. (1950) *Veterinariya*, **27**, 53.
Fernelius, A. L. (1968) *Arch. ges. Virusforsch.*, **25**, 211.
Fernelius, A. L. (1969) *Arch. ges. Virusforsch.*, **27**, 1.
Fernelius, A. L., Lambert, G., & Packer, R. A. (1969) *Am. J. vet. Res.*, **30**, 1541.
Fontanelli, E., Menascè, I., & d'Ascani, E. (1959) *Zooprofilassi*, **14**, 467.
Frantsi, C., & Gregory, K. F. (1969) *Virology*, **37**, 145.
Fuccillo, D. A., Gitnick, G. L., Traub, R., Wong, K., Sever, J. L., & Huebner, R. J. (1967) *Proc. Soc. exp. Biol. Med., U.S.A.*, **125**, 1015.
Fuchs, F. (1968) in *Handbuch der Virusinfektionen bei Tieren.* Band III. Ed.: Rohrer. Jena: Gustav Fischer.
Furukawa, T., Plotkins, S., Sedwick, D., & Profeta, M. (1967) *Nature, Lond.*, **215**, 172.
Gillespie, J. H., & Baker, J. A. (1959) *Cornell Vet.*, **49**, 439.
Gillespie, J. H., Baker, J. A., & McEntee, K. (1960a) *Cornell Vet.*, **50**, 73.
Gillespie, J. H., Coggins, L., Thompson, J., & Baker, J. A. (1961) *Cornell Vet.*, **51**, 155.
Gillespie, J. H., Sheffy, B. E., & Baker, J. A. (1960b) *Proc. Soc. exp. Biol. Med., U.S.A.*, **105**, 679.
Gregg, N. McA. (1945) *Med. J. Aust.*, **1**, 313.
Gustafson, D. P., & Pomerat, C. M. (1957) *Am. J. vet. Res.*, **18**, 473.
Gutekunst, D. E., & Malmquist, W. A. (1965) *Arch. ges. Virusforsch.*, **15**, 159
Habel, K. (1942) *Publ. Hlth Rep., Wash.*, **57**, 1126.
Hafez, S. M., Petzoldt, K., & Reczko, E. (1968) *Acta virol., Prague*, **12**, 471.

Halonen, P. E., Casey, H. L., Stewart, J. A., & Hall, A. D. (1967a) *Proc. Soc. exp. Biol. Med., U.S.A.*, **125**, 167.
Halonen, P. E., Ryan, J. M., & Stewart, J. A. (1967b) *Proc. Soc. exp. Biol. Med., U.S.A.*, **125**, 162.
Hamvas, J. J., Ugovsek, S., Iwakata, S., & Labzoffsky, N. A. (1969) *Arch ges. Virusforsch.*, **26**, 287.
Hanson, R. P. (1957) *J. Am. vet. med. Ass.*, **131**, 211.
Harding, J. D. J., Done, J. T., & Darbyshire, J. H. (1966) *Vet. Rec.*, **79**, 388.
Hobbins, T. E., & Smith, K. O. (1968) *Proc. Soc. exp. Biol. Med., U.S.A.*, **129**, 407.
Holmes, I. H., Wark, M. C., & Warburton, M. F. (1969) *Virology*, **37**, 15.
Horzinek, M. (1966) *J. Bact.*, **92**, 1723.
Horzinek, M. (1967) *Arch. ges. Virusforsch.*, **21**, 447.
Horzinek, M., Reczko, E., & Petzoldt, K. (1967) *Arch. ges. Virusforsch.*, **21**, 475.
Hovi, T., & Vaheri, A. (1970a) *Virology*, **42**, 1.
Hovi, T., & Vaheri, A. V. (1970b) *J. gen. Virol.*, **6**, 77.
Huck, R. A. (1957) *J. comp. Path.*, **67**, 267.
Huck, R. A., & Aston, F. W. (1964) *Vet. Rec.*, **76**, 1151.
Huck, R. A., & Cartwright, S. F. (1964) *J. comp. Path.*, **74**, 346.
Huygelen, C., Peetermans, J., & Prinzie, A. (1969) *Progr. med. Virol.*, **11**, 107.
Hyllseth, B. (1969) *Arch. ges. Virusforsch.*, **28**, 26.
Hyllseth, B. (1970) *Arch. ges. Virusforsch.*, **30**, 97.
Hylsop, N. St. G. (1966) *Vet. Rec.*, **78**, 858.
Inaba, Y., Tanaka, Y., Kumagai, T., Omori, T., Ito, H., & Matumoto, M. (1968) *Jap. J. Microbiol.*, **12**, 35.
Ingalls, T. H., Babbott, F. L., Hampson, K. W., & Gordon, J. E. (1960) *Am. J. med. Sci.*, **239**, 363.
Ishii, S. (1963) *Adv. vet. Sci.*, **8**, 263.
Ito, Y., Kono, Y., & Kobayashi, K. (1969) *Arch. ges. Virusforsch.*, **28**, 411.
Johnson, A. W. (1966) *Vet. Bull.*, **36**, 465.
Johnston, K. G. (1959) *Aust. vet. J.*, **35**, 101.
Karasszon, D., & Bodon, L. (1963) *Acta microbiol. acad. sci. hung.*, **10**, 287.
Keast, J. C., Littlejohn, I. R., & Helwig, D. M. (1962) *Aust. vet. J.*, **38**, 129.
Kenny, M. T., Albright, K. L., Emery, J. B., & Brittle, J. L. (1969) *J. Virol.*, **4**, 807.
Kobayashi, K., & Kono, Y. (1967) *Natn. Inst. Anim. Hlth Q., Tokyo*, **1**, 1 and 8.
Kono, R., Hibi, M., Hayakawa, Y., & Ishi, K. (1969) *Lancet*, **1**, 343.
Kono, Y., & Kobayashi, K. (1966) *Natn. Inst. Anim. Hth. Q., Tokyo*, **6**, 194.
Kono, Y., Kobayashi, K., & Fukunaga, Y. (1971) *Arch. ges. Virusforsch.*, **34**, 202.
Kono, Y., Yoshino, T., Fukunaga, Y. (1970) *Arch. ges. Virusforsch.*, **30**, 252.
Koprowski, H., James, T., & Cox, H. (1946) *Proc. Soc. exp. Biol. Med., U.S.A.*, **63**, 178.
Korn, G. (1964) *Zentbl. VetMed.*, **11b**, 119.
Korn, G., & Hecke, F. (1964) *Zentbl. VetMed.*, **11b**, 40.
Korn, G., Nishimura, Y. (1963) *Mh. Tierheilk*, **15**, 328.
Krugman, S. (1965) *Arch. ges. Virusforsch.*, **16**, 477.
Kumagai, T., Shimizu, T., & Matumoto, M. (1958) *Science*, **128**, 366.
Laufs, R., & Thomssen, R. (1968a) *Arch. ges. Virusforsch.*, **24**, 164.
Laufs, R., & Thomssen, R. (1968b) *Arch. ges. Virusforsch.*, **24**, 181.
Liebhaber, H. (1970) *J. Immunol.*, **104**, 818.
Loan, R. W. (1964) *Am. J. vet. Res.*, **25**, 1366.
Loan, R. W. (1965) *Am. J. vet. Res.*, **26**, 1110.
Loan, R. W., & Gustafson, D. P. (1964) *Am. J. vet Res.*, **25**, 1120.
Loan, R. W., & Storm, M. M. (1968) *Am. J. vet. Res.*, **29**, 807.
London, W. T., Fucillo, D. A., & Sever, J. L. (1969) *Proc. Internat. Symp. on Rubella Vaccine*, **11**, 121. Basel/New York: Karger.
McBryde, C. N., & Cole, C. G. (1936) *J. Am. vet. med. Ass.*, **89**, 652.
McCarthy, K., Taylor-Robinson, C. H., & Pillinger, S. E. (1963) *Lancet*, **2**, 593.
McCollum, W. H. (1969) *J. Am. vet. med. Ass.*, **155**, 318.

McCollum, W. H., Doll, E., Wilson, J., & Johnson, C. B. (1961) *Am. J. vet. Res.*, **22**, 731.
McCombs, R. M., & Rawls, W. E. (1968) *J. Virol.*, **2**, 409.
McKercher, D. G., Saito, J. K., Crenshaw, G. L., & Bushnell, R. B. (1968) *J. Am. vet. med. Ass.*, **152**, 1621.
Maes, R., Vaheri, A., Sedwick, D., & Plotkin, S. (1966) *Nature, Lond.*, **210**, 384.
Maess, J., & Reczko, E. (1970) *Arch. ges. Virusforsch.*, **30**, 39.
Maess, J., Reczko, E., & Bohm, H. O. (1970) *Arch. ges. Virusforsch.*, **30**, 47.
Magnusson, P., Hyllseth, B., & Marusyk, H. (1970) *Arch. ges. Virusforsch.*, **30**, 105.
Magnusson, P., & Skaaret, P. (1967) *Arch. ges. Virusforsch.*, **20**, 374.
Malmquist, W. A., Fernelius, A. L., & Gutekunst, D. E. (1965) *Am. J. vet. Res.*, **26**, 1316.
Mansi, W. (1957) *J. comp. Path.*, **67**, 297.
Marcus, S. J., & Moll, T. (1968) *Am. J. vet. Res.*, **29**, 817.
Matumoto, M., Shimizu, T., & Ishizaki, R. (1965) *C. r. Séanc. Soc. Biol.*, **159**, 1262.
Mayr, A., Bachmann, P. A., Sheffy, B. E., & Sigel, G. (1967) *Arch. ges. Virusforsch.*, **21**, 113.
Mayr, A., & Mahnel, H. (1964) *Zenbl. Bakt. ParasitKde, I Abt. Orig.*, **195**, 157.
Mengeling, W. L., Pirtle, E. C., & Torrey, J. P. (1963) *Can. J. comp. Med.*, **27**, 249.
Menser, M. A., Harley, J. D., Hertzberg, R., Dorman, D. C., & Murphy, A. M. (1967) *Lancet*, **2**, 387.
Mergenhagen, S. E., Notkins, A. L., & Dougherty, S. F. (1967) *J. Immunol.*, **99**, 576.
Mettler, N. E., Petrelli, R., & Casals, J. (1968) *Virology*, **36**, 503.
Mills, J. H. L., Nielsen, S. W., & Luginbuhl, R. E. (1965) *J. Am. vet. Med. Ass.*, **146**, 691.
Molnár, I. (1954) *Magy. Állatorv. Lap.*, **9**, 146.
Monif, G. R. G., Avery, G. B., Korones, S. B., & Sever, J. L. (1965) *Lancet*, **1**, 723.
Moore, R. W., Livingston, C. W., Jr., & Redmond, H. E. (1966) *SWest Vet.*, **19**, 187.
Moore, R. W., Redmond, H. E., Katada, M., & Wallace, M. (1970) *Am. J. vet. Res.*, **31**, 1569.
Murphy, F. A., Halonen, P. E., & Harrison, A. K. (1968) *J. Virol.*, **2**, 1223.
Nakajima, H., & Obara, J. (1964) *Natn. Inst. anim. Hlth Q., Tokyo*, **4**, 129.
Nakajima, H., Tajima, M., Tanaka, S., & Ushimi, C. (1969a) *Arch. ges. Virusforsch.*, **28**, 348.
Nakajima, H., Tanaka, S., & Ushimi, C. (1969b) *Arch. ges. Virusforsch.*, **26**, 389.
Nakajima, H., Tanaka, S., & Ushimi, C. (1969c) *Arch. ges. Virusforsch.*, **26**, 395.
Nakajima, H., Tanaka, S., & Ushimi, C. (1970) *Arch. ges. Virusforsch.*, **31**, 273.
Nakajima, H., & Ushimi, C. (1971) *Inf. Immunity*, **3**, 373.
Neva, F. A., Alford, C. A., & Weller, T. H. (1964) *Bact. Rev.*, **28**, 444.
Noice, F., & Schipper, I. A. (1959) *Proc. Soc. exp. Biol. Med., U.S.A.*, **100**, 84.
Notkins, A. L. (1965) *Bact. Rev.*, **29**, 143.
Notkins, A. L., Mage, M., Ashe, W. K., & Mahar, S. (1968) *J. Immunol.*, **100**, 314.
Notkins, A. L., Mahar, S., Scheele, C., & Goffman, J. (1966) *J. exp. Med.*, **124**, 81.
Notkins, A. L., Mergenhagen, S. E., Rizzo, A. A., Scheele, C., & Waldmann, T. A. (1966) *J. exp. Med.*, **123**, 347.
Notkins, A. L., & Scheele, C. (1963a) *Virology*, **20**, 640.
Notkins, A. L., & Scheele, C. (1963b) *J. exp. Med.*, **118**, 7.
Notkins, A. L., & Scheele, C. (1964) *J. natn. Cancer Inst.*, **33**, 741.
Oxford, J. S. (1967) *J. Immunol.*, **98**, 697.
Ohshima, K., McGuire, T. C., Henson, J. B., & Gorham, J. R. (1970) *Res. vet. Sci.*, **11**, 405.
Parkman, D., Buescher, E. L., & Arnstein, E. S. (1962) *Proc. Soc. exp. Biol., U.S.A.*, **111**, 225.
Parkman, D. D., Buescher, E. L., Artenstein, M. S., McGown, J. M., Mundon, F. K., & Druzd, A. D. (1964) *J. Immunol.*, **93**, 595.
Parkman, P. D., & Meyer, H. M. (1969) *Progr. med. Virol.*, **11**, 80.
Parkman, P. D., Meyer, H. M., Kirschstein, R. L., & Hopps, H. E. (1966) *New Engl. J. med.*, **275**, 569.
Parkman, P. D., Phillips, P. E., Kirschstein, R. L., & Meyer, H. M., Jr. (1965) *J. Immunol.*, **95**, 743.

Petters, J. T. (1924) *Presse med.*, **32**, 105.
Pirtle, E. C. (1964) *Can. J. comp. Med.*, **28**, 193.
Pirtle, E. C., & Kniazeff, A. J. (1968) *Am. J. vet. Res.*, **29**, 1033.
Plagemann, P. G. W., & Swim, H. E. (1966) *Proc. Soc. exp. Biol. Med., U.S.A.*, **121**, 1147.
Plotkin, S. A., Farquhar, J., Katz, M., & Ingalls, T. H. (1967) *Am. J. Epidemiol.*, **86**, 468.
Pope, J. H., & Rowe, W. F. (1964) *Proc. Soc. exp. Biol. Med., U.S.A.*, **116**, 1015.
Pritchard, W. R. (1963) *Adv. vet. Res.*, **8**, 1.
Prosser, P. R., & Evans, R. (1967) *J. gen. Virol.*, **1**, 419.
Rawls, W. E. (1968) *Progr. med. Virol.*, **10**, 238.
Rawls, W. E., Desmynter, J., & Melnick, J. L. (1967a) *Proc. Soc. exp. Biol. Med., U.S.A.*, **124**, 167.
Rawls, W. E., & Melnick, J. L. (1966) *J. exp. Med.*, **123**, 795.
Rawls, W. E., et al. (19 authors) (1967b) *Bull. Wld Hlth Org.*, **37**, 79.
Rhim, J. S., Schell, K., & Huebner, R. J. (1967) *Proc. Soc. exp. Biol. Med., U.S.A.*, **125**, 1271.
Riley, V. (1963) *Ann. N.Y. Acad. Sci.*, **100**, 762.
Riley, V. (1964) *Science*, **146**, 921.
Riley, V., Lilly, F., Huerto, E., & Bardell, D. (1960) *Science*, **132**, 545.
Ritchie, A. E., & Fernelius, A. L. (1969) *Arch. ges. Virusforsch.*, **28**, 369.
Rodionov, M., & Oleynik, H. K. (1946) *Veterinariya*, **23**, 11.
Romvary, J. (1965) *Acta vet. hung.*, **15**, 451.
Rosner, S. F. (1968) *J. Am. vet. med. Ass.*, **152**, 898.
Rowson, K. E. K., Mahy, B. W. J., & Bendinelli, M. (1966) *Virology*, **28**, 775.
Rowson, K. E. K., Parr, I. B., & Alper, T. (1968) *Virology*, **36**, 157.
Russell, B., Selzer, G., & Goetze, H. (1967) *J. gen. Virol.*, **1**, 305.
Salmi, A. A. (1970) *Arch. ges. Virusforsch.*, **32**, 91.
Sampson, G. R., Sauter, R. A., Williams, L. M., & Marshall, V. (1965) *J. Am. vet. Med. Ass.*, **146**, 836.
Sato, U., Hanaki, T., & Nobuto, K. (1969) *Arch. ges. Virusforsch.*, **26**, 1.
Sato, V., Nishimura, Y., Hanaki, T., & Nobuto, K. (1964) *Arch. ges. Virusforsch.*, **14**, 394.
Saurino, V. R., Waddell, G. H., Flynn, J. H., & Teigland, M. B. (1966) *J. Am. vet. med. Ass.*, **149**, 1416.
Schiff, G. M., & Sever, J. L. (1966) *Progr. med. Virol.*, **8**, 30.
Schmidt, N. J., Dennis, J., & Lennette, E. H. (1968a) *Arch. ges. Virusforsch.*, **25**, 308.
Schmidt, N. J., Lennette, E. H., & Dennis, J. (1969) *Proc. Soc. exp. Biol. Med., U.S.A.*, **132**, 128.
Schmidt, N. J., Lennette, E. H., Gee, P. S., & Dennis, J. (1968b) *J. Immunol.*, **100**, 851.
Sedwick, W. D., & Sokol, F. (1970) *J. Virol.*, **5**, 478.
Segre, D. (1964) *Proc. Soc. exp. Biol. Med., U.S.A.*, **117**, 567.
Sever, J. L., Huebner, R. J., Castellaro, G. A., Sarma, P. S., Fabiyi, A., Schiff, G. M., & Casumano, C. L. (1965) *Science*, **148**, 385.
Sever, J., & White, L. R. (1968) *Ann. Rev. Med.*, **19**, 471.
Sheffy, B. E., Coggins, L., & Baker, J. A. (1962) *Proc. Soc. exp. Biol. Med., U.S.A.*, **109**, 349.
Shope, R. E. (1958) *J. exp. Med.*, **107**, 609, and **108**, 159.
Sigurdardottir, B., Givan, K. F., Rozee, E. R., & Rhodes, A. J. (1963) *Can. med. Ass. J.*, **88**, 128.
Simonyi, E., Bognar, K., Biro, J., & Palatka, Z. (1968) *Acta vet. hung.*, **18**, 237.
Sirbu, Z., Ieremia, D., & Bona, C. (1964) *Ref. Zootech. Med. vet.* (Bucarest), **14**, 59.
Slavin, G. (1938) *J. comp. Path.*, **51**, 213.
Snowdon, W. A., & French, E. L. (1968) *Aust. vet. J.*, **44**, 179.
Soekawa, M., & Isawa, H. (1960) *Kitasato Arch. expl. Med.*, **33**, 25.
Stair, E. L., Rhodes, M. B., Aiker, J. M., Underdahl, N. R., & Young, D. A. (1963) *Proc. Soc. exp. Biol. Med., U.S.A.*, **113**, 656.
Steck, W. (1967) *Vet. Rec.*, **81**, 205.
Stein, C. D., & Gates, D. W. (1952) *Am. J. vet. Res.*, **13**, 195.

Stein, C. D., & Mott, L. O. (1944) *Vet. Med.*, **39**, 408.
Stewart, G. L., Parkman, P. D., Hopps, H. E., Douglas, R. D., Hamilton, J. P., & Mayer, H. M. (1967) *New Engl. J. Med.*, **276**, 554.
Swan, C., Tostevin, A. L., & Black, G. H. B. (1946) *Med. J. Aust.*, **2**, 889.
Symposium on Rubella Vaccines (1969) Symp. Series in Immun. Stand. Vol. 11. Eds.: R. H. Regamey, A. de Barbieri, W. Hennesseu, D. Ikic, & F. T. Perkins. Basel/New York: Karger.
Tajima, M., Nakajima, H., & Ito, Y. (1969) *J. Virol.*, **4**, 521.
Tanaka, Y., Inaba, Y., Omori, T., & Matumoto, M. (1968) *Jap. J. Microbiol.*, **12**, 201.
Tanaka, K., & Sakaki, K. (1962) *Natn. Inst. Anim. Hlth Q.*, Tokyo, **2**, 128.
Taylor, D. O. N., Gustafson, D. P., & Claflin, R. M. (1963) *Am. J. vet. Res.*, **24**, 143.
Thomsen, R., Laufs, R., & Miller, J. (1968) *Arch. ges. Virusforsch.*, **23**, 332.
Torrey, J. P., & Prather, J. K. (1964) *Proc. 67th Ann. Meet. U.S. Livestock Sanitary Ass.*, 414.
Tuchinda, P., Nii, S., Sasada, T., Naito, T., Ono, N., & Chatiyanon, K. (1969) *Biken's J.*, **12**, 201.
Vaheri, A., Sedwick, W. D., & Plotkin, S. A. (1967) *Proc. Soc. exp. Biol. Med., U.S.A.*, **125**, 1086.
Vaheri, A., Sedwick, W. D., Plotkin, S. A., & Maes, R. (1965) *Virology*, **27**, 239.
Watanabe, S. (1960) *Jap. J. vet. Sci.*, **22**, 65 and 87.
Weibel, R. E., Stokes, J., Jr., Buynak, E. B., Whitman, J. E., Jr., Leagus, M. B., & Hilleman, M. R. (1968) *J. Am. med. Ass.*, **205**, 554.
Weller, T. H., Alford, C. A., & Neva, F. A. (1965) *Yale J. Biol. Med.*, **37**, 455.
Weller, T. H., & Neva, F. (1962) *Proc. Soc. exp. Biol. Med., U.S.A.*, **111**, 215.
Wilsdon, A. J. (1958) *Vet. Rec.*, **70**, 3.
Wilson, J. C., Doll, E. R., McCollum, W. H., & Cheatham, J. (1962) *Cornell Vet.*, **52**, 200.
Woods, W. A., Johnson, R. T., Hostetler, D. D., Lepow, M. L., & Robbins, F. C. (1966) *J. Immunol.*, **96**, 253.
Yaffe, D. (1962) *Cancer Res.*, **22**, 573.
Young, G. A., Kitchell, R. L., Luedke, A. J., & Sautter, J. H. (1955) *J. Am. vet. med. Ass.*, **126**, 165.

PART II

DNA Viruses

12

Parvovirus

The name 'Picodnavirus', suggested for the group, was not approved by the International Committee for Nomenclature of Viruses.

Review: Toolan (1968).

The following properties characterize the genus: Virions are un-enveloped, 18 to 22 nm in diameter, with icosahedral symmetry, and probably 32 capsomeres. Contain single-stranded, circular DNA of MW $1 \cdot 2 \times 10^6$ to $1 \cdot 8 \times 10^6$. Two subgroups have been suggested: in one (adeno-associated viruses), virions contain complementary (+ or −) DNA strands which come together *in vitro* to form a double strand. Virus replication in the nucleus. Ether-resistant, relatively heat-stable.

LATENT RAT VIRUS

Description: Kilham & Olivier (1959).
Distinct from the K virus (p. 301).

Review: Kilham (1966).

Synonyms: H1, H3 etc. (Toolan).

Morphology. Virions are 17 to 25 nm in diameter with icosahedral symmetry and probably 32 capsomeres (Toolan et al., 1964; Breese et al., 1964; Vasquez & Brailowsky, 1965).

Chemical composition. Thirty-four per cent of the virion weight is made up of single-stranded DNA (Robinson & Hetrick, 1969) with MW $1 \cdot 7 \times 10^6$ (Salzman & Jori, 1970). Nearest neighbour frequency pattern similar to that of host cells and base composition A 25·5–26·8; T 29·3–29·6; G 20·6–22·6; C 22·6–22·9 (McGeogh et al., 1970).

Physico-chemical characters. Density of virion 1·43 g/cm^3 and of DNA 1·72 (Robinson & Hetrick, 1969).

Haemagglutination. Agglutinates guinea-pig RBCs at 23°C–40°C; strains differ in ability to agglutinate rat and human cells. Virus does not elute spontaneously from guinea-pig cells; receptor-destroying enzyme destroys guinea-pig red cell receptors. The haemagglutinin has not been separated from the virion; it is stable between pH 2 and 11 (Greene, 1965). Haemadsorption is useful. Toolan (1967) describes the agglutination by various strains of the RBCs of a number of species.

Antigenic properties. Antihaemagglutinins and neutralizing antibodies (tested for in tissue culture) are present in sera of some normal laboratory and wild rats (*R. norvegicus*). One of Toolan's (1960a) strains is identical with Kilham's; another is serologically distinct with at most a little 1-way crossing (Moore, 1962). According to Portella (1964) and Greene & Karasaki (1965) there are 3 serotypes.

Cultivation. Growth occurs in rat- but not mouse-embryo tissue cultures, also (Toolan & Ledinko, 1965) in various cell-lines and other cultures of primate and hamster origin. Dense masses of cells form in the cultures, with thin radiating extensions. These CPEs are seen after 9–12 days at 36°C, at which time haemagglutinins appear in the supernatant fluids. Growth occurs also in hamster embryo, but without this characteristic CPE (Moore, 1962). Intranuclear inclusions like those of polyoma can be found in epithelial and connective tissue cells of rat embryo (Dawe et al., 1961). Growth is enhanced by co-cultivation with adenovirus type 12 (Chany & Brailovsky, 1967; Ledinko & Toolan, 1968).

Pathogenicity. Toolan et al. (1960) produced a fulminating infection in suckling hamsters with large doses of virus. Smaller doses have caused dwarfism and so-called 'mongoloid' appearance (Toolan, 1960b; Kilham, 1961). As a result of studying the basis of this appearance, Dalldorf (1960) has described this as an osteolytic virus. The virus also infects pregnant hamsters and suckling and pregnant rats (Kilham, 1966), especially those of an unusually susceptible strain (Moore & Nicastri, 1965). Hepatitis may also be produced in very young rats. A naturally occurring strain caused congenital lesions in rats (Kilham & Margolis, 1966b). Congenital abnormalities may occur also in offspring of hamsters infected in pregnancy. The virus can cause destruction of cerebellar cells in very young hamsters and kittens (Kilham & Margolis, 1965). Matsuo & Spencer (1969) reported pathogenicity for new-born mice.

Ecology. Virus has been recovered from rat tumours and from rats in which human tumours have been growing. It is presumably a virus latent in rats. Its direct isolation from human tumours has been considered as possible, but one awaits evidence that this has been achieved in a laboratory where the virus was not already under study. It has been recovered by swabbing a laboratory bench.

HAEMORRHAGIC ENCEPHALOPATHY IN RATS

A virus serologically related to the Kilham rat virus but not to the H1 strain caused haemorrhage and necrosis of the brains and cords of rats; its activity was potentiated by immunosuppressive drugs (Nathanson et al., 1970).

MINUTE MOUSE VIRUS

Description: Crawford (1966).

Morphology. Virions are hexagonal in outline, sometimes in closely packed array with centre-to-centre spacing of 19 to 26 nm.

Chemical composition. Contain single-stranded DNA with nearest-neighbour frequency pattern and base composition similar to that of latent rat virus (p. 281) (McGeogh et al., 1970).

Physico-chemical characters. Density in CsCl 1·43 g/ml.

Haemagglutination. Agglutinates guinea-pig RBCs and, to slightly lower titre, those of hamster, rat and mouse; the temperature could, without harm, be varied between 4° and 37°, the pH between 7 and 8·5.

Antigenic properties. Distinct from the latent rat virus, though there were slight cross-reactions in the HAI test. Wild rat sera had more antibodies against this than against rat parvoviruses (Kilham & Margolis, 1970).

Cultivation. Multiplied in cultures of cells from rat or mouse embryo; no plaques or CPE.

Pathogenicity. Multiplies in many organs of baby mice and may cause runting and cerebrellar lesions, but not ataxia. (Kilham & Margolis, 1970). Fatal to suckling rats.

Ecology. The virus was carried by 79 per cent of mouse colonies studied by Parker et al. (1970) and was found as a contaminant of many transplanted tumours or leukaemias. It was present in urine and faeces.

PANLEUCOPENIA OF CATS

Synonyms: Feline infectious enteritis. Feline agranulocytosis. Cat fever. Cat distemper. Cat plague. Show fever. Enteritis of mink. Ataxia of cats.

Review: Johnson (1969).

The virus is only provisionally included with the parvoviruses as it may not have single-stranded DNA. Johnson & Cruickshank (1966) held that the DNA was double-stranded.

Morphology. Diameter about 20–25 nm.

Physico-chemical characters. Stable between pH 3 and pH 9. Evidence as to resistance to desiccation is conflicting: it is said to withstand lyophilization, but to be readily inactivated when dried at room temperature. On the other hand, fomites seem to retain infectivity for months under natural conditions. Inactivated by 0·2 per cent formalin. Fairly stable at 50°, inactivated in 30 minutes between 80° and 85°.

Haemagglutination. Some strains, especially from mink, have weak haemagglutinins for pig RBCs at 4°.

Antigenic properties. Neutralizing antibodies are demonstrable; this is most readily done in tissue culture. Fluorescent antibodies can also be detected (in tissue culture): if present at more than 1 in 4 dilution, their presence was correlated with resistance to challenge (King & Crogham, 1966). Hyperimmune sera produced in cats will passively protect kittens.

Cultivation. No multiplication occurs in fertile eggs. The virus was cultivated in monolayers of kitten-kidney cells, the first isolation being from a leopard (Johnson, 1964). Nuclear inclusions developed similar to those seen in infected cats. Their appearance and that of CPE was transient (Johnson, 1967).

Distribution. World-wide.

Pathogenicity. Chiefly affects young cats in endemic areas, but cats of any age where virus is newly introduced. There is a fever, the animals becoming very ill at the second peak. The cat is depressed and dehydrated often with vomiting and profuse, sometimes blood-stained, diarrhoea. Discharge from eyes and nose occurs. Mortality when the disease has declared itself is from 65 to 90 per cent, but there are probably many unrecognized subclinical infections which result in immunity. The incubation period after contact is rarely over 6 days.

All the *Felidae* are susceptible, even lions and tigers, though in zoological gardens the virus rather affects the smaller species. Raccoons, of the *Procyonidae*, are also attacked (Goss, 1948), but of *Mustelidae* only mink naturally and young ferrets experimentally. A fatal virus enteritis in mink was shown by Wills (1952) to be caused by this virus; symptoms are similar to those in cats.

A virus from spontaneous ataxia of cats was found to be antigenically identical with panleucopenia (Johnson et al., 1967). It was serially transmitted IC in kittens and ferrets, the latter being easier to work with. Cerebellar ataxia developed after 4 or 5 weeks. Spontaneous transmission occurred among kittens, but not ferrets (Kilham & Margolis, 1966a). The panleucopenia virus can apparently give rise to absorption of foeti, early neonatal deaths or ataxia in surviving kittens. (Kilham et al., 1967).

Experimentally, the disease is transmissible to cats and mink, by injection by various routes; the incubation period may be 48 hours. Animals of other families are resistant.

After an initial leucocytosis, white blood cells, both lymphocytes and polymorphonuclears, progressively disappear from the circulation and may, finally, be almost absent (Lawrence & Syverton, 1938; Hammon & Enders, 1939).

Pathological lesions. Post mortem are found acute enteritis, particularly of the lower ileum, spleen enlargement, swollen mesenteric lymph nodes and a 'semi-liquid' aplastic bone marrow (Jennings, 1947). In intestinal epithelium, also in lymph nodes, intranuclear inclusions are found (Hammon & Enders, 1939). Though they are at times eosinophilic, study of their whole course of development suggests that they differ from Cowdry's type A inclusions (Lucas & Riser, 1945), occurring in clustered granular and diffuse granular forms and becoming basophilic with ageing. Tuomi & Kangas (1963), using fluorescent antibody, detected antigen in cytoplasm of intestinal cells of mink with enteritis.

Ecology. All secretions and excretions contain virus; recovered mink, perhaps also cats, may excrete virus for a year. Rugs and other fomites in houses where cats have died may carry infection to other cats introduced later. Recovered animals are highly immune and maternal antibody may protect kittens for a while. If they encounter the virus during this period, they may undergo a subclinical immunizing infection. Natural transmission through infected arthropods, especially fleas, or through worms has been suspected, but never proved.

Control. A vaccine made by treated infected tissue suspensions with 0·2 per cent formalin has been successfully used for immunizing kittens (Pridham & Wills, 1959). An attenuated virus vaccine may protect cats from panleucopenia and mink from enteritis. (Gorham et al., 1965; King & Gutekunst, 1970).

PORCINE PARVOVIRUS

Morphology. Like that described for parvoviruses (Mayr et al., 1968).

Chemical composition. DNA, but strandedness not determined.

Physico-chemical properties. Density 1·37–1·38. Relatively heat-resistant: survived 70° for 2 hours. Ether- and acid-(pH3) resistant.

Haemagglutination: Agglutinates RBCs of chick, rat, guinea-pig, cat, rhesus, patas and human O cells (Cartwright et al., 1969).

Antigenic properties. Johnson & Collings (1969) observed antigenic responses, using HAI and VN tests.

Cultivation. Grew in pig-kidney cultures; nuclear inclusions developed in infected cells (Mayr et al., 1968; Cartwright et al., 1969).

Pathogenicity. The virus was recovered from swine with infertility and from aborted piglets. Surviving piglets had virus in various organs up to 9 weeks of age (Johnson & Collings, 1969). The virus may be the same as the pig infertility virus described by Cartwright & Huck (1967).

BOVINE PARVOVIRUS

A haemadsorbing virus isolated from the genito-urinary tract of cattle may be a parvovirus (Storz & Warren, 1970). The virus described by Abinanti & Warfield (1961) may be the same.

CANINE PARVOVIRUS

A 'minute virus of canines' was isolated from dog faeces by Binn et al. (1970). It was cultivable only in a line of dog cells and produced large intranuclear inclusions in the cultures.

AVIAN PARVOVIRUS

Particles resembling parvoviruses have been found together with adenoviruses in materials derived from quail bronchitis (Dutta & Pomeroy, 1967).

ADENO-ASSOCIATED (SATELLITE) VIRUSES (AAV)

Descriptions: Atchison et al. (1966). Hoggan et al. (1966). Rose et al. (1966).

Small particles may be found in tissue cultures of adenoviruses. They are apparently virions not genetically related to the adenoviruses, defective in that they cannot replicate in the absence of multiplying adenoviruses.

Morphology: 22–24 nm across with hexagonal profiles (Mayor et al., 1965). Smith et al. (1966) describe a 'net-like capsid' similar to that of reoviruses.

Chemical composition. Virions contain complementary (+ or −) DNA strands which come together when extracted to form double-stranded DNA with MW of 3×10^6 daltons and 54·2 per cent G+C (Mayor et al., 1969a, b; Rose et al., 1966; 1970).

Physico-chemical characters. Thermostability greater than that of adenoviruses: half life at 60° 30′. Ether-stable. Buoyant density of DNA in CsCl 1·717. Density of virion 1·25–1·33 (Smith et al., 1966). Other estimates give 1·27–1·28, and for type 4, 1·43 (Mayor et al., 1969b; Parks et al., 1967; Rose et al., 1966).

Haemagglutination. Type 4 agglutinates human O RBCs at 4° (Ito & Mayor, 1968). No haemagglutinin was separable from the virus.

Antigenic properties. At least 4 serotypes exist, separable by neutralization, complement fixation and precipitation tests; some strains, however, show cross-relationships in the CF test. Antibodies may be found in human sera. They have also been found (Rapoza & Atchison, 1967) in antisera against certain adenoviruses, even though no AAV were directly demonstrable. They may therefore exist as defective passengers.

Cultivation. Only replicate in the presence of multiplying adenoviruses and therefore in cultures supporting the growth of the latter. May reach much higher titres than do the adenoviruses. Can be transferred from a culture of 1 adenovirus type to that of another. Have a depressing effect on the growth of some adenoviruses which grow to a lower titre in their presence. They have been found in cultures of adenoviruses from man, monkeys, dogs, mice and chicks (Hoggan et al., 1968). The various serotypes are permitted to grow to varying extents by different adenovirus types (Boucher et al., 1969). In the presence of herpes simplex virus, AAV can produce antigens, presumably coat proteins, but not complete infectious viruses (Atchison, 1970; Blacklow et al., 1970).

Pathogenicity. None known. Kirschstein et al. (1968) found that AAV partly inhibited the oncogenicity of adenovirus 12, but Gilden et al. (1968) found no such effect.

Ecology. Field adenovirus isolates are usually free from AAV. Adenoviruses cultures may perhaps become contaminated by them in the laboratory. Blacklow et al. (1968) studied the incidence of AAV in a nursery population.

Control. Adenoviruses cultures can be freed from AAV by treatment with specific antisera.

REFERENCES

Abinanti, F. R., & Warfield, M. S. (1961) *Virology*, **14**, 288.
Atchison, R. W. (1970) *Virology*, **42**, 155.
Atchison, R. W., Casto, B. C., & Hammon, W. McD. (1966), *Virology*, **29**, 353.
Binn, L. N., Lazar, E. C., Eddy, G. A., & Kajime, M. (1970). *Infect. Immunol.*, **1**, 503.
Blacklow, N. R., Hoggan, M. D., & McClanahan, M. S. (1970) *Proc. Soc. exp. Biol. Med., U.S.A.*, **134**, 952.
Blacklow, N. R., Hoggan, M. D., & Rowe, W. P. (1968) *J. natn. Cancer Inst.*, **40**, 319.
Boucher, D. W., Parks, W. P., & Melnick, J. L. (1969) *Virology*, **39**, 932.
Breese, S. S., Jr., Howatson, A. F., & Chany, C. (1964) *Virology*, **24**, 603.
Cartwright, S. F., & Huck, R. A. (1967) *Vet. Rec.*, **81**, 196.
Cartwright, S. F., Lucas, M., & Huck, R. A. (1969) *J. comp. Path.*, **79**, 371.
Chany, C., & Brailovsky, C. (1967) *Proc. natn. Acad. Sci., U.S.A.*, **57**, 87.
Crawford, L. V. (1966) *Virology*, **29**, 605.
Dalldorf, G. (1960) *Bull. N.Y. Acad. Med.*, **36**, 795.
Dawe, C. J., Kilham, L., & Morgan, C. D. (1961). *J. natn. Cancer Inst.*, **27**, 221.
Dutta, S. K., & Pomeroy, B. S. (1967) *Am. J. vet. Res.*, **28**, 296.
Gilden, R. V., Kern, J., Beddow, T. G., & Huebner, R. J. (1968) *Nature, Lond.*, **219**, 80.
Gorham, J. R., Hartsough, G. H., Burger, D., Lusle, S., Sato, N. (1965) *Cornell Vet.*, **55**, 559.
Goss, L. J. (1948) *Am. J. vet. Res.*, **9**, 65.
Greene, E. L. (1965) *Proc. Soc. exp. Biol. Med., U.S.A.*, **118**, 973.
Greene, E. L., & Karasaki, S. (1965) *Proc. Soc. exp. Biol. Med., U.S.A.*, **119**, 918.
Hammon, W. McD., & Enders, J. F. (1939) *J. exp. Med.*, **69**, 327.
Hoggan, M. D., Blacklow, N. R., & Rowe, W. P. (1966) *Proc. natn. Acad. Sci., U.S.A.*, **55**, 1467.
Hoggan, M., Shatkin, A. J., Blacklow, W., Kazot, F., & Rose, J. (1968) *J. Virol.*, **2**, 850.
Ito, M., & Mayor, H. D. (1968) *J. Immunol.*, **100**, 61.
Jennings, A. R. (1947) *Br. vet. J.*, **105**, 89.
Johnson, R. H. (1964) *Vet. Rec.*, **76**, 1008.
Johnson, R. H. (1967) *Res. vet. Sci.*, **8**, 250.

Johnson, R. H. (1969) *Vet. Rec.*, **84**, 338.
Johnson, R. H., & Collings, D. F. (1969) *Vet. Rec.*, **85**, 446.
Johnson, R. H., and Cruickshank, J. G. (1966) *Nature, Lond.*, **212**, 622.
Johnson, R. H., Margolis, G., & Kilham, L. (1967) *Nature, Lond.*, **214**, 175.
Kilham, L. (1961) *Proc. Soc. exp. Biol. Med., U.S.A.*, **106**, 825.
Kilham, L. (1966) In *Viruses of laboratory rodents. Natn. Cancer Inst. monogr. No.* 20, p. 117.
Kilham, L., & Margolis, G. (1965) *Science*, **148**, 244.
Kilham, L., & Margolis, G. (1966a) *Am. J. Path.*, **48**, 991.
Kilham, L., & Margolis, G. (1966b) *Arch. Path.*, **49**, 457.
Kilham, L., & Margolis, G. (1970) *Proc. Soc. exp. Biol. Med., U.S.A.*, **133**, 1447.
Kilham, L., Margolis, G., & Colby, E. D. (1967) *Lab. Invest.*, **13**, 465.
Kilham, L., & Olivier, L. J. (1959) *Virology*, **7**, 428.
King, D. A., & Crogham, D. L. (1966) *Can. J. comp. Med.*, **29**, 85.
King, D. A., & Gutekunst, D. E. (1970) *Vet. med. small anim. Clin.*, **65**, 377 and 380.
Kirschstein, K. O., Smith, K. O., & Peters, E. A. (1968) *Proc. Soc. exp. Biol. Med., U.S.A.*, **128**, 670.
Lawrence, J. S., & Syverton, J. T. (1938) *Proc. Soc. exp. Biol. Med., U.S.A.*, **38**, 914.
Ledinks, N., & Toolan, H. W. (1968) *J. Virol.*, **2**, 155.
Lucas, A. M., & Riser, W. H. (1945) *Am. J. Path.*, **21**, 435.
McGeogh, D. J., Crawford, L. V., & Follett, E. A. C. (1970) *J. gen. Virol.*, **6**, 33.
Matsuo, Y., & Spencer, H. J. (1969) *Proc. Soc. exp. Biol. Med., U.S.A.*, **130**, 294.
Mayor, H. D., Jamison, R. M., Jordan, L. E., & Melnick, J. L. (1965) *J. Bact.*, **90**, 235.
Mayor, H. D., Jordan, L., & Ito, M. (1969a) *J. virol.*, **4**, 191.
Mayor, H. D., Torikai, K., Melnick, J. L., & Mandel, M. (1969b) *Science*, **166**, 1280.
Mayr, A., Bachmann, P. A., Siegl, G., Mahnel, H., & Sheffy, B. E. (1968) *Arch. ges. Virusforsch.*, **25**, 38.
Moore, A. E. (1962) *Virology*, **18**, 182.
Moore, A. E., & Nicastri, A. D. (1965) *J. natn. Cancer Inst.*, **35**, 937.
Nathanson, N., Cole, G. A., Santos, G. W., Squire, R. A., & Smith, K. O. (1970) *Am. J. epidemiol.*, **91**, 328.
Parker, J. C., Cross, S. S., Collins, M. J., & Rowe, W. P. (1970) *J. natn. Cancer Inst.*, **45**, 297.
Parks, W. P., Green, M., Pina, M., & Melnick, J. L. (1967) *J. virol.*, **1**, 980.
Portella, O. B. (1964) *Arch. ges. Virusforsch.*, **14**, 277.
Pridham, J., & Wills, C. G. (1959) *J. Am. vet. med. Ass.*, **135**, 279.
Rapoza, N. P., & Atchison, R. W. (1967) *Nature, Lond.*, **215**, 1186.
Robinson, D. M., & Hetrick, F. M. (1969) *J. gen. Virol.*, **4**, 269.
Rose, J. A., Berns, K. I., Hoggan, M. P., & Koczot, F. J. (1970) *Proc. natn. Acad. Sci., U.S.A.*, **64**, 863.
Rose, J. A., Hoggan, M. D., & Shatkin, A. J. (1966) *Proc. natn. Acad. Sci., U.S.A.*, **56**, 68.
Salzman, L. A., & Jori, L. A. (1970) *J. Virol.*, **5**, 114.
Smith, K. O., Gehle, W. D., & Thiel, J. F. (1966) *J. Immunol.*, **97**, 754.
Storz, J., & Warren, G. S. (1970) *Arch. ges. Virusforsch.*, **30**, 271.
Toolan, H. W. (1960a) *Science*, **131**, 1446.
Toolan, H. W. (1960b) *Proc. natn. Acad. Sci., U.S.A.*, **46**, 1256.
Toolan, H. W. (1967) *Proc. Soc. exp. Biol. Med., U.S.A.*, **124**, 144.
Toolan, H. W. (1968) *Int. Rev. exp. Path.*, **6**, 135.
Toolan, H. W., Dalldorf, G., Barclay, M., Chandra, S., & Moore, A. E. (1960) *Proc. natn. Acad. Sci., U.S.A.*, **46**, 1256.
Toolan, H. W., & Ledinko, N. (1965) *Nature, Lond.*, **208**, 812.
Toolan, H. W., Sanders, E. L., Greene, E. L., & Fabrizio, D. P. A. (1964) *Virology*, **22**, 286.
Tuomi, J., & Kangas, J. (1963) *Arch. ges. Virusforsch.*, **13**, 430.
Vasquez, C., & Brailowsky, C. (1965) *Exp. molec. Path.*, **4**, 130.
Wills, C. G. (1952) *Can. J. comp. Med.*, **16**, 419.

13

Papovaviridae

Review: Allison (1965).

The international committee for the nomenclature of viruses has divided the papovaviruses into 2 genera, papillomavirus and polyomavirus; they are to be included together in the family *Papovaviridae*. This name covers the 2 first letters of PApilloma and POlyoma, with VA for vacuolating agent.

The viruses contain double-stranded DNA which is often circular and supercoiled. The nature and extent of supercoiling are discussed by Crawford (1969). Crawford & Crawford (1963) suggest that a diameter of 52–54 nm is likely for the true wart viruses in contrast to 43 nm for polyoma. The ratio would be 1:1·2 for diameter and 1:1·8 for volume. Form basically icosahedral with 5–3–2 symmetry, but filamentous forms occur with several. The number of capsomeres has been variously estimated as 42, 60, 70 and 92, but there is now general acceptance of the contention of Finch & Klug (1965) that the capsid contains 72 morphological units and that it is an icosahedron of skew form. Papovaviruses are ether-stable, and most withstand heating for 30 minutes at 56°–65°; they survive well in 50 per cent glycerol and when frozen and dried. Most of them are potentially oncogenic. Their development is mainly intranuclear.

PAPILLOMAVIRUSES

Diameter 52–54 nm. MW $5{\cdot}3 \times 10^6$ daltons. These viruses cause proliferation of epithelium of skin or mucous membranes; this is usually innocent, but may proceed to malignancy. The viruses are, in general, host specific, and also differ antigenically (Le Bouvier et al., 1966).

RABBIT PAPILLOMA

Synonym: Shope papilloma.

Reviews: Shope (1933) (original description). Bryan & Beard (1940).

Morphology and development. The icosahedral virions are of left-handed skew form (Klug & Finch, 1968). Filamentous forms occur

(Williams et al., 1960). Virus apparently begins to multiply in the nucleolus, particles then seen being 33 nm in diameter; the rest of the nucleus is soon involved and mature virus particles can be found in large quantity. They may be arranged in orderly patterns, but are not seen in such regular crystalline arrangements as with adenoviruses (Moore et al., 1959). Some preparations may contain incomplete virus.

Chemical composition. DNA, which is double-stranded, forms 6–8 per cent of the total weight (Watson & Littlefield, 1960). It apparently exists in a cyclic form, altogether 2·3 to 2·8 μm long, a number of endless loops being visible (Kleinschmidt et al., 1965). Extraction with hot or cold phenol yielded an active material, sensitive to deoxyribonuclease, possibly representing infective DNA (Rogers, 1959). GC content 47 per cent (Crawford & Crawford, 1963). Much more arginase is present in the warts than in normal skin (Rogers & Moore, 1963), but this is not necessarily a component of the virus (Orth et al., 1967).

Physico-chemical characters. Heat-resistant: inactivated in 30′ at 70° not at 67°. Survives well in 50 per cent glycerol—up to 20 years (Fischer & Green, 1947). Much more resistant to X-radiation than are most viruses (Syverton et al., 1941). Density in aqueous suspension 1·133 (Sharp et al., 1946), but Breedis et al. (1962) found that particles containing different amounts of DNA had densities varying from 1·29 to 1·34. Stable between pH 3 and 7. Iso-electric point about pH 5·0 (Sharp et al., 1942). Can be purified by means of fluorocarbon or by precipitation with methanol (Fischer, 1949). Methods of purification have been compared by Beaty & Hodes (1966).

Haemagglutination. Though rabbit RBCs adsorb the virus, they are not agglutinated (Barabadze, 1960).

Antigenic properties. Activity is readily neutralized by sera of recovered rabbits. Warts from domestic rabbits, which may contain no demonstrable infective virus, may nevertheless induce antibodies when inoculated IP. Complement-fixing antigen is readily extracted from cotton-tail warts, much less readily from warts in domestic rabbits; it is more stable to UV irradiation than infectivity, but is not mechanically separable from the virus particle (Kidd, 1938). No cross-immunity was demonstrable between this virus and those causing warts in cattle and dogs nor with the rabbit oral papilloma.

Yoshida and Ito (1968) describe an antigen which may be a T-antigen like that of polyoma (see p. 299).

Cultivation. Successful serial propagation of the virus has not been reported. Kreider et al. (1967) observed no changes in cells of rabbit skin to which virus had been adsorbed, but these gave rise to papilloma when transferred to hamster cheek pouches. Shiratori et al. (1969) established a cell-line from wart material. Specific antigen, revealed in nuclei with fluorescent antibody, disappeared from cultures held at 37°; it reappeared when temperature was lowered to 30°, but this time in cytoplasm.

Pathogenicity. Only reported to occur naturally in *Sylvilagus floridanus* (cotton-tail rabbit) in North America. There is, however, one record of a small growth in a jack-rabbit (*Lepus californicus*) (Beard & Rous, 1935). The natural tumours are often tall, thin black or grey horns.

Experimentally, domestic rabbits (*Oryctolagus*), also several species of *Lepus*, can be infected by rubbing virus into scarified skin. In domestic rabbits growth is 'relatively exuberant and fleshy and may be pink or sooty, low or projecting, dry or succulent' (Rous & Beard, 1934). Warts may regress, but much more often than in cotton-tails go on to become malignant. From domestic rabbit—unlike cotton-tail—warts, virus is usually recoverable with difficulty, though transplantation through 14 domestic rabbits in series has been reported. There has been much dispute as to whether 'masked virus' in tame rabbits is qualitatively or only quantitatively different from that in cotton-tails. Shope has suggested that it may in the former be present as incomplete virus, possibly infective DNA. The papillomata and resulting cancers have been the basis of much important work on the relation between viruses and tumours (Rous & Beard, 1934; 1935; Greene, 1955, etc.).

The proliferative epithelial changes leading to superficial warts in wild and tame rabbits are described by E. W. Hurst (in Shope, 1933), and those of transplanted warts by Rous & Beard (1934). Noyes (1959) and Stone et al. (1959) found virus mainly in keratinizing, not in proliferating, layers of epithelium.

Ecology. Natural transmission is probably by direct contact, though mosquitoes and reduviid bugs can transmit the virus experimentally (Dalmat, 1958), and a possible role of nematodes as transmitting agents has been suggested by Rendtorff & Wilcox (1957).

Control. Evans et al. (1962) report that vaccination with tissue suspensions may cause some regressions of growths.

ORAL PAPILLOMATOSIS
(of rabbits)

Description: Parsons & Kidd (1943).

Morphology. Virions 50–52 nm across, or with 40 nm centre-to-centre spacing when in crystalline masses in nuclei (Rdzok et al., 1966).

Physico-chemical characters. Survives at least 2 years in 50 per cent glycerol at 4° C; stable when freeze-dried. Survives heating for 30′ at 65°; some infectivity persists after heating to 70° to 30′.

Antigenic properties. Recovered animals are immune for at least some months; there is no cross-immunity with the Shope papilloma.

Pathogenicity. Frequent in domestic rabbits in New York. Papillomas up to 5 mm across are found in the mouth, usually beneath the tongue. They usually persist only for a month or two, but may do so for longer.

On inoculation into the oral mucosa, papillomata appear after 6–38 days. *Sylvilagus* and *Lepus* spp. are susceptible though only tame rabbits (*Oryctolagus*) have been found naturally infected. The virus will not 'take' on the skin.

Basophilic inclusions have been found near the nucleus of superficial cells of the papillomas of tame rabbits.

Ecology. Infection does not spread readily in animal quarters, but may do so from doe to sucklings. Virus has been recovered from mouth-washings of rabbits free from papillomas.

INFECTIOUS WARTS
(of man)

Review: Rowson & Mahy (1967).

Synonyms: Verruca vulgaris. Myrmecia. Papilloma. Common wart. Condyloma.

Various clinical types of wart are described: juvenile (plane), digitate, filiform, plantar, genital, laryngeal. Despite suggestions (Lyell & Miles, 1951) that 2 agents are concerned, it is increasingly believed that there is only 1 human wart virus.

Morphology. Particles may be packed into crystalline masses within nuclei, in which alone multiplication seems to take place. The virion, an icosahedron with 72 capsomeres (Klug & Finch, 1965; 1968), is of right-handed skew form. Howatson (1962) describes cylindrical forms, and, as with many other viruses, virions may be intact or empty (Noyes, 1965).

Chemical composition. GC content 41 per cent. Three components of the DNA separated by band centrifugation may represent circular, linear and supercoiled forms (Crawford, 1965). As with the rabbit virus (see p. 291) arginase is present in large amounts.

Physico-chemical characters. Survives heating 30′ at 50°; also, for some time, in 50 per cent glycerol.

Antigenic characters. Antibodies, demonstrable by gel diffusion and other methods, are present in many human sera. (Almeida & Goffe, 1965; Goffe et al., 1966). Regression of warts may not depend on humoral antiviral antibodies; warts may regress at one site while new warts are appearing elsewhere. T antigens may be concerned, but have not been demonstrated.

Cultivation. Claims to have grown the virus in eggs have not been confirmed. Morgan & Balduzzi (1964) cultivated fragments from vaginal warts and observed focal lesions after 2 months: transmission to fresh cultures of human embryonic skin and kidney was possible. Noyes (1965) was successful in infecting primary cultures of human embryonic skin and muscle; and these showed 'transformation', with piling up of epithelial cells and loss of contact-inhibition. There have been reports, also, of propagation of the virus in cultures of rhesus kidney (Mendelson & Kligman, 1961) and green monkey epithelial cells (Macpherson, 1962).

Pathogenicity. Warts in man tend to persist for many months, but ultimately regress. Genital warts or condylomata acuminata, however, may become malignant. Various names for clinical types, mentioned above, testify the varied appearances in different sites. Experimental transmission to man has been reported, the incubation period varying between 6 weeks and 8 months. A laryngeal papilloma inoculated on to the skin produced flat cutaneous warts (Ullman, 1923). Isolated reports of transmissions to a baboon and a dog lack confirmation.

Histologically the essential change is of proliferation of the Malpighian layer of the skin, but appearances vary greatly in different types. The eosinophilic intranuclear inclusions have been studied by Bunting et al. (1952) and Bloch & Godman (1957); they are of Cowdry's type B; they appear to be products of abnormal keratinization and not to contain much virus. Virus particles form a large part of the basophilic inclusions (Almeida et al., 1962). Masses in cytoplasm are vacuolated.

Howatson (1962) finds that the mode of development of virus in the skin is like that described for rabbit papilloma; most virus is in keratinized layers.

Ecology. Transmission is by direct or indirect contact; in the case of genital warts it may be, but is probably not always, venereal. The tendency to spontaneous regression makes assessment of therapeutic claims difficult.

Control. Warts may be cured by X-rays, cautery, nitric acid or other directly destructive means.

BOVINE PAPILLOMATOSIS

Morphology. Virions are found in the nuclei of the stratum granulosum and corneum of the skin, in crystalline or random arrangement (Tajima et al., 1968). The capsomeres are $7{\cdot}5 \times 5$–6 nm in size with a possible longitudinal central cavity (Boiron et al., 1964).

Chemical composition. Contains 10 per cent DNA (Allison, 1965). The main component of the DNA is circular and with a right-handed twist (Bujard, 1967).

Physico-chemical characters. Density 1·334.

Antigenic properties. All strains are alike as shown by gel-diffusion tests; antibodies are present in affected cattle and in immunised rabbits and chicks (Lee & Olson, 1969).

Cultivation on the CAM of fertile eggs was reported by Olson et al. (1960), but has not been confirmed. In tissue culture the virus causes transformation of embryonic cells of cattle (Thomas et al., 1964), mice (Black et al., 1963) or hamsters (Geraldes, 1969). Extracted DNA acted in a similar way (Boiron et al., 1965). Serial passage was possible in the hamster system; even adult hamsters are susceptible (Robl, 1968).

Pathogenicity. Bovine warts, sometimes in immense numbers, occur on head and neck of calves, less commonly elsewhere on the skin. They regress after some months. Warts on teats, penis and vagina are less common. A similar agent may cause bladder-growths in cattle (Olson et al., 1959; 1965; Pamukcu, 1963). The virus differs from other wart-viruses in that connective tissue proliferation is an important component of the warts. When inoculated into calves IC the virus gives rise to meningeal fibroma (Gordon & Olson, 1968). Though these growths resemble equine sarcoids (see p. 297), immunological evidence indicates that 2 different agents are concerned (Ragland & Spencer, 1968).

The virus has produced connective tissue tumours on subcutaneous inoculation into new-born hamsters and a particular strain of mice (C_3H/EB) (Boiron et al., 1964); the tumours grew very slowly, but were ultimately fatal.

The agent has been transmitted to horses causing connective tissue tumours (Olson & Cook, 1951; Segre et al., 1955). These produced warts on inoculation back to calves.

Ecology. Transmission is believed to be by direct contact.

Control. Wart-suspensions, autogenous or otherwise, made up in glycerol saline or formalinized, have been used as vaccines, either for prophylaxis or cure. Several groups report success, but in a disease which clears up spontaneously judgement is difficult.

CANINE ORAL PAPILLOMATOSIS

Affects mucous membranes, especially of young dogs.

Morphology. As for other wart-viruses (Crawford & Crawford, 1963; Watrach et al., 1969).

Physico-chemical characters. The virus survives heating for an hour to 45°, not 58°.

Antigenic properties. Recovered dogs are immune and have neutralizing antibodies in sera (Chambers et al., 1960).

Pathogenicity. Warts, which may become cauliflower-like, usually begin on lips and spread to the inside of the mouth and pharynx; they disappear after 4–21 weeks. They rarely become malignant. They can be transferred to other young dogs, the incubation period being 4 to 8 weeks; only oral mucosa and neighbouring skin could be infected. Other species were insusceptible (McFadyean & Hobday, 1898; De-Monbreun & Goodpasture, 1932; Chambers & Evans, 1959). Dermal papillomata in dogs may be caused by a different virus (Allison, 1965).

Ecology. The disease readily spreads in kennels.

Control. Wart suspensions given intramuscularly along with adjuvants had prophylactic but no curative value (Chambers et al., 1960).

CANINE DERMAL PAPILLOMATOSIS

Morphology. Like that of the oral virus (Watrach, 1969). Most authors agree that the oral virus will not infect skin. The 2 viruses are probably distinct (Allison, 1965), but immunological evidence for this is not reported.

EQUINE PAPILLOMATOSIS

Affects nose and lips, but is transferable to skin of neck. Warts usually numerous, not more than 1 cm. across. Readily transmitted to horses, not to other species. Agent survived 75 days at 4° and frozen at −35° for 185 days. Recovered animals are immune (Cook & Olson, 1951).

Inclusions are found in nuclei of affected cells and virions like those of other papovaviruses can also be seen. Similar virions have been seen in 'sarcoids', relatively benign connective-tissue tumours which also occur in horses. Experimentally produced sarcoids contain bands of epidermal cells as well as whorls of fibroblasts (Voss, 1969). The sarcoids are probably caused by a virus distinct from that of bovine papillomatosis, though this can produce similar lesions in horses (Ragland & Spencer, 1968).

PAPILLOMATOSIS OF MONKEYS

Lucké et al. (1950) described warts in South American *Cebus* monkeys. They transferred infection with glycerolated material to both Old and New World monkeys.

PAPILLOMATOSIS OF GOATS AND SHEEP

Affects some herds (Davis & Kemper, 1936), but is uncommon. The warts may develop into carcinomata. Transmission experiments have not been reported.

Papilloma of chamois is orf (see p. 389).

DEER FIBROMA

Multiple skin nodules occurring in skins of white-tailed deer in North America are histologically fibromata. They are caused by a virus morphologically like that of bovine papilloma (Tajima et al., 1968); tumour cells contain intranuclear inclusions of papilloma type (Allison, 1965). Experimentally the incubation period of the disease is about 7 weeks. The agent was stable in 50 per cent glycerol in the cold; it was not transmissible to a calf, rabbits, guinea-pigs or sheep (Shope et al., 1958).

PAPILLOMATOSIS OF HAMSTERS

Graffi et al. (1968) observed skin tumours in 5 per cent of Syrian hamsters more than 6 months old; the papillomata contained papilloma-virus-like particles. When inoculated SC into rats or hamsters they gave rise not to epithelial tumours but to abdominal lymphomata and reticular-cell sarcomata. These contained 'C-particles' like those of murine leukaemia (see p. 149); the papilloma virus appears to have activated a latent leukovirus.

POLYOMAVIRUSES

The viruses of this genus are smaller than the papilloma viruses, having a diameter of about 43–45 nm and a MW of about $3{\cdot}2 = 10^6$ daltons. All are normally present in their hosts as latent infections; they have only revealed their oncogenic powers under artificial conditions in the laboratory. Only the rabbit vaccuolating virus (p. 304) has not as yet been shown to be potentially oncogenic. Host DNA may become enclosed in virus capsids; the product is called a 'pseudovirus' (Trilling & Axelbrod, 1970).

POLYOMA

Morphology and development. Most estimates of diameter agree upon 45 nm. Mattern et al. (1967) suggest that the virions are constructed, from within out, of a 22 nm shell with 12 capsomeres, a thin shell of DNA, a 38 nm shell of 32 capsomeres and a 48 nm shell of 72 capsomeres. A layer of phospholipid covers some particles. Filamentous forms are present, varying in diameter, some containing DNA, others not (Mattern & DeLeva, 1968). Horne & Wildy (1961) and Howatson (1962) figured filamentous forms; Stoker (1962) has suggested how these might have been built up from the subunits which normally form icosahedra. Particles are first found in the nucleus where they may be more or less regularly packed; they are sometimes contained within vacuoles. Later they reach the cytoplasm and here they seem to be larger (Negroni et al., 1959). Release from cell is gradual. In tumour cells antigen stainable by fluorescent antibody (? incomplete virus) may appear in the cytoplasm without evidence of earlier growth in nucleus (Sachs & Fogel, 1960). Neither external membrane nor dense nucleoid could be demonstrated by Bernhard et al. (1959).

Chemical composition. DNA molecule is circular; MW $3{\cdot}5 \times 10^6$. GC content 48 per cent (Crawford, 1963). The sequential appearance of various components is described by Gershon & Sachs (1964). Isolation of an infectious DNA was reported by Di Mayorca et al. (1959), and confirmed by Harris et al. (1961). Progressive degradation and separation of nucleic acid and protein occur in the presence of alkaline buffers (Perry et al., 1969). Thorne & Warden (1967) adduced evidence for presence of a single protein component in the capsid. The protein composition was also studied by Murukami et al. (1968) and Fine et al. (1968) who suggest the existence of a basic internal polypeptide.

Physico-chemical characters. Survives many months at 4°C or −70°C and 8 weeks at 37°C. Heating 30′ at 60° did not affect the titre: at 70° it was usually inactivated. It was resistant to ether, 2 per cent phenol and 50 per cent ethanol (Brodsky et al., 1959). Winocour (1963) separated full (DNA-containing) particles (specific gravity 1·339) from empty ones (sp. gr. 1·297).

Haemagglutination. The virus agglutinates cells of many species at 4° over a pH range of 5·4–8·4: guinea-pig cells, being most reliable, are chiefly used. There is nonenzymatic elution at 37°C (Eddy et al., 1958). Many animal sera and cell extracts contain nonspecific haemagglutinin inhibitors; these have been removed by heating and in other ways (Deinhardt et al., 1960).

Antigenic properties. Infected mice and hamsters develop antihaemagglutinins, neutralizing and CF antibodies. It has not been possible to separate haemagglutinin or CF antigen from the virus by mechanical means (Rowe et al., 1958). Minor antigenic differences between isolates have been described by Hare & Chan (1966). Cells transformed or rendered malignant by polyoma virus develop a nonstructural (T) antigen revealed in complement-fixation tests (Habel, 1965). There are also superficial, transplantation antigens (Sjögren et al., 1961; Habel, 1962), which are of importance in tumour immunity.

Cultivation. Multiplies in cultures of embryo and other mouse tissues, also in mouse tumours and in other rodent tissues. In infected cultures nuclei are enlarged and chromatin coarsened. The early reports of growth in monkey and chick tissues have not been confirmed. In monolayers, plaque-formation occurs; 2 types of plaque have been described, due to variants with slightly different biological properties (Gotlieb-Stematsky & Leventon, 1960).

Some multiplication occurs in rat tumour tissue *in vitro* (Sachs & Fogel, 1960). There may be a destructive effect on cultivated mouse cells or a stable cell-virus association may be set up: embryonic cells of mouse or hamster may thus be rendered malignant *in vitro* (Vogt & Dulbecco, 1960; Stoker & MacPherson, 1961). Cells of rats, rabbits, guinea-pigs, dogs, cattle, monkeys and man may also be transformed in culture (Black, 1968). Infectious virus cannot as a rule be recovered from cells transformed in culture, but Fogel & Sachs (1970) succeeded in inducing synthesis of complete virus by treatment with UV irradiation or mitomycin C (see reviews by Stoker, 1963; and Black, 1968). Many papers have been published on transformation of cells in culture by polyoma virus and on the new antigens which appear in transformed cells.

Pathogenicity. Normally a completely inapparent infection affecting some stocks of laboratory mice, also wild mice (*Mus musculus*) from both urban and rural environments. Spontaneous tumours caused by the virus are exceedingly rare, but contact infection in thymectomized rats may be followed by development of tumours (Law, 1965).

Experimentally, virus quantitatively exalted in tissue culture and inoculated into suckling mice, rats, guinea-pigs or hamsters leads to production of tumours of many histological types, parotid tumours being amongst the most frequent in mice and sarcomata in hamsters. Apart from tumours, the virus may cause in suckling mice dwarfism, nephritis, anaemia and conjunctivitis (Stewart et al., 1958); sometimes hydrocephalus. In inoculated suckling hamsters foci of sarcomatous cells may be found in the kidneys after a few days, and the animals may die after only 1 or 2 weeks (Stoker, 1960). Some die with haemorrhagic lesions affecting particularly the liver (Negroni et al., 1959; Defendi & Lehman, 1964). The tumours caused by polyoma virus may be transplantable, but it may be impossible to recover virus from the transplanted tumours. Inoculated new-born rabbits develop multiple fibromata which regress (Eddy et al., 1959), but a few rabbits develop metastasizing fibrosarcomata (Lehman & Defendi, 1970). Fibrosarcomata have also been produced in new-born ferrets (Harris et al., 1961).

Ecology. Transmission of silent infection among mice is probably by the intranasal route. Virus is excreted in faeces and urine and may be present in the litter of the nests. Huebner and his colleagues (1962) have described the ecology of the virus in wild mice.

In laboratories working with the virus, the environment may

become heavily contaminated, so that uninoculated mice become infected and there is every chance of an accidental 'pick-up' of virus in the course of experiments (Rowe et al., 1961).

K VIRUS

Description: Kilham & Murphy (1953). Fisher & Kilham (1953).

There has been confusion in some people's minds between this virus and the latent rat virus, also described by Kilham (see p. 281). One worker (Brailovsky, 1966) has, to add to the confusion, referred to the latter as the rat K virus.

Morphology. Spheres 40–50 nm across.

Chemical composition. DNA, 7 per cent by weight (Mattern et al., 1963).

Physico-chemical characters. Resisted heating to 70°C for 3 hours but not 4½; withstood 60°C for 4 hours. Ether-stable. Not readily inactivated by 0·5 per cent formalin.

Haemagglutination. Sheep RBCs are agglutinated by tissue suspensions from suckling mice at room temperature or 37°. Heating at times unmasks the haemagglutinin (Kilham, 1961).

Cultivation. Cultures of embryonic mouse lung were infected, and foci of transformation appeared after 6 weeks. One line of such transformed cells produced tumours in new-born or X-rayed mice (Takemoto & Fabisch, 1970).

Pathogenicity. Causes fatal pneumonia and sometimes liver lesions after inoculation by various routes into mice less than 10 days old. The incubation period is 6–15 days, varying with the route of inoculation. Virus in high titre is found in all tissues. Nuclear inclusions occurred in endothelium of lung arterioles.

Ecology. Virus was found in urine and faeces up to 4 weeks after inoculation. It has been recovered from wild mice.

SIMIAN VACUOLATING VIRUS

Synonym: SV40. Vacuolating agent.

Description: Sweet & Hilleman (1960).

Morphology. Spherical, 40–45 nm in diameter. Filamentous forms also occur. The icosahedral surface lattice is of the right-handed form. Smaller particles are also present (Anderer et al., 1967). The capsomere diameters are given as 6·5 × 6·5 nm (Bernhard et al., 1962). The virions may be aggregated into crystalline masses in nuclei. Prunieras et al. (1964) have described virus development in nucleus and subsequent transfer to cytoplasm by way of vacuoles near the nuclear membrane. Incomplete virus may appear when serial passages are carried out in grivet kidney cells, with undiluted inocula (Uchida et al., 1966).

Chemical composition. MW of DNA $2{\cdot}7 \times 10^6$ (Crawford & Black, 1964) or $2{\cdot}25 \times 10^6$ (Anderer et al., 1967). MW of the virion $17{\cdot}3 \times 10^6$. Alkaline degradation of the virion separated 3 polypeptides (Schlumberger et al., 1968), while Girard et al. (1970) identified 6. An infectious DNA has been extracted by Boiron et al. (1962) and Gerber (1962).

Physico-chemical characters. Stable on storage at −20°C and −70°C. Withstands heating for 1 hour, perhaps longer, at 56°C. Ether-resistant. Inactivation by 1:4,000 formalin at 37°C is slower than for poliovirus, and some active vacuolating virus has probably been present in some formolized poliovirus and adenovirus vaccines. It is, however, more readily inactivated by the photodynamic action of toluidine blue. Density in CsCl 1·30 (Mayor et al., 1963).

Haemagglutination. No haemagglutinins detected for guinea-pig, chick or human RBCs.

Antigenic properties. Neutralizing antibodies, revealed in tests in tissue culture, were present in sera of 12/18 rhesus sera. They have not been found in 'normal' human sera, but are present in sera of numerous people who have been given adenovirus or polio vaccines (ostensibly inactivated). Antibodies occur in human sera in parts of India where there is close contact between man and rhesus monkeys; they may be found to higher titre in monkey-handlers (Shah, 1966), they have since been found in some people in the United States giving no history of contact with monkeys (Shah et al., 1971). As in the case of polyoma, nonstructural (T) antigens and transplantation antigens are produced in transformed cultures and in tumours (Pope & Rowe, 1964; Diamandopoulos & Dalton-Tucker, 1969). The T antigen may be transiently present in the absence of tumour development.

There is some evidence for existence of a normal fetal antigen related to one induced by SV40 (Coggin et al., 1970; Baranska et al., 1970).

Cultivation. Multiplies in a variety of cells in culture, but when first discovered was cytopathic only for grivet (*Cercopithecus aethiops*) kidney cultures (Sweet & Hilleman, 1960), and to a less extent patas kidney and rhesus testis cultures. In the grivet cultures it produces vacuolation of cytoplasm, beginning in 3 or 4 days, and it forms plaques on monolayers of grivet or baboon cells.

A stable line of rhesus cells shows a CPE after 12 days (Meyer et al., 1962). Foci of transformation with piling up of cells may appear in infected cultures of human, bovine, porcine, hamster, rabbit and mouse cells (Shein & Enders, 1962; Koprowski et al., 1962; Black & Rowe, 1963a, b). Virus DNA produced similar effects. The transformed cells may produce tumours in hamsters. These are usually sarcomata, but there may be carcinomata when the virus has been grown in cultures of skin (Diamandopoulos & Dalton-Tucker, 1969). Infectious virus commonly ceases to be demonstrable. The viral genome must, however, persist because infectious virus may reappear when apparently virus-free transformed cells are co-cultivated with grivet cells. (Black & Rowe, 1963b; Gerber, 1964). Presence of inactivated Sendai virus helps this rescue by permitting fusion of normal and transformed cells. (Koprowski et al., 1967).

In cells doubly infected with SV40 and some adenoviruses a 'hybrid' virus appears; some adenoviruses grow well in certain cultures only in its presence. The oncogenic effects may be greater than that of either adenovirus or SV40 'parent'. The phenomena are reviewed by Rapp & Melnick (1966): they refer to the hybrid particles as PARA (Particles Assisting-and-assisted by Replication of Adenoviruses). The SV40 element is present only in the genome, the capsid always being that of the adenovirus.

Pathogenicity. Virus has been recovered from cultures of 'normal' rhesus and less frequently from cynomolgus and cercopithecus kidneys. It can produce a silent infection in man, leading to antibody formation, especially when given by the respiratory route. Virus may be excreted in stools of children for several weeks. It can produce fibrosarcomata when given to suckling hamsters (Eddy et al., 1961), grivets, baboons or rhesus monkeys; in the grivets there may be signs of trivial renal damage (Fabiyi et al., 1967). The tumours are serially transplantable, and virus may or may not be recovered from them. A chronic latent infection follows inoculation of grivets or baboons.

Control. Repeated injection of the virus into hamsters during the latent period before tumours could appear has the effect of inhibiting their appearance (Potter et al., 1969). The virus can exist as a contaminant in cultures of monkey kidney used to make vaccines for man. It can be eliminated by adequate formolinization.

RABBIT-KIDNEY VACUOLATING VIRUS

Description: Hartley & Rowe (1964).

Morphology. Like that of polyoma and SV40. Virus in nucleus at least to begin with. Diameter of virion 47 nm, of core 38 nm. Filaments were also present (Chambers et al., 1966).

Chemical composition. Viral DNA is circular, double-stranded and super-coiled with GC content of 43 per cent and MW about 3×10^7 (Crawford & Follett, 1967). An infectious DNA was extracted by Ito et al. (1966).

Physico-chemical characters. Infectivity unaffected after 30 minutes at 60° and not destroyed after 30 minutes at 70°.

Haemagglutination. Of guinea-pig cells at 4° or 20°. Receptors destroyed by RDE.

Antigenic properties. Unrelated to rabbit papilloma virus.

Cultivation. Produces cell-vacuolation in cultures of domestic and cotton-tail rabbit kidneys, similar to that seen in grivet-kidney cells infected with SV40. Plaques formed in monolayers. Cells of other species insusceptible.

Pathogenicity. Present as a latent infection in cotton-tail (*Sylvilagus*) rabbits. Not known to be pathogenic for any species.

HUMAN POLYOMAVIRUSES

Virions morphologically identical to polyomavirus have been described in oligodendrocytes from patients with progressive multifocal encephalopathy (see review by Zu Rhein, 1969). Propagation of this virus in tissue cultures has recently been reported by Padgett et al. (1971). A similar virus showing slight antigenic relationship with SV40 has also been found in the urinary tract of a patient with a kidney transplant and propagated in a monkey (Vero) cell line producing haemagglutinin to human RBC (Gardner et al., 1971).

Polyoma-like viruses have also been demonstrated in cultures of human Wilm's tumour (Smith et al., 1969), and there are suggestions that a papovavirus may be involved in subacute sclerosing panencephalitis (Koprowski et al., 1970).

REFERENCES

Allison, A. C. (1965) in *Comparative physiology & pathology of the skin*. Oxford: Blackwell.

Almeida, J. D., & Goffe, A. P. (1965) *Lancet*, **2**, 1205.

Almeida, J. D., Howatson, A. F., & Williams, M. G. (1962) *J. invest. Dermatol.*, **38**, 337.

Anderer, F. A., Schlumberger, H. D., Hoch, M. A., Frank, H., & Eggers, H. J. (1967) *Virology*, **32**, 511.

Barabadze, Y. M. (1960) *Probl. Virol.*, **5**, 103 and 105.

Baranska, W., Koldovsky, P., & Koprowski, H. (1970) *Proc. natn. Acad. Sci., U.S.A.*, **67**, 193.

Beaty, L. E., & Hodes, M. E. (1966) *J. natn. Cancer Inst.*, **36**, 375.

Beard, J. W., & Rous, P. (1935) *Proc. Soc. exp. Biol. Med., U.S.A.*, **33**, 191.

Bernhard, W., Febvre, H. L., & Cramer, R. (1959) *C. r. hebd. Séanc. Acad. Sci., Paris*, **249**, 483.

Bernhard, W., Vasquez, C., & Tournier, P. (1962) *J. Microscopie*, **1**, 343.

Black, P. H. (1968) *Ann. Rev. Microbiol.*, **22**, 391.

Black, P. H., Hartley, J. W., Rowe, W. P., & Huebner, R. J. (1963) *Nature, Lond.*, **199**, 1016.

Black, P. H., & Rowe, W. P. (1963a) *Proc. natn. Acad. Sci., U.S.A.*, **50**, 606.

Black, P. H., & Rowe, W. P. (1963b) *Proc. Soc. exp. Biol. Med., U.S.A.*, **114**, 721.

Bloch, D. P., & Goodman, G. C. (1957) *J. exp. Med.*, **105**, 161.

Boiron, M., Levy, J. P., Thomas, M., Friedmann, J. C., & Bernard, J. (1964) *Nature, Lond.*, **201**, 423.

Boiron, M., Paoletti, C., Thomas, M., Rebière, J. P., & Bernard, J. (1962) *C. r. hebd. Séanc. Acad. Sci., Paris*, **254**, 2097.

Boiron, M., Thomas, M., & Chenaille, P. (1965) *Virology*, **26**, 150.

Brailowsky, C. (1966) *Annls Inst. Pasteur, Paris*, **110**, 49.

Breedis, C., Berwick, L., & Anderson, T. F. (1962) *Virology*, **17**, 84.

Brodsky, I., Rowe, W. P., Hartley, J. W., & Lane, W. T. (1959) *J. exp. Med.*, **109**, 439.

Bryan, W. R., & Beard, J. W. (1940) *J. natn. Cancer Inst.*, **1**, 607.

Bujard, H. (1967) *J. Virol.*, **1**, 1135.

Bunting, H., Strauss, M. J., & Banfield, W. G. (1952) *Am. J. Path.*, **28**, 985.

Chambers, V. C., & Evans, C. A. (1959) *Cancer Res.*, **19**, 1188.

Chambers, V. C., Evans, C. A., & Weiser, R. S. (1960) *Cancer Res.*, **20**, 1083.

Chambers, V. C., Hsia, S., & Ito, Y. (1966) *Virology*, **29**, 32.

Coggin, J. H. (1970) *J. Virol.*, **6**, 524.

Cook, R. H., & Olson, C. (1951) *Am. J. Path.*, **27**, 1087.

Crawford, L. V. (1963) *Virology*, **19**, 279.

Crawford, L. V. (1965) *J. molec. Biol.*, **13**, 362.

Crawford, L. V. (1969) *Int. Virol.*, **1**, 20.

Crawford, L. V., & Black, P. H. (1964) *Virology*, **24**, 388.

Crawford, L. V. & Crawford, E. M. (1963) *Virology*, **21**, 258.

Crawford, L. V., Follett, E. A. C. (1967) *J. gen. Virol.*, **1**, 19.

Dalmat, H. T. (1958) *J. exp. Med.*, **108**, 9.

Davis, C. L., & Kemper, H. E. (1936) *J. Am. vet. med. Ass.*, **88**, 175.

Defendi, V., & Lehman, J. M. (1964) *Cancer Res.*, **24**, 329.

Deinhardt, F., Henle, G., & Marks, M. (1960) *J. Immunol.*, **84**, 599.

De Monbreun, W. A., & Goodpasture, E. W. (1932) *Am. J. Path.*, **8**, 43.

Diamandopoulos, G. T., & Dalton-Tucker, M. F. (1969) *Am. J. Path.*, **56**, 59.

Di Mayorca, G. A., Eddy, B. E., Stewart, S. E., Hunter, W. S., Friend C., & Bendick, A. (1959) *Proc. natn. Acad. Sci., U.S.A.*, **45**, 1805.

Eddy, B., Borman, G. S., Berkeley, W. H., & Young, R. D. (1961) *Proc. Soc. exp. Biol. Med., U.S.A.*, **107**, 191.

Eddy, B. E., Rowe, W. P., Hartley, J. W., Stewart, S. E., & Huebner, R. J. (1958) *Virology*, **6**, 290.

Eddy, B., Stewart, S. E., Kirschstein, R. L., & Young, R. D. (1959) *Nature, Lond.*, **183**, 766.
Evans, C. A., Gorman, L. R., Ito, Y., & Weiser, R. S. (1962) *J. natn. Cancer Inst.*, **29**, 277 and 287.
Fabiyi, A., Calcagno, P. L., Sever, J. L., Antonovych, T., & Wolman, F. (1967) *Nature, Lond.*, **215**, 88.
Finch, J. T., & Klug, A. (1965) *J. molec. Biol.*, **13**, 1.
Fine, R., Mass, M., & Murakami, U. T. (1968) *J. molec. Biol.*, **36**, 167.
Fischer, R. G. (1949) *Proc. Soc. exp. Biol. Med., U..SA.*, **72**, 323.
Fischer, R. G., & Green, R. G. (1947) *Proc. Soc. exp. Biol. Med., U.S.A.*, **64**, 452.
Fisher, E. R., & Kilham, L. (1953) *Arch. Path.*, **55**, 14.
Fogel, M., & Sachs, L. (1970) *Virology*, **40**, 174.
Gardner, S., Field, A., Coleman, D., & Hulme, B. (1971) *Lancet*, **1**, 1253.
Geraldes, A. (1969) *Nature, Lond.*, **222**, 1283.
Gerber, P. (1962) *Virology*, **16**, 96.
Gerber, P. (1964) *Science*, **145**, 833.
Gershon, D., & Sachs, L. (1964) *Virology*, **24**, 604.
Girard, M., Marty, L., & Suarez, F. (1970) *Biochem. biophys. Res. Comm.*, **40**, 97.
Goffe, A. P., Almedia, J. D., & Brown, F. (1966) *Lancet*, **2**, 607.
Gordon, D. E., & Olson, C. (1968) *Cancer Res.*, **28**, 2423.
Gotlieb-Stematsky, T., & Leventon, S. (1960) *Br. J. exp. Path.*, **41**, 507.
Graffi, A., Schramm, I., Graffie, D., Bierwolf, D., & Bender, E. (1968) *J. natn. Cancer Inst.*, **40**, 867.
Greene, H. S. N. (1955) *Cancer Res.*, **15**, 748.
Habel, K. (1962) *Virology*, **18**, 553.
Habel, K. (1965) *Virology*, **25**, 55.
Hare, J. D., & Chan, J. C. (1966) *Virology*, **30**, 62.
Harris, R. J. C., Chesterman, F. C., & Negroni, G. (1961) *Lancet*, **1**, 788.
Hartley, J. W., & Rowe, W. P. (1964) *Science*, **143**, 258.
Horne, R. W., & Wildy, P. (1961) *Virology*, **15**, 348.
Howatson, A. F. (1962) *Br. med. Bull.* **18**, 193.
Howatson, A. F., Nagai, M., & Zu Rhein (1965) *Can. med. Ass. J.*, **93**, 379.
Huebner, R. J., Rowe, W. P., Hartley, J. W., & Lane, W. T. (1962) in *Tumour viruses of murine origin* (Ciba Foundation symposium), p. 314. London: Churchill.
Ito, Y., Hsia, S., & Evans, C. A. (1966) *Virology*, **29**, 26.
Kidd, J. (1938) *J. exp. Med.*, **68**, 703.
Kilham, L. (1961) *Virology*, **15**, 389.
Kilham, L., & Murphy, H. W. (1953) *Proc. Soc. exp. Biol. Med., U.S.A.*, **82**, 133.
Kleinschmidt, A. K., Kass, S. J., Kleinschmidt, R. C., & Knight, C. A. (1965) *J. molec. Biol.*, **13**, 749.
Klug, A., & Finch, J. T. (1965) *J. molec. Biol.*, **11**, 403 and 424.
Klug, A., & Finch, J. T. (1968) *J. molec. Biol.*, **31**, 1.
Koprowski, H., Barbanti-Brodiano, G., & Katz, M. (1970) *Nature, Lond.*, **225**, 1045.
Koprowski, H., Jenson, F. C., & Steplewski, Z. (1967) *Proc. natn. Acad. Sci., U.S.A.*, **58**, 127.
Koprowski, H., Ponten, J. A., Jensen, F., Ravdin, R. G., Moorhead, P., & Saksela, E. (1962) *J. cell. comp. Physiol.*, **59**, 281.
Kreider, J. W., Breedis, C., & Curran, J. S. (1967) *J. natn. Cancer Inst.*, **38**, 921.
Law, L. W. (1965) *Nature, Lond.*, **205**, 672.
Le Bouvier, G. L., Sussman, M., & Crawford, L. V. (1966) *J. gen. Microbiol.*, **45**, 497.
Lee, K. P., & Olson, C. (1969) *Am. J. vet. Res.*, **30**, 725.
Lehman, J. M., & Defendi, V. (1970) *J. natn. Cancer Inst.*, **44**, 125.
Lucké, B., Ratcliffe, H., & Breedis, C. (1950) *Fed. Proc.*, **9**, 336.
Lyell, A., & Miles, J. A. R. (1951) *Br. med. J.*, **1**, 922.
McFadyean, J., & Hobday, F. (1898) *J. comp. Path.*, **11**, 341, 912.
Macpherson, I. A. (1962) Wistar Institute biennial res. rep. 1962–63, p. 52.

Mattern, C. F. T., Allison, A. C., & Rowe, W. P. (1963) *Virology*, **20**, 413.
Mattern, C. F. T., & DeLeva, A. M. (1968) *Virology*, **36**, 683.
Mattern, C. F. T., Takemoto, K. K., & DeLeva, A. M. (1967) *Virology*, **32**, 378.
Mayor, H. D., Jamison, R. M., & Jordan, L. E. (1963) *Virology*, **19**, 359.
Mendelson, C. G., & Kligman, A. M., (1961) *Arch. Dermatol.*, **83**, 559.
Meyer, H. M., Hobbs, H. E., Rogers, N. G., Brooks, B. E., Bernheim, B. C., Jones, W. P., Nisalak, A., & Douglas, R. D. (1962) *J. Immunol.*, **88**, 796.
Moore, D. H., Stone, R. S., Shope, R. E., & Gelber, D. (1959) *Proc. Soc. exp. Biol. Med., U.S.A.*, **101**, 575.
Morgan, H. R., & Balduzzi, P. C. (1964) *Proc. natn. Acad. Sci., U.S.A.*, **52**, 1561.
Murakami, W. T., Fine, R., Harrington, M. R., & Ben Sassan, Z. (1968) *J. molec. Biol.*, **36**, 153.
Negroni, G., Dourmashkin, R., & Chesterman, F. C. (1959) *Br. med. J.*, **2**, 1359.
Noyes, W. F. (1959) *J. exp. Med.*, **109**, 423.
Noyes, W. F. (1965) *Virology*, **25**, 358.
Olson, C., & Cook, R. H. (1951) *Proc. Soc. exp. Biol. Med., U.S.A.*, **77**, 281.
Olson, C., Pamukcu, A. M., Brobst, D. F., Kowalczyk, T., Satter, E. J., & Price, J. M. (1959) *Cancer Res.*, **19**, 779.
Olson, C., Pamukcu, A. M., & Brobst, D. F. (1965) *Cancer Res.*, **25**, 840.
Olson, C., Segre, D., & Skidmore, L. V. (1960) *Am. J. vet. Res.*, **21**, 233.
Orth, G., Vielle, F., & Changeux, J. P. (1967) *Virology*, **31**, 729.
Padgett, B. L., Walker, D. L., Zu Rhein, G. M., Eckrvade, R. J., & Dessel, B. H. (1971) *Lancet*, **1**, 1257.
Pamukcu, A. M. (1963) *Ann. N.Y. Acad. Sci.*, **108**, 938.
Parsons, R. J., & Kidd, J. (1943) *J. exp. Med.*, **77**, 233.
Perry, J. L., To, C. M., & Consigli, R. A. (1969) *J. gen. Virol.*, **4**, 403.
Pope, J. H., & Rowe, W. P. (1964) *J. exp. Med.*, **120**, 121.
Potter, C. W., Oxford, J. S., & Hoskins, J. M. (1969) *Arch. ges. Virusforsch.*, **27**, 87.
Prunieras, M., Chardonnet, T., & Sohier, R. (1964) *Annls Inst. Pasteur, Paris*, **106**, 1.
Ragland, W. L., & Spencer, G. R. (1968) *Am. J. vet. Res.*, **29**, 1363.
Rapp, F., & Melnick, J. L. (1966) *Progr. med. Virol.*, **8**, 349.
Rdzok, E. S., Shipkowitz, N. L., & Richter, W. L. (1966) *Cancer Res.*, **26**, 160.
Rendtorff, R. C., & Wilcox, A. (1957) *J. infect. Dis.*, **100**, 119.
Robl, M. G. (1968) *Cancer Res.*, **28**, 1596.
Rogers, S. (1959) *Nature, Lond.*, **183**, 1815.
Rogers, S., & Moore, M. (1963) *J. exp. Med.*, **117**, 521.
Rous, P., & Beard, J. W. (1934) *J. exp. Med.*, **60**, 701.
Rous, P., & Beard, J. W. (1935) *J. exp. Med.*, **62**, 523.
Rowe, W. P., Hartley, J. W., Brodsky, I., & Huebner, R. J. (1958) *Science*, **128**, 1339.
Rowe, W. P., Huebner, R. J., & Hartley, J. W. (1961) in *Perspectives in virology*, **II**, 177. Minneapolis: Burgess.
Rowson, K. E. K., & Mahy, B. W. J. (1967) *Bact. Rev.*, **31**, 110.
Sachs, L., & Fogel, M. (1960) *Virology*, **11**, 722.
Schlumberger, H. D., Anderer, F. A., & Koch, M. A. (1968) *Virology*, **36**, 42.
Segre, D., Olson, C., & Hoerlein, A. B. (1955) *Am. J. vet. Res.*, **16**, 517.
Shah, K. V. (1966) *Proc. Soc. exp. Biol. Med., U.S.A.*, **121**, 303.
Shah, K. V., Ozer, H. L., Pond, H. S., Palma, L. D., & Murphy, G. O. (1971) *Nature Lond.*, **231**, 448.
Sharp, D. G., Taylor, A. R., & Beard, J. W. (1946) *J. biol. Chem.*, **163**, 289.
Sharp, D. G., Taylor, A. R., Beard, D., & Beard, J. W. (1942) *J. biol. Chem.*, **142**, 193.
Shein, H. M., & Enders, J. F. (1962) *Proc. Soc. exp. Biol. Med., U.S.A.*, **109**, 495.
Shiratori, O., Osato, T., & Ito, Y. (1969) *Proc. Soc. exp. Biol. Med., U.S.A.*, **130**, 115.
Shope, R. E. (1933) *J. exp. Med.*, **58**, 607.
Shope, R. E., Mangold, R., Macnamara, L. G., & Dumbell, K. R. (1958) *J. exp. Med.*, **108**, 797.
Sjögren, H. O., Hellstrom, I., Klein, G. (1961) *Cancer Res.*, **21**, 329.

Smith, J. W., Pinkel, D., & Dabrowski, S. (1969) *Cancer*, **24**, 527.
Stewart, S. E., Eddy, B. E., & Borgese, N. G. (1958) *J. natn. Cancer Inst.*, **20**, 1223.
Stoker, M. G. P. (1960) *Br. J. Cancer*, **14**, 679.
Stoker, M. G. P. (1962) in *Tumour viruses of murine origin* (Ciba foundation symposium), p. 52. London: Churchill.
Stoker, M. G. P. (1963) *Br. med. J.*, **1**, 1305.
Stoker, M. G. P., & MacPherson, I. (1961) *Virology*, **14**, 359.
Stone, R. S., Shope, R. E., & Moore, D. H. (1959) *J. exp. Med.*, **110**, 543.
Sweet, H., & Hilleman, M. R. (1960) *Proc. Soc. exp. Biol. Med., U.S.A.*, **105**, 420.
Syverton, J. T., Berry, G. P., & Warren, S. L. (1941) *J. exp. Med.*, **74**, 223.
Tajima, M., Gordon, D. E., & Olson, C. (1968) *Am. J. vet. Res.*, **29**, 1185.
Takemoto, K. K., & Fabisch, P. (1970) *Virology*, **40**, 135.
Thomas, M., Boiron, M., Tanzer, J., Levy, J. P., & Bernard, J. (1964) *Virology*, **202**, 709.
Thorne, H. V., Warden, D. (1967) *J. gen. Virol.*, **1**, 135.
Trilling, D. M., & Axelrod, D. (1970) *Science*, **168**, 268.
Uchida, S., Watanabe, S., & Kato, M. (1966) *Virology*, **28**, 135.
Ullman, E. V. (1923) *Acta oto-laryng.* (*Stockh.*), **5**, 317.
Vogt, M., & Dulbecco, R. (1960) *Proc. natn. Acad. Sci., U.S.A.*, **46**, 365.
Voss, J. L. (1969) *Am. J. vet. Res.*, **30**, 183.
Watrach, A. M. (1969) *Cancer Res.*, **29**, 2079.
Watrach, A. M., Hanson, L. E., & Meyer, R. C. (1969) *J. natn. Cancer Inst.*, **43**, 453.
Watson, J. D., & Littlefield, J. W. (1960) *J. molec. Biol.*, **2**, 161.
Williams, R. C., Kass, S. J., & Knight, C. A. (1960) *Virology*, **12**, 48.
Winocour, E. (1963) *Virology*, **19**, 158.
Yoshida, T. O., & Ito, Y. (1968) *Proc. Soc. exp. Biol. Med., U.S.A.*, **128**, 587.
Zu Rhein, G. M. (1969) *Progr. med. Virol.*, **11**, 185.

14

Adenoviruses

Reviews: Huebner et al. (1958). Brandon & McLean (1962). Pereira et al. (1963b). Sohier et al. (1965). Schlesinger (1969).

Adenoviruses have been isolated from man, monkeys, cattle, swine, dogs, mice, sheep and birds. Serological evidence suggests that other animal species (horse, goat and deer) may also harbour these viruses (Darbyshire & Pereira, 1964). A strain tentatively identified as adenovirus has been isolated from horses (Todd, 1969).

Morphology and development. Adenoviruses of human (Horne et al., 1959), simian (Archetti & Steve-Bocciarelli, 1963), bovine (Tanaka et al., 1968), porcine (Chandler, 1965), ovine (McFerran et al., 1969), murine (Hashimoto et al., 1966), canine (Davies et al., 1961) and avian (McPherson et al., 1961; Dutta & Pomeroy, 1963) origins are 70 to 80 nm in diameter, unenveloped, with icosahedral symmetry, made up of 252 capsomeres.

The 240 capsomeres making up the faces and part of the sides of the triangles (hexons) are spherical or prismatic, approximately 7 nm in diameter (Wilcox & Ginsberg, 1963a; Wilcox et al., 1963) whereas the apical capsomeres (pentons) are complex structures consisting of a base, 7 nm in diameter and a fibre with a terminal swelling. This is projected outwards from each of the 12 apices of the capsid (Valentine & Pereira, 1965; Norrby, 1966). The length of the fibres varies in different serotypes of human (Norrby, 1968) and canine (Marusyk et al., 1970) adenoviruses. Within the capsid is a dense nucleoprotein core or nucleoid 30 nm across in which a filamentous structure is visible (Epstein et al., 1960; Valentine, 1960). Isolation of cores from purified virus has been described by Laver et al. (1968). Intranuclear virions are often arranged in a crystalline structure, forming a cubic lattice with centre-to-centre spacing of 60–65 nm (Low & Pinnock, 1956). Besides crystals of virus particles, infected nuclei may contain protein crystals (Morgan et al., 1956; Valentine, 1960) and striated structures of uncertain nature (Martinez-Palomo et al., 1967).

The virus penetrates either by phagocytosis (Dales, 1962) or directly through the cell membrane (Morgan et al., 1969). Synthesis of virus-specific RNA has been described in both productive (Kohler & Odaka, 1964; Flanagan & Ginsberg, 1964; Rose et al., 1965) and nonproductive infection (Fujinaga & Green, 1967). Viral DNA synthesis is detected a few hours before the production of infectious virus (Wilcox & Ginsberg, 1963b; Polasa & Green, 1965; Mantyjarvi & Russell, 1969). Virus-specific proteins synthesized at early stages of lytic infection (Hoggan et al., 1965) or in virus-induced tumour or transformed cells (Huebner et al., 1963) may represent nonstructural viral proteins (T or neo-antigen) or internal components (P antigen) of the virion (Russell et al., 1967). Capsid proteins are synthesized in the cytoplasm and rapidly transferred to the nucleus (Thomas & Green, 1966; Velicer & Ginsberg, 1969; Horwitz et al., 1969). The sequential synthesis of viral proteins has been described by Russell et al. (1967). Virus maturation takes place in the nucleus and is dependent on the synthesis of an arginine-rich component (Rouse & Schlesinger, 1967; Russell & Becker, 1968). Virus is released by rupture of the nuclear membrane (Morgan et al., 1956) or by the accumulation of virions in nuclear protrusions which eventually separate from the main body of the nucleus (Fong et al., 1965). The effect of adenovirus replication on host macromolecular synthesis is reviewed by Schlesinger (1969).

Chemical composition. Virions contain 11·3 to 13·5 per cent DNA (Green & Piña, 1964) in the form of a single, noncircular, double-stranded molecule 11 to 13 microns in length, with a MW of 20 to 25 × 10^6 daltons (Green et al., 1967). From buoyant density (1·708 to 1·720) and thermal denaturation temperature ($Tm^{(0)}$ – 89·4 to 94·1), the G + C content of viral DNA has been estimated by Green and co-workers (see review by Schlesinger, 1969) to vary between 47 and 60 per cent, there being an inverse correlation between G + C and oncogenicity of different serotypes (Piña & Green, 1965). However, the DNA of an oncogenic simian adenovirus (SA7) has a GC content of 58 to 60 per cent (Burnett & Harrington, 1968a; Piña & Green, 1968). Oncogenic adenovirus DNAs were found by Kubinski & Rose (1967) to contain fewer dC-rich sequences and more dA-rich sequences than DNAs of nononcogenic serotypes. Distinctions correlated with oncogenicity have also been demonstrated by DNA-DNA hybridization (Lacy & Green, 1964; 1965; 1967). Nearest-neighbour base sequence analysis reveals only limited resemblance between the DNA of human adenovirus type 2 and that of the host cell (Morrison et al., 1967). Evidence for a unique distribution of adenine-thymine-rich clusters in the DNA of human adenovirus type 2 has been presented by Doerfler

& Kleinschmidt (1970). In contrast with human adenoviruses, phenol-extracted DNA from a simian adenovirus is infectious and induces tumours in hamsters (Burnett & Harrington, 1968a, b).

Proteins make up 86·5 to 88·7 per cent of virions (Green & Piña, 1964). Analysis of the structural proteins by polyacrylamide gel electrophoresis (Maizel et al., 1968; Russell et al., 1968; Laver et al., 1968) revealed the presence of up to 10 polypeptides. Five of these are clearly related to hexon (MW 120,000), penton base (MW 70,000) fibre (MW 62,000) and 2 core proteins (MW 44,000 and 24,000) (Maizel et al., 1968), and the remainder may represent aggregation artefacts or break-down products (Pereira & Skehel, 1970). Amino acid composition of purified virions (Polasa & Green, 1967; Boulanger et al., 1969) and of isolated structural proteins (Biserte et al., 1964; Pettersson et al., 1967; 1968; Russell et al., 1971; Laver, 1970) reveals as an outstanding finding, a high arginine content of one of the internal proteins. Laver (1970) demonstrated alanine as the N-terminal amino acid of this arginine-rich adenovirus core protein. The internal proteins make up about 18 to 20 per cent of the total viral protein (Laver et al., 1967; Russell et al., 1968).

Physico-chemical characters. Virions have buoyant density of 1·34 g/ml in CsCl (Allison & Burke, 1962). Resistant to lipid solvents and trypsin. Heat-sensitivity is either unaffected (murine—Hashimoto et al., 1966) or enhanced (human—Wallis et al., 1962; simian—Casto & Hammon, 1966; avian—Burke et al., 1968) by divalent ions. Human (Kjellén, 1965), bovine (Tanaka et al., 1968), porcine (Haig et al., 1964) and canine strains are acid-stable. Human (Hiatt et al., 1960) and avian (Petek et al., 1963) strains are sensitive to photodynamic inactivation. Inactivation by chloramine T and permanganate takes place at a rate dependent on oxidation potential with activation energy about 10,000 cal per mol between 0°C and 37°C (Lund, 1966). Infectivity in aerosols is stable at high relative humidity (Miller & Artenstein, 1967). Different human serotypes vary in sensitivity to ultraviolet and to nitrous acid (Wasserman, 1962). Canine (Larin, 1959) and avian (Petek et al., 1963) strains are relatively UV-resistant.

HUMAN ADENOVIRUSES

Synonyms: Adenoid-degeneration (AD) agents. Adenoidal-pharyngeal-conjunctival (APC) agents. Diseases caused include: acute pharyngitis or febrile catarrh, pharyngo-conjunctival fever and epidemic kerato-conjunctivitis.

Haemagglutination. RBCs of a number of species are agglutinated

(Rosen, 1958; 1960; Nász et al., 1962), rat cells most commonly. Human O, rhesus, *Cercopithecus* and mouse RBCs are agglutinated by some strains. As a rule several isolates belonging to a single type behave alike but some discrepancies are recorded.

Rosen (1960) has pointed out that human adenoviruses can be grouped according to their behaviour in the HA test. A first subgroup including types 3, 7, 11, 14, 16, 20, 21, 25 and 28 agglutinate rhesus but not rat RBCs. A second subgroup including types 8, 9, 10, 13, 15, 19, 22, 23, 24, 26, 27, 29 and 30 agglutinate rat cells, but rhesus cells are affected not at all or to lower titre. A third subgroup including types 1, 2, 3, 4, 5, and 6 partially agglutinate rat but not rhesus cells. Agglutination by these types is enhanced by heterotypic sera to types within the subgroup. Types 12, 18 and 31 were first reported to lack haemagglutinins, but subsequently shown to cause incomplete agglutination of rat cells as in members of Rosen's third subgroup (Schmidt et al., 1965). Modifications and elaborations of the above scheme for subgrouping human adenoviruses according to haemagglutination have been proposed by several workers (see review by Wadell, 1970). Haemagglutinating activity has been associated with virions (Bauer et al., 1964) or with pentons or fibres in different states of aggregation (Norrby, 1969a). Haemagglutination behaviour is well correlated with other biological properties characteristic of different subgroups.

Human O cells treated with receptor-destroying enzyme, by influenza virus or by a factor present in some adenovirus cultures are less readily agglutinated (Kasel et al., 1961). An erythrocyte receptor for adenovirus type 7 haemagglutinin has been described (Neurath et al., 1969).

Antigenic properties. Virus neutralization tests distinguish 33 serotypes of human adenoviruses (Blacklow et al., 1969), although cross-reactions are sometimes recorded (Rafajko, 1964; Wigand et al., 1965). Intratypic antigenic heterogeneity has been described within types 3 (Omura & Mutai, 1966) and 7 (Pereira & Kelly, 1957b). Serotypes may also be distinguished by haemagglutination-inhibition, although discrepancies between HI and VN tests have been described. Most of these discrepancies can be explained by different specificities ascribed to antigens either present in soluble form or incorporated into virions (Wigand & Fliedner, 1968). Intermediary forms with capsids composed of antigenic mosaics have been described (Norrby, 1969b). The type-specific antigenic component involved in virus neutralization has been shown to reside in the hexon (Kjellén & Pereira, 1968) which also contains a group-specific antigen. Distinct type- and group- or subgroup-specific antigens are also found in the penton, fibre and base. The specificities of the capsid antigens have been extensively investigated by Norrby and co-workers (see reviews by Norrby, 1968; 1969a; Wadell, 1970). Broadly reactive internal antigens including a group-specific 'p' antigen have been described (Russell & Knight, 1967; Hayashi & Russell, 1968). Virus-specific antigens found in adenovirus-induced tumours or transformed cells (Huebner et al., 1963) or in early stages of lytic infections (Hoggan et al., 1965) are generally considered to be nonstructural virus-coded proteins. They are broadly reactive allowing, however, the distinction of 2 groups of oncogenic serotypes (Riggs et al., 1968). Transplantation antigens are also broadly reactive, but allow the distinction of subgroups correlated with oncogenic activity (Ankerst & Sjogren, 1970).

Cultivation. Continuous human cell-lines have been chiefly used in studies of human adenoviruses, but most cultures of human cells are suitable for propagation. Epithelioid cells are more susceptible than fibroblasts and primary human-kidney cultures are particularly sensitive. There is an early cytopathic effect due to the penton antigen and a late CPE associated with virus multiplication. Intranuclear inclusions show characteristics peculiar to certain groups of serotypes (Barski, 1956; Boyer et al., 1959). Plaque assays (Rouse et al., 1963) may be used for infectivity and virus-neutralization assays.

A number of cells other than human support productive or abortive infection by human adenoviruses. Some types can be serially propagated in cultures of rabbit, hamster, swine and monkey cells, whereas others tend to replicate abortively in nonhuman cells (see reviews by Rapp & Melnick, 1966; Schlesinger, 1969). Multiplication of some human serotypes in monkey cell cultures is promoted by concomitant infection with SV40 (Rabson et al., 1964; Beardmore et al., 1965) or with simian adenovirus SV15 (Naegele & Rapp, 1967). A defective particle (MAC, or Monkey-cell Adapting Component) capable of promoting replication of human adenovirus type 7 in monkey-kidney cells has been described (Butel & Rapp, 1967). A strain of adenovirus type 7 which had been propagated together with SV40 in African green monkey-kidney cultures was shown to have incorporated at least part of the SV40 genome, thus acquiring the capacity to produce SV40 tumours in hamsters and to induce the synthesis of SV40 T antigen in cultured cells (Rowe & Baum, 1964; Rapp et al., 1964). The particles responsible for these activities are defective hybrids consisting of adenovirus capsids containing portions of adenovirus and SV40 genomes linked by an alkali-stable bond (Rowe & Pugh, 1966; Baum et al., 1966). The defective genomes containing either SV40 or MAC components can be transferred to capsids of other adenovirus serotypes (transcapsidation) by co-cultivation (Rowe & Pugh, 1966; Butel & Rapp, 1967). A nondefective hybrid has been isolated from adeno 2-SV40 hybrid populations by Lewis et al. (1969). Complementation of human by simian adenoviruses leading to the formation of phenotypically mixed virions with antigens from both parents and to adaptation for growth in green monkey-kidney cells has been described (Alstein & Dodonova, 1968). *In vitro* cell transformation by human adenoviruses, originally described by McBride & Wiener (1964), was subsequently observed by many workers (see review by Black, 1968). Infective virus is recovered with difficulty (Mart et al., 1968) or not at all from transformed or tumour cells.

Pathogenicity. Infection of man may be inapparent or take the form of one of the minor respiratory infections of childhood. Types 1, 2, 5 and 6 are particularly concerned, also in sporadic infections in adults. These, too, are the types commonly found latent in human adenoids and tonsils (Rowe et al., 1955). Types 3, 4, 7, 14 and 21 in particular cause outbreaks of fever and pharyngitis, especially in service recruits and boarding schools. Occurrence of conjunctivitis in some outbreaks has led to the use of the term 'pharyngo-conjuctival fever'. Adenoviruses may cause 1 form of atypical pneumonia, unassociated with development of cold agglutinins. Some serotypes have been reported to cause fatal pneumonia in infants, occurring sporadically (Chany et al., 1958) or in extensive epidemics (Jen et al., 1962). Type 8 is commonly associated with epidemic kerato-conjunctivitis (Jawetz, 1959) and also with outbreaks of swimming-pool conjunctivitis (Fukumi et al., 1958). Other adenovirus types have been recovered from sporadic cases of follicular conjunctivitis, particularly in Arabia (Bell et al., 1960). Huebner (1959) gives the types commonly associated with various syndromes; types less commonly isolated are in parentheses. Acute respiratory disease: 4, 7 (3, 14). Pharyngo-conjunctival fever: 3, 7a (1, 2, 5, 6, 14). Acute febrile conjunctivitis: (1, 2, 3, 5). Follicular conjunctivitis: 3, 7a (1, 2, 5, 6, 14). Epidemic kerato-conjunctivitis: 8 (3, 7a, 9). Virus pneumonia in infants: (7a, 1, 3). Pneumonia in adults: 4, 7 (3). An extensive study by Brandt et al. (1969) revealed that approximately 7 per cent of respiratory illness among 18,000 infants was associated with adenovirus infections, the types most commonly represented being 1, 2, 3, 5, 6 and 7.

Experimentally. Laboratory animals are generally insusceptible to human strains. However, Pereira & Kelly (1957) produced latent infections with type 5 in rabbits, and the good antibody response to a single dose of virus in guinea-pigs, cotton-rats and hamsters has suggested that virus may multiply in them (Rowe et al., 1955; Ginsberg, 1956).

Types 1, 2, 5 and 6 have produced pneumonia in young pigs deprived of colostrum (Betts et al., 1962). Experimental infection of calves by adenovirus type 1 has been reported by Bettinotti & Straub (1966). Inapparent infection of dogs inoculated with types 2, 3, 4 and 7 has been reported (Sinha et al., 1960). Type 5 causes fatal infections in very young hamsters (Pereira et al., 1963a). Type 12 causes a wasting syndrome in thymectomized immature baboons (Kalter et al., 1967a).

Human adenoviruses have been divided according to oncogenicity into 4 subgroups (Huebner, 1967; McAllister *et al.*, 1969). Subgroups A (types 12, 18 and 31) and B (types 3, 7, 14, 16 and 21) are characterized by high and low oncogenic activity respectively and by the capacity to transform cells *in vitro*; whereas subgroups C (types 1, 2, 5 and 6) and D (types 9, 10, 13, 15, 17, 19 and 26) are incapable of inducing tumours *in vivo*, but can transform rat and hamster cells *in vitro* and are distinguishable from each other by T antigen specificity.

Pathological lesions. In a large number of human autopsies carried out by Teng (1960) the main findings consisted of necrotic and proliferative bronchitis and peribronchial pneumonia with necrosis of the bronchial and alveolar walls and marked oedema. Basophilic intranuclear inclusions similar to those seen in tissue cultures were frequently observed in epithelial cells of the alveoli and bronchi (see also Chany et al., 1958). Tumours induced by human adenoviruses in hamsters are characterized as densely cellular, embryonal-type small-cell sarcomata (Berman, 1967).

Ecology. Infection is probably mainly air-borne from respiratory secretions (Couch et al., 1966), but virus is frequently present in stools and may also be recovered from urine (Gutekunst & Heggie, 1961; Numazaki et al., 1968). Types above No. 9, except 14 and 21, have been recovered almost wholly from the intestinal tract. Virus latent in tonsils and adenoids is believed to persist after an infection. Respiratory outbreaks in recruits are recorded mainly in winter months, while 'swimming-pool conjunctivitis' is a summer disease.

Control. Formalinized vaccines have proved of value in controlling epidemics in recruits (see review by Hilleman, 1966). Vaccination with attenuated strains of types 3, 4, 5, 6 and 7 has been carried out by Selivanov et al. (1964). More recently, human subjects have been immunized against adenovirus type 4 by oral administration of living virus in enteric-coated capsules (Chanock et al., 1966; Gutekunst et al.,

1967). The possible use of subunit adenovirus vaccine (Kasel et al., 1964) deserves further investigation.

SIMIAN ADENOVIRUSES

Review: Hull (1968).

A number of adenoviruses are included in the SV (Simian Virus) series of Hull and his colleagues (1956; 1957; 1958; 1968). These have been renumbered by Pereira et al. (1963b). Certain strains of the SA series of simian viruses described by Malherbe & Harwin (1957; 1963) have also been identified as adenoviruses. More recently, adenoviruses antigenically related or not to previously described simian or human serotypes have been isolated from baboons (Kalter et al., 1967b), vervet monkeys (Kim et al., 1967) and chimpanzees (Hillis & Goodman, 1969).

Haemagglutination. Rapoza (1968) has classified simian adenoviruses into 4 subgroups according to haemagglutination behaviour. Subgroup 1 is characterized by the capacity to agglutinate rhesus-monkey erythrocytes both at 4°C and at 37°C and includes a single serotype—M11 (SV36). Subgroup 2 strains agglutinate rat, rhesus and guinea-pig erythrocytes at 4°C, but only rat erythrocytes at 37°C; it includes serotypes M2 (SV23), M3 (SV32, SV37, SV39), M4 (SV15), M6 (SV17) and M9 (SV27 and SV31). Subgroup 3 is characterized by negative or incomplete agglutination of rat erythrocytes and includes serotypes M1 (SV1), M5 (SV11), M7 (SV20), M8 (SV25) and M10 (SV30, SV34 and SV38).

Adenoviruses isolated from chimpanzees were divided into 3 haemagglutination subgroups (Hillis & Goodman, 1969). Subgroup 1 represented by strains of serotype C1, agglutinate rhesus- or vervet-monkey erythrocytes. Subgroup 2 includes strains antigenically related to human serotypes 2 and 5, agglutinate rat erythrocytes in the presence of heterotypic sera of Rosen's subgroup 3 (see p. 312). Subgroup 3 includes 4 isolates unrelated to each other or to known adenovirus serotypes and is characterized by failure to agglutinate monkey, rat, guinea-pig or human erythrocytes.

Antigenic properties. Simian strains share the CF antigen with other strains of mammalian origin but can be divided into several serotypes by virus neutralization and haemagglutination-inhibition.

Pereira et al. (1963b) proposed the distinction of serotypes isolated from monkeys (mainly rhesus and cynomolgus isolates of Hull and co-workers) and 1 serotype from chimpanzees. Subsequent work by

Rapoza (1968) established the antigenic relationships between these isolates by virus neutralization and haemagglutination-inhibition. The distribution of strains of Hull's SV series into the proposed simian serotypes is (Hull's SV numbers in parentheses): M1 (SV1, SV34); M2 (SV23, SV39); M3 (SV32, SV37); M4 (SV15); M5 (SV11); M6 (SV17); M7 (SV20); M8 (SV25); M9 (SV27, SV30, SV31); M10 (SV33, SV38); M11 (SV36). In addition to these, new serotypes may be proposed for strain SA7 of Malherbe & Harwin (1957) for strain V340 isolated from vervet monkeys (Kim et al., 1967) and from baboons (Engster et al., 1969) and for 4 unrelated strains from chimpanzees (Hillis & Goodman, 1969). Other isolates from baboons or vervet monkeys are related to previously described simian serotypes, and several isolates from chimpanzees are antigenically related, though on the whole, not identical to human serotypes 14, 18, 5 and 2 or to the chimpanzee serotype C1.

Cultivation. Simian strains grow better in monkey-kidney tissue than in human cells; a chimpanzee strain (Rowe et al., 1958) is said to grow equally well in both. M4 (SV15) has been adapted to grow indefinitely in human HEp-2 cell-line (Hammon et al., 1963). Prier & Le Beau (1958) report that M9 (SV27) has a CPE similar to that of human adenoviruses 3 and 4. So had M6 (SV17); this strain also grew, though poorly, in calf-kidney cells. A detailed study by light and electron microscopy of productive and abortive infection in monkey and hamster cells respectively has been reported by Fong et al. (1968).

Pathogenicity. Several serotypes have been associated with respiratory and enteric infections Rhesus (Bullock, 1965), *Erythrocebus* (Tyrrell et al., 1960), *Cercopithecus* (Kim et al., 1967) monkeys and in baboons (Engster et al., 1969).

Five of 17 adenovirus strains of rhesus or cynomolgus monkey origin (SV20, SV23, SV34, SV37 and SV38) and 1 strain from an African green monkey (SA7) are oncogenic for new-born hamsters (Hull et al., 1965). One of these strains (SV20) has also produced a tumour in a new-born mouse and another (SA7) induced tumours in suckling rats. Earlier results by Huebner et al. (1962) suggested that M1 (SV1) might also be oncogenic for hamsters.

BOVINE ADENOVIRUSES

Haemagglutinin. Types 1 to 3 agglutinate rat erythrocytes. Types 2 and 3 are also capable of agglutinating mouse and monkey erythrocytes respectively (Klein et al., 1960). Erythrocytes from a number of other mammalians (cattle, horse, sheep, goat, guinea-pig, hamster) are agglutinated by some strains, sometimes the reaction being temperature dependent (Inaba et al., 1968).

Antigenic properties. Nine serotypes have been described: types 1 and 2 by Klein et al. (1959; 1960), type 3 by Darbyshire et al. (1965), types 4 and 5 by Bartha & Aldasi (1966), type 6 by Rondhuis (1968), type 7 by Matumoto et al. (1969), type 8 by Bartha et al. (1970) and type 9 by Guenov et al. (1971). All types share the adenovirus group-specific antigen and are distinguishable from each other and from human serotypes by virus neutralization. There are discrepancies in serotype numbers given by different workers. Cross-reactions have been revealed by virus neutralization between some of the newly proposed serotypes (Matumoto et al., 1969).

Cultivation. Types 1 and 2 grow preferentially in bovine-kidney tissue cultures with CPE typical of adenoviruses. Type 2 grows poorly in monkey-kidney cells. Bartha & Csontos (1969) distinguishes 2 groups of strains: 1) readily propagated in bovine-kidney cultures with production of irregularly shaped intranuclear inclusions; 2) unable to grow in bovine-kidney cultures but growing in bovine testis cultures with formation of multiple, regularly shaped inclusions.

Pathogenicity. Serological evidence obtained by Darbyshire and Pereira (1964) suggests that adenoviruses play a part in the aetiology of bovine respiratory disease. Bovine adenoviruses have been associated with epidemic pneumoenteritis of calves (Aldasi et al., 1965) and with conjunctivitis and kerato-conjunctivitis (Wilcox, 1969). Experimental infection of young calves (Aldasi et al., 1965; Mohanty & Lillie, 1965; Darbyshire et al., 1966; 1969) leads mainly to respiratory and sometimes intestinal symptoms. Type 3 is oncogenic in hamsters (Darbyshire, 1966; Gilden et al., 1967).

Ecology. In bovine respiratory and intestinal tracts. Serological evidence suggests high incidence of infection (Darbyshire & Pereira, 1964). Several strains have been isolated from uninoculated cultures of bovine testis (Bartha & Csontos, 1969) and kidney (Schpov et al., 1968).

PORCINE ADENOVIRUSES

Haemagglutination. Type 1 agglutinates monkey, rat, human O, guinea-pig and mouse erythrocytes, but not those of rabbit, fowl, ox, sheep or pig. Types 2 and 3 fail to agglutinate any of these cells (Clarke et al., 1967). A strain isolated by Kasza (1966) agglutinates chicken and rat erythrocytes strongly and guinea-pig, mouse and hamster erythrocytes weakly.

Antigenic properties. Three distinct serotypes have been isolated in Great Britain (Haig et al., 1964; Clarke et al., 1967). A strain described by Kasza (1966) may represent a fourth serotype. Rasmussen (1969) independently described 4 serotypes. Presence of the group-specific adenovirus antigen has been demonstrated by complement fixation and immunodiffusion (Haig et al., 1964). Serotypes distinguished by virus neutralization. No cross-reactions by virus neutralization with adenoviruses from other hosts (Clarke et al., 1967; Kasza, 1966).

Cultivation. Wide range of cells susceptible to infection, including pig, calf, dog, hamster and human (Kasza, 1966). Some strains grow better in calf- than in pig-kidney cultures (Mayr et al., 1967). Ultrastructural changes in infected cells have been described by Koestner et al. (1968), Mahnel & Sigel (1968) and Derbyshire et al. (1968).

Pathogenicity. Types 2 and 3 inoculated intranasally into colostrum-deprived new-born pigs give rise to infection of the tonsils and lower intestinal tract with serological response, but no symptoms or gross lesions (Sharpe & Jesset, 1967). Pig embryos aborted after inoculation *in utero* with type 1 and showed intranuclear inclusions in intestinal epithelial cells (Sharpe, 1967).

Ecology. Commonly found in the digestive tract of pigs. Isolation from primary kidney cultures of slaughter pigs has been reported (Mahnel & Bibrack, 1966; Kohler & Apodaca, 1966).

OVINE ADENOVIRUSES

McFerran et al. (1969) reported the isolation of 8 adenovirus strains from sheep. A representative strain had typical dimensions and morphology by negative staining, was stable at pH 3·0, inactivated at 60°C for 30 minutes and agglutinated rat erythrocytes. Multiplication in sheep-kidney cells, inhibited by BUDR.

MURINE ADENOVIRUSES

Haemagglutination. Several strains reported to lack haemagglutinins for human, monkey, rabbit, mouse, guinea-pig, cattle, pig, sheep and fowl erythrocytes.

Antigenic properties. Murine adenoviruses described by Hartley & Rowe (1960) and by Hashimoto et al. (1966) are probably 2 distinct serotypes. Their relationship to strains isolated by Missal (1969) are unknown.

Cultivation. Growth with typical CPE in mouse-embryo tissue cultures, but not in mouse cell-lines or in rat, monkey or human cells (Hartley & Rowe, 1960; Hashimoto et al., 1966). Adaptation to grow in human cells has been reported (Sharon & Pollard, 1964). The strain described by Missal (1969) grows in primary and established cell cultures and also in hamster- (BHK-21), calf- and pig-kidney cells, producing CPE in the last.

Pathogenicity. Hartley & Rowe (1960) describe fatal disease in suckling mice inoculated by various routes; whereas Sugiyama et al. (1967) and Missal (1969) report asymptomatic infections. Adaptation to grow intraperitoneally in mice with Ehrlich ascites tumours leads to generalized infection with high mortality rate (Schmitt-Ruppin, 1968). Ginder (1964) described persistent experimental infection in mice.

Ecology. Frequently found in mouse colonies without association with illness.

CANINE ADENOVIRUSES

Synonyms: Infectious canine hepatitis. Rubarth's disease. Fox encephalitis. Hepatitis infectiosa canis. Canine laryngotracheitis. Kennel cough.

Review: Rubarth (1947) original description. Cabasso (1962). Identification with fox encephalitis (Green et al., 1930).

The virus was shown to be an adenovirus by Kapsenberg (1959).

Haemagglutination. According to Fastier (1957) the virus agglutinates fowl RBCs at 4°C and pH 7·5–8, but Espmark & Salenstedt (1961) could not confirm this and found agglutination of rat and human O cells at pH 6·5–7·5, temperature being immaterial. Soluble haemagglutinins resembling those of human adenoviruses (Rosen's Group III) have been described (Marusyk et al., 1970).

Antigenic properties. A strain associated with canine laryngotracheitis ('kennel cough') is distinguished from infectious canine hepatitis virus by haemagglutination-inhibition and by virus neutralization although the latter test reveals some cross-reaction (Ditchfield et al., 1962).

An antigenic relationship with human adenovirus type 8 has been described by Smith et al. (1970).

Cultivation. The virus grows readily in cultures of dog, ferret, raccoon and pig kidneys or testis producing a CPE and, in monolayers, readily counted plaques. The sequence of ultrastructural changes associated with virus multiplication has similarities with other adenoviruses (Garg et al., 1967; Yamamoto, 1969). Intranuclear inclusions and protein crystals of various configurations have been demonstrated in infected cells (Leader et al., 1960; Matsui & Bernhard, 1967).

Pathogenicity. The disease in dogs may be inapparent. Symptoms are commonest in newly weaned puppies; these may show high fever, apathy, oedema, vomiting and diarrhoea. Transient corneal opacity is common. The mortality is 10–25 per cent. The incubation period after contact is 6–9 days, but is less (3–6 days) after experimental infection. These adenoviruses may also cause 'kennel cough'.

In foxes the disease presents as acute encephalitis with convulsions passing into paralysis and coma; death usually occurs within 24 hours. There may be nasal and ocular discharge and diarrhoea. Transmission of infection from dogs to bears has been described (Kapp & Lehoczki, 1966). Serological evidence of natural human infections has been presented (Smith et al., 1970).

Experimentally. Dogs and foxes may be infected by any route, including the oral. Other canines—coyotes and wolves—are susceptible, also raccoons. Grey foxes (*Urocyon*) are, however, resistant (Green et al., 1934). Inoculation on to the cornea may lead to opacity, and this may be partly due to a toxic action seen also in inoculated kittens and ferrets which are not otherwise susceptible and in which the virus has not been passed in series (Cabasso et al., 1954). Toxic effects may also be shown in suckling mice (Kapsenberg, 1959). Salenstedt (1958) reports that when guinea-pigs were inoculated subcutaneously, there were fever, viraemia, viruria and lymphocytosis and 'up to 25 per cent died', often showing perihepatitis; there is no mention of serial transmission. Strains associated with both hepatitis and respiratory infections have been shown to be oncogenic for hamsters (Sarma et al., 1967; Dulac et al., 1970).

Pathological lesions. In dogs dying of the disease there are seen subcutaneous oedema, ascites, often blood-stained, haemorrhages in various viscera, a swollen, pale liver with thickened, often haemorrhagic, gall-bladder. In the fatal fox disease, the symptoms are seen to have been due to cerebral haemorrhages—the virus is not truly neurotropic. Lesions, especially in the liver, show numerous nuclear inclusions resembling those of other adenovirus infections.

Ecology. Spread is probably from respiratory tract or via the urine, since recovered animals may excrete virus in the urine for many months. On fur-ranges the infection may be enzootic over long periods.

Control. Formalinized and attenuated live virus vaccines both have been used. Virus attenuated by growth in pig-tissue culture has been successfully used in a bivalent vaccine, combined with attenuated distemper virus (Cabasso et al., 1958; Piercy & Sellers, 1960). The ocular route has sometimes been used for immunization, success being judged by the appearance of opacity.

AVIAN ADENOVIRUSES

Synonyms: GAL (Gallus Adeno-Like) virus. CELO (Chicken Embryo Lethal Orphan) virus. Quail bronchitis virus.

Antigenic properties. Eight unnumbered serotypes have been described by Kawamura et al. (1964), including GAL (Burmester et al., 1960), CELO (Yates & Fry, 1957) and 6 additional types isolated in Japan. Sharpless (1962) distinguished 3 or possibly 4 antigenic subgroups among 35 strains isolated in the United States, all of which showed some relation to GAL. Burke et al (1968) describe 2 serotypes distinguishable by virus neutralization and represented by CELO and GAL 1 strains respectively. A strain isolated from geese (Csontos, 1967) showed no cross-reactions with isolates from chickens. DuBose & Grumbles (1959) failed to differentiate CELO from quail bronchitis virus by virus neutralization.

A soluble antigen cross-reacting with adenoviruses of mammalian hosts has not been demonstrated, but all avian strains share a common antigen revealed by immunodiffusion and less clearly by complement fixation (Kawamura et al., 1964).

Cultivation. Grow in tissue cultures of embryonic chick liver, spleen, kidney and other tissues. Plaque-formation occurs in monolayers (Stoker, 1959). Some strains produce red plaques (Kawamura et al., 1964). Virus proteins are synthesized in the cytoplasm and transferred to the nucleus (Ishibashi, 1970) where maturation takes place with release only when cell membranes disintegrate (Sharpless et al., 1961). Optimal growth at 40°C (Moureal, 1968), but temperature-sensitive mutants growing only at lower temperatures have been described (Ishibashi, 1970). The CPE is typical of that for adenoviruses. Multiplication in chick embryos leads to death within 3 or 4 days. CELO virus has also been grown in turkey embryos (DuBose & Grumbles, 1959). CELO virus transforms hamster and human cells *in* *vitro* (Anderson et al., 1969) with induction of T antigen distinct from that of human adenovirus type 12 (Oxford & Potter, 1970).

Pathogenicity. GAL virus has not been associated with any natural disease, but may cause necrosis of the liver with basophilic intranuclear inclusions when inoculated into chicks (Sharpless & Jungherr, 1961).

CELO virus has been shown to be closely related to quail bronchitis virus first described by Olson (1950) and to reproduce this condition in experimentally infected animals (DuBose & Grumbles, 1959). CELO virus is oncogenic in hamsters (Sarma et al., 1965).

Ecology. Widely distributed in some countries (Cook, 1970). Frequently found in uninoculated chick-embryo tissues and in the faeces of apparently normal chickens.

POSSIBLE ADENOVIRUSES FROM OTHER HOSTS

Viruses morphologically identical to adenoviruses have been isolated from horses with respiratory disease (Todd, 1969) and from frog (*Rana pipiens*) kidneys with granulomas (Clark & Zeigel, quoted by Granoff, 1969).

REFERNCES

Aldasi, P., Csontos, L., & Bartha, A. (1965) *Acta vet. hung.*, **15**, 167.

Allison, A. C., & Burke, D. (1962) *J. gen. Microbiol.*, **27**, 181.

Alstein, A. D., & Dodonova, N. N. (1968) *Virology*, **35**, 248.

Ankerst, J., & Sjögren, H. O. (1970) *Int. J. Cancer*, **6**, 84.

Anderson, J., Yates, V. J., Jasty, V., & Mancini, L. O. (1969) *J. natn. Cancer Inst.*, **42**, 1 and **43**, 575.

Archetti, I., & Steve-Boccianelli, D. (1963) *Virology*, **20**, 399.

Barski, G. (1956) *Annls Inst. Pasteur, Paris*, **91**, 614.

Bartha, A., & Aldasy, P. (1966) *Acta vet. hung.*, **16**, 107.

Bartha, A., & Csontos, L. (1969) *Acta vet. hung.*, **19**, 323.

Bartha, A., Mathe, S., & Aldsay, P. (1970) *Acta phys. hung.*, **20**, 399.

Bauer, H., Wigand, R., & Adam, W. (1964) *Z. Naturforsch.*, **19b**, 587.

Baum, S. G., Reich, P. R., Hybner, C. J., Rowe, W. P., & Weissman, S. M. (1966) *Proc. natn. Acad. Sci. U.S.A.*, **56**, 1509.

Beardmore, W. B., Havlick, M. J., Serafini, A., & McLean, I. W., Jr., (1965) *J. Immunol.*, **95**, 422.

Bell, S. D., Rota, T. R., & McComb, S. T. (1960) *Am. J. trop. Med.*, **9**, 523.

Berman, L. D. (1967) *J. natn. Cancer Inst.*, **39**, 847.

Bettinotti, C. M., & Straub, O. C. (1966) *Zentbl. Bakt. ParasitKde, I. Abt. Orig.*, **199**, 427.

Betts, A. O., Jennings, A. R., Lamont, P. H., & Page, Z. (1962) *Nature, Lond.*, **193**, 45.

Biserte, G., Havez, R., Samaille, J., Desmet, G., & Waroquier, R. (1964), *Biochim. biophys. Acta*, **93**, 361.

Black, P. H. (1968) *Ann. Rev. Micriobiol.*, **22**, 391.

Blacklow, N., Hoggan, M., Austin, J., & Rowe, W. (1969) *Am. J. Epidemiol.*, **90**, 501.

Boulanger, P. A., Flamencourt, P., & Biserte, G. (1969) *Europ. J. Biochem.*, **10**, 116.

Boyer, G. S., Denny, F. W., & Ginsberg, H. S. (1959) *J. exp. Med.*, **110**, 827.
Brandon, F. B., & McLean, I. W., Jr. (1962) *Adv. Virus Res.*, **9**, 157.
Brandt, C. D., Kim, H. W., Vargosko, A. J., Jeffries, B. C., Arrobio, J. O., Rindge, B., Parrott, R. H., & Chanock, R. M. (1969) *Am. J. Epidemiol.*, **90**, 484.
Bullock, G. (1965) *J. Hyg., Camb.*, **63**, 383.
Burke, C. N., Luginbuhl., R. E., & Williams, L. F. (1968) *Avian Dis.*, **12**, 483.
Burmester, B. R., Sharpless, G. R., & Fontes, A. K. (1960) *J. natn. Cancer Inst.*, **24**, 1443.
Burnett, J. P., & Harrington, J. R. (1968a) *Proc. natn. Acad. Sci. U.S.A.*, **60**, 1023.
Burnett, J. P., & Harrington, J. A. (1968b) *Nature, Lond.*, **21**, 1245.
Butel, J. S., & Rapp. F. (1967) *Virology*, **31**, 573.
Cabasso, V. J. (1962) *Ann. N.Y. Acad. Sci.*, **101**, 489.
Cabasso, V. J., Stebbins, M. R., & Avampato, J. M. (1958) *Proc. Soc. exp Biol. Med., U.S.A.* **99**, 46.
Cabasso, V. J., Stebbins, M. R., Norton, T. W., & Cox, H. R. (1954) *Proc. Soc. exp. Biol. Med., U.S.A.*, **85**, 239.
Casto, B. C., & Hammon, W. McD. (1966) *Proc. Soc. exp. Biol. Med., U.S.A.*, **122**, 1216.
Chandler, R. L., Haig, D. A., & Smith, K. (1965) *Jl R. microsc. Soc.*, **84**, 133.
Chanock, R. M., Ludwig, W., Huebner, R. J., Cate, T. R., & Chu, L.-W. (1966) *J. Am. med. Ass.*, **195**, 445.
Chany, C., Lepine, P., Lelong, M., Le-Tan-Vinh, Satge, P., & Virat, J. (1958) *Am. J. Hyg.*, **67**, 367.
Clarke, M. C., Sharpe, H. B. A., & Derbyshire, J. B. (1967) *Arch. ges. Virusforsch.*, **21**, 91.
Cook, J. K. A. (1970) *Res. Vet. Sci.*, **11**, 343.
Couch, R. B., Cate, T. R., Fleet, W. F., Gerome, P. J., & Knight, V. (1966) *An. Rev. resp. Dis.*, **93**, 529.
Csontos, L. (1967) *Acta vet. hung.*, **17**, 217.
Dales, S. (1962) *J. Cell. Biol.*, **13**, 303.
Darbyshire, J. H. (1966) *Nature, Lond.*, **211**, 102.
Darbyshire, J. H., Dawson, P. S., Lamont, P. H., Ostler, D. C., & Pereira, H. G. (1965) *J. comp. Path.*, **75**, 327.
Darbyshire, J. H., Jennings, A. R., Dawson, P. S., Lamont, P. H., & Omar, A. R. (1966) *Res. Vet. Sci.*, **7**, 81.
Darbyshire, J. H., Kinch, D. A., & Jennings, A. R. (1969) *Res. vet. Sci.*, **10**, 39.
Darbyshire, J. H., & Pereira, H. G. (1964) *Nature, Lond.*, **201**, 895.
Davies, M. C., Englert, M. E., Stebbins, M. R., & Cabasso, V. J. (1961) *Virology*, **15**, 87.
Derbyshire, J. B., Chandler, R. L., & Smith, K. (1968) *Res. vet. Sci.*, **9**, 300.
Ditchfield, J., Macpherson, L. W., & Zbitnew, A. (1962) *Can. vet. J.*, **3**, 238.
Doerfler, W., & Kleinschmidt, A. K. (1970) *J. molec. Biol.*, **50**, 579.
DuBose, R. T., & Grumbles, L. C. (1959) *Avipa Dis.*, **3**, 321.
Dulac, G. C., Swango, L. J., & Burnstein, T. (1970) *Can. J. Microbiol.*, **16**, 391.
Dutta, S. K., & Pomeroy, B. S. (1963) *Proc. Soc. exp. Biol. Med., U.S.A.*, **114**, 539.
Engster, A. K., Kalter, S. S., Kim, C. S., & Pinkerton, M. E. (1969) *Arch. ges. Virusforsch.*, **26**, 260.
Epstein, M. A., Holt, S. J., & Powell, A. K. (1960) *Br. J. exp. Path.*, **41**, 559.
Espmark, J. A., & Salenstedt, C. R. (1961) *Arch. ges. Virusforsch.*, **11**, 61.
Fastier, L. B. (1957) *J. Immunol.*, **78**, 413.
Flanagan, J. F., & Ginsberg, H. (1964) *J. Bact.*, **87**, 977.
Fong, C. K. Y., Bensch, K. G., & Hsiung, G. D. (1968) *Virology*, **35**, 297.
Fong, C. K. Y., Bensch, K. G., Miller, L. R., & Hsiung, G. D. (1965) *J. Bact.*, **90**, 1786.
Fujinaga, K., & Green, M. (1967) *J. Virol.*, **1**, 576.
Fukumi, H., Nishikawa, F., Kurimoto, U., Inoue, H., Usui, J., & Hirayama, T. (1958) *Jap. J. med. Sci. Biol.*, **11**, 467.
Garg, S. P., Moulton, J. E., & Sekhri, K. K. (1967) *Am. J. vet. Res.*, **28**, 725.
Gilden, R. V., Kern, J., Beddow, T. G., & Huebner, R. J. (1967) *Virology*, **31**, 727.
Ginder, D. R. (1964) *J. exp. Med.*, **120**, 1117.
Ginsberg, H. S. (1956) *J. Immunol.*, **77**, 271.

Granoff, A. (1969) *Curr. Topics Microbiol., Immunol.*, **50**, 107.
Green, M., Piña, M., Kimes, R., Wensink, P. C., McHattie, L. A., & Thomas, C. A., Jr. (1967) *Proc. natn. Acad. Sci. U.S.A.*, **57**, 1302.
Green, M., & Piña, M. (1964) *Proc. natn. Acad. Sci. U.S.A.*, **51**, 1251.
Green, R. G., Ziegler, N. R., & Carlson, W. C. (1934) *Am. J. Hyg.*, **19**, 343.
Green, R. G., Ziegler, N. R., Green, B. B., & Dewey, E. T. (1930) *Am. J. Hyg.*, **12**, 109.
Guenov, I., Sartmadshiev, K., Schopov, I., Shlhabinkov, Z., & Fjodorov, W. (1917) *Zentbl. VetMed.*, **16B**, 1064.
Gutekunst, R. R., & Heggie, A. D. (1961) *New Engl. J. Med.*, **264**, 347.
Gutekunst, R. R., White, R. J., Edmondson, W. P., & Chanock, R. M., (1967) *Am. J. Epidemiol.*, **86**, 341.
Haig, D. A., Clarke, M. C., & Pereira, M. S. (1964) *J. comp. Path.*, **74**, 81.
Hammon, W. McD., Yohn, D. S., Casto, B. C., & Atchison, R. W. (1963) *J. natn. Cancer Inst.*, **31**, 329.
Hartley, J. W., & Rowe, W. P. (1960) *Virology*, **11**, 645.
Hashimoto, K., Sugiyama, T., & Sasaki, S. (1966) *Jap. J. Microbiol.*, **10**, 115.
Hayashi, K., & Russell, W. C. (1968) *Virology*, **34**, 470.
Hiatt, C. W., Kaufman, C., Helprin, J. J., & Baron, S. (1960) *J. Immunol.*, **84**, 480.
Hilleman, M. R. (1966) in *Viruses inducing cancer*. Ed.: Burdett. Salt Lake City: Univ. of Utah Press.
Hillis, W. D., & Goodman, R. (1969) *J. Immunol.*, **103**, 1089.
Hoggan, M. D., Rowe, W. P., Black, P. H., & Huebner, R. J. (1965) *Proc. natn. Acad. Sci. U.S.A.*, **53**, 12.
Horne, R. W., Brenner, S., Waterson, A. P., & Wildy, P. (1959) *J. molec. Biol.*, **1**, 89.
Horwitz, M. S., Scharff, M. D., & Maizel, J. V., Jr. (1969) *Virology*, **39**, 682.
Huebner, R. J. (1959) *Publ. Hlth Rep., Wash.*, **74**, 6.
Huebner, R. J. (1967) in *Perspectives in virology*. Ed.: Pollard. **5**, 147–66. New York: Academic Press.
Huebner, R. J., Rowe, W. P., & Chanock, R. M. (1958) *Ann. Rev. Microbiol.*, **12**, 49.
Huebner, R. J., Rowe, W. P., & Lane, W. T. (1962) *Proc. natn. Acad. Sci. U.S.A.*, **48**, 2051.
Huebner, R. J., Rowe, W. P., Turner, H. C., & Lane, W. T. (1963) *Proc. natn. Acad. Sci. U.S.A.*, **50**, 379.
Hull, R. N. (1968) *Virology monogr.*, **2**, 1. Vienna/New York: Springer-Verlag.
Hull, R. N., Johnson, I. S., Culbertson, C. G., Reimer, C. B., & Wright, H. F. (1965) *Science*, **150**, 1044.
Hull, R. N., & Minner, J. R. (1957) *Ann. N.Y. Acad. Sci.*, **67**, 413.
Hull. R. N., Minner, J. R., & Mascoli, C. C. (1958) *Am. J. Hyg.*, **68**, 31.
Hull, R. N., Minner, J. R., & Smith, J. W. (1956) *Am. J. Hyg.*, **63**, 204.
Inaba, Y., Tanaka, T., Sato, K., Ito, H., Ito, Y., Omori, T., & Matumoto, M. (1968) *Jap. J. Microbiol.*, **12**, 219.
Ishibashi, M. (1970) *Proc. natn. Acad. Sci. U.S.A.*, **65**, 304.
Jawetz, E. (1959) *Br. med. J.*, **1**, 873.
Jen, K. F., Tai, Y., Lin, Y. C., & Wang, H. Y. (1962) *Chin. med. J.*, **81**, 141.
Kalter, S. S., Kim, C. S., & Sueltenfuss, E. A. (1967b) *J. infect. Dis.*, **117**, 301.
Kalter, S. S., Ratner, J. A., Britton, H. A., Viu, T. E., Eugster, A. K., & Rodriguez, A. R. (1967a) *Nature, Lond.*, **213**, 610.
Kapp, P., & Lehoczki, Z. (1966), *Acta vet. hung.*, **16**, 429.
Kapsenberg, J. G. (1959) *Proc. Soc. exp. Biol. Med., U.S.A.*, **101**, 611.
Kasel, J. A., Huber, M., Loda, F., Banks, P. A., & Knight, V. (1964) *Proc. Soc. exp. Biol. Med., U.S.A.*, **117**, 186.
Kasel, J. A., Rowe, W. P., & Nemes, J. L. (1961) *J. exp. Med.*, **114**, 717.
Kasza, L. (1966) *Am. J. vet. Res.*, **27**, 751.
Kawamura, H., Shimizu, F., & Teuhanara, K. (1964) *Natn. Inst. Anim. Hlth Q., Tokyo*, **4**, 183.
Kim, S. S., Sueltenfuss, E. A., & Kalter, S. S. (1967) *J. infect. Dis.*, **117**, 292.
Kjellén, L. (1965) *Virology*, **27**, 580.

Kjellén, L., & Pereira, H. G. (1968) *J. gen. Virol.*, **2**, 177.
Klein, M., Earley, E., & Zellat, J. (1959) *Proc. Soc. exp. Biol. Med., U.S.A.*, **102**, 1.
Klein, M., Zellat, J., & Michaelson, T. C. (1960) *Proc. Soc. exp. Biol. Med., U.S.A.*, **105**, 340.
Koestner, A., Kasza, L., Kindig, O., Shadduck, J. A. (1968) *Am. J. Path.*, **53**, 651.
Kohler, H., & Apodaca, J. (1966) *Zentbl. Bakt. ParasitKde, I. Abt. Orig.*, **199**, 338.
Kohler, K., & Odaka, T. (1964) *Z. Naturforsch.*, **19b**, 331.
Kubinski, H., & Rose, J. A. (1967) *Proc. natn. Sci. U.S.A.*, **57**, 1720.
Lacy, S., & Green, M. (1964) *Proc. natn. Acad. Sci. U.S.A.*, **52**, 1053.
Lacy, S., & Green, M. (1965) *Science*, **150**, 1296.
Lacy, S., & Green, M. (1967) *J. gen. Virol.*, **1**, 413.
Larin, N. M. (1959) *Br. vet. J.*, **115**, 35.
Laver, W. G. (1970) *Virology*, **41**, 488.
Laver, W. G., Pereira, H. G., Russell, W. C., & Valentine, R. C. (1968) *J. molec. Biol.*, **37**, 379.
Laver, W. G., Suriano, J. R., & Green, M. (1967) *J. Virol.*, **1**, 723.
Leader, R. W., Pomerat, C. M., & Lefebre, C. G. (1960) *Virology*, **10**, 268.
Lewis, A. M., Jr., Levin, M. J., Wiese, W. H., Crumpacker, C. S., & Henry, P. H. (1969) *Proc. natn. Acad. Sci. U.S.A.*, **63**, 1128.
Low, B., & Pinnock, P. R. (1956) *J. biophys. biochem. Cytol.*, **2**, 483.
Lund, E. (1966) *Arch. ges. Virusforsch.*, **19**, 32.
McAllister, R. M., Nicolson, M. O., Reed, G., Kern, J., Gilden, R. V., & Huebner, R. J. (1969) *J. natn. Cancer Inst.*, **43**, 917.
McBride, W. D., & Wiener, A. (1964) *Proc. Soc. exp. Biol. Med., U.S.A.*, **115**, 870.
McFerran, J. B., Nelson, R., McCracken, J. M., & Ross, J. G. (1969) *Nature, Lond.*, **221**, 194.
McIntosh, K., Payne, S., & Russell, W. C. (1971) *J. gen. Virol.*, **10**, 251.
Macpherson, I., Wildy, P., Stoker, M. G. P., & Horne, A. W. (1961) *Virology*, **13**, 146.
Mahnel, H., & Bibrack, B. (1966) *Zentbl. Bakt. ParasitKde, I. Abt. Orig.*, **199**, 329.
Mahnel, H., & Sigel, G. (1968) *Zentbl. Bakt. ParasitKde, I. Abt. Orig.*, **206**, 149.
Maizel, J. V., Jr., White, D. O., & Scharff, M. D. (1968) *Virology*, **36**, 115 and 126.
Malherbe, H., & Harwin, R. (1957) *Br. J. exp. Path.*, **38**, 539.
Malherbe, H., & Harwin, R. (1963) *S. Afr. J. med. Sci.*, **37**, 407.
Mantyjarvi, R., & Russell, W. C. (1969) *J. gen. Virol.*, **5**, 339.
Mart, A., Connor, J. D., & Sigel, M. M. (1968) *J. natn. Cancer Inst.*, **40**, 243.
Martinez-Palomo, A., LeBuis, J., & Bernhard, W. (1967) *J. Virol.*, **1**, 817.
Marusyk, R. G., Norrby, E., & Lindquist, U. (1970) *J. Virol.*, **5**, 507.
Matsui, K., & Bernhard, W. (1967) *Annls Inst. Pasteur, Paris*, **112**, 773.
Matumoto, M., Inaba, Y., Tanaka, Y., Sato, K., Ito, H., & Omori, T. (1969) *Jap. J. Microbiol.*, **13**, 131.
Mayr, A., Bibrack, B., & Bachmann, P. (1967) *Zentbl. Bakt. ParasitKde, I. Abt. Orig.*, **203**, 59.
Miller, W. S., & Artenstein, M. S. (1967) *Proc. Soc. exp. Biol. Med., U.S.A.*, **125**, 222.
Missal, O. (1969) *Arch. Exp. vet. Med.*, 23, 783.
Mohanty, S. B., & Lillie, M. G. (1965) *Proc. Soc. exp. Biol. Med., U.S.A.*, **120**, 679.
Monreal, G. (1968) *Zentbl. VetMed.*, **15b**, 685, 717.
Morgan, C., Howe, C., Rose, H. M., & Moore, D. H. (1956) *J. biophys. biochem. Cytol.*, **2**, 351.
Morgan, C., Rosenkranz, H. S., & Mednis, B. (1969) *J. Virol.*, **4**, 777.
Morrison, J. M., Keir, H. M., Subak-Sharpe, H., & Crawford, L. V. (1967) *J. gen. Virol.*, **1**, 101.
Naegele, R. F., & Rapp, F. (1967) *J. Virol.*, **1**, 838.
Nász, I., Lengyel, A., Dan, P., & Kalcsar, G. (1962) *Acta microbiol. hung.*, **9**, 69.
Neurath, A. R., Hartzell, R. W., & Rubin, B. A. (1969) *Nature, Lond.*, **221**, 1069.
Norrby, E. (1966) *Virology*, **29**, 236.
Norrby, E. (1968) *Curr. Topics in Microbiol. & Immunol.*, **43**, 1.
Norrby, E. (1969a) *J. gen. Virol.*, **5**, 221.

Norrby, E. (1969b) *J. Virol.*, **4**, 657.
Numazaki, Y., Shigeta, S., Kumusaka, T., Miyazawa, T., Yamanaka, M., Nano, N., Takai, S., & Ishida, N. (1968) *New Engl. J. Med.*, **278**, 700.
Olson, N. O. (1950) *Proc. 54th Ann. Meeting U.S. Livestock Sant. Ass.*, p. 171.
Omura, S., & Mutai, M. (1966) *Biken's J.*, **9**, 231.
Oxford, J. S., & Potter, C. W. (1970) *J. gen. Virol.*, **8**, 33.
Pereira, H. G., Allison, A. C., & Niven, J. S. F. (1963a) *Nature, Lond.*, **196**, 244.
Pereira, H. G., Huebner, R. J., Ginsberg, H. S., & van der Veen, J. (1963b) *Virology*, **20**, 613.
Pereira, H. G., & Kelly, B. (1957a) *Nature, Lond.*, **180**, 615.
Pereira, H. G., & Kelly, B. (1957b) *Proc. r. Soc. Med.*, **50**, 754.
Pereira, H. G., & Skehel, J. J. (1970) *J. gen. Virol.*, **12**, 13.
Petek, M., Felluga, B., & Zoletto, R. (1963) *Avian Dis.*, **7**, 38.
Pettersson, U., Philipson, L., & Hoglund, S. (1967) *Virology*, **33**, 575.
Pettersson, U., Philipson, L., & Hoglund, S. (1968) *Virology*, **35**, 204.
Piercy, S. E., & Sellers, R. F. (1960) *Res. vet. Sci.*, **1**, 84.
Piña, M., & Green, M. (1965) *Proc. natn. Acad. Sci. U.S.A.*, **43**, 547.
Piña, M., & Green, M. (1968) *Virology*, **36**, 321.
Polasa, H., & Green, M. (1965) *Virology*, **25**, 68.
Polasa, H., & Green, M. (1967) *Virology*, **31**, 565.
Prier, J. E., & Le Beau, R. W. (1958) *J. Lab. clin. Med.*, **41**, 495.
Rabson, A. S., O'Connor, G. T., Berezesky, I. K., & Paul, F. J. (1964) *Proc. Soc. exp. Biol. Med., U.S.A.*, **116**, 187.
Rafajko, R. R. (1964) *Am. J. Hyg.*, **79**, 310.
Rapp, F., & Melnick, J. L. (1966) *Progr. med. Virol.*, **8**, 349.
Rapp, F., Melnick, J. L., Butel, J. S., & Kitahara, T. (1964) *Proc. natn. Acad. Sci. U.S.A.*, **52**, 1348.
Rapoza, N. P. (1968) *Am. J. Epidemiol.*, **86**, 736.
Rasmussen, P. G. (1969) *Acta vet. scand.*, **10**, 10.
Riggs, J. C., Takemori, N., & Lennette, E. H. (1968) *J. Immunol.*, **100**, 384.
Rondhuis, P. R. (1968) *Arch. ges. Virusforsch.*, **25**, 235.
Rose, J. A., Reich, P. R., & Weissman, S. M. (1965) *Virology*, **27**, 571.
Rosen, L. (1958) *Virology*, **5**, 574.
Rosen, L. (1960) *Am. J. Hyg.*, **71**, 120.
Rouse, H. C., Bonifas, V. H., & Schlesinger, R. W. (1963) *Virology*, **20**, 357.
Rouse, H. C., & Schlesinger, R. W. (1967) *Virology*, **33**, 513.
Rowe, W. P., & Baum, S. G. (1964) *Proc. natn. Acad. Sci. U.S.A.*, **52**, 1340.
Rowe, W. P., Hartley, J. W., & Huebner, R. J. (1958) *Proc. Soc. exp. Biol. Med., U.S.A.*, **97**, 465.
Rowe, W. P., Huebner, R. J., Hartley, J. W., Ward, T. G., & Parrott, R. H. (1955) *Am. J. Hyg.*, **61**, 197.
Rowe, W. P., & Pugh, W. E. (1966) *Proc. natn. Acad. Sci. U.S.A.*, **55**, 1126.
Rubarth, S. (1947) *Acta path. microbiol. scand.* (suppl.), **69**, 222.
Russell, W. C., & Becker, Y. (1968) *Virology*, **35**, 18.
Russell, W. C., Hayashi, K., Sanderson, P. J., & Pereira, H. G. (1967) *J. gen. Virol.*, **1**, 495.
Russell, W. C., & Knight, B. E. (1967) *J. gen. Virol.*, **1**, 523.
Russell, W. C., Laver, W. G., & Sanderson, P. J. (1968) *Nature, Lond.*, **219**, 1127.
Russell, W. C., McIntosh, & Skehel, J. J. (1971) *J. gen. Virol.*, **11**, 35.
Salenstedt, C. F. (1958) *Arch. ges. Virusforsch.*, **8**, 600.
Sarma, P. S., Huebner, R. J., & Lane, W. T. (1965) *Science*, **149**, 1108.
Sarma, P. S., Vass, W., Huebner, R. J., Igel, H., Lane, W. T., & Turner, H. C. (1967) *Nature, Lond.*, **215**, 293.
Schlesinger, R. W. (1969) *Adv. Virus Res.*, **14**, 1–61.
Schmidt, N. J., King, C. J., & Lennette, E. H. (1965) *Proc. exp. Biol. Med., U.S.A.*, **118**, 208.
Schmitt-Ruppin, K. H. (1968) *Arch. ges. Virusforsch.*, **25**, 126.

Schpov, I., Genov, I., & Fjodorov, W. (1968) *Arch. Exp. vet. Med.*, **22**, 487.
Selivanov, A. A., Pleshanova, R. A., Skriabina, E. A., & Smorodinstev, A. A. (1964) *Acta virol., Prague*, **8**, 263.
Sharpe, H. B. (1967) *J. Path. Bact.*, **93**, 353.
Sharpe, H. B., & Jessett, D. M. (1967) *J. comp. Path.*, **67**, 45.
Sharpless, G. R. (1962) *Ann. N.Y. Acad. Sci.*, **101**, 513.
Sharpless, G. R., & Jungherr, E. L. (1961) *Am. J. vet. Res.*, **22**, 986.
Sharpless, G. R., Levine, S., Davies, M. C., & Englert, M. E. (1961) *Virology*, **13**, 315.
Sharon, N., & Pollard, M. (1964) *Nature, Lond.*, **202**, 1139.
Sinha, S. K., Fleming, L. W., & Scholes, S. (1960) *J. Am. vet. med. Ass.*, **136**, 481.
Smith, K. O., Gehle, W. D., & Kniker, W. T. (1970) *J. Immunol.*, **105**, 1036.
Sohier, R., Chardonnet, Y., & Prunieras, M. (1965) *Progr. med. Virol.*, **7**, 253.
Stoker, M. G. P. (1959) *Virology*, **8**, 250.
Sugiyama, T., Hashimoto, K., & Sasaki, S. (1967) *Jap. J. Microbiol.*, **11**, 33.
Tanaka, Y., Inaba, Y., Ito, Y., Omori, T., & Matumoto, M. (1968) *Jap. J. Microbiol.*, **12**, 77.
Teng, C. H. (1960) *Chin. med., J.*, **80**, 331.
Thomas, D. C., & Green, M. (1966) *Proc. natn. Acad. Sci. U.S.A.*, **56**, 243.
Todd, J. D. (1969) *J. Am. vet. Med. Ass.*, **155**, 387.
Tyrrell, D. A. J., Buckland, F. E., Lancaster, M. C., & Valentine, R. C. (1960) *Br. J. exp. Path.*, **41**, 610.
Valentine, R. C. (1960) *Fourth internat. Congr. Electron Microscopy*, 577.
Valentine, R. C., & Pereira, H. G. (1965) *J. molec. Biol.*, **13**, 13.
Velicer, L. F., & Ginsberg, H. S. (1969) *Proc. natn. Acad. Sci. U.S.A.*, **61**, 1264.
Wadell, G. (1970) *Structural and biological properties of capsid components of human adenoviruses.* Stockholm: David Brobey.
Wallis, C., Yang, C. S., & Melnick, J. L. (1962) *J. Immunol.*, **89**, 41.
Wasserman, F. E. (1962) *Virology*, **11**, 335.
Wigand, R., Bauer, H., Lang, F., & Adam, W. (1965) *Arch. ges. Virusforsch.*, **15**, 188.
Wigand, R., & Fliedner, D. (1968) *Arch. ges. Virusforsch.*, **24**, 245.
Wilcox, G. E. (1969) *Aust. vet. J.*, **45**, 265.
Wilcox, W. C., & Ginsberg, H. S. (1963a) *J. exp. Med.*, **118**, 295.
Wilcox, W. C., & Ginsberg, H. S. (1963b) *Virology*, **20**, 269.
Wilcox, W. C., Ginsberg, H. S., & Anderson, T. F. (1963) *J. exp. Med.*, **118**, 307.
Yates, V. J., & Fry, D. E. (1957) *Am. J. vet. Res.*, **18**, 657.
Yamamoto, T. (1969) *J. gen. Virol.*, **4**, 397.

15

Herpesvirus

Reviews: Plummer (1967). Burrows (1970). Roizman (1969).

Members of the Herpesvirus group are relatively large ether-sensitive DNA viruses. Nuclear inclusions of Cowdry's (1934) type A are characteristically present in infected cells. The herpes simplex virus and some related to it are readily released from infected cells. Others, e.g. varicella and malignant catarrh, are so closely associated with cells that infectivity of cell-free preparations is hard to demonstrate.

A suggestion that the genus should be subdivided on this basis is hard to support, since the distinction is not a clear one; nor are the cytomegaloviruses sharply separable from other herpesviruses (Plummer, 1967).

The observations on which the following account is based have mainly been made on herpes simplex virus, but there is every likelihood that they apply to the whole genus, perhaps with minor differences. Some differences may be associated with growth in cells of various kinds.

Morphology. The complete virus (Wildy et al., 1960) when negatively stained is seen to have an icosahedral core with 162 capsomeres on its surface. These are hollow prisms $12{\cdot}5 \times 9{\cdot}5$ nm in diameter. Outside the capsid is a membrane appearing up to 180 nm across. Particles may be found with and without nucleoids and with and without membranes (Watson et al., 1963). There is dispute as to whether naked particles are infectious. They are certainly much less so than those with envelopes. The nonenveloped particles are said to have 3 shells round the core: with envelopes they acquire a further 2 shells (Roizman et al., 1969). Some strains, e.g. equine rhinopneumonitis (Reczko & Mayr, 1963) and canine tracheobronchitis (Strandberg & Carmichael, 1965), show nucleoids shaped like crosses or stars. Aberrant forms, including tubules bearing capsomeres on their outer surfaces have been described especially in herpesvirus from amphibians (Stackpole & Mizel, 1968). Also dense, homogeneous spherical bodies 300–500 nm within vacuoles or larger bodies 2–2·8 μm in diameter have been observed in cells infected with cytomegaloviruses (Luse & Smith, 1958).

Morgan et al. (1968) have described the processes concerned in the entry of virions into the cell and the early events which follow. Development begins in the nucleus, where at an early stage, dense particles 30–40 nm across appear. They then acquire single membranes and a diameter of 70–100 nm. Later, particles with double membranes penetrate the nuclear membrane and are found in the cytoplasm; these have now a diameter of 120–130 nm and as mature particles leave the cell altogether (Morgan et al., 1954). Apparently they leave the nucleus by passing along preformed tubules (Schwartz & Roizman, 1969). Virus infection of adjacent cells may take place by direct transfer. The different elements added to the virus and identified by immunofluorescence are apparently formed in different compartments of nucleus or cytoplasm (Roizman et al., 1967). Release of virus from cells is probably gradual without involving cell destruction.

Chemical composition. Virions contain double-stranded DNA with MW in the range of 60×10^6 to 90×10^6 in the majority of strains although lower values (e.g. 32×10^6 for Allerton virus and 54×10^6 for infectious bovine rhinotracheitis) have been recorded. Comparative estimations of the MW, base composition and buoyant density of the DNA of different members of the group (Russell & Crawford, 1964; Plummer et al., 1969a, b; Goodheart, 1970) or individual strains (Soehner et al., 1965—Equine herpes 1; Aurelian, 1969—canine tracheobronchitis; Lee et al., 1970—Marek's disease virus; Weinberg & Becker, 1969—EB virus; Crawford & Lee, 1964—Cytomegalovirus). GC contents vary from values as low as 45 to 50 per cent (avian laryngotracheitis, feline rhinotracheitis, cercopithecus virus) to as high as 68 to 74 per cent (herpes simplex, pseudorabies and infectious bovine rhinotracheitis) and buoyant densities, from 1·692 to 1·731 g/ml. DNA makes up about 10 per cent of particle weight. At least 25 proteins are found in the cytoplasm of cells infected by herpes simplex. Some of these enter the nucleus (Spear & Roizman, 1968); others are glycoproteins which may become bound to cytoplasmic membranes and be found in the envelope of partly purified virus. An alkaline DNA-exonuclease (Morrison & Keir, 1968) and a thymidine kinase (Klemperer et al., 1967) have been recognised in herpesvirus infected cells. Infectious virus is not formed in the absence of arginine (Becker et al., 1967). Asher et al. (1969) studied the incorporation of lipids into virus particles.

Physico-chemical characters. The density is estimated at 1·27–1·29, and the iso-electric point at pH 7·2–7·6. The virus is fairly heat-labile, being inactivated in 30′ at 50–52°C, in 50–80 hours at 41·5°, but it withstood 90°C for 30′ when dried (Holden, 1932). According to Scott et al. (1961) several biologically different strains were alike in having a half-life of $1\frac{1}{2}$ hours at 37°C, $3\frac{3}{4}$ hours at 30–31°C. Heat-stability at 50° is stabilized by molar Na_2SO_4 and Na_2HPO_4, but not, as with picornaviruses by $MgCl_2$ (Wallis & Melnick, 1965). Inactivated by phosphatases (Amos, 1953). Inactivation by various other chemicals is described by Holden (1932). Differences in stability of various herpesviruses are summarized by Plummer (1967). Plummer et al. (1969b) have compared the densities of the nucleic acids of 14 herpesviruses: these range from 1·692 to 1·731.

Antigenic properties. Specific complement fixation and neutralization are readily shown. A plaque assay is commonly applicable in tissue culture. There is a soluble antigen separate from the virus particles (Hayward, 1949); this is largely destroyed in 1 hour at 56°C, but a little survives boiling for 5′. Virus-treated tanned sheep RBCs are agglutinated by specific antisera. The reaction is best carried out at 25°C and a pH of 6·4 (Scott et al., 1957).

Sabin (1934) found cross-reactions among the trio herpes simplex, pseudorabies and B virus. Watson et al. (1967) confirmed this as regards herpes simplex and B, but found pseudorabies related only by virtue of a common antigen, evident in gel-diffusion tests. Crossing has been reported between H. simplex and canine herpesvirus. Cross-reactions between infectious bovine rhinotracheitis and equine rhinopneumonitis are reported, and even minor ones between varicella and herpes simplex.

Cultivation. The viruses all grow in tissue culture and some of them also do so in fertile eggs. Those most closely related to herpes simplex produce pocks on chorioallantoic membranes.

Pathogenicity. Lesions produced are at most briefly proliferative, soon necrotic. Many members grow in the central nervous system of their hosts or of experimental animals and some travel along nerves to reach the CNS from a peripheral lesion. The characteristic intranuclear inclusions of Cowdry's type A (1934) are formed by all the agents included here.

HERPES SIMPLEX

Synonyms: Herpes febrilis, genitalis, labialis, etc. Fever blisters. *Herpesvirus hominis.*

Review: Kaplan (1969).

Antigenic properties. Watson et al. (1966) observed 12 lines of precipitation in gel-diffusion tests while Tokumaru (1970) found evidence that there were as many as 18 antigens, some of them very labile. Several workers have reported serological differences among strains when various tests were used. It is now apparent that there are 2 main types (Schneweis, 1962; Plummer et al., 1970) and these are related to the site of origin, whether oral or genital (Dowdle et al., 1967). Most strains from the nasal and oral regions belong to type 1, those from the ano-genital region to type 2. Roizman et al. (1970), however, argue that the differences are not stable and that intermediate strains occur. Ashe & Notkins (1966) report that infectious virus-antibody complexes occur and that these can be inactivated by anti-γ-globulins. McKenna et al. (1966) described an early antigen (neo-antigen) and such has also been reported by Sabin (1968). An antigen has been prepared giving specific skin tests in susceptible persons (Nagler, 1944; 1946). Studies on human sera reveal the curious fact that those susceptible to recurrent herpes have good titres of neutralizing and CF antibody while those not subject to it have no antibodies; attacks of recurrent herpes do not usually affect the titre, while primary attacks lead to a rise.

Cultivation. *In fertile eggs*. Herpes virus will grow after inoculation by all the usual routes. On the CAM plaques are considerably smaller than those of vaccinia—only 1–2 mm across. Examination is best made 36–48 hours after incubation at 36°C. Growth also occurs in de-embryonated eggs and on the blastoderm of 1-day eggs (Yoshino, 1956).

In tissue culture. Cultivation is reported in many types of cell—rabbit, chick embryo and human (HeLa, amnion and others). Cytopathic effects include inclusion body and giant-cell formation, later cell destruction. 35°C seems to be the optimal temperature for growth. In monolayers, countable plaques appear and variants producing macro- and micro-plaques are described (Hoggan & Roizman, 1959a). There have been several careful studies of various stages of growth—adsorption, eclipse multiplication, release (Kaplan, 1957; Stoker & Ross, 1958; Hoggan & Roizman, 1959b). An early effect in HeLa cells is inhibition of mitosis (Stoker & Newton, 1959).

Nahmias & Dowdle (1968) found that freshly isolated genital strains (type 2) produced larger plaques in culture than did type 1 strains: they observed other differences between the strains in tissue cultures and on the CAM of eggs. Blackmore & Morgan (1967) and Nii (1969) describe the growth of the virus, sometimes cyclical, in chronically infected L cells.

Pathogenicity. In man, primary infection is mainly in young children; it may be subclinical or may take the form of acute stomatitis. Many people, including such children, although developing antibodies are subject to recurrent herpes for much of their lives. Such latent infection may be activated by various stimuli such as certain infections, fever, menstruation (cf. Doerr, 1924–25). Primary infection may, however, occur later and reactivation after immunosuppressive measures. Very rarely the virus causes acute hepatitis in man. It has also been associated with upper respiratory tract infections in children. The commonest sites for 'fever-blisters' are the external nares and lips. The lesions, at first vesicular, soon scabbing, heal within a few days without a scar. There are other forms of the disease; Kaposi's varicelliform eruption (eczema herpeticum) is a primary infection especially in children, in which there is widespread herpetic infection of eczematous skin. 'Traumatic herpes' may affect burns and other superficial lesions. Herpetic kerato-conjunctivitis may lead to serious corneal opacity. A generalized fatal infection may occur in new-born or undernourished children, and a herpetic meningo-encephalitis in children or adults may be fatal.

Experimental. Man-to-man transmission has frequently been carried out; it is more often successful in people subject to herpes.

Rabbits are very susceptible to infection by inoculating cornea, skin, brain or testis. Neurotropic strains travel along nerves to the CNS from cornea or other peripheral sites and cause convulsions and death from encephalitis. Some strains produce necrotic foci in adrenals. A latent infection in rabbits may be activated by severe allergic reactions or injection of adrenalin. Guinea-pigs, mice, particularly suckling mice, monkeys of several genera and other species, including even tortoises, are susceptible. Some strains cause myelitis in mice after IP or ID injection, but the virus is readily modified in the laboratory to produce different effects in different species. Spontaneous infections, sometimes fatal, have been reported in captive gibbons, owl monkeys (*Aotus*) and skunks.

Pathological lesions. Proliferation, then ballooning degeneration of epithelial cells, precede vesicle formation in skin; numerous giant cells may occur and contain many nuclear inclusions. These type A intranuclear inclusions contain little or no virus and represent a scar—a sequel to virus activity (Lebrun, 1956) (cf. p. 331). At an early stage tissues stained with haematoxylin and eosin show nuclei filled with homogeneous purple-staining material. Later basophilic material goes to the periphery and eosinophilic matter to form the central inclusion. The inclusions are found in almost all affected tissues. Recombination of strains with different genetic markers has been described by Wildy (1955). A virus component, described as a toxin, may act as a pyrogen when given IV to rabbits; this or other substances give rise to syncytia in cultures (Tokumaru, 1968).

Important recent work indicates a possible relationship between infection with type 2 virus and cervical cancer in women (Naib et al., 1966; Rawls et al., 1968). Antibodies to this virus were much higher (79 per cent) in the cancer patients than in matched controls. Moreover, many women with genital herpes were found to have premalignant or malignant cervical lesions.

Finally exfoliated cells from patients with cervical cancer contain antigens related to herpes subtype 2 (Royston & Aurelian, 1970); and viruses of this subtype can transform hamster cells in culture (Duff & Rapp, 1971).

Ecology. Recurrent herpes and associated high antibody levels occur more frequently at lower social-economic levels. Young children are more susceptible than are adults; so these are often infected from recurrent lesions in a mother, to acquire themselves a persistent latent infection (Burnet & Williams, 1939). Epidemics amongst children may occur.

Control. Vaccination of man has been attempted without success. Some immunity may be induced experimentally in rodents.

Kaufman (1962) has been successful in treating rabbits with herpetic kerato-conjunctivitis by means of 5-iodo-2′-deoxyuridine and there are encouraging reports of its use in similar conditions in man; some related compounds show similar activity.

B VIRUS OF MONKEYS

Synonym: *Herpesvirus simiae.*

Antigenic properties. Neutralization tests reveal a group relationship to herpes simplex and pseudorabies. Immune B sera neutralize herpes simplex very well, but the contrary is not true; virus B may be antigenically more complex. Not all B strains are antigenically identical (Prier & Goulet, 1961). Search for evidence of active immunity in cross-protection tests showed little if any between *H. hominis* and *H. simiae*, but a definite small amount between *H. simiae* and pseudorabies (Sabin, 1934). One of the recorded fatal infections in man was in someone previously known to have antibodies to herpes (Nagler & Klotz, 1958).

Of newly caught rhesus monkeys, 10 per cent have antibodies; the percentage rises to 60–70 per cent when they are confined in 'gang-cages' (Hull & Nash, 1960).

Cultivation. *In fertile eggs.* The virus has been serially passed 30 times on chorioallantoic membranes (Burnet et al., 1939).

In tissue culture. It grows readily in cultures of monkey, human and rabbit tissues and has been frequently isolated from cultures of apparently normal rhesus kidneys. The cell-to-cell spread of virus within cultures was studied by Black & Melnick (1955).

Distribution. A natural herpes-like infection of normal Asiatic monkeys—*Macaca mulatta* (rhesus), *irus* (cynomolgus), *cyclopis* and *fuscata*—(Endo et al., 1960).

Pathogenicity. In monkeys (*Macaca* spp.), natural infection, occurring especially in monkeys crowded together, leads to vesicular lesions on tongue and lips, sometimes skin. Lesions in CNS may be found histologically (Keeble et al., 1958).

Infection may occur in human beings 10 to 20 days after a monkey bite, or even in monkey handlers or people working with monkey-tissue cultures in absence of history of bite. There may be local inflammation at the site of a bite, followed by ascending myelitis (Sabin, 1949), or there may be acute encephalitis or encephalomyelitis. In one reported case there was early infection of the respiratory tract (Love & Jungherr, 1962). Almost all cases have been fatal; there has been much residual disability in the few survivors (Breen et al., 1958).

Experimentally, cortisone-treated monkeys have been infected intraspinally. Rabbits are readily infected by various routes and usually die with encephalitis or encephalomyelitis. Skin lesions are more necrotic than those of herpes simplex and may not appear for 10–12 days. The first symptoms of CNS involvement may be severe pruritus. Virus probably reaches the CNS along nerves (Sabin, 1934). Mice under 3 weeks old can be infected IC, also day-old chicks. The virus is irregularly fatal for guinea-pigs.

Pathological lesions. Besides local necrotic lesions in the skin of rabbits and inflammatory and destructive lesions in brain and cord, there may be necrotic foci in liver, spleen and adrenals. Typical type-A nuclear inclusions may be found whenever the virus is active in any of its hosts.

Ecology. The virus spreads by direct contact between monkeys especially in captivity. Incidence of antibodies under various conditions has already been mentioned. The virus has not been found in stools or urine of infected monkeys.

Control. Precautions are possible to avoid infection of monkey handlers. Monkeys are best caged by 1s or 2s, not in gangs, and quarantined before use in experiments. Really safe gloves are difficult to devise and some workers only inoculate anaesthetized animals.

Attempts have been made to produce formolized vaccines for those working with monkeys; their efficiency is unproved. Some immunity can be produced in rabbits (Benda, 1966).

OTHER SIMIAN HERPESVIRUSES

Marmoset Virus

Review: Hunt & Melendez (1969).

Synonyms: Herpes-T. (Tamarindus). *Herpesvirus platyrhinae.*

A herpes-like virus has been repeatedly isolated from throat-swabs and autopsy material of marmosets (Holmes et al., 1964; Melnick et al., 1964). It multiplied in marmoset, rabbit and human cells, poorly in rhesus kidney. Pocks were produced on the CAM. It was pathogenic for rabbits and mice IC and for suckling mice and hamsters IP. Herpes-like intranuclear inclusions appeared in them. Daniel & Melendez (1970) found that it was only a large-plaque variant which was pathogenic for these species. In contrast to reports from the above-mentioned workers, Tischendorf (1969) found some cross-protection between this and herpes simplex virus. The virus has been recovered from owl-monkeys (*Aotus*) and squirrel monkeys (*Saimiri*) and one or other of these may be the natural hosts. The squirrel monkeys had labial and oral lesions.

Patas Virus

McCarthy et al. (1968) recovered a virus from patas (*Erythrocebus*) monkeys. It caused rashes and some deaths in these monkeys. It was cultivated in various primate, but not other, cells, giving rise to inclusion bodies and giant cells.

Viruses from Cercopithecus and Papio

Malherbe & Harwin (1958) recovered a virus (SA 8) from kidneys of vervets (*Cercopithecus aethiops*), and later it was obtained from a baboon (*Papio ursinus*). Its GC content was 51 per cent. Density of RNA 1·710.

A second virus (SA 15) from a baboon was probably a herpesvirus (Malherbe & Strickland-Cholmley, 1969; 1970). SA 8 was serologically related to herpes simplex; unlike B virus it was fatal only to very young rabbits. A virus causing severe exanthematous disease in captive *C. aethiops* (Clarkson et al., 1967) is distinct from SA 8; *Cercopithecus* may not have been the natural host.

Herpesvirus Saimiri

Besides the marmoset virus, another has been isolated from squirrel monkeys (*Saimiri sciureus*). GC content 67 per cent. Density of RNA 1·726. This virus is of great interest, as it has produced malignant lymphomata in all inoculated marmosets or owl monkeys; they die in 13–48 days after inoculation (Hunt et al., 1970). Lymphocytosis and reticular-cell invasion of various viscera were observed.

VIRUS III OF RABBITS

Synonym: *Herpesvirus cuniculi.*

This virus is so-called as the discoverers (Rivers & Tillett, 1923, 1924) mainly studied the third strain they isolated. It exists as a latent infection of some stocks of rabbits.

Antigenic properties. Recovered rabbits are solidly immune and their sera contain neutralizing antibodies. Such antibodies occur in sera of many of the rabbits of stocks in which the virus is endemic. Complement-fixing antibodies also are described. Partly immune animals show an allergic response.

Cultivation. The virus has not been grown in eggs, but multiplies in tissue culture of rabbit testis, producing nuclear inclusions (Andrewes, 1929—this was perhaps the earliest report of production of a specific virus change in tissue culture). Growth occurs also in other rabbit tissues (Ivanovics & Hyde, 1936) and poorly in kidney tissues of grivet monkeys (Nesburn, 1969).

Distribution. It has been found in some but not all stocks of rabbits in the United States, Britain and Switzerland, but isolation was not reported for many years. Nesburn (1969), however, recovered what is almost certainly the same virus in Boston.

Pathogenicity. Virus III is not known to produce any natural disease in rabbits, being only activated after serial ‘blind passages’ of tissues. Once exalted in this way it produces lesions after inoculation by various routes. Given intradermally it gives rise to slightly raised erythematous lesions; intratesticular injections produce acute orchitis and fever; intrathoracic inoculation leads to pericarditis (Miller et al., 1924) and IC injection to encephalitis (Rivers & Stewart, 1928). In all these tissues the acute inflammatory lesions are associated with type A intranuclear inclusions, those in the testis being largely in the interstitial tissues. Cowdry (1930) has reported on minor differences between the inclusions of Virus III and herpes. While only persisting for a few days after inoculation into normal rabbits, the virus may be carried for some time in transplanted tumours (Rivers & Pearce, 1925; Andrewes, 1940). Species other than rabbits are insusceptible.

PSEUDORABIES

Synonyms: Aujeszky's disease. Mad itch. Infectious bulbar paralysis. *Herpesvirus suis.*

Review: Kaplan (1969).

Chemical composition and physical characters. GC content 74 per cent. Density of RNA 1·731 and MW 68×10^6 (Russell & Crawford, 1964).

Antigenic properties. Antisera have usually been prepared in pigs since the virus is so lethal to most other species and inactivated virus does not readily immunize. Neutralizing antibodies have been tested for in guinea-pigs rather than in the very sensitive rabbit. Tissue-culture techniques are, however, now available. Sabin (1934) found evidence of a group antigen common to this and herpes. There is also evidence suggesting that monkeys with antibodies to B virus may be resistant to pseudorabies (Hurst, 1936).

All strains, including the American Mad Itch, seem alike antigenically.

Cultivation. In fertile eggs, the virus grows on the chorioallantoic membrane producing white plaques (Bang, 1942) or a haemorrhagic reaction and death of the embryo by the fourth day (Glover, 1939). It can also be cultivated in the yolk sac.

The virus can be grown in cultures of chick, rabbit, guinea-pig, pig and dog tissue and doubtless those of other species. It causes cell-destruction, and plaques are produced in monolayers of pig-kidney or chick-embryo cells. Less virulent strains are reported to produce larger plaques. Of 2 strains cultivated by Tokumaru (1957) one caused rounding-up of cells, the other giant-cell formation.

Intranuclear inclusions are found both in infected eggs and in tissue cultures.

Distribution. The virus is prevalent in most of Europe, in South Africa, North and South America; it is rare in Britain, but sporadic in Northern Ireland.

Pathogenicity. Cattle, sheep, pigs, dogs, cats, foxes and mink are naturally affected. In pigs the infection is commonly inapparent, but at times 5 to 10 per cent may have nervous symptoms with fever, convulsions and paralysis; even so recovery is the rule. Sabo et al. (1968a; b) have followed the course of infection from primary lesions in tonsils in cats and piglets. In cattle, sheep and carnivores the disease is usually fatal, the predominant symptom being intense pruritus causing the animal to gnaw or scratch part of the body—usually head or hind quarters—until great tissue destruction is caused. Other symptoms of encephalomyelitis—violent excitement, fits and paralysis—precede death, which usually occurs within a few days. Horses are doubtfully susceptible. There are 3 reports of infection of laboratory workers: in the best authenticated instance, local pruritus occurred, there were aphthae in the mouth and virus was recovered from the blood (Aksel & Tuncman, 1940).

Experimentally, the disease can be reproduced in many species of several orders. Rabbits are extremely susceptible to infection by various routes. Intracerebral injection leads to fatal meningitis or encephalitis without pruritus, but after peripheral inoculation, intense pruritus is the rule, death following very soon. Guinea-pigs are slightly more resistant, rats and mice even more so; chicks and ducks can also be infected. Virus spreads centripetally from primary peripheral lesions but spread by blood-stream may occur also.

Pathological lesions. In all species pruritus and consequent self-mutilation result in severe destructive oedematous lesions. The pruritus is apparently due to lesions of spinal ganglia and posterior horn cells in the cord. Here, and elsewhere in the CNS, A-type intranuclear inclusions are found (Hurst, 1933; 1936). Focal necrotic lesions occur in cerebrum, cerebellum, adrenals and other viscera in rabbits, and in this species severe pulmonary oedema may be a cause of death (Shope, 1931).

Ecology. The disease does not spread amongst infected sheep or cattle; these often contract infection from pigs with inapparent infections, having virus in their nasal secretions (Shope, 1935a). Pigs in their turn may contract infection through eating carcases of infected rats (Shope, 1935b). It is suggested that mice might act as reservoirs (Fraser & Ramachandran, 1969). Animals can certainly be infected by eating infected material. Possibly cattle may pick up infection directly from rats. Most workers have failed to recover virus from urine; some have succeeded, especially with pigs.

Control. Attempts to make attenuated or inactivated vaccines have given little encouragement. One promising report is from Beladi & Ivanovics (1954), who inactivated virus grown in chick embryos with UV irradiation and were able to immunize guinea-pigs and mice. These are, of course, not the most susceptible species. It has lately been tested in cattle (Skoda, 1962) and has given encouraging results in pigs and sheep (Bartha & Kojnok, 1963). The virus has also been attenuated by growth in chick- or calf-tissue cultures, but it has been hard to obtain in a wholly avirulent yet still immunogenic state.

VARICELLA

Synonyms: *Herpesvirus varicellae. Varicella-zoster (V-Z) virus.* The virus causes disease of 2 clinical types: Varicella, chickenpox or Windpocken; and Herpes zoster, or zona.

Review: Downie (1959).

Antigenic properties. Viruses obtained from varicella and zoster are identical by all serological tests. These include agglutination of partly purified virus suspensions (Amies, 1934), complement fixation (Netter & Urbain, 1926), precipitation and neutralization tests. CF antigen is better shown with vesicle fluid than with material from tissue cultures. An antigen separable from the virus particles is concerned (Taylor-Robinson & Downie, 1959; Caunt et al., 1961). Gel-diffusion tests in agar reveal the presence of 3 lines of precipitation (Taylor-Robinson & Rondle, 1959). Virus antigen can be revealed also by the aid of fluorescent antibody and in neutralization tests in tissue culture; the latter was best achieved by incorporating antisera in the culture system (Weller, 1958). The earliest antigen detected by immunofluorescence by Slotnik & Rosanoff (1963) was in the perinuclear region of the cytoplasm. Convalescent zoster sera tend to have higher antibody titres than post-varicella sera, probably because a secondary antigenic stimulus is concerned. A small degree of antigenic relation to herpes simplex has been shown (Schmidt et al., 1969).

Cultivation. The virus has not been cultivated in eggs except that Goodpasture & Anderson (1944) reported development of inclusions in human skin grafted on the chick CAM. Weller (1953; 1958) succeeded in propagating strains from varicella and zoster in tissue cultures of human-embryo skin and muscle and of prepuce. The virus has since been grown in HeLa cells and various monkey tissues. Cytopathic effects appear in 2–7 days, are focal in character, involving the cell sheet gradually. Virus, in culture in contrast to that in vesicle fluid, is mostly cell-bound, and serial cultivation is ordinarily only achieved by passing material containing tissue; this is probably because virus outside cells is very labile (Cook & Stevens, 1968). Infection apparently passes to contiguous cells rather than through the fluid medium. In cultures of thyroid tissue, however, virus is demonstrable in the fluid phase of the culture (Caunt & Taylor-Robinson, 1964). The same may be possible with some lines of monkey-kidney cells (Geder et al., 1964–65) or in organ cultures of human embryonic tissue (Caunt, 1969).

Pathogenicity. Varicella affects chiefly children, causing a papular rash, which soon becomes vesicular and then scabby. The incubation period is 14–16 days, rarely up to 21 days. Pneumonitis, sometimes fatal, may occur, especially in adults. Encephalitis is a rare complication. The disease may be fatal in children on cortisone therapy. The infection may also be activated by immunosuppressive measures.

Zoster is a painful local condition, usually in adults; skin lesions resemble those of varicella. Zoster ophthalmicus may be very serious. Zoster may follow exposure to varicella, but usually behaves as if it were a latent infection, activated by such stimuli as arsenic treatment or tumour growth; virus is suspected to lie dormant in the body after an attack of varicella. Zoster rarely complicates varicella in children, and a generalized rash rarely occurs in zoster in adults. Typical varicella may be contracted from exposure to a zoster case, and zoster material has experimentally produced varicella in children (Kundratitz, 1925).

Animals other than man seem to be generally insusceptible. There is, however, a report (Rivers, 1926) of production of nuclear inclusions in testes of inoculated *Cercopithecus* monkeys.

Pathological lesions. Ballooning degeneration of skin epithelium precedes the formation of vesicles. Here and elsewhere are found the characteristic type A intranuclear inclusions. In zoster, the skin lesions are preceded by inflammatory changes in posterior root ganglia and posterior columns of the cord. Anterior columns are not commonly affected, and paralytic symptoms are rare, though they do occur. In fatal cases of varicella in children, focal lesions, with inclusions, are found in various viscera.

Ecology. Spread of varicella is probably mainly through the air. Hope-Simpson (1965) has suggested that the recrudescence of active infection in the form of zoster may help in the survival of the virus, since case-to-case transmission of varicella may fail.

Control. No method of active immunization is yet available. Convalescent serum has not been successful in prevention nor treatment.

INFECTIOUS BOVINE RHINOTRACHEITIS

Synonyms: IBR. Necrotic rhinitis. Red nose. The causative virus can also give rise to infectious pustular vulvovaginitis or bovine coital exanthema.

Review: McKercher (1959).

Antigenic properties. Neutralization tests in tissue culture show that the virus is antigenically homogeneous, only minor differences existing between strains (York et al., 1957). The virus is antigenically distinct from herpes simplex, but shows some relation to that of equine rhinopneumonitis in complement-fixation and gel-diffusion tests (Carmichael and Barnes, 1961), oral, genital and conjunctival strains are antigenically alike.

Cultivation. No growth in fertile eggs. The virus has been cultivated in bovine-embryo tissues—kidney, testis, lung, skin, producing CPE within 1–2 days (Madin et al., 1956). The disease was reproduced with culture material after 21 but not 40 passages (Schwarz et al., 1957). Intranuclear inclusions were produced in culture. Growth also occurs in pig-, sheep-, goat- and horse-kidney tissue cultures, in rabbit spleen, in human amnion and, after adaptation, in HeLa cells. Similar results are reported with a strain from coital exanthema (Greig, 1958). Plaques may be found in bovine-kidney monolayers. In organ cultures the virus inhibited ciliary activity (Shroyer & Easterday, 1968).

Distribution. Infectious rhinotracheitis occurs as such chiefly in the western half of the United States, also in Britain, Germany, Central Africa, Australia and New Zealand. As pustular vulvovaginitis or coital exanthema it is recorded from the eastern United States and Canada, several European countries and North Africa.

Pathogenicity. 'IBR' may be mild and unrecognized or very acute, involving the whole respiratory tract with acute inflammation, exudate and mucosal necrosis. There are fever, depression and often also bloody diarrhoea. The natural incubation period is 4–6 days; experimentally it may be as little as 18 hours. The natural course is 10–14 days. In severe outbreaks mortality may be 75 per cent, but is usually much lower.

Only cattle are known to be naturally susceptible; antibodies have, however, been found in mule deer (*Odontocoileus*) (Chow & Davis, 1964). Virus from cases of rhinotracheitis or vulvovaginitis may reproduce either disease according to the route of inoculation (Brown & Bjornson, 1959). Infection produced by cultivated virus is milder than the natural disease.

In calves the virus may cause only conjunctivitis (Quin, 1961). Other workers have offered evidence that the infection may cause abortions in cows (McKercher & Wade, 1964), a fatal disease in new-born calves, pustular balano-posthitis in bulls, pustular conjunctivitis or encephalitis (cf. Studdert et al., 1964 for references). Cattle inoculated IC may have a fatal infection (Straub & Böhm, 1965).

Experimentally, young goats may be infected and develop fever (McKercher, 1959).

Rabbits inoculated intradermally or intratesticularly with tissue-culture material developed local lesions, but inclusions were not found and attempts at serial passage were unsuccessful (Armstrong et al., 1961). The virus can also cause meningo-encephalitis in rabbits with paralysis of the hind legs (Persechino et al., 1965).

Pathological lesions are those of acute inflammation and necrosis of affected mucous membranes, which may be covered with glairy muco-purulent exudate. There may be patchy pneumonia and ulceration of the abomasum is frequent.

Ecology. Transmission is by contact, especially where there is overcrowding. Virus may still be present in nasal secretions for several months after an attack. McKercher & Theilen (1963) suggest that the virus, known mainly as a cause of disease of the genital tract in Europe, was introduced into North America about 1930 and was exalted in virulence in large concentrations of 'feed-lot' animals, acquiring the power to spread by the respiratory route.

Control. Unmodified virus will immunize without producing disease if given intramuscularly. However, it is better to use virus which has been attenuated by cultivation in bovine (Schwarz et al., 1957) or porcine (Schwarz et al., 1958) tissue culture.

MALIGNANT CATARRH

Synonyms: Malignant catarrhal fever. Snotsiekte (South Africa). Bovine epitheliosis.

Review: Berkman et al. (1960)—pathology.

This virus was found by Plowright et al. (1963) to be morphologically identical to herpesviruses.

Antigenic properties. Immunity of cattle which survive an attack persists for at least a few months (Piercy, 1954). The virus is neutralized by sera of recovered cattle.

Cultivation. No continued cultivation in eggs is convincingly reported. The virus may, however, be grown in cultures of thyroids or adrenals from infected cattle (Plowright et al., 1960); syncytia and Cowdry A-type inclusions form in cultures. Once growth has been thus initiated cultivation has proved possible in sheep thyroid, calf testis or adrenal, rabbit and wildebeest kidney. Cultures up to the nineteenth passage in calf kidney were still infectious. In thyroid cultures transmission by means of cell-free fluids was possible.

Distribution. What is called malignant catarrh occurs sporadically in every continent.

Pathogenicity. An extremly fatal, sporadic disease. After 1 or 2 days' fever, acute inflammation of nasal and oral mucous membranes begins. There occur ulceration, exudation and nasal obstruction, and the lesions may spread to pharynx and lungs. Keratitis and various other eye lesions are common. Most cases show nervous symptoms, usually stupor, but there may be excitement. Death may occur within 24 hours, but the disease may last a fortnight or more. Mortality is over 90 per cent.

These severe symptoms are seen in cattle, but wildebeests and sheep may have inapparent infections. Huck et al. (1961) report a similar disease amongst Père David's deer; an agent was transmitted to other deer, calves and rabbits.

Experimental transmission has been achieved almost wholly by workers in Africa, a fact suggesting the possibility that more than one disease is concerned. Piercy (1952) readily transmitted the disease to cattle, also to rabbits (Piercy, 1955; Plowright, 1953); the latter showed little apart from fever, but after 9 rabbit passages virus still infected cattle. The agent transmitted to various species by Daubney & Hudson (1936) seems to have been something different.

Pathological lesions are those of destructive inflammation of mucous membranes of the respiratory and upper intestinal tracts (Berkman et al., 1960). Bronchopneumonia may occur, and the brain shows evidence of meningo-encephalitis. Cytoplasmic inclusions have been seen in various tissues, but the striking intranuclear inclusions which are seen in tissue cultures have not been found in animals infected *in vivo*.

Ecology. Much circumstantial evidence suggests that infection is transmitted when cattle are kept in contact with sheep, which may have symptomless infections and that in Africa it may follow contact with wildebeest (*Gorgon taurinus*). Plowright (1963) detected virus in blue wildebeests of all ages; there was viraemia in 40 per cent of wildebeest calves 1 to 2 months old in the wild. Some seemed to be infected congenitally and to transmit infection to others by contact. All 181 wildebeest examined had antibodies in sera (Plowright, 1967). Contact transmission amongst cattle seems not to occur. Arthropod transmission seems unlikely, at least in North America, where it is a winter disease.

BOVINE ULCERATIVE MAMMILLITIS—ALLERTON VIRUS

Descriptions: Allerton virus—Alexander et al. (1957) Kipps & Polson (1966). Bovine ulcerative mammillitis—Martin et al. (1966).

A virus causing mammillitis in cattle in Britain (Martin et al., 1966) is not distinguishable antigenically nor in physical properties from the Allerton virus; this was isolated from lumpy skin disease in South Africa, but is not now thought to be the cause of that condition. The initials M and A in the following account indicate which virus is the subject of particular observations.

Antigenic properties. Neutralizing antibodies are present in sera of recovered animals (A, M).

Cultivation. Multiplies in calf-kidney tissue culture, forming large syncytia; these form as early as 8 hours after inoculation or as late as 8 days; once they form, cell destruction is soon complete. Numerous large inclusions of Cowdry's type A are present (A). Grows also in BHK (hamster cells) (A, M) and lamb testis (A).

Pathogenicity. Infection in South Africa (A) at one time confused with lumpy skin disease. In Scotland (M) the virus causes lesions on the teats, less frequently the udders of milking cows; they take the form of a deep slowly healing ulcer; mastitis may follow.

When inoculated into cattle the Allerton virus produces fever and an eruption of skin nodules all over the body; they become necrotic. Lymphadenitis is also present (M, A). One strain of virus (A) produces similar lesions when inoculated into suckling mice and causes transient lesions when given intradermally to rabbits. In day-old rats, mice and chinese hamsters the virus may produce rashes, stunting and death (M) (Rweyemamu et al., 1968).

The histological changes include oedema, vesiculation, mononuclear infiltration, formation of syncytia and nuclear inclusions (Martin et al., 1969; Rweyemamu et al., 1969). It will also infect sheep and probably goats (P. B. Capstick, personal communication, 1966).

Huygelen (1960) isolated in Ruanda-Urundi a virus similar to Allerton from cattle with extensive erosion of the teats.

Ecology. Allerton virus may be mechanically transmitted by insects (Capstick). Virus is also excreted in urine and faeces.

Control. Formalinized virus gave poor protection; nor could the virus be attenuated in tissue cultures. Better results were obtained by giving living unattenuated virus intramuscularly (M) (Rweyemamu & Johnson, 1969).

EQUINE RHINOPNEUMONITIS

Synonyms: Mare abortion. Equine abortion. Equine influenza (in part). Equine herpesvirus 1.

The name 'equine rhinopneumonitis' proposed by Doll et al. (1959) seems preferable since another virus, that of equine arteritis (cf. p. 268), can also cause abortion in mares.

Haemagglutination. Horse RBCs are agglutinated between 4° and 37°C by tissues of affected hamsters (McCollum et al., 1956). The haemagglutinin is labile at 56°C. It is neutralized not by convalescent horse sera but by sera of horses hyperimmunized with infected hamster tissue. Semerdjiev (1962) described agglutination of horse, and also guinea-pig cells, preferably after treatment with formalin.

Antigenic properties. Specific CF antigens occur in lungs of infected hamsters and in infected eggs, and there are CF antibodies in sera of infected horses, though these only persist for a few months. Neutralization tests may be carried out in hamster and, presumably, also in tissue culture. Equine herpesviruses antigenically differing from equine rhinopneumonitis are referred to on p. 348.

Complement-fixation and gel-diffusion, but not neutralization, tests indicate that an antigen is shared by this and infectious bovine rhinotracheitis virus (p. 342) (Carmichael & Barnes, 1961).

Cultivation. *In fertile eggs.* The virus has been adapted to growth on the chorioallantoic membrane, in the yolk sac and amnion (Doll & Wallace, 1954; Randall, 1955).

In tissue culture it grows in fetal horse tissue, also in HeLa cells, human amnion, sheep, pig, cattle, cat and chick tissue. Inclusion bodies are formed *in vitro* (McCollum et al., 1962; Mayr et al., 1965).

In roller tube cultures of horse-kidney cells a specific CPE is produced (Shimizu et al., 1958).

Distribution. The disease is reported from several European countries, South Africa and the United States; probably occurs elsewhere under the guise of equine influenza.

Pathogenicity. In mares, the disease is usually inapparent, but abortion occurs, especially in the eighth, ninth and tenth months of pregnancy. Symptoms affecting the central nervous system have been recorded, but rarely. Genital vesicular exanthema or pustular vulvo-vaginitis of horses may be caused by this virus (Petzoldt, 1970) (but see p. 348); it is distinct from the virus affecting cattle.

Only equines are naturally affected.

Experimentally, suckling Syrian hamsters are readily infected IP with production of fatal hepatitis (Anderson & Goodpasture, 1942; Doll et al., 1953). Later it was found possible to adapt the virus to produce fatal infection in adult hamsters; also to go intracerebrally in baby mice (Kaschula et al., 1957). The virus will also cause abortion in pregnant guinea-pigs (Doll et al., 1953).

No lesions have been found in aborting mares, but aborted fetuses show multiple necrotic foci in the liver and petechiae elsewhere. Type A intranuclear inclusions occur abundantly in infected epithelial cells, in liver and elsewhere; in infected hamsters 99 per cent of hepatic cells may contain them.

Ecology. Probably transmitted by the respiratory route. Transmission by 'carrier' stallions has been alleged, but is not generally accepted.

Control. It is safe and effective to give living virus to mares when neither they nor their stable companions are pregnant (Doll et al.,

1955); use of virus attenuated by intracerebral mouse-passage has also been suggested (Byrne et al., 1958).

OTHER EQUINE HERPESVIRUSES

Other viruses from horses have been referred to as equine herpesviruses 2, 3, 4.

Equine herpesvirus 2

Plummer & Waterson (1963) isolated a herpesvirus from a horse with catarrh. It was serologically quite distinct from equine rhinopneumonitis (virus 1).

Equine herpesvirus 3

Another serotype was recovered from foal-kidney cultures by Karpas (1966).

A fourth virus has been described as an equine cytomegalovirus (Plummer et al., 1969a). Viruses 2 and 3 and 4 are more closely related to each other serologically than to virus 1. Like it, and unlike cytomegaloviruses, they will grow in tissue cultures of various species; on the other hand they are more closely cell-associated than is virus 1.

Other herpesviruses from horses have been described as being distinct from equine herpesvirus 1, but it is not stated whether they have been compared with the others. One was recovered from mares with coital exanthema (Girard et al., 1968); another was obtained from leucocytes of 88·7 per cent of apparently normal horses (Kemenry & Pearson, 1970).

FELINE VIRAL RHINOTRACHEITIS

Cultivation. *In tissue culture* grows in cultures of cat kidney, lung and testis, producing focal lesions in 40–48 hours; the whole cell sheet was affected and fell off the glass in 120 hours. Plaques were formed in monolayers. No CPE in cultures of bovine, human and monkey cells. Intranuclear inclusions and giant cells were formed in culture (Crandell & Maurer, 1958; Crandell et al., 1960). Abortive infections were produced in human cells; inclusion bodies were formed, but no infectious virus (Tegtmeyer & Enders, 1969).

Pathogenicity. Causes fever, lacrimation and nasal discharge in kittens. Incubation period 1–3 days. Described also as endemic viral coryza (Lindt et al., 1965).

Ecology. Recovered cats may become carriers and cause clinical disease in kittens in contact (Povey & Johnson, 1967).

CANINE HERPESVIRUS

Synonym: Canine tracheo-bronchitis virus.

Antigenic properties. A plaque-reduction test revealed a low degree of crossing with herpes simplex (Aurelian, 1968).

Cultivation. In dog-kidney cultures, causing CPE, not in human, bovine or porcine tissues.

Pathogenicity. The virus can cause a disease in new-born puppies often fatal (Carmichael et al., 1965; Stewart et al., 1965a). There are necrotizing rhinitis and pneumonia with necrosis and haemorrhages in various organs (Cornwell & Wright, 1969). Nonsuppurative meningo-encephalomyelitis may occur. The virus has been recovered from a pup with malignant lymphoma, but may have been only a 'passenger' (Kakuk et al., 1969). A virus causing tracheo-bronchitis (kennel cough) in older dogs is probably the same (Karpas et al., 1968). ('Kennel cough' may also be caused by an adenovirus, see p. 321). Genital lesions caused by a similar virus are described (Poste & King, 1971).

Ecology. The virus has been recovered from pups obtained by caesarean section (Stewart, 1965). The tracheo-bronchitis virus has been isolated from apparently normal dogs, which could be carriers.

AVIAN INFECTIOUS LARYNGO-TRACHEITIS

Not to be confused with infectious bronchitis of chicks (p. 183).

Haemagglutination. Agglutination of fowl erythrocytes reported by Prokof'eva & Babkin (1965).

Antigenic properties. Neutralization by antisera is readily measured in tests on the chorioallantoic membrane (Burnet, 1936). No serological types are described, but some strains though fully antigenic are poorly neutralized by antisera, a situation recalling the P–Q phase variation of influenza viruses (Pulsford, 1953). A neutralization test involving counting plaques on tissue-culture monolayers is described. The gel-diffusion test is also useful in diagnosis (Woernle & Brunner, 1961).

Cultivation. *In fertile eggs.* On the CAM plaques of 2 kinds appear; larger ones with an opaque white periphery and necrotic centre, from more virulent strains; and smaller ones without necrosis, from less virulent viruses (Burnet, 1936). Amniotic inoculation causes production of lesions with intranuclear inclusions in trachea and bronchi (Burnet & Foley, 1941).

In tissue culture, growth occurs in kidney and muscle of day-old chicks with CPE, also in HeLa cells without CPE (Webster, 1959). A long eclipse phase is followed by sudden virus release after 18 hours (Pulsford, 1960). Atherton & Anderson (1957) grew the virus in chick-embryo tissue and found activity particularly associated with the tissue in the culture; transmission was thought to be largely from cell to cell. Nuclear inclusions are formed in tissue cultures, as in the CAM; giant cells may appear also, and these may have as many as 80–100 nuclei (Watrach & Hanson, 1963). Crystalline aggregates of virions were seen in infected nuclei (Watrach et al., 1968).

Distribution. Occurs in every continent.

Pathogenicity. Naturally affects mainly fowls and pheasants causing haemorrhagic tracheitis with gasping and coughing. Virus is mainly in the respiratory tract, in smaller quantity in liver and spleen (Beach, 1931). In some outbreaks haemorrhagic conjunctivitis occurs. The incubation period lasts for 2–6 days. Mortality may be anything up to 70 per cent, the course of infection in survivors being 2–3 weeks. Egg-production suffers.

A less severe disease with nothing worse than coughing or sneezing is caused by a less virulent strain occurring in Australia and the United States.

Most accounts say that only fowls and pheasants are susceptible, but occasional infection of ducks, pigeons and turkeys is reported. No mammals have been infected.

Haemorrhagic inflammation and oedema of larynx, trachea, bronchi and sometimes conjunctiva occur in severe forms, or there may be a caseous exudate in the air passages and air sacs. Type A intranuclear inclusions occur in groups of epithelial cells and may be seen as early as 12 hours after inoculation. Though these are homogeneous as seen after ordinary staining, silver staining may reveal argentophilic granules (Seifried, 1931). Bang & Bang (1967) followed the course of infection histologically. Inclusion bodies and syncytia appeared at 21 hours; sloughing occurred at 3 to 7 days and repair from 8 to 21 days.

Ecology. Transmission is by the respiratory route. The disease is often highly infectious. It may be endemic in some areas, apparently because some recovered birds continue to excrete virus for long periods. Indirect transmission through human agency is also suspected, and transmission occurs through the egg, or perhaps by contamination of the outside of the egg-shell.

Control. *Vaccination.* Living virus has been successfully applied as a vaccine to the bursa of Fabricius off the cloaca (Beaudette & Hudson, 1933). It is best to use a strain attenuated by cultivation in eggs (Cox, 1952). Churchill (1965) describes an attenuated virus to be given intra-ocularly.

Since carriers exist, it is clearly rash to mix recovered birds with clean stock.

Pacheco's Disease of Parrots

A virus isolated from South American parrots especially those of the genus *Amazona* (Pacheco, 1930–31) is tentatively included in the herpesvirus group. It can be propagated on CAM of 10-day chick embryos producing opaque white lesions and killing a high proportion of embryos.

Symptoms in affected parrots consist of weakness, wing drooping, indifference to stimuli, diarrhoea, finally coma and death. Budgerigars (*Melopsittacus*) are highly susceptible to experimental infection, young chicks less so. Lesions include necrotic foci in liver and spleen and accumulations of fluid in serous cavities. Histologically one sees proliferation of mesenchymatous cells, sometimes syncytia. Type A nuclear inclusions occur in mononuclear cells.

Disease of Owls

Green & Shillinger (1936) have described a fatal disease of wild horned owls (*Bubo virginianus*). There were small necroses of liver and spleen. The disease was transmitted with material which was unfiltered but apparently free from bacterial pathogens to another horned owl, a screech owl (*Otus*), but not to a barred owl (*Strix*). Large eosinophilic nuclear inclusions were found in the hepatic cells. Burtscher (1968) cultivated a herpes-like virus on the CAM and in the allantois of eggs and could reproduce the disease in owls, but not in other birds.

Inclusion Disease of Pigeons

Smadel et al. (1945), in the course of studies of ornithosis in pigeons, encountered another agent, which produced intranuclear inclusions in infected cells. The virus caused focal necrosis of parenchymatous tissue of the liver in pigeons, but was not transmitted to any other species. In eggs, titres of 10^7/ml were reached, embryos dying 4 days after inoculation. It could be stored for several months at −20°C. Complement-fixing but not neutralizing antibodies could be demonstrated. The virus has been recognised in Britain, Denmark and the United States. Cornwell & Wright (1970) record that conjunctivitis and dullness are the main symptoms in infected birds. Infection could apparently be maintained in adult carriers. It resembled other herpesviruses by electron-microscopy. It differed from infectious laryngotracheitis in the character of the lesions produced on the CAM and in tissue culture (Cornwell & Weir, 1970).

Cormorant Virus

French (personal communication, 1954) has described a virus which may fall into the same family as the avian viruses just described. It was isolated on the CAM of a fertile hen's egg inoculated with the blood of a little pied cormorant (*Phalacrocorax melanoleucos*) 3 weeks old. It produced pocks on the membrane and could also be grown in the amniotic and allantoic cavities. There was no regular lethal effect. Neither chicks, pigeons, parrots nor laboratory rodents could be infected. The virus passed a 0·6 μm gradocol membrane without loss of titre. It survived storage at −70°C and lyophilization. Thirty minutes at 55°C inactivated it. No neutralizing antisera could be prepared.

Lesions on the CAM were at first proliferative, later necrotic. Eosinophilic intranuclear inclusions were found, but these were thought to represent enlarged nucleoli rather than Cowdry type A inclusions.

There is no evidence that the virus is pathogenic for cormorants.

MAREK'S DISEASE

Synonyms: Neurolymphomatosis. Fowl paralysis.

Review: Biggs (1967).

This disease was formerly included with the fowl leukosis viruses, but has now been shown to be etiologically quite distinct.

Haemagglutination. A claim that plasma of infected chicks would agglutinate sheep RBCs has not been confirmed (Payne & Rennie, 1970).

Antigenic properties. The techniques of double diffusion in agar gel (Chubb & Churchill, 1968) and direct immunofluorescence (Purchase & Burgoyne, 1970) can be used for diagnosis.

Cultivation. The virus has been cultivated in chicken-kidney cells, producing foci or micro-plaques of cell destruction, sometimes preceded by proliferation (Churchill, 1968). Virus was cell-associated, supernatant fluids of cultures being noninfectious.

Pathogenicity. In the nervous form of the disease, there is progressive paralysis, usually of a wing or leg. Quite young chicks may be affected, but it begins most commonly in birds 2–8 months old. Lymphoid tumours may be seen. Virions are frequently found in feather follicles (Calnek et al., 1970).

The disease has been recorded in pheasants, turkeys and quails (Andrewes & Glover, 1939; Jungherr, 1939; Wright, 1963), also in ducks, swans, geese, pigeons and budgerigars (Baxendale, 1969).

Of recent years an acute disease has appeared and increased in America and Europe. This is characterized by enlargement and lymphocytic infiltration into liver and other viscera, rather than into nerves; nevertheless the properties of the causative agent suggest that it is related to neurolymphomatosis and it has been called Acute Marek's disease. The first report was probably that of Benton & Cover (1957).

Marek's disease has been serially transmitted to chicks by IP inoculation of blood or plasma; the incubation period was about 4 weeks (Biggs & Payne, 1967); and virus from the acute form of the disease was more virulent for inoculated birds (Purchase & Biggs, 1967). Cell-free transmission was possible if virus from feather follicle epithelium was used (Calnek et al., 1970; Nazerian & Wither, 1970).

Lesions in the nerves are first those of inflammation, with oedema and infiltration with lymphocytes, lymphoblasts and plasma cells. At length the nerves become grossly thickened as a result of this infiltration, but the axons themselves are only secondarily affected. Payne & Biggs (1967) described 3 types of nerve lesion, differing in the extent of cell infiltration and oedema. Viral antigen was predominantly present in cells of feather follicles and Bursa of Fabricius.

'Grey eye', a disease involving loss of pigment in the iris and going on to blindness, has been considered as an ocular form of the disease, but its true etiology is obscure.

Ecology. Marek's disease was first described in 1907, became troublesome, especially in the United States, in 1925, but has since become less common, while visceral lymphomatosis has increased.

The infection is spread by contact; virus is present in oral secretions, but not in faeces. Dust from poultry houses remains infectious for as long as 4 weeks (Beaseley et al., 1970). Biggs et al. (1968) have studied genetic resistance to the disease.

Control. Live attenuated viruses have been obtained by passing the virus in cultures, and these seem to have value in protecting against challenge or contact infection (Churchill et al., 1969, Kottaridis & Luginbuhl, 1969). Protection by vaccination with the herpesvirus of turkeys has been reported (Okasaki et al., 1970).

Herpesvirus of Turkeys

A cell-associated virus antigenically related to that of Marek's disease was recovered from turkeys by Witter et al. (1970). It could produce tumours of various organs in chicks (Purchase et al., 1971).

Duck Plague

Synonym: Duck virus enteritis.

Review: Jansen (1964).

Resembles herpesviruses in morphology and in chemical and physical properties (Hess & Dardiri, 1968).

Haemagglutination. Negative with chicken, duck, sheep and horse RBCs.

Antigenic properties. Indian, American and Dutch strains are antigenically alike. Unrelated to fowl plague, NDV or duckling hepatitis.

Cultivation. On the CAM of 12-day-old duck eggs; embryos die with extensive haemorrhages after 4 days. Adaptation to chick fibroblasts was achieved; the adapted virus was nonpathogenic for ducks, but it immunized (Dardiri, 1969).

Distribution. Outbreaks have occurred in Holland since 1949 and in India since 1963. Serological evidence suggests that it may be present in wild mallard in Britain (Asplin, 1970).

Pathogenicity. Course very rapid with nasal and ocular discharge and diarrhoea. Post-mortem: multiple petechiae; sometimes diphtheritic inflammation of oesophagus and cloaca. Mortality up to 97 per cent, but lower recently.

Natural infection seen only in domestic ducks. Only *Anserinae* (ducks, geese and swans) can be infected experimentally, except that day-old chicks can be infected IM (Dardiri & Gailunas, 1969).

Ecology. Outbreaks very local, usually between January and July, only in areas where ducks have free access to water.

Control. Virus attenuated by growth in hens' eggs forms an effective vaccine (Toth, 1970).

INFECTIOUS MONONUCLEOSIS

Synonyms: Glandular Fever. Monocytic Angina.

Review: Evans & Paul (1965).

There is strong evidence that the Epstein-Barr (EB) virus, first isolated from Burkitt tumours, is the causative agent of infectious mononucleosis (IM), and this view, though not finally proved, is provisionally accepted here.

Chemical composition. Eight proteins were identified in virions by electrophoretic techniques (Weinberg & Becker, 1969).

Physico-chemical characters. Density 1·2–1·3 (Minowada et al., 1969) MW 100×10^8 daltons. Density of DNA 1·72 (Weinberg & Becker, 1969).

Antigenic properties. Specific EB antigens can be detected in cultures by fluorescent antibody and gel-diffusion tests. These may appear in the course of the infection with infectious mononucleosis, but are also present in many persons with no history of the disease. At least 1 antigen, detected by gel diffusion, is shared by this and the Lucké frog virus (Fink et al., 1968).

Many patients develop heterophile antibodies agglutinating sheep RBCs (Paul & Bunnell, 1932). These antibodies, in contrast to those developing in some other infections, can be absorbed by bovine RBCs, but not by guinea-pig kidney. If the sheep cells are treated with a plant-protease, the haemagglutination test is said to become more specific (Springer & Callahan, 1965). The Paul-Bunnell test is not positive in all cases, and it is likely that some of these are infections with cytomegalovirus. This may produce a clinical picture very like that of infectious mononucleosis. A haemolysin for bovine RBCs is also described (Mikkelsen et al., 1958). Human red cells modified by treatment with an Australian strain of Newcastle disease virus are agglutinated by many recent mononucleosis sera (Burnet & Anderson, 1946).

Cultivation. A virus having the morphology of a herpesvirus and recognizable by specific immunological tests has been cultivated from cultures of Burkitt's tumour, a malignant tumour affecting particularly children in parts of Africa (Epstein et al., 1965; Stewart et al., 1965b; Hummeler et al., 1966). It is cell-associated and not transferable in series in the ordinary way. However, when EB-infected cells were lethally irradiated and co-cultivated with leucocytes of normal infants it induced these to grow progressively; EB antigen and characteristic chromosomal changes appeared in them (Henle et al., 1967; Gerber et al., 1970). Similar transfer was effected to human leukaemia cells (Miller et al., 1969). Leucocytes from normal persons will not grow indefinitely in culture, but those from patients with IM will do so, changing into blastoid cells and containing EB antigen (Diehl et al., 1968; Glade et al., 1968). Growth of the EB virus in culture is enhanced if arginine-deficient media are used (Henle & Henle, 1968).

Pathogenicity. The onset of the disease is insidious and its course chronic; symptoms include fever, swollen lymph nodes, especially cervical ones at first, enlarged spleen and sore throat. Rashes and jaundice may occur; some hepatitis is apparently the rule. The blood shows a great increase in mononuclear cells, probably abnormal lymphocytes. Maximal incidence is in the age-group 15–25. The incubation period is said to be 34–49 days in adults (Hoagland, 1964) or under 2 weeks in children (Hobson et al., 1958). Experimental transmission to human volunteers and to monkeys has been reported by a few workers, but the results are unconvincing and most attempts have been negative. Claims to have transmitted the virus to new-born thymectomized hamsters and mice (Stewart, 1969) await confirmation. 'EB virus' has been cultivated from apparently normal chimpanzees (Landon et al., 1968), but according to Stevens et al. (1970) the agent isolated differs antigenically from strains of human origin.

Pathological lesions in biopsy material from man have shown various changes in lymph nodes up to great distortion from lymphocytic or reticulo-endothelial hyperplasia; there was also hepatitis. Similar pictures have been seen in the few fatal cases (Custer & Smith, 1948).

Ecology. Most cases are sporadic, rare outbreaks being particularly in closed communities. The disease seems not to be highly infectious. Hoagland (1955) thinks 'intimate oral contact' important in transmission.

Henle & Henle (1970) record that primary seroconversions occur especially under the age of 6 or over 7.

Relation to Burkitt's and Other Tumours

The virus has been recognised in most cultures of malignant lymphoblasts from Burkitt's tumour, even when not detected in the tumour itself. EB antigen has, however, been demonstrated in cells lacking morphological evidence of presence of virus. The virus has also been found in cultures of nasopharyngeal tumours and leukaemia. Patients with Burkitt's lymphoma and nasopharyngeal tumours have EB antibodies in higher proportion than do normal persons (Henle et al., 1969; 1970). There is evidence, from nucleic acid hybridization experiments, that EB genome may be present even when virus is not directly demonstrable (Hausen et al., 1970). Since EB infection is world-wide and Burkitt's tumour mainly prevalent in tropical areas, EB virus cannot be the only causative agent of the tumour. A possible contributory role of endemic malaria has been suggested by several workers (review by Epstein, 1970).

HERPESVIRUSES ASSOCIATED WITH RENAL CARCINOMA OF THE LEOPARD FROG

Reviews: Lucké (1934, 1938). Rafferty (1964).

Adenomata or adenocarcinomata occur in the kidneys of *Rana pipiens* from certain areas. Though a virus is regularly present, doubts have been raised as to the role of this and of other frog viruses (see p. 417) in causing the tumours. It now seems fairly certain that this virus is the one properly associated with the tumour. A virus causing lymphosarcomata in clawed toads (Xenopus) appears to be of a similar nature.

Morphology. As for other herpesviruses. Aberrant forms of virus are frequently seen; these include tubules bearing capsids on their outer surfaces (Stackpole & Mizell, 1968).

Cultivation. The virus grows in cultures of frog tissue, preferably at 35°C; nuclear inclusions were not formed *in vitro* (Lucké, 1939; Lucké et al., 1953) but CPE and plaque formation have been observed (Gravell et al., 1968).

Distribution. Natural occurrence is mainly in *R. pipiens* from parts of Vermont and adjacent areas of Canada, also in Wisconsin and possibly elsewhere in North America.

Pathogenicity. The naturally occurring kidney tumours are found in 2 per cent of Vermont frogs, especially those 30–100 g in weight.

They are commonly bilateral and may involve almost all the kidney tissue. They are usually progressive and fatal metastases are recorded, but regression can occur. Strains from Wisconsin may, in large doses, produce acute disease and death in tadpoles (Rafferty, 1965).

Attempted subcutaneous transplantation of the tumour apparently fails, but many inoculated frogs develop renal tumours several months later. Other frog species and even another subspecies of *R. pipiens* proved insusceptible (Lucké, 1938). However, Rose & Rose (1952) inoculated young salamanders (*Triturus*) or regenerating limb-buds of older ones and produced osteochondromata.

The tumour cells characteristically contain intranuclear inclusion bodies, often in large numbers. Their appearance is, however, greatly affected by temperature. At summer temperatures these are not seen, but when inclusion-free frogs were brought into the cold, inclusions developed (Mizell et al., 1968). In contrast, tumour-development was favoured by warmer weather.

Ecology. There is a suggestion (Rafferty & Rafferty, 1961) that cross-infection may occur within a laboratory, but probably not between adult frogs. Virus may be excreted in the urine and transmitted when frogs congregate for spawning in March. Rafferty (1963) raises the question of whether injection of tumour extracts may not act merely by accelerating the occurrence of tumours which would develop anyway. The facts have to be considered in the light of the possibility that there are (a) a tumour-causing virus with unknown properties and (b) a 'passenger' herpesvirus which is, like some other herpesviruses, closely cell-associated.

JAAGSIEKTE

Synonyms: Pulmonary adenomatosis. Chronic progressive pneumonia. Lungers. (Jaagsiekte means 'driving sickness' from the reduced tolerance of affected sheep to exercise).

Its relation to Laikipia lung disease of East Africa is doubtful.

Reviews: Marsh (1958). Mackay & Nisbet (1966).

It is probable that the herpes-like virus described by Smith & Mackay (1969) is the causative agent, but this is not definitely proved.

Cultivation. Cultures of affected lung tissue showed proliferation of cells, particularly macrophages; these contained intranuclear inclusions. Transmission with cell-free material proved possible, and, again, macrophages were particularly involved (Mackay, 1969). No transmission of cultivated virus to animals is reported.

Distribution. The disease is known from several parts of Europe including Britain, Iceland, South Africa, Peru, the United States, possibly also Israel and India.

Pathogenicity. A chronic lung infection usually involving sheep 4 years old or more. Symptoms as with Maedi are those of emaciation and dyspnoea. Cuba-Caparo et al. (1961) report 2 cases in goats in Peru. Transmission to sheep is reported by Sigurdsson (1958) after inoculation intranasally and intratracheally; lung lesions were found when the animals were killed 14 months later.

Pathological lesions are quite different from those of Maedi, being those of widespread adenomatosis which gradually replaces the lung tissue (Robertson, 1904; Dungal et al., 1938). Nobel et al. (1969) found that metastases were frequently seen in Israel, though apparently rare elsewhere.

Ecology. Infection was apparently by the respiratory route, but most contact experiments failed (Dungal, 1946).

Control. The disease was apparently eliminated from Iceland by a slaughter policy.

CARP POX

Synonym: Epithelioma of carp.

Morphology. Schubert (1966) describes the virus as morphologically indistinct from herpesviruses.

Cultivation. Grutzner (1956) inoculated tissue cultures from *Lebistes* and observed specific changes.

Pathogenicity. The pock-like lesions were described by Keysselitz (1908) and Loewenthal (1907).

CYTOMEGALOVIRUSES

There is a group of agents affecting various animals and all highly species-specific. Many of them have a particular affinity for salivary glands or kidneys. The names 'salivary gland virus' or 'submaxillary virus' have been used for these agents and for the human infection cytomegalic inclusion disease, visceral disease, inclusion-body disease, salivary-gland virus disease, giant-cell pneumonia and other names. Weller et al. (1960) suggest the name Cytomegalovirus for the human agents, but the term, if to be of use, will have to include related viruses attacking other species.

Cytomegalic Inclusion Disease of Man

Synonyms: See above.

Reviews: Smith (1959). Rowe (1960). Nelson & Wyatt (1959). Hanshaw (1968).

Morphology. As for other herpesviruses. There have also been seen dense homogeneous spherical bodies 300–500 nm across. Both types occurred within vacuoles. Still larger bodies 2–2·8 μm in diameter had superficial blebs, often containing the smaller particles just described (Luse & Smith, 1958).

Chemical composition. GC content 58 per cent (Crawford & Lee, 1964). Virions in cytoplasm but not those in nucleus have an acid mucopolysaccharide 'as a structural feature' (Martin & Kuntz, 1966).

Physico-chemical characters. Rather sensitive to freezing and thawing and to prolonged storage at −70°C. Ether-sensitive. Inactivated at 56°C in 10 or 20 minutes. Fairly stable between pH 9 and 5, quickly inactivated at pH 4.

Haemagglutinin. None found.

Antigenic properties. CF and neutralizing antibodies (the latter tested in tissue culture) are recorded. Neutralizing antibodies alone may be present in infants' sera, but older children's and adult sera are commonly positive by both tests. Antibodies are recorded from 80 per cent of normal adult sera. Strains may differ antigenically (Weller et al., 1960). Chiang et al. (1970) distinguished 5 subgroups by immunofluorescence, but not by CF tests. The virus is relatively insensitive to interferon (Glasgow et al., 1967).

Cultivation. *In tissue culture.* Cytomegaloviruses of various species grow preferably in fibroblasts, usually only in tissues from natural hosts (but cf. Black et al., 1963). Human strains have been cultivated in fibroblasts from human myometrium, adenoid tissue, embryonic skin-muscle tissue and foreskin. Focal lesions consisting of a few enlarged cells appear, sometimes not for many weeks: it may be two months before the whole culture degenerates. Effects develop more quickly after passage, but tend to be erratic. Virus is closely associated with tissue and is often not demonstrable in the fluid phase. Inclusion bodies are abundantly formed in cells in the culture. Titres of virus remain low—$10^{3.5}$/ml (References in Smith, 1959).

Distribution. Probably world-wide.

Pathogenicity. Four types of disease in man are recognized:

1. inapparent infection especially of salivary glands.
2. overwhelming neonatal infection (cytomegalic inclusion disease).
3. more chronic disease in older children or adults, associated with other diseases such as tumours, leukaemia and pertussis.
4. very rarely localized granulomata.

In severe, usually fatal, neonatal infections, liver and spleen are much enlarged and anaemia and haemorrhages occur. More chronic infections lead to microcephaly, intracerebral calcification and hydrocephalus (Weller & Hanshaw, 1962). Medearis (1964a) has reviewed the clinical aspects of infections in children. People given immunosuppressive therapy in connection with kidney transplantation (Hedley-Whyte & Craighead, 1965) may develop fatal pneumonia or other serious infections including a mononucleosis-like syndrome (Lang et al., 1968). It is uncertain whether these represent activation of latent virus or fresh infections. Symptoms resembling infectious mononucleosis may also occur apart from immunosuppressive treatments. Infected adults occasionally develop hepatitis.

Large cells up to 40 μm across bear intranuclear inclusions 8–10 μm across. These are less acidophilic than those of the herpesviruses, but the amount of basophilia varies. Small basophilic granules may occur in cytoplasm. These are probably aggregates of lysosomes (McGavran & Smith, 1965). The nuclear inclusions are usually in epithelial cells (in contrast to what one might expect from tissue-culture findings). In their most characteristic form they occur in salivary glands, but may be in almost any organ. In the rare cases in adults they are commonest in lungs. They may also be found in the brains of sporadic cases of encephalitis of which they are the apparent cause. Cellular infiltrations surround the inclusion-bearing tissues.

Ecology. A number of sera studies from various countries suggest that the infection is commoner where socio-economic conditions are poor (Rowe, 1960). Stern (1968) found that 10 per cent of children between 2 months and 5 years excreted virus, while Numazaki et al. (1970) obtained virus from oral swabs or urine of 60 per cent of infants 5 to 9 months old. Virus has been found in the urine for up to 2 years in recovered cases. It has also been recovered from the urine of apparently normal children. Infection may be contracted *in utero* (Rowe et al., 1958; Weller et al., 1957).

Control. Tobin (1968) reported favourably on the use of iodo-deoxyuridine in 2 congenitally infected infants.

Salivary Gland Virus of Guinea-Pigs

Synonym: Submaxillary virus.

Review: Smith (1959).

Morphology. Patrizi et al. (1967) describe needle-like structures in nuclei of infected cells, possibly concerned with passage of virus into cytoplasm.

Antigenic properties. Virus neutralized by sera of infected guinea-pigs including apparently normal animals of infected stocks. A complement-fixing antigen is present in infected tissue cultures. Serologically unrelated to salivary viruses of other species.

Cultivation. Multiplied in fibroblasts derived from embryonic guinea-pig muscle (Hartley et al., 1957). Small foci of enlarged cells appeared in 10 days, the whole cell sheet being affected in 28 days. After serial passage, using fluid together with ground cells, cytopathic effects were seen in 1–2 days. Nuclear inclusions appeared in most cells. Titres of $10^{5 \cdot 8}$/ml were obtained.

Pathogenicity. In the natural host the disease is usually inapparent, the inclusions in salivary ducts only being discovered on histological examination. What may correspond to the generalized disease caused by the human virus was described in 2 guinea-pigs by Pappenheimer & Slanetz (1942).

The disease can normally be passed in series by inoculating the virus peripherally, allowing it to localize in the salivary glands and using these for transfer a few weeks later. Intracerebral inoculation will produce fatal meningitis in susceptible young guinea-pigs, but serial IC passage has been only once recorded and then only for 2 or 3 passages (Hudson & Markham, 1932). Intratracheal inoculation has produced a pneumonia with inclusions and an unusually virulent strain killed with generalized infection when inoculated by various routes, though serial transmission was no better than usual (Rosenbusch & Lucas, 1939). Generalized disease was produced by inoculating fetuses (Markham & Hudson, 1936)—another analogy with disease in man.

Pathological lesions. The very large nuclear inclusions in much enlarged salivary duct cells resemble those of the human virus. Cytoplasmic bodies are also seen. Nuclear inclusions produced in mononuclear cells of meningeal exudates after intracerebral injection are not particularly large and, being acidophilic, are more like typical type A inclusions.

Ecology. Many but not all stocks of guinea-pigs are latently infected. Virus is normally present in salivary glands; also in the kidneys and probably urine of young animals.

Salivary Gland Virus of Mice

Review: Smith (1959).

Antigenic properties. Unrelated to other salivary viruses. No neutralizing or CF antibodies are found in latently infected mice, but these may be produced by hyperimmunizing rabbits (Mannini & Medearis, 1961).

Cultivation. Grows in fibroblasts in cultures of mouse embryonic tissue (Smith, 1954) and of conventional (Grand, 1958) and organ cultures (Raynaud, 1967) of mouse salivary glands. Focal destruction occurs in the cultures, degeneration being diffuse in 9–12 days. After passage this only took 3 or 4 days. A plaque-assay method has been described (Field & Fong, 1964). Inclusion bodies are formed in the cultures *in vitro*. Virus is released into the fluid phase of cultures much more freely than with the human and guinea-pig viruses, and titres of $10^{7.5}$/ml may be recorded.

Pathogenicity. Normally a wholly latent infection.

Experimentally, young mice of a clean stock can be infected by any route, as with guinea-pig virus, the agent localizing in salivary glands, with which tissue alone serial passage is possible (Kuttner & Wang, 1934). However, large doses given intraperitoneally will kill in 4 to 7 days; inclusion bodies are to be found in various viscera and there is evidence of limited multiplication in liver and spleen, though insufficient to make serial passage possible (McCordock & Smith, 1936). Smaller doses produce a focal hepatitis. Intracerebral inoculation of suckling mice with quite small doses is fatal. Infection of pregnant females leads to infection of the placenta and fetal losses (Johnson, 1969). Medearis (1964b) has studied the course of chronic infection and methods of transmission. Ruebner et al. (1966) studied virus development in salivary gland and liver.

Ecology. Probably ubiquitous in wild mice, but present perhaps only in a minority of stocks of laboratory mice. There is prolonged excretion of virus in saliva (Brodsky & Rowe, 1958).

Cytomegaloviruses in Other Species

An allied virus may cause abortions of *laboratory rats* (Lyon et al., 1959). It occurs in salivary glands (Thompson, 1932; Kuttner & T'ung, 1935—these did do some transmissions). Production of a haemagglutinin is reported by Ashe (1969).

Rattus norvegicus. In kidneys of wild rats (Syverton & Larson, 1947). Kuttner & Wang (1934) transmitted an agent from wild to tame white rats.

Chinese hamsters (*Cricetulus griseus*). In salivary glands (Kuttner & Wang, 1934). Transmission from hamster to hamster was reported. The results of intracerebral and subcutaneous injections were the same as have been described for the guinea-pig virus.

Australian opossums (*Trichiurus*) (Hurst et al., 1943). Inclusions in kidneys were found only in animals which had been some while in the laboratory, not in wild ones.

Intranuclear inclusions likely to have been caused by related viruses have also been described in various organs of chimpanzees (Vogel & Pinkerton, 1955), in monkeys of several genera (Covell, 1932; Cowdry & Scott, 1935), field mice (*Apodemus*) (Raynaud & Raynaud, 1945), ground-moles (Lucas, 1936), moles (Rector & Rector, 1933), dogs (Haberman et al., 1960), sheep (Hartley & Done, 1963) and chicks (Lucas, 1947). Viruses have been isolated from ground-squirrels (*Citellus*) (Diosi et al., 1967) and sand-rats (*Psammomys*) (Melendez et al., 1967); in these 2 instances the virus grew in epithelial cells rather than fibroblasts.

INCLUSION-BODY RHINITIS OF PIGS

Review: Gwatkin (1948).

Cultivation. Grows in pig tissues, but, unlike most other cytomegalic disease viruses, does so better in epithelial than in fibroblastic cells (J. T. Done, personal communication, 1961; L'Ecuyer & Corner, 1966).

Distribution. North America and Europe.

Pathogenicity. Affects particularly 2-week-old piglets; mortality may be very high. Symptoms include sneezing, nose-bleeding and distortion of the snout from atrophy of turbinate bones; this last feature is characteristic of atrophic rhinitis, the relation of which to inclusion-body rhinitis is obscure. *Pasteurella multocida* occurs as a secondary invader.

Transmission is possible only to very young pigs. A fatal generalized cytomegalic disease of young pigs has occurred in outbreaks. Inclusion bodies are found in many organs (Corner et al., 1964).

There occur inclusions in a swollen nucleus; these resemble those of other cytomegalic diseases.

Bakos et al. (1960) describe an association between rhinitis and pneumonia in piglets.

OTHER POSSIBLE MEMBERS OF THE HERPESVIRUS GROUP

Hsiung & Kaplow (1969) obtained a virus from a strain of guinea-pigs. This appeared to be different from the cytomegalovirus: it grew readily in rabbit-tissue cultures.

Monroe et al. (1968) found bodies morphologically resembling herpesviruses in the venom of a cobra (*Naja naja*).

Hinze (1971) has described a herpesvirus from cotton-tail rabbits, producing in them a lymphoproliferative disease; it was distinct from virus III of rabbits.

A herpesvirus causing respiratory symptoms in calves is said to be distinct from other bovine herpesviruses (Mohanty et al., 1971).

A herpesvirus from iguanas has been described by Clark et al. (1968), and an agent associated with an acute viral disease of channel catfish has been shown by K. Wolf & R. W. Darlington (A. Granoff, personal communication) to be morphologically identical to herpesviruses.

REFERENCES

Aksel, I. S., & Tuncman, Z. (1940) *Z. ges. Neurol. Psychiat.*, **169**, 598.
Alexander, R. A., Plowright, W., & Haig, D. A. (1957) *Bull. epizoot. Dis.*, **5**, 489.
Amies, C. R. (1934) *Br. J. exp. Path.*, **15**, 314.
Amos, H. (1953) *J. exp. Med.*, **98**, 365.
Anderson, K., & Goodpasture, E. W. (1942) *Am. J. Path.*, **18**, 555.
Andrewes, C. H. (1929) *Br. J. exp. Path.*, **10**, 188.
Andrewes, C. H. (1940) *J. Path. Bact.*, **50**, 227.
Andrewes, C. H., & Glover, R. E. (1939) *Vet. Rec.*, **51**, 934.
Armstrong, J. A., Pereira, H. G., & Andrewes, C. H. (1961) *Virology*, **14**, 276.
Ashe, W. K. (1969) *J. gen. Virol.*, **4**, 1.
Ashe, W. K., & Notkins, A. L. (1966) *Proc. natn. Acad. Sci., U.S.A.*, **56**, 447.
Asher, Y., Heller, M., & Becker, Y. (1969) *J. gen. Virol.*, **4**, 65.
Asplin, F. D. (1970) *Vet. Rec.*, **87**, 182.
Atherton, J. G., & Anderson, W. (1957) *Aust. J. exp. Biol. med. Sci.*, **35**, 335.
Aurelian, L. (1968) *Proc. Soc. exp. Biol. Med., U.S.A.*, **127**, 485.
Aurelian, L. (1969) *J. Virol.*, **4**, 197.
Bakos, K., Obel, A-L., Swahn, O., & Walzl, H. (1960) *Zentbl. VetMed.*, **7**, 262.
Bang, B. G., & Bang, F. B. (1967) *J. Exp. Med.*, **125**, 409.

Bang, F. B. (1942) *J. exp. Med.*, **76**, 263.
Bartha, A., & Kojnok, J. (1963) *Proc. 17th Wld vet. Congress Hanover*, **1**, 531.
Baxendale, W. (1969) *Vet. Rec.*, **85**, 341.
Beach, J. R. (1931) *J. exp. Med.*, **54**, 809.
Beaseley, J. N., Patterson, L. T., & McWade, D. H. (1970) *Am. J. vet. Res.*, **31**, 339.
Beaudette, F. R., & Hudson, C. B. (1933) *J. Am. vet. med. Ass.*, 82, 460.
Becker, Y., Olshevsky, U., & Levitt, J. (1967) *J. gen. Virol.*, **1**, 471.
Beladi, J., & Ivanovics, G. (1954) *Acta microbiol. hung.*, **2**, 151.
Benda, R. (1966) *J. Hyg. Epidem. Microbiol. Immun., Moscow*, **10**, 105.
Benton, W. J., & Cover, M. S. (1957) *Avian Dis.*, **1**, 320.
Berkman, R. N., Barner, R. D., Morrill, C. C., & Langham, R. F. (1960) *Am. J. vet. Res.*, **21**, 1015.
Biggs, P. M. (1967) *Vet. Rec.*, **81**, 583.
Biggs, P. M., & Payne, L. N. (1967) *J. natn. Cancer Inst.*, **39**, 237.
Biggs, P. M., Thorpe, R. J., & Payne, L. N. (1968) *Br. Poultry Sci.*, **9**, 37.
Black, F. L., & Melnick, J. L. (1955) *J. Immunol.*, **74**, 236.
Black, P. H., Hartley, J. W., & Rowe, W. P. (1963) *Proc. Soc. exp. Biol. Med., U.S.A.*, **112**, 601.
Blackmore, R. V., & Morgan, H. R. (1967) *Acta virol.*, **11**, 1.
Breen, G. E., Lamb, S. G., & Otaki, A. T. (1958) *Br. med. J.*, **2**, 22.
Brodsky, I., & Rowe, W. P. (1958) *Proc. Soc. exp. Biol. Med., U.S.A.*, **99**, 654.
Brown, A. L., & Bjornson, C. B. (1959) *Am. J. vet. Res.*, **20**, 985.
Burnet, F. M. (1936) *J. exp. Med.*, **63**, 685.
Burnet, F. M., & Anderson, S. G. (1946) *Br. J. exp. Path.*, **27**, 236.
Burnet, F. M., & Foley, M. (1941) *Aust. J. exp. Biol. med. Sci.*, **19**, 235.
Burnet, F. M., Lush, D., & Jackson, A. V. (1939) *Aust. J. exp. Biol. med. Sci.*, **17**, 35.
Burnet, F. M., & Williams, S. W. (1939) *Med. J. Aust.*, **1**, 637.
Burrows, R. (1970) *Proc. 2nd int. Conf. equine Inf. Diseases* (1969) Paris: Karger, p. 154.
Burtscher, H. (1968) *Zentbl. VetMed.*, **15B**, 540.
Byrne, R. J., Quan, A. L., & Kaschula, V. R. (1958) *Am. J. vet. Res.*, **19**, 655.
Calnek, B. W., Ubertini, T., & Adldinger, H. K. (1970) *J. natn. Cancer Inst.*, **45**, 341.
Carmichael, L. E., & Barnes, J. D. (1961) *Proc. 65th ann. Meeting U.S. sanitary livestock Ass.*, 384.
Carmichael, L. E., Strandberg, J. D., & Barnes, F. D. (1965) *Proc. Soc. exp. Biol. Med., U.S.A.*, **120**, 644.
Caunt, A. E. (1969) *Br. J. exp. Path.*, **50**, 26.
Caunt, A., Rondle, C. J. M., & Downie, A. W. (1961) *J. Hyg. Camb.*, **59**, 249.
Caunt, A. E., & Taylor-Robinson, D. (1964) *J. Hyg. Camb.*, **62**, 413.
Chiang, W-T., Wentworth, R. B., & Alexander, E. R. (1970) *J. Immunol.*, **104**, 992.
Chow, T. L., & Davis, R. W. (1964) *Am. J. vet. Res.*, **25**, 518.
Chubb, R. C., & Churchill, A. E. (1968) *Vet. Rec.*, **83**, 4.
Churchill, A. E. (1965) *Vet. Rec.*, **77**, 1227.
Churchill, A. E. (1968) *J. natn. Cancer Inst.*, **41**, 939 and 951.
Churchill, A. E., Payne, L. N., & Chubb, R. C. (1969) *Nature, Lond.*, **221**, 744.
Clark, H. F., Zeigel, R. F., Fabian, F., & Karzon, D. T. (1968) *Bact. Proc.*, p. 149, abst. V34.
Clarkson, M. J., Thorpe, E., & McCarthy, K. (1967) *Arch. ges. Virusforsch.*, **22**, 219.
Cook, M. L., & Stevens, J. G. (1968) *J. Virol.*, **2**, 1458.
Corner, A. H., Mitchell, D., Julian, R. J., & Meads, E. B. (1964) *J. comp. Path.*, **74**, 192.
Cornwell, H. J. C., & Weir, A. R. (1970) *J. comp. Path.*, **80**, 509 and 517.
Cornwell, H. J. C., & Wright, N. G. (1969) *Vet. Rec.*, **84**, 2.
Cornwell, H. J. C., & Wright, N. G. (1970) *J. comp. Path.*, **80**, 221 and 229.
Covell, W. P. (1932) *Am. J. Path.*, **8**, 151.
Cowdry, E. V. (1930) *Arch. Path.*, **10**, 23.
Cowdry, E. V. (1934) *Arch. Path.*, **18**, 527.
Cowdry, E. V., & Scott, G. H. (1935) *Am. J. Path.*, **11**, 647.

Cox, H. R. (1952) *Ann. N.Y. Acad. Sci.*, **55**, 236.
Crandell, R. A., Ganaway, J. R., Nieman, W. H., & Maurer, F. D. (1960) *Am. J. vet. Res.*, **21**, 504.
Crandell, R. A., & Maurer, F. D. (1958) *Proc. Soc. exp. Biol. Med., U.S.A.*, **97**, 487.
Crawford, L. V., & Lee, A. J. (1964) *Virology*, **23**, 105.
Cuba-Caparo, A., de la Vega, E., & Copaira, M. (1961) *Am. J. vet. Res.*, **22**, 673.
Custer, R. P., & Smith, E. B. (1948) *Blood*, **3**, 830.
Daniel, M. D., & Melendez, L. V. (1970) *Arch. ges. Virusforsch.*, **32**, 45.
Dardiri, A. H. (1969) *Arch. ges. Virusforsch.*, **27**, 55.
Dardiri, A. H., & Gailunas, P. (1969) *Bull. Wildl. Dis. Ass.*, **5**, 235.
Daubney, R., & Hudson, J. R. (1936) *J. comp. Path.*, **49**, 63.
Diehl, V., Henle, G., Henle, W., & Kohn, G. (1968) *J. Virol.*, **2**, 663.
Diosi, P., Balusceae, L., & David, C. (1967) *Arch. ges. Virusforsch.*, **30**, 383.
Doerr, R. (1924–5) *Z. Haut- u. Geschl.-Kr.*, **13**, 417, **15**, 1, 129, 239, **16**, 481.
Doll, E. R., Crowe, M. E. W., Bryans, J. T., & McCollum, W. H. (1955) *Cornell Vet.*, **45**, 387.
Doll, E. R., Crowe, M. E., McCollum, W. H., & Bryans, J. T. (1959) *Cornell Vet.*, **59**, 49.
Doll, E. R., Richards, M. G., & Wallace, M. E. (1953) *Cornell Vet.*, **43**, 551.
Doll, E. R., & Wallace, M. E. (1954) *Cornell Vet.*, **44**, 453.
Dowdle, W. R., Nahmias, A. J., Harwell, R. W., & Pauls, F. P. (1967) *J. Immunol.*, **99**, 974.
Downie, A. W. (1959) *Br. med. Bull.*, **15**, 197.
Duff, R., & Rapp, F. (1971) *Nature new Biol.*, **233**, 48.
Dungal, N. (1946) *Am. J. Path.*, **22**, 737.
Dungal, N., Gislason, G., & Taylor, E. L. (1938) *J. comp. Path.*, **57**, 46.
Endo, M., Kanimura, T., Aoyama, Y., Hayashida, T., Kinjo, T., Ono, Y., Kotera, S., Suzuki, K., Tajima, Y., & Ando, K. (1960) *Jap. J. exp. Med.*, **30**, 227.
Epstein, M. A. (1970) *Adv. Cancer Res.*, **13**, 383.
Epstein, M. A., Henle, G., Achong, B. G., & Barr, Y. M. (1965) *J. exp. Med.*, **121**, 761.
Evans, A. S., & Paul, J. R. (1965) in *Viral rickettsial diseases of man.* Eds.: Horsfall and Tamm. 4th Ed. p. 94. Philadelphia: Lippincott.
Field, A. K., & Fong, J. (1964) *J. Bact.*, **87**, 1238.
Fink, M. A., King, G. S., & Mizell, M. (1968) *J. natn. Cancer Inst.*, **41**, 1477.
Fraser, G., & Ramachandran, S. P. (1969) *J. comp. Path.*, **79**, 435.
Geder, L., Jeney, E., & Gönczöl, E. (1964/65) *Acta microbiol. hung.*, **11**, 361.
Gerber, P., Whang-Pen, J., & Monroe, J. H. (1970) *Proc. natn. Acad. Sci., U.S.A.*, **63**, 740.
Girard, A., Greig, A. S., & Mitchell, D. (1968) *Can. J. comp. Med.*, **32**, 603.
Glade, P. R., Kasel, J. A., Moses, H. L., Whang-Peng, J., Hoffman, P. F., Kammermeyer, J. K., & Chassin, L. V. (1968) *Nature, Lond.*, **217**, 564.
Glasgow, L. A., Hanshaw, J. W., Merigan, T. C., & Petrelli, J. K. (1967) *Proc. Soc. exp. Biol. Med., U.S.A.*, **125**, 843.
Glover, R. E. (1939) *Br. J. exp. Path.*, **20**, 150.
Goodheart, C. R. (1970) *J. Am. med., Ass.*, **211**, 91.
Goodpasture, E. W., & Anderson, K. (1944) *Am. J. Path.*, **20**, 447.
Grand, N. G. (1958) *Am. J. Path.*, **34**, 775.
Gravell, M., Granoff, A., & Darlington, R. W. (1968) *Virology*, **36**, 467.
Green, R. G., & Shillinger, J. E. (1936) *Am. J. Path.*, **12**, 405.
Greig, A. S. (1958) *Can. J. Microbiol.*, **4**, 487.
Grutzner, L. (1956) *Zentbl. Bakt. ParasitKde, 1, Abt Orig.*, **165**, 81.
Gwatkin, R. (1948) *Adv. vet. Sci.*, **4**, 211.
Haberman, R. I., Williams, F. P., & Fite, G. L. (1960) *J. Am. vet. Med. Ass.*, **137**, 161.
Hanshaw, J. B. (1968) *Virology monogr.*, 3, 1. Berlin: Springer.
Hartley, J. W., Rowe, W. P., & Huebner, R. J. (1957) *Proc. Soc. exp. Biol. Med., U.S.A.*, **96**, 281.
Hartley, W. J., & Done, J. T. (1963) *J. comp. Path.*, **73**, 84.

Hausen, H. zur, Schulte-Holthausen, H., Klein, G., Henle, W., Henle, G., Clifford, P., & Santesson, L. (1970) *Nature, Lond.*, **228**, 1056.
Hayward, M. E. (1949) *Br. J. exp. Path.*, **30**, 520.
Hedley-Whyte, E. T., & Craighead, J. E. (1965) *New Engl. J. Med.*, **272**, 473.
Henle, G., & Henle, W. (1970) *J. infect. Dis.*, **121**, 303.
Henle, G., Henle, W., & 10 others (1969) *J. natn. Cancer Inst.*, **43**, 1147.
Henle, W., Diehl, V., Kohn, G., zur Hausen, H., & Henle, G. (1967) *Science*, **151**, 1064.
Henle, W., & Henle, G. (1968) *J. Virol.*, **2**, 182.
Henle, W., Henle, G., & 8 others (1970) *J. natn. Cancer Inst.*, **44**, 225.
Hess, W. R., & Dardiri, A. H. (1968) *Arch. ges. Virusforsch.*, **24**, 148.
Hinze, H. C. (1971) *Infect. Immunol.*, **3**, 350.
Hoagland, R. J. (1955) *Am. J. med. Sci.*, **229**, 262.
Hoagland, R. J. (1964) *Am. J. publ. Hlth*, **54**, 1699.
Hobson, F. G., Lawson, B., & Wigfield, M. (1958) *Br. med. J.*, **1**, 845.
Hoggan, M. D., & Roizman, B. (1959a) *Virology*, **8**, 508.
Hoggan, M. D., & Roizman, B. (1959b) *Am. J. Hyg.*, **70**, 208.
Holden, M. (1932) *J. infect. Dis.*, **50**, 218.
Holmes, A. W., Caldwell, R. G., Dedmon, R. E., & Deinhardt, F. (1964) *J. Immunol.*, **92**, 602.
Hope-Simpson, R. E. (1965) *Proc. r. Soc. Med.*, **58**, 9.
Hsiung, G. D., & Kaplow, L. S. (1969) *J. Virol.*, **3**, 355.
Huck, R. A., Shand, A., Allsop, P. J., & Paterson, A. B. (1961) *Vet. Rec.*, **73**, 457.
Hudson, N. P., & Markham, F. S. (1932) *J. exp. Med.*, **55**, 405.
Hull, R. N., & Nash, J. C. (1960) *Am. J. Hyg.*, **71**, 15.
Hummeler, K., Henle, G., & Henle, W. (1966) *J. Bact.*, **91**, 1366.
Hunt, R. D., & Melendez, L. V. (1969) *Lab. Animal Care*, **19**, 221.
Hunt, R. D., Melendez, L. V., King, N. W., Gilmore, C. E., Daniel, M. D., Williamson, M. E., & Jones, T. C. (1970) *J. natn. Cancer Inst.*, **44**, 447.
Hurst, E. W. (1933) *J. exp. Med.*, **58**, 415.
Hurst, E. W. (1936) *J. exp. Med.*, **63**, 449.
Hurst, E. W., Cooke, B. T., Mawson, J., & Melvin, P. (1943) *Aust. J. exp. Biol. med. Sci.*, **21**, 149.
Huygelen, C., (1960) *Zentbl. VetMed.*, **7**, 664.
Ivanovics, G., & Hyde, R. R. (1936) *Am. J. Hyg.*, **23**, 55.
Jansen, J. (1964) *Indian vet. J.*, **41**, 309.
Johnson, J. (1969) *Indian vet. J.*, **41**, 309.
Jungherr, E. (1939) *J. Am. vet. med. Ass.*, **94**, 49.
Kakuk, T. J., Conner, G. H., Langham, R. F., Moore, J. A., & Mitchell, J. R. (1969) *Am. J. vet. Res.*, **30**, 1951.
Kaplan, A. S. (1957) *Virology*, **4**, 435.
Kaplan, A. S. (1969) *Virology monogr.*, **5**, 4. Berlin: Springer.
Karpas, A. (1966) *Annls Inst. Pasteur, Paris*, **110**, 688.
Karpas, A., Garcia, F. G., Calvo, F., & Cross, R. E. (1968) *Am. J. vet. Res.*, **29**, 1251.
Kaschula, V. R., Beaudette, F. R., & Byrne, R. J. (1957) *Cornell Vet.*, **47**, 137.
Kaufman, H. E. (1962) *Proc. Soc. exp. Biol. Med., U.S.A.*, **109**, 251.
Keeble, S. A., Christofinis, G. J., & Wood, W. (1958) *J. Path. Bact.*, **76**, 189.
Kemeny, L., & Pearson, T. E. (1970) *Can. J. comp. Med.*, **34**, 59.
Keysselitz, G. (1908) *Arch. Protistenk.*, **11**, 326.
Kipps, A., & Polson, A. (1966) *S. Afr. med. J.*, **40**, 127.
Klemperer, H. G., Haynes, G. R., Shedden, W. I. H., & Watson, D. H. (1967) *Virology*, **31**, 120.
Kottaridis, S. D., & Luginbuhl, R. E. (1969) *Nature, Lond.*, **221**, 1258.
Kundratitz, K. (1925) *Mschr. Kinderheilk.*, **29**, 516.
Kuttner, A. G., & T'ung, T. (1935) *J. exp. Med.*, **62**, 805.
Kuttner, A. G., & Wang, S. H. (1934) *J. exp. Med.*, **60**, 773.

Landon, J. C., Ellis, L. B., Zeve, V. H., & Fabrizio, D. P. A. (1968) *J. natn. Cancer Inst.*, **40**, 181.
Lang, D. J., Scolnick, E. M., & Willerson, J. T. (1968) *New Engl. J. med.*, **278**, **1147**.
Lebrun, A. C. (1956) *Virology*, 2, 496.
L'Ecuyer, C., & Corner, A. H. (1966) *Can. J. comp. Med.*, **30**, 321.
Lee, L. F., Roizman, B., Spear, P. G., Kieff, G. E. D., Burmester, B. R., & Nazerian, K. (1970) *Proc. natn. Acad. Sci., U.S.A.*, **64**, 952.
Lindt, S., Mühlethaler, E., & Bürki, F. (1965) *Schweiz. Arch. Tierheilk.*, **107**, 91.
Loewenthal, W. (1907) *Z. Krebsforsch.*, **5**, 197.
Love, F. M., & Jungherr, E. (1962) *J. Am. med. Ass.*, **179**, 804.
Lucas, A. M. (1936) *Am. J. Path.*, **12**, 933.
Lucas, A. M. (1947) *Am. J. Path.*, **23**, 1005.
Lucké, B. (1934) *Am. J. Cancer*, **20**, 352.
Lucké, B. (1938) *J. exp. Med.*, **68**, 457.
Lucké, B. (1939) *J. exp. Med.*, **70**, 270.
Lucké, B., Berwick, L., & Nowell, P. (1953) *J. exp. Med.*, **97**, 505.
Luse, S. A., & Smith, M. G. (1958) *J. exp. Med.*, **107**, 623.
Lyon, H. W., Christian, J. J., & Miller, C. W. (1959) *Proc. Soc. exp. Biol. Med., U.S.A.*, **101**, 164.
McCarthy, K., Thorpe, E., Laursen, A. C., Heymann, C. S., & Beale, A. J. (1968) *Lancet*, 2, 856.
McCollum, W. H., Doll, E. R., & Bryans, J. T. (1956) *Am. J. vet. Res.*, **17**, 267.
McCollum, W. H., Doll, E. R., Wilson, J. C., & Johnson, C. B. (1962) *Cornell Vet.*, **52**, 164.
McCordock, H. A., & Smith, M. G. (1936) *J. exp. Med.*, **63**, 303.
McGavran, M. H., & Smith, M. H. (1965) *Exp. & Mol. Path.*, **4**, 1.
Mackay, J. M. K. (1969) *J. comp. Path.*, **79**, 141 and 147.
Mackay, J. M. K., & Nisbet, D. I. (1966) *Vet. Rec.*, 78, 18.
McKenna, J. M., Davis, F. E., Prier, J. E., & Kleger, B. (1966) *Nature, Lond.*, **212**, 1602.
McKercher, D. G. (1959) *Adv. vet. Sci.*, **5**, 299.
McKercher, D. G., & Theilen, G. H. (1963) *Proc. 7th Wld vet. Congr., Hanover*, **1**, 625.
McKercher, D. G., & Wade, E. M. (1964) *J. Am. vet. Med. Ass.*, **144**, 136.
Madin, S. H., York, C. J., & McKercher, D. G. (1956) *Science*, **124**, 721.
Malherbe, H., & Harwin, R. (1958) *Lancet*, 2, 530.
Malherbe, H., & Strickland-Cholmley, M. (1969) *Lancet*, 2, 1300.
Malherbe, H., & Strickland-Cholmley, M. (1970) *Lancet*, **1**, 785.
Mannini, A., & Medearis, D. N. (1961) *Am. J. Hyg.*, 73, 329.
Markham, F. S., & Hudson, N. P. (1936) *Am. J. Path.*, **12**, 175.
Marsh, H. (1958) *Adv. vet. Sci.*, **4**, 163.
Martin, A. M., & Kuntz, S. M. (1966) *Arch. Path.*, 82, 27.
Martin, W. B., James, Z. H., Lauder, I. M., Murray, M., & Pirie, H. M. (1969) *Am. J. vet. Res.*, **30**, 2151.
Martin, W. B., Martin, B., Hay, D., & Lauder, I. M. (1966) *Vet. Rec.*, **78**, 494.
Mayr, A., Böhm, H. O., Brill, J., & Woyciechowska, S. (1965) *Arch. ges. Virusforsch.*, **17**, 216.
Medearis, D. N. (1964a) *Am. J. Hyg.*, **80**, 103.
Medearis, D. N. (1964b) *Bull. Johns Hopkins Hosp.*, **114**, 181.
Melendez, L. V., Hunt, R. D., King, N. W., Garcia, F. G., Like, A. A., & Miki, E. (1967) *Lab. Animal Care*, **17**, 302.
Melnick, J. L., Midulla, M., Wimberley, I., & Barrera-Oro, J. G. (1964) *J. Immunol.*, 92, 596.
Mikkelsen, W., Tupper, C. J., & Murray, J. (1958) *J. Lab. clin. Med.*, **52**, 648.
Miller, C. P., Andrewes, C. H., & Swift, H. F. (1924) *J. exp. Med.*, **40**, 773.
Miller, G., Enders, J. F., Lisco, H., & Kohn, H. I. (1969) *Proc. Soc. exp. Biol. Med., U.S.A.*, **132**, 247.
Minowada, J., Chai, L., & Moore, G. E. (1969) *Cancer*, **23**, 300.
Mizell, M., Stackpole, C. W., & Halperen, S. (1968) *Proc. Soc. exp. Biol. Med., U.S.A.*, **127**, 808.
Mohanty, S. B., Hammond, R. C., & Lillie, M. G. (1971) *Arch. ges. Virusforsch.*, **34**, 394.

Monroe, J. H., & 6 others (1968) *J. natn. Cancer Inst.*, **40**, 135.
Morgan, C., Ellison, S. A., Rose, H. M., & Moore, D. H. (1954) *J. exp. Med.*, **100**, 195.
Morgan, C., Rose, H. M., & Mednis, B. (1968) *J. virol.*, **2**, 507.
Morrison, J. M., & Keir, H. M. (1968) *J. gen. virol.*, **3**, 337.
Nagler, F. P. O. (1944) *J. Immunol.*, **48**, 213.
Nagler, F. P. O. (1946) *Aust. J. exp. Biol. med. Sci.*, **24**, 103.
Nagler, F. P. O., & Klotz, M. (1958) *Can. med. Ass. J.*, **79**, 743.
Nahmias, A. J., & Dowdle, W. R. (1968) *Progr. med. Virol.*, **10**, 110.
Naib, Z. M., Nahmias, A. J., & Josey, W. E. (1966) *Cancer*, **19**, 1026.
Nazerian, K., & Wither, R. L. (1970) *J. Virol.*, **5**, 388.
Nelson, J. S., & Wyatt, J. P. (1959) *Medicine*, **38**, 223.
Nesburn, A. (1969) *J. virol.*, **3**, 59.
Netter, A., & Urbain, A. (1926) *C.r. Séanc Soc. Biol.*, **94**, 98.
Nii, S. (1969) *Biken's J.*, **12**, 45.
Nobel, T. A., Neumann, F., & Klopfer, G. (1969) *J. comp. Path.*, **79**, 537.
Numazaki, Y., Yano, N., Morizuka, T., Takai, S., & Ishida, N. (1970) *Am. J. Epidemiol.*, **91**, 410.
Okasaki, W., Purchase, H. G., & Burmester, B. R. (1970) *Avian dis.*, **141**, 413.
Pacheco, G. (1930–31) *C.r. Séanc. Soc. Biol.*, **105**, 109, **106**, 372.
Pappenheimer, A. M., & Slanetz, C. A. (1942) *J. exp. Med.*, **76**, 299.
Patrizi, G., Middelkamp, J. N., & Reed, C. A. (1967) *Am. J. Path.*, **50**, 779.
Paul, J. R., & Bunnell, W. W. (1932) *Am. J. med. Sci.*, **183**, 90.
Payne, L. N., & Biggs, P. M. (1967) *J. natn. Cancer Inst.*, **39**, 281.
Payne, L. N., & Rennie, M. (1970) *Vet. Rec.*, **87**, 109.
Persechino, A., Merucci, P., & Orfei, Z. (1965) *Nuova. Vet.*, **41**, 213.
Petzoldt, K. (1970) *Dtsch. tierarztl, Wschr.*, **77**, 162.
Piercy, S. E. (1952) *Br. vet. J.*, **108**, 35, 214.
Piercy, S. E. (1954) *Br. vet. J.*, **110**, 87.
Piercy, S. E. (1955) *Br. vet. J.*, **111**, 484.
Plowright, W. (1953) *J. comp. Path.*, **63**, 318.
Plowright, W. (1963) *Bull. epizoot. Dis. Africa*, **11**, 149.
Plowright, W. (1967) *Res. vet. Sci.*, **8**, 129.
Plowright, W., Ferris, R. D., & Scott, G. R. (1960) *Nature, Lond.*, **188**, 1167.
Plowright, W., Macadam, R. F., & Armstrong, J. A. (1963) *J. gen. Microbiol.*, **39**, 253.
Plummer, G. (1967) *Progr. med. Virol.*, **9**, 302.
Plummer, G., Bowling, C. P., & Goodheart, C. R. (1969a) *J. virol.*, **4**, 738.
Plummer, G., Goodheart, C. R., Henson, D., & Bowling, C. P. (1969b) *Virology*, **39**, 134.
Plummer, G., & Waterson, A. P. (1963) *Virology*, **19**, 412.
Plummer, G., Waner, J. L., Phuangsaab, A., & Goodheart, C. R. (1970) *J. virol.*, **5**, 51.
Poste, G., & King, N. (1971) *Vet. Rec.*, **88**, 229.
Povey, R. C., & Johnson, R. H. (1967) *Vet. Rec.*, **81**, 686.
Prier, J. E., & Goulet, N. R. (1961) *Am. J. vet. Res.*, **22**, 1112.
Prokof'eva, M. T., & Babkin, V. F. (1965) *Vetorirariya, Moscow*, **42**, 24.
Pulsford, M. F. (1953) *Nature, Lond.*, **172**, 1193.
Pulsford, M. F. (1960) *Aust. J. exp. Biol. med. Sci.*, **38**, 153.
Purchase, H. G., & Biggs, P. M. (1967) *Res. vet. sci.*, **8**, 440.
Purchase, H. G., & Burgoyne, G. H. (1970) *Am. J. vet. Res.*, **31**, 117.
Purchase, H. G., Burmester, B. R., & Cunningham, C. H. (1971) *Infect. Immunity*, **3**, 295.
Quin, A. H. (1961) *Vet. Med.*, **56**, 192.
Rafferty, K. A. (1963) *J. natn. Cancer Inst.*, **30**, 1103.
Rafferty, K. A. (1964) *Cancer Res.*, **24**, 169.
Rafferty, K. A. (1965) *Ann. N.Y. Acad. Sci.*, **126**, 3.
Rafferty, K. A., & Rafferty, N. S. (1961) *Science*, **133**, 702.
Randall, C. C. (1955) *Proc. Soc. exp. Biol. Med., U.S.A.*, **90**, 176.
Rawls, W. E., Tompkins, W., Figueroa, M. E., & Melnick, J. L. (1968) *Science*, **161**, 1255.
Raynaud, A., & Raynaud, J. (1945) *Annls Inst. Pasteur, Paris*, **71**, 344.

Raynaud, J. (1967) *Annls Inst. Pasteur, Paris*, **113**, 460.
Rector, L. E., & Rector, E. J. (1933) *Proc. Soc. exp. Biol. Med., U.S.A.*, **31**, 192.
Reczko, E., & Mayr, A. (1963) *Arch. ges. Virusforsch.*, **13**, 591.
Rivers, T. M. (1926) *J. exp. Med.*, **43**, 275.
Rivers, T. M., & Pearce, L. (1925) *J. exp. Med.*, **42**, 523.
Rivers, T. M., & Stewart, F. W. (1928) *J. exp. Med.*, **48**, 603.
Rivers, T. M., & Tillet, W. S. (1923) *J. exp. Med.*, **38**, 673.
Rivers, T. M., & Tillet, W. S. (1924) *J. exp. Med.*, **39**, 777.
Robertson, W. (1904) *J. comp. Path.*, **17**, 221.
Roizman, B. (1969) *Curr. topics in Microbiol. and Immunol.*, **49**, 1.
Roizman, B., Keller, J. M., Spear, P. G., Terni, M., Nahmias, A., & Dowdle, W. (1970) *Nature, Lond.*, **227**, 1253.
Roizman, B., Spring, S. S., & Roane, P. R. (1967) *J. Virol.*, **1**, 181.
Roizman, B., Spring, S. S., & Schwartz, J. (1969) *Fed. Proc.*, **28**, 1890.
Rose, S. M., & Rose, F. C. (1952) *Cancer Res.*, **12**, 1.
Rosenbusch, C. T., & Lucas, A. M. (1939) *Am. J. Path.*, **15**, 303.
Rowe, W. P. (1960) in *Viral infections of infancy and childhood*. Ed.: Rose. Symposium No. 10. *N.Y. Acad. med.* p. 205. New York: Hoeber–Harper.
Rowe, W. P., Hartley, J. W., Waterman, S., Turner, H. E., & Huebner, R. J. (1958) *Proc. Soc. exp. Biol. Med., U.S.A.*, **92**, 418.
Royston, J., & Aurelian, L. (1970) *Am. J. epidemiol.*, **91**, 531.
Ruebner, B. H., Hirano, T., Slusser, R., & Osborn, J., & Medearis, D. N. (1966) *Am. J. Path.*, **48**, 971.
Russell, W. C., & Crawford, L. V. (1964) *Virology*, **22**, 288.
Rweymamu, M. M., & Johnson, R. H. (1969) *Res. vet. Sci.*, **10**, 419.
Rweymamu, M. M., Johnson, R. H., & McCrea, M. R., (1968) *Br. vet. J.*, **124**, 317.
Rweymamu, M. M., Osborne, A. D., & Johnson, R. H. (1969) *Res. vet. Sci.*, **10**, 203.
Sabin, A. B. (1934) *Br. J. exp. Path.*, **15**, 248, 268, 321, 372.
Sabin, A. B. (1949) *J. clin. Invest.*, **28**, 808.
Sabin, A. B. (1968) *Int. virol.*, **1**, 189.
Sabo, A., Rajčáni, J., & Bláskovič, D. (1968a) *Acta virol.*, **12**, 214.
Sabo, A., Rajčáni, J., Raus, J., & Karelová, E. (1968b) *Arch. ges. Virusforsch.*, **25**, 288.
Schmidt, N. J., Lennette, E. H., & Magoffin, R. L. (1969) *J. gen. Virol.*, **4**, 321.
Schneweis, K. E. (1962) *Z. Immun. Forsch.*, **124**, 24.
Schubert, G. H. (1966) *Bull. Office Int. epizoot.*, **65**, 1011.
Schwartz, J., & Roizman, B. (1969) *Virology*, **38**, 42.
Schwarz, A. J. F., York, C. J., Zirbel, L. W., & Estela, L. A. (1957) *Proc. Soc. exp. Biol. Med., U.S.A.*, **96**, 453.
Schwarz, A. J. F., Zirbel, L. W., Estela, L. A., & York, C. J. (1958) *Proc. Soc. exp. Biol. Med., U.S.A.*, **97**, 680.
Scott, L. V., Felton, F. G., & Barney, J. A. (1957) *J. Immunol.*, **78**, 211.
Scott, T. F. McN., Macleod, D. L., & Tokumaru, T. (1961) *J. Immunol.*, **86**, 1.
Seifried, O. (1931) *J. exp. Med.*, **54**, 817.
Semerdjiev, B. (1962) *Zentbl. Bakt. ParasitKde, I. Abt. Orig.*, **185**, 316.
Shimizu, T., Ishizaki, R., Kono, Y., Ishii, S., & Matumoto, M. (1958) *Jap. J. exp. Med.*, **27**, 175.
Shope, R. E. (1931) *J. exp. Med.*, **54**, 233.
Shope, R. E. (1935a) *J. exp. Med.*, **62**, 85.
Shope, R. E. (1935b) *J. exp. Med.*, **62**, 101.
Shroyer, E. L., & Easterday, B. C. (1968) *Am. J. vet. Res.*, **29**, 1355.
Sigurdsson, B. (1958) *Arch. ges. Virusforsch.*, **8**, 51.
Skoda, R. (1962) *Acta virol., Prague*, **6**, 189.
Slotnick, V. B., & Rosanoff, E. (1963) *Virology*, **19**, 589.
Smadel, J. E., Jackson, E. B., & Harman, J. W. (1945) *J. exp. Med.*, **81**, 385.
Smith, M. G. (1954) *Proc. Soc. exp. Biol. Med., U.S.A.*, **86**, 435.

Smith, M. G. (1959) *Progr. med. Virol.*, **2**, 171.
Smith, W., & Mackay, J. M. K. (1969) *J. comp. Path.*, **79**, 421.
Soehner, R. L., Gentry, G. A., & Randall, C. C. (1965) *Virology*, **26**, 394.
Spear, P. G., & Roizman, B. (1968) *Virology*, **36**, 545.
Springer, G. F., & Callahan, H. J. (1965) *J. Lab. clin. Med.*, **65**, 617.
Stackpole, C. W., & Mizell, M. (1968) *Virology*, **36**, 63.
Stern, H. (1968) *Br. med. J.*, **1**, 665.
Stevens, D. A., Pry, T. W., Blackham, E. A., & Manaker, R. A. (1970) *Proc. Soc. exp. Biol. Med., U.S.A.*, **133**, 678.
Stewart, S. E. (1965) *Science*, **148**, 1341.
Stewart, S. E. (1969) *Ann. Rev. Microbiol.*, **15**, 291.
Stewart, S. E., David-Ferreira, J., Lovelace, E., Landor, J., & Stock, J. (1965a) *Science*, **148**, 1341.
Stewart, S., Lovelace, E., Whang, J., & Ngu, V. A. (1965b) *J. natn. Cancer Inst.*, **34**, 319.
Stoker, M. G. P., & Newton, A. (1959) *Virology*, **7**, 438.
Stoker, M. G. P., & Ross, R. W. (1958) *J. gen. Microbiol.*, **19**, 250.
Strandberg, J. D., & Carmichael, L. E. (1965) *J. Bact.*, **90**, 1790.
Straub, O. C., & Böhm, H. O. (1965) *Dtsch. tierärztl. Wschz.*, **72**, 124.
Studdert, M. J., Barker, C. A. V., & Savan, M. (1964) *Am. J. vet. Res.*, **25**, 303.
Syverton, J. T., & Larson, C. L. (1947) *Arch. Path.*, **43**, 541.
Taylor-Robinson, D., & Downie, A. W. (1959) *Br. J. exp. Path.*, **40**, 398.
Taylor-Robinson, D., & Rondle, C. J. M. (1959) *Br. J. exp. Path.*, **40**, 517.
Tegtmeyer, P., & Enders, J. F. (1969) *J. Virol.*, **3**, 469.
Thompson, J. (1932) *J. infect. Dis.*, **50**, 162.
Tischendorf, P. (1969) *Zentbl. Bakt. ParasitKde, I. Abt. Orig.*, **211**, 1.
Tobin, J. O'H. (1968) *Int. Virology*, **1**, 255. Basel: Karger.
Tokumaru, T. (1957) *Proc. Soc. exp. Biol. Med., U.S.A.*, **96**, 55.
Tokumaru, T. (1968) *Arch. ges. Virusforsch.*, **24**, 104.
Tokumaru, T. (1970) *Arch. ges. Virusforsch.*, **29**, 295.
Toth, T. E. (1970) *Am. J. vet. Res.*, **31**, 1275.
Vogel, F. S., & Pinkerton, H. (1955) *Endocrinology*, **60**, 251.
Wallis, C., & Melnick, J. L. (1965) *J. Bact.*, **90**, 1632.
Watrach, A. M., & Hanson, L. E. (1963) *Proc. Soc. exp. Biol. Med., U.S.A.*, **112**, 230.
Watrach, A. M., Hanson, L. E., & Watrach, M. A. (1968) *Virology*, **35**, 321.
Watson, D. H., Russell, W. C., & Wildy, P. (1963) *Virology*, **19**, 250.
Watson, D. H., Shedden, W. I. H., Elliot, A., Tetsuka, T., Wildy, P., Bourgaux-Ramoisy, D., & Gold, E. (1966) *Immunology*, **11**, 399.
Watson, D. H., Wildy, P., Harvey, B., & Shedden, W. (1967) *J. gen. virol.*, **1**, 139.
Webster, R. G. (1959) *N.Z. vet. J.*, **7**, 67.
Weinberg, A., & Becker, Y. (1969) *Virology*, **39**, 312.
Weller, T. H. (1953) *Proc. Soc. exp. Biol. Med., U.S.A.*, **83**, 340.
Weller, T. H. (1958) Harvey lectures. New York: Academic Press.
Weller, T. H., & Hanshaw, J. B. (1962) *New Engl. med. J.*, **266**, 1233.
Weller, T. H., Hanshaw, J. B., & Scott, D. E. (1960) *Virology*, **12**, 130.
Weller, T. H., MacCauley, J. E., Craig, J. M., & Wirth, P. (1957) *Proc. Soc. exp. Biol. Med., U.S.A.*, **94**, 4.
Wildy, P. (1955) *J. gen. Microbiol.*, **13**, 346.
Wildy, P., Russell, W. C., & Horne, R. W. (1960) *Virology*, **23**, 419.
Witter, R. L., Nazerian, K., Purchase, H. G., & Burgoyne, G. H. (1970) *Am. J. vet. Res.*, **31**, 525.
Woernle, H., & Brunner, A. (1961) *Tierärztl. Umsch.*, **16**, 245.
Wright, P. A. L. (1963) *Vet. Rec.*, **75**, 685.
York, C. J., Schwarz, A. J. F., & Estela, L. A. (1957) *Proc. Soc. exp. Biol. Med., U.S.A.*, **94**, 740.
Yoshino, K. (1956) *J. Immunol.*, **76**, 301.

16

Poxviruses

Reviews: Joklik (1966; 1968). Woodson (1968).

Morphology and development. Virions brick-shaped or oval 250–300 × 200 nm in diameter. There is an electron-dense pepsin-resistant central body, 100–200 nm across—the nucleoid. This may be in form of a biconcave disc with thickenings in the surrounding layer forming 'lateral bodies'. The nucleoid is filled with dense filamentous material, which may be in a tight S-shape (Peters & Müller, 1963), but there is no helical or cubical nucleocapsid. Immediately surrounding the nucleoid is a palisaded structure and finally an outer membrane containing protein threads, differing in arrangement in the various sub-groups. These impart to the virion a mulberry-like surface. According to Westwood et al. (1964) they are formed from loops of filaments 9 nm wide which are themselves double helices, probably protein rather than nucleo-protein in nature. Some virions do not show this structure, probably because phospho-tungstic acid used in preparing the specimen has penetrated more deeply and masked it. Multiplication is in cytoplasm; virions mature in cytoplasmic foci and bud from cellular membranes.

Chemical composition. Contain double-stranded DNA of MW 120–240 × 10^6 daltons. GC content 35–36 per cent. Two per cent lipid may be present but this is not 'essential lipid', for the members vary in sensitivity to ether. Fuller details of chemical composition available especially for vaccinia.

Physico-chemical characters. Relatively resistant to heat, very stable at −75°, less so at −10°. Density of DNA about 1·7.

The poxviruses can be tentatively divided into 6 groups as shown in Table 8.

A haemagglutinin separable from the virus particle itself is described for some members.

TABLE 8

POXVIRUSES

	Subgroup I Viruses closely related to variola	*Subgroup II* Viruses related to orf	*Subgroup III* Other viruses affecting ungulates	*Subgroup IV* Avian poxes	*Subgroup V* Viruses related to myxoma	*Subgroup VI* Unclassified poxviruses
Members of group	Variola Alastrim Vaccinia Rabbit pox Monkey pox Ectromelia Cow pox Buffalo pox	Orf Milker's nodes (paravaccinia) Bovine papular stomatitis	Sheep pox Goat pox Lumpy skin disease	Fowl pox Canary pox and other bird poxes	Rabbit myxoma Rabbit fibroma Hare fibroma Squirrel fibroma	Molluscum contagiosum Swine pox Tanapox virus Yaba virus Horse pox Camel pox
Special characters of group	All close related antigenically, similar morphologically and ether-resistant	Similar morphologically (woven pattern) Moderately ether-sensitive	Some are partly ether-sensitive	Large virus particles contained in a matrix, probably lipo-protein Ether-resistant Often transmitted by insects	Ether-sensitive Normally mechanically transferred by insects	Contains those pox viruses which cannot yet be placed in another group

Antigenic properties. (See review by Wilcox & Cohen, 1969.) All the viruses contain a common nucleo-protein antigen demonstrable by complement fixation or fluorescent antibody (Takahashi et al., 1959; Woodroofe & Fenner, 1962).

Several viral antigens can be revealed by complement fixation, precipitation, virus neutralization and haemagglutinin-inhibition. Some of the viruses, though distinct in biological behaviour, stand very close together antigenically. When 2 closely related poxviruses grow in the same cells, hybrids or recombinants having characters derived from both 'parents' may be formed (Woodroofe & Fenner, 1960). Also, poxviruses inactivated by heat may have their activity restored within cells in which another poxvirus is multiplying (Joklik et al., 1960). Only other poxviruses have this effect; other viruses tested have been ineffective (Fenner, 1962). Here may be a character useful in defining the group.

Cultivation. *In eggs:* most but not all poxviruses have been cultivated in developing chick embryos; those which do so produce 'pocks' or focal lesions on chorioallantoic membranes.

Tissue culture of most poxviruses has proved possible when seriously attempted.

Habitat. Many mammals and birds are susceptible to one or more poxviruses; there is no evidence that carp 'pox' is a member of the group.

Pathology. The epidermis, or at least the superficies of the animal, is characteristically attacked, with formation of focal lesions often proliferative in character. Such proliferative lesions are often followed by necrosis. Generalized infection occurs with most of the group. Many form cytoplasmic inclusions in epidermal or other cells.

Transmission is by the respiratory route or through the skin. Some are mechanically transmitted by arthropods but a developmental cycle in arthropods is unproved.

Prevention. Living attenuated viruses are useful for control of many poxviruses, inactivated vaccines generally much less so.

SUBGROUP I. VIRUSES RELATED TO VACCINIA

Viruses in this group are so closely related that it is fair to assume that facts elicited about vaccinia are equally true for the other members.

Those included in the group are vaccinia, variola, monkey pox, rabbit pox, cow pox, infectious ectromelia and buffalo pox.

VACCINIA

The origin of the vaccinia used for protection against smallpox is obscure. A derivation from variola has been claimed but many consider an origin from cowpox more probable. Bedson & Dumbell (1964) suggest that it may have arisen as a hybrid between those two viruses.

Morphology and development. Internal structure of the virion as for other poxviruses. After virus has been taken into the cell by a process of engulfment, the outer coat is removed by a stripping or uncoating enzyme (Dales, 1963; Joklik, 1962b). This seems to be a 2-stage process, part occurring immediately, part after a lag (Joklik, 1966). Further development is within the cytoplasm; liberation from the cell need not entail its disruption. Easterbrook (1966) used nonionic detergents and mercaptoethanol to take the virus gradually to pieces, first removing the outer coat, then the lateral bodies. The 'nucleocapsid' within had 2 layers, the outer of cylindrical subunits 10×5 nm, and an inner one 5 nm thick. Long threads of nucleic acid could be released from the core. Before liberation from the cell, double membranes may be seen (Morgan et al., 1954) and stalked forms on the surface of the cell have been observed (Robinow, 1950). Negative-staining techniques reveal subunits on the surface of the particle but within the outer membrane (Noyes, 1962). Treatment with various chemicals throws light on their structure (McCrea et al., 1962).

Chemical composition. The central body contains deoxyribonucleoprotein; the outer parts probably consist largely of protein. The DNA molecules may be up to 83 μm long. MW 80×10^6. GC content 36 per cent (Joklik, 1962a). MW $150–170 \times 10^6$ daltons (Sarov & Becker, 1967). Holowczak & Joklik (1967) studied the composition of virions and concluded that there were probably 17 components. Three 'early' proteins were probably associated with formation of the virus core, ones formed later with more superficial structures (Wilcox & Cohen, 1967; Cohen & Wilcox, 1968). The DNA may exist as a single molecule, possibly circular (Easterbrook, 1967). A number of enzymes have been identified in the virions including a DNA-dependent RNA polymerase (Kates & McAuslan, 1967). There have been numerous studies of the RNA which can be detected during virus synthesis (Kates & Beeson, 1970); some of the RNA is apparently double-stranded (Colby & Duesberg, 1969).

Physico-chemical characters. *Physical properties.* The specific gravity is variously reported as between 1·10 and 1·33. The virus carries a negative charge between pH 5·5 and pH 8·4 (Douglas et al., 1929).

Inactivation by physical agents. The radio-sensitive volume of the unit carrying infective properties is $1{\cdot}9 \times 10^{-17}$ cm^3, or about 10 per cent of the whole viral DNA (Deeker et al., 1969). The virus can be kept indefinitely at $-75°$ in the presence of peptone or other protective colloids and freeze-dried virus survives *in vacuo* or in nitrogen for many years. Ability of dried vaccine to withstand storage at 37° or 45° for many weeks or months. Cross et al. (1957) is of great practical importance in vaccine campaigns in the tropics. Suspensions are inactivated in 10 minutes at 60°C, but dried virus withstands 100°C for 10 minutes. Kaplan (1958), Woodroofe (1960) and Sharp et al. (1964) have studied the kinetics of heat inactivation; apparently virus becomes more resistant to heating on storage. It loses activity in 1 hour at pH 3, but it is fairly stable between pH 5 and 9. It can survive for weeks when dried on cloths or glass. Methods of preserving vaccine intended for immunization have been reviewed by Collier (1954) from the historical and practical aspects.

Inactivation by chemical agents. The virus is very resistant to ethyl ether in the cold and to sodium desoxycholate, but is inactivated by chloroform (Wittman & Matheka, 1958). It resists 1 per cent phenol at 4°C, but not at 37°C and is on the whole more resistant to disinfectants than are most bacteria. It is most readily destroyed by oxidizing agents such as potassium permanganate or by ethylene oxide. It is inactivated by *p*-iodo-acetamide and other SH-reactive compounds (cf. Dunham & MacNeal, 1943).

Haemagglutination. After incubation for an hour at 37°C preparations of virus agglutinate the red blood cells of turkeys (Datt, 1964) and of some but not all fowls (Nagler, 1942). There may be partial agglutination of cells of some other species (Clark & Nagler, 1943). The haemagglutinin, which may be a lipoprotein (Burnet & Stone, 1946), is separable from virus particles by centrifugation, withstands boiling and has an estimated diameter of 65 nm (Chu, 1948) though there may be 2 components, the larger of which is more heat stable (Gillen et al., 1950). Some vaccinial strains (also a strain of rabbit pox) do not form haemagglutinins (Fenner, 1958). In some such negative preparations a haemagglutinin may be revealed by sonication; possibly an inhibitor is destroyed (Rondle, 1969). Antibodies to the haemagglutinin, distinct from other vaccinial antibodies, appear in immunized animals and in man.

Antigenic properties. The virus surface contains a complex (LS) antigen with 2 serologically reactive components L (heat labile), S (heat stable). A third, NP (nucleoprotein) antigen is also recognized (Craigie & Wishart, 1934; Smadel et al., 1942). These are concerned in reactions of precipitation, agglutinins for elementary bodies and complement-fixation. The L and S components of the LS antigen may be degraded separately. These antigens are distinct from the haemagglutinin. Immunodiffusion tests in agar indicate the presence of 8 antigens extractable from purified virus (Zwartouw et al., 1965) while extracts from infected rabbit skin show as many as 17 lines of precipitate. Some of them may represent substances not incorporated into virus (Westwood et al., 1965). Some success in separating various immunoprecipitins is reported by Marquardt et al. (1967). Neutralizing antibody is probably unrelated to any of them; this may be titrated by mixing with virus, titrating intradermally in the rabbits' skin, on the chorioallantoic membrane of hens' eggs or in tissue culture. The last, using a plaque-counting technique, is said to be the best for some purposes (Cutchins et al., 1960).

All tests show that the viruses of vaccinia, variola, cow pox and ectromelia are very closely related serologically. Differences can, however, be shown by neutralization tests on chorioallantoic membranes (Downie & McCarthy, 1950), by indirect complement fixation (Downie & Macdonald, 1950) and by double-diffusion in agar (Gispen, 1955). There is of course cross-immunity, when it can be tested, between these viruses in experimental animals and in man. Variola is said to protect less well against vaccinia in monkeys than the other way round (Horgan & Hasseeb, 1939). The practical use of vaccinia is discussed under variola. An antigen particularly concerned with immunity has been described by Appleyard (1961). Turner et al. (1970) have compared methods of inactivation in an endeavour to find an effective killed vaccine.

Cultivation. *In developing eggs.* Vaccinia grows well on the chorioallantoic membrane, especially of 7- to 13-day-old eggs; dilute suspensions produce isolated focal lesions or pocks and, thus, is available a method of virus-titration. Pocks are formed even up to a temperature of 40·5° (Bedson & Dumbell, 1961). Possibilities of increasing the accuracy of this method of titration have been studied by Westwood et al. (1957). Different vaccinia strains produce varying types of lesion; these—with neurovaccinia—may be haemorrhagic. The death rate of the embryo also varies with strain and conditions of test from 100 per cent to something quite low. Inoculation into the allantoic cavity and yolk sac may also be used, but these are less sensitive than the chorioallantoic membrane (Cabasso & Moore, 1957).

In tissue-culture. Vaccinia was the first virus to be grown in tissue culture. It can be grown in cultures of chick embryo, rabbit kidney, testis, and other tissues, bovine embryo, continuous cell-lines such as HeLa and L and in various other cells. Cytopathic effects are produced as early as 48 hours and include formation of giant cells and reticulum formation from lengthening of cytoplasmic processes. Basophilic and eosinophilic inclusions are formed in the cytoplasm. Some of the effects seem to be due to a 'toxic' action for which living virus is not necessary (Bernkopf et al., 1959); these may be brought about by synthesis of some virus-induced proteins (Bablanian, 1968). Ultimately there is cell destruction. With suitable techniques this can be localized so that plaques are formed and can be counted (Noyes, 1953; Porterfield & Allison, 1960). Culture in bovine embryo has been advocated as a practical measure for vaccine production (Wesslén, 1956).

Pathogenicity. *Disease in man* is usually local. A primary vaccinial reaction is a papule developing into a vesicle and then a pustule, the lesion being maximal at the eighth to tenth day. Other reactions are referred to under variola (p. 383). Generalized vaccinia may occur, especially in children and in persons with pre-existing dermatoses: it may be fatal. An even rarer condition, progressive vaccinia, in hypogammaglobulinaemics, is usually fatal.

Disease in experimental animals. Local skin lesions are produced in calves and sheep; these are used for routine production of vaccine. Much work has been done with rabbits. Strains of virus passed by inoculating scarified skin produce abundant papulo-vesicles; those passed by intradermal inoculation give rise to large swellings necrosing centrally; passage by either method soon modifies the virus so that it takes better when inoculated by the accustomed route. Virulence is exalted by intratesticular or intracerebral passage, resulting in production of neurovaccinia. Corneal inoculation leads to keratitis and a vaccinial pneumonia may be produced by intranasal instillation under anaesthesia. With neurovaccinia and some other strains, generalization may occur, giving rise to multiple lesions in skin, mucous membranes and adrenals. Guinea-pigs react much as rabbits do. Strains vary in ease of adaptation to mice. Most produce local lesions on inoculation into skin or brain or intranasally, but only some can be passed in series or kill. Big doses given IV to suckling mice will kill without necessarily multiplying (Zakay-Roness & Bernkopf, 1962). Fenner (1958) describes the biological characters of 24 different strains used in his genetic studies.

Morbid anatomy and histopathology. Skin lesions induced in rabbits vary with the strain of virus. Levaditi & Nicolau (1923) and others emphasize the affinity for epidermal and other epiblastic tissues with proliferation followed by necrosis, but Ledingham (1924) describes the lesions as 'of the order of an infective granuloma' . . . with 'no evidence of an elective affinity for epiblastic tissue'. Douglas et al. (1929) describe the lesions of generalized vaccinia in rabbits. Presence of cytoplasmic inclusions, Guarnieri bodies, usually juxta-nuclear, has been used in diagnosis, particularly their occurrence in the cornea of inoculated rabbits or guinea-pigs. Kato et al. (1959) describe in poxvirus infections cytoplasmic inclusions of 2 types A and B (not to be confused with Cowdry's A and B intranuclear inclusions). The A inclusions contain protein, the B inclusions are Feulgen-positive (DNA containing) and also contain different antigens as shown by staining by fluorescent antibody (Loh & Riggs, 1961). Virus particles—elementary bodies or Paschen bodies—can be stained in preparations of infected tissues by Giemsa's stain or Victoria blue.

Ecology. *Transmission.* The virus may be accidentally transmitted to contacts of vaccinated persons or to laboratory workers, usually through the skin. Generalized vaccinia is said to be unusually common in such cases. Normal rabbits housed alongside infected ones may undergo an immunizing inapparent infection.

Control. Use of vaccinia for immunization is discussed under variola.

Chemotherapy. A number of substances have proved active in the laboratory. Amongst the most promising are thiosemicarbazones (Hamre et al., 1950; Bauer, 1955) and rifampicin (Subak-Sharpe et al., 1969). IDUR (2-iodo-2′-deoxyuridine) is effective against vaccinial kerato-conjunctivitis in rabbits (Kaufman et al., 1962).

RABBIT POX

Epidemics amongst laboratory rabbits, often with high fatality, are described from Holland, the United States and elsewhere. The virus, rabbit pox (Greene, 1934) or rabbit plague (Jansen, 1941) is immunologically identical with vaccinia or almost so, but differs in virulence for rabbits. After infection by the respiratory route it not only multiplies in the rabbits' lungs but causes generalized infections, as vaccinia and variola did not (Westwood et al., 1966). Christensen et al. (1967) have described a 'pockless' rabbit pox. A comparison of its other biological characters with those of a number of vaccinia strains (Fenner, 1958) reveals no other firm basis for separation from vaccinia. It has not been reported amongst wild rabbits and may well be a laboratory 'sport' from vaccinia.

MONKEY POX

An epidemic disease has also been described amongst captive monkeys (Magnus et al., 1959; Prier et al., 1960): there is a generalized variola-like rash. Infected monkeys are not seriously ill except, at times, very young animals. The sporadic appearance amongst stocks of monkeys suggests that the virus may exist as a silent infection (Magnus et al., 1959). The virus is very close to vaccinia. It is distinct from the Yaba virus (p. 406) and Tanapox (p. 407).

These reports of a pox disease among monkeys describe an illness with a generalized variola-like rash which is nevertheless not very severe (Magnus et al., 1959; Prier et al., 1960).

More recently there have been accounts of epidemics in zoos (Peters, 1966; Gispen et al., 1967; Wenner et al., 1968). Cynomolgus monkeys were more seriously affected than rhesus, while the disease caused deaths among orang-outans and severe illness in gorillas; the severity of the disease in chimpanzees was variable. An ant-eater (*Myrmecophaga*) was also affected. The virus was readily transmissible to rabbits, but resembled variola in producing only small pocks on the CAM (Gispen et al., 1967). Rouhandeh et al. (1967) describe the properties of the virus. Immunization with vaccinia gave good protection.

Arita & Henderson (1968) describe outbreaks among *Cebus* monkeys in South America. Spread to man may occasionally occur (WHO, 1971).

VARIOLA

Synonyms: Smallpox.
Less virulent forms: variola minor, alastrim, amaas.

Reviews: cf. Poxvirus. Bedson & Dumbell (1967).

Physico-chemical characters. The virus is very stable and has survived in crusts kept at room temperature for over a year. Vesicle fluid remains active for years if kept in sealed tubes in the cold. Most strains are inactivated by heating for 30′ at 55°C; resists most disinfectants, as does vaccinia, and is most readily destroyed by oxidizing agents such as KMnO4.

Haemagglutination. Haemagglutinins are like those of vaccinia, but may not be readily obtained in such high titre. In tissue cultures haemadsorption of sensitive fowls RBCs occurs (Vieuchange, 1959). Anthony et al. (1970) describe 3 haemagglutinins.

Antigenic properties. Very similar to vaccinia and cow pox; minor differences can be shown (cf. Vaccinia, p. 378). Precipitation, complement fixation (Gordon, 1925; Craigie & Wishart, 1934) or agar gel-diffusion may be used for diagnosis if smallpox crusts are available. Sera used for diagnosis should be able to detect the heat-stable (S) antigen (cf. Vaccinia, p. 377). Soluble antigen may be present in the blood of patients with fulminating haemorrhagic smallpox (Downie et al., 1969).

Cultivation. *In developing eggs.* 'Pocks' on infected chorioallantoic membranes are smaller than those of vaccinia, more dome-shaped, less necrotic and not haemorrhagic; embryos are usually not killed unless the infecting dose is large. Limited growth is possible after inoculation into the yolk sac or amniotic cavity; allantoic inoculation is unsuccessful (Hahon et al., 1958). Variola major will produce pocks when incubated at 38–38·5°, unlike variola minor (alastrim) which will only do so at temperatures below 38°C (Bedson & Dumbell, 1961).

In tissue culture. Grows, producing CPE, in tissue cultures of many kinds—chick embryo, HeLa cells, bovine tissues, etc. Foci of proliferation may be seen before cell destruction sets in. Virus may be isolated directly from human lesions in tissue cultures of human-embryo skin and muscle (Marennikova et al., 1959).

Distribution. Variola major has been a widespread and lethal disease, especially in parts of Asia and Africa, and in Central America. In other parts of the world variola minor (alastrim) is endemic. Strains of intermediate virulence may occur in Africa (WHO, 1964). Recent outbreaks in Britain have mostly been caused by introduction of variola major; they have been rapidly brought under control.

Pathogenicity. The extensive, often confluent, rash produced in man by variola major is familiar. With alastrim, pocks may be very few and may be missed altogether. In vaccinated contacts infection may take the form of pneumonia, or conjunctivitis may be the only symptom of the disease (Kempe et al., 1969).

Experimentally, various species of monkeys can be infected with production of a generalized rash, and infection can be propagated in series. Dumbell & Bedson (1966) after preliminary passages in baby rabbit-kidney cultures, adapted 2 strains to grow in series in skins of rabbits inoculated by scarification. There was no change in virus properties towards those of vaccinia. Suckling mice can be infected intracerebrally, intraperitoneally or intranasally. In general, attempted serial propagation in rodents has failed, but the virus has been adapted to go serially in suckling mouse brains (Brown et al., 1960). The production of keratitis and Guarnieri bodies in eyes of inoculated rabbits (Paul's test) was formerly used in diagnosis.

The classical account of the pathology is by Councilman et al. (1904), and there are good reviews by Lillie (1930) and Bras (1952). First lesions in the skin of man are proliferative, all layers of the skin being involved; later, necrosis and fibrosis lead to the familiar 'pitting'. Similar lesions occur in mucous membranes. Cytoplasmic inclusions (Guarnieri) bodies occur in epithelial cells in man and experimental animals. They probably consist of collections of virus particles together with matrix. Intranuclear inclusions also are found from time to time, but their significance is obscure.

Ecology. Infection is probably air-borne. The source of infection, early in the disease, is probably from mouth and nose, later from dried crusts. Contaminated bed-clothes can infect.

Control. The whole question of control and the possibility of eradication of the disease are discussed in a WHO report (1968). Isolation of patients, quarantine of contacts and vaccination are all-important. Vaccination is most frequently carried out with living vaccinia virus propagated on the skin of calves or sheep. Great advantages would accrue from the substitution of virus propagated in a medium free from bacteria such as chick-embryo or bovine-embryo tissue culture (Wesslén, 1956). Such vaccines are on trial with promising results. Virus propagated wholly in chick embryo tends to lose immunogenicity after numerous passes; it may be revived, however, by rabbit passage and become effective once more (Noordaa et al., 1967). Scarification and multiple pressure are the techniques recommended; intradermal inoculation is not so good. Lymph is commonly preserved in glycerol, but freeze-dried vaccine (Collier, 1955) has far better keeping properties especially for use in the tropics. Official requirements for smallpox vaccine have been laid down by WHO (1959).

Encephalitis following vaccination is an infrequent complication, but an undeniable small risk. It is associated with demyelination and is probably an auto-immune phenomenon analogous to other post-infection encephalitides. It is less frequent when vaccination is carried out in infancy.

Vaccinated persons may show: (a) when fully susceptible, a primary reaction with vesiculation, maximal at 8–10 days; (b) with partial immunity, a vaccinoid or accelerated reaction, maximal at 3–7 days, with presence of a smaller vesicle; (c) an early or immediate reaction at 2–3 days, usually only papular; this last may be elicited by live or killed virus and may occur both in people who are immune and in those who are susceptible.

Chemotherapy. Bauer et al. (1963; 1969) and do Valle et al. (1965) have shown that N-methyl isatin-β-thiosemicarbazone (methisazone) has prophylactic value when given to contacts.

COW POX

Now agreed to be a disease distinct from both vaccinia and from milkers' nodes (often called natural cow pox in the United States).

Haemagglutination. Though to lower titre, is against the same range of fowl red cells as with vaccinia.

Antigenic properties. Very close to vaccinia, but distinguishable by refined tests using complement-fixation, agar gel-diffusion and antibody-adsorption (Downie, 1939; Downie & McCarthy, 1950; Gispen, 1955; Rondle & Dumbell, 1962).

Cultivation. *In eggs*. Haemorrhagic pocks are produced on chorio-allantoic membranes, but not above 40° (Bedson & Dumbell, 1961); with large doses the embryo may be killed. Several variant strains have been isolated causing white pocks on the membrane; these owe their colour to the more rapid necrosis produced and absence of haemorrhage (Downie & Haddock, 1952).

In tissue culture. Plaques are produced on monolayers of chick-embryo cultures, as with vaccinia (Porterfield & Allison, 1960); also in human and bovine cells.

Pathogenicity. *Natural host range*. Affects the skin, particularly teats and udders of cows. Frequently spreads to milkers affecting their hands, sometimes arms and face, and even the eye. In cattle, papules develop into vesicles on a firm inflamed base. Crusting follows and may not clear up for several weeks. Lesions in man resemble those of primary vaccination.

Readily infects rabbits, guinea-pigs, mice and monkeys—other species have probably not been tested; skin and testis have been used as sites for infection. There is less rapid epithelial necrosis than with vaccinia, more invasion of mesodermal tissue and more haemorrhage (Downie, 1939). Mice inoculated intraperitoneally are killed more regularly than by vaccinia (Moritsch, 1956). Lesions on rabbit cornea are smaller.

The cytoplasmic inclusions in the lower epidermal layers are larger than the Guarnieri bodies of vaccinia and variola (Downie, 1947). Kato et al. (1959) suggest that the larger inclusions are of the nature of their protein A inclusions (cf. Vaccinia, p. 380 and not analogues of Guarnieri bodies.

Ecology. Infection is readily spread amongst a herd of cattle, probably by the hands of milkers, who may in turn be infected themselves. Man-to-man infection is rare. Occasionally cattle may be infected with true vaccinia virus coming from a recently vaccinated subject (Dekking quoted by Downie, 1959).

BUFFALO POX

A virus was isolated from a severe outbreak in India (Baxby & Hill, 1969). It was related to cow pox and even more closely to vaccinia (Kataria & Singh, 1970). It produced pocks on the CAM up to a temperature of 38·1°–38·2°.

INFECTIOUS ECTROMELIA

Synonym: Mouse pox.

Reviews: Fenner (1949), Briody (1959).

Haemagglutination. Agglutinates the same range of fowl and pigeon cells as vaccinia. Reported agglutination of mouse cells (Burnet & Stone, 1946) has not been generally reproducible. A heat-labile haemagglutinin-inhibitor occurs in the sera of most normal mice (Briody, 1959).

Antigenic properties. Very close to vaccinia but separable by indirect complement fixation, antibody adsorption and gel diffusion (cf. Vaccinia). Some tests suggest a closer relation to cow pox than to vaccinia and variola. Antihaemagglutinins may or may not be present in sera of latently infected mice. Recovered mice are solidly immune.

Cultivation. *In eggs.* The virus grows on the chorioallantoic membrane, but only below 30°C. From the membrane it reaches the whole embryo and embryonic fluids.

In tissue culture. Grows in various types of culture—HeLa cells, L cells, mouse fibroblasts, chick embryo (on which it will form plaques—Porterfield & Allison, 1960). There is evidence in culture of cell-to-cell spread and of giant-cell formation (Nii, 1959). An attenuated strain, unlike a virulent one, grew poorly in cultivated mouse macrophages. (Roberts, 1964).

Distribution. An endemic infection of laboratory mice in Europe and Asia; more recently recognized in the United States. The only known recovery from wild mice was in a laboratory where the disease was under study, but latent infection in wild mice would almost certainly be overlooked.

Pathogenicity. *In mice.* The infection is latent in many stocks of laboratory mice and is activated by various stresses, such as transport, X-radiation and especially when in the course of experiments serial passages are made of tumour cells or tissues. The disease then elicited has many forms and often remains undiagnosed for some time. It may present as conjunctivitis, pneumonia, meningitis or hepatitis rather than in the classical form of oedema and necrosis followed by loss of limb or tail.

With established virus intradermal inoculation into a footpad leads to this oedema and necrosis, signs first appearing about 5 days after inoculation. With virulent strains generalization occurs with a 'rash' of pocks or, more often, fatal necrosis of liver and spleen. With less virulent strains, though viraemia occurs, visible lesions are only seen locally. Intraperitoneal inoculations lead to death from hepatitis; pancreatic necrosis and ascites also occur. Results after inoculation by other routes are variable. The course of infection is fully described by Fenner (1948a). There are genetic differences in susceptibility between different strains of mice.

In other hosts. Inapparent infection is produced in rats inoculated intranasally (Burnet & Lush, 1936b); virus multiplies in olfactory bulbs. Local lesions are produced in the skins of rabbits, guinea-pigs and cotton-rats, but serial transmission is not reported. No ill effects have followed inoculation into man; ectromelia was the active agent in a vaccine ostensibly containing attenuated typhus rickettsiae and given to very many people.

Morbid anatomy. In fatal cases the liver may be an even yellow-grey or may show varying amounts of mottling. The spleen often has large white necrotic areas. White patches over the peritoneum are caused by fat necrosis following pancreatic damage. Intestinal haemorrhages may occur. Early lesions on feet and tail are those of oedema. Whole limbs may fall off after severe damage or only single toes or scabs from the tail. Roberts (1962) described the course of infection in skin and respiratory tract.

Histopathology. There are characteristic cytoplasmic inclusions, often multiple. They may be found in epithelial cells of many organs, and in fibroblasts. Only exceptional virus strains produce them in the liver;

they are not found in the spleen. Intestinal lesions are probably common and important, but are not often looked for. The necrotic lesions are secondary to gross damage to blood vessels. A and B inclusions both are seen (Kameyama et al., 1959), virus antigen being revealed by fluorescent-antibody 'staining' in the B inclusions.

Ecology. *Transmission* occurs especially from skin lesions (Fenner, 1947b), but the respiratory route may also be important (Briody, 1959). Virus is also present in urine and faeces. Transmission by mosquitoes has been demonstrated in the laboratory, but arthropod-transmission is of doubtful importance in nature. Regularity of transmission by contact is variable with different strains. Gledhill (1962) has shown that virus may be recovered from tail skin and faeces for several months after recovery from infection.

The epizootic behaviour of the disease has been studied by Greenwood et al. (1936) and by Fenner (1948a and b).

Control. *Vaccines.* Immunization with vaccinia virus given either intranasally (Fenner, 1947a) or into the tail (Salaman & Tomlinson, 1957) is effective in preventing deaths—more doubtfully in eliminating the infection from a stock.

Chemotherapy. Isatin-β-thiosemicarbazone, while experimentally effective against vaccinia in mice, is useless against ectromelia. On the other hand, several derivatives of the drug (Bauer & Sadler, 1961) act against ectromelia, not vaccinia.

Wholesale slaughter may be necessary to get rid of infection.

SUBGROUP II. VIRUSES RELATED TO ORF

Orf, milkers' nodes and stomatitis papulosa forms a group of closely related viruses (Peters et al., 1964). They can be divided into 2 groups of ovine and bovine origin, respectively (Nagington, 1968). Viruses from milkers' nodes and stomatitis papulosa could not be distinguished by Nagington et al. (1967).

Properties common to the whole group will be described and letters will indicate that particular findings have been described for orf (O), milkers' nodes (MN) and stomatitis papulosa (SP).

Morphology. Diameter 296 × 190 nm (MN—Friedman-Kien et al., 1963) or 252 × 158 nm (O—Abdussalam & Coslett, 1957). Particles narrower than those of vaccinia and having an appearance in electron-micrographs suggesting a ball of yarn (Nagington & Horne, 1962); this seems to be caused by a left-handed spiral coil of a single thread (Nagington et al., 1964).

Physico-chemical characters. Inactivated by chloroform; reports on ether-resistance are divergent; the viruses are apparently intermediate in behaviour between ether-sensitive and resistant ones. Dried scabs (O) can remain virulent out of doors for months, or the virus may survive at room temperature for 15 years (Hart et al., 1949). Inactivated in 30′ at 58–60° not 55°. Labile at pH 3 (Castrucci et al., 1970).

Antigenic properties. Complement-fixation, precipitation and neutralization tests have been used. Nagington et al. (1965) found it difficult to separate milkers' nodes from orf by serological tests. However Horgan & Hasseeb (1947) thought all strains of orf might not be identical serologically, though in practice there seem to be good cross-immunity between strains from different continents. According to some authors there is a little crossing with members of the vaccinia group, but this may be explicable by possession of the common pox antigen of Takahashi et al. (1959).

Cultivation. There is one report of cultivation on the CAM (SP—Mason & Neitz, 1940), but other workers have been unsuccessful (MN, O, SP).

The viruses have been grown in tissue culture, mainly of ovine or bovine, but also (MN) in human cells. A strain from sheep grew initially only in sheep cells, but the same virus after isolation from lesions in man had lost that specificity. Plaques are produced in culture, but the ovine and bovine strains are distinguishable by the type of CPE (Nagington, 1968).

MILKER'S NODES

Synonyms: Pseudo-cow pox. Paravaccinia.

Review: Lipschütz (1920).

This pock-like disease occurs on udders of cows and is often confused with cow pox, especially as, like it, it may spread to milkers and cause lesions on their hands.

Pathogenicity. Lesions in cattle and men take the form of hemispherical cherry-red papules. The incubation period in man is 5 days. Histologically the characteristic lesion is one of endothelial proliferation with formation of new small blood vessels. Juxtanuclear cytoplasmic inclusions occur and more rarely inclusions in nuclei. Lipschütz failed to transmit infection to rabbits, nor has attempted propagation in mice, guinea-pigs or chick embryos been successful. Mayr (1966) found that, in contrast to vaccinia, this virus would not infect young chicks by feather-follicle inoculation.

ORF

Synonyms: Contagious pustular dermatitis (of sheep) (CPD). Contagious ecthyma of sheep. Sore mouth. Scabby mouth. Contagious pustular stomatitis. Infectious labial dermatitis.
(Contagious pustular dermatitis or stomatitis of the horse is probably a different disease, cf. p. 402.)

Chemical composition. The internal component consists of DNA with a MW of $2{\cdot}85 \times 10^{-6}$ g/particle; it may be arranged in an S-curve (Nagington et al., 1964).

Haemagglutination. Abdussalam (1958) tested cells of a number of species, but found none sensitive.

Antigenic properties. Goat pox will immunize against orf, but not vice versa (Bennett et al., 1944). Goat-pox antisera were found by Shama & Bhatia (1959) to neutralize orf, but the reverse neutralization test was negative.

Distribution. The disease occurs in all continents.

Pathogenicity. In the *natural hosts*, the disease mainly affects young lambs and kids, though adults are not resistant. Lips and mouth are mainly involved with vesicles followed by pustules and ulcers or by proliferative wart-like lesions. Malignant aphtha is probably a severe form. Exceptionally there may be considerable mortality amongst lambs. Other parts of the skin may be affected including eyes, anus and, in suckling ewes, the udder (Glover, 1930). Lesions usually regress in 3–4 weeks.

Natural host range. So-called papilloma of chamois is orf (Grausgruber, 1964). The only other species to be naturally infected is man, who may contract a local skin lesion from affected animals—usually on face or hands.

Experimentally, sheep, goats and man can be readily infected by scarification, lesions appearing in 2–5 days, being maximal (in lambs) on the eighth day. There is dispute as to whether calves are susceptible; most workers have failed to infect them. Evidence as to susceptibility of rabbits is confused as some published work describes the reaction to a fungus infection rather than orf. However, Abdussalam (1957a) apparently did propagate the virus in series in rabbits.

The ulcerative dermatosis virus of Trueblood (1966) is probably a strain of orf.

There is a report that dogs have been infected after eating unskinned sheep carcases (Wilkinson & Prydie, 1970).

Pathological lesions. Lesions in epithelium show ballooning of cells leading to degeneration and vesicle formation. Later the lesion may be granulomatous. No specific inclusion bodies are described (Wheeler & Cawley, 1956; Abdussalam, 1957b).

Ecology. Virus persists in the soil of an affected pasture for some months, and infection is probably by direct contact of the mouth with the virus.

Control. A living virus vaccine has been applied by scarification to immunize lambs and kids.

STOMATITIS PAPULOSA OF CATTLE

Probably the same disease as ulcerative stomatitis, papular stomatitis, pseudo-aphthous stomatitis, erosive stomatitis of cattle, but it is possible from discrepancies in accounts that more than one agent is concerned.

Pathogenicity. Lesions in mouths of cattle somewhat resemble those of foot-and-mouth disease. There may be crateriform ulcers up to 1 cm across. There is no generalization.

The agent that caused ulcerative stomatitis in Rwanda-Burundi (Huygelen et al., 1958) was reported to infect sheep and goats, but a virus from Holland (Jansen & Kunst, 1959) failed to do so. Transmission to man was reported by Carson & Kerr (1967) and transfer back from man to cattle by Carson et al. (1968).

Pathological lesions. A number of writers describe cytoplasmic inclusions. According to Plowright & Ferris (1959) there is a paranuclear acidophilic part, partly surrounded by a hoop-like basophilic part.

BALANO-POSTHITIS OF SHEEP

A filtrable agent causing venereal infection of sheep in America was described by Tunnicliff & Matishek (1941) and by Tunnicliff (1949). It could cause ulcerative dermatitis as well as balanitis and ulcerative vulvitis. Similar conditions are described from Australia, South Africa and Britain (Glover, personal communication, 1962), but their relation to each other and to known viruses is obscure.

SUBGROUP III. VIRUSES RELATED TO SHEEP POX

GOAT POX

This seems to be an entity, but little is known of the properties of the virus.

The virus is ether-resistant (Bennett et al., 1944); the Kedong strain of 'sheep pox' (really goat pox) is sensitive (Plowright & Ferris, 1959). Elementary body suspensions are agglutinated by immune sera.

Reported to immunize against orf (see p. 389) and against sheep pox (Rafyi & Ramyan, 1959), but the reverse protection is said to operate in neither case. Protects cattle against lumpy skin disease (see p. 393) (P. B. Capstick, personal communication, 1963).

Antigenic properties. Gel-diffusion tests indicate that there are some specific antigens and others shared with sheep pox (Bhambani & Murty, 1963). Hyperimmune sera prepared in rabbits and tested in tissue culture showed cross-neutralization between goat pox, sheep pox and lumpy skin disease viruses. There was some crossing also with vaccinia and cow pox (P. B. Capstick, personal communication 1963).

Cultivation on the chorioallantoic membrane of 12-day eggs produces opaque pocks; the virus is said to be attenuated for goats after 4 to 8 passes (Rafyi & Ramyan, 1959).

Grows, with CPE in lamb and kid kidney and testis cultures (Pandey & Singh, 1970). Cytoplasmic inclusions are produced.

Distribution. Prevalent particularly in North Africa and the Middle East—in fact where there are most goats. Also from a few European countries and Australia. See p. 392 for a possible relationship to African strains of sheep pox.

Pathogenicity. It produces generalized pocks on mucous membranes and skin. Is transmissible to sheep and allegedly to calves, rabbits and other species, but other workers deny this and it requires confirmation. A virus dermatitis in goats (Haddow & Idnani, 1948) is said to be more severe than ordinary goat pox. Cytoplasmic inclusion bodies are like those of sheep pox.

Control. Virus attenuated by cultivation in eggs has been used to immunize (Rafyi & Ramyan, 1959).

SHEEP POX

Synonyms: Clavelée. Variola ovina.

Morphology. 194 × 115 nm by electron microscopy—said to be more elongated than other poxviruses (Abdussalam, 1957b).

Physico-chemical characters. Inactivated in 15′ by 2 per cent acid or 2 per cent phenol (Angeloff et al., 1956).

Antigenic properties. All strains are serologically alike (Ježić, 1932). For cross-protection by goat-pox vaccine and for antigenic relations with goat pox and lumpy skin disease cf. Goat pox (p. 391).

Neutralization by antisera was studied by Borrel (1903) and incompletely neutralized virus used for immunisation (sero-clavelization). Specific complement fixation is reported.

Cultivation. *Developing eggs.* Adapted, apparently with some difficulty, to growth on chorioallantoic membranes (Yuan et al., 1957; Sabban, 1957). Sabban found no change in properties, but the Chinese workers report attenuation after 90 passes, so that the virus could be used for immunization. Others report failure to cultivate the virus in eggs.

In tissue culture. Growth is reported in skin, kidney and testis of sheep, goats and calves with complete CPE in most tissues in 4–12 days and a peak titre (10^6/0·2 ml) in 4 to 5 days. Virulence for sheep unchanged (Plowright & Ferris, 1958; Boué et al., 1957). Aygün (1955) on the other hand reports attenuation after 15 passages in sheep-embryo cultures and that this attenuated virus has been used for immunization in Turkey.

Distribution. Prevalent in parts of Africa, Asia, the Middle East: in Europe only in the Southeast and in the Iberian peninsula. Strains of sheep pox from eastern and southern Africa are said to be more closely related to goat pox than to classical sheep pox (P. G. Capstick, personal communication, 1963).

Pathogenicity. Produces a generalized pock disease in sheep, with often tracheitis and caseous nodules in the lungs. Breeds of sheep vary greatly in susceptibility; mortality varies from 5 to 50 per cent.

Only sheep are naturally infected.

Local lesions can be produced experimentally in goats and gazelle and, with some strains, in cattle.

Subcutaneous inoculation of sheep produces oedematous swellings and on passage a virus is obtained adapted to growth in mesodermal tissues. There are many contradictory claims to have succeeded—or otherwise—in infecting other species; success should be regarded as unproved.

Pathological lesions. The characteristic cytoplasmic inclusion body, not unlike the Guarnieri body, occurs in 'cellules claveleuses'; there is also condensation of chromatin on the nuclear membrane. These changes are seen also in cells in tissue culture.

Control. As this is an economically very important disease, many attempts at producing a vaccine have been made since the 'seroclavelization' of Borrel (1903). That widely used in Egypt is the dried Roumanian virus given intradermally under the tail. Reactions are slight and immunity good for 14 months (Sabban, 1960). An Iranian strain is similarly used. Similarly favourable results are reported by Ramyar & Hessami (1967) and by Borisovich et al. (1966). Virus attenuated in eggs or tissue culture (see above) may in time prove better.

LUMPY SKIN DISEASE (Neethling Strain)

Review: Weiss (1968).

Morphology. Similar by electron-microscopy to vaccinia (Munz & Owen, 1966).

Of several viruses isolated from lumpy skin disease of cattle in Africa, the Neethling virus proves to be closely related to African sheep pox, which itself lies near goat pox (P. B. Capstick, personal communication, 1963).

The virus withstands 3 cycles of freezing and thawing. It is sensitive to 20-per cent ether (Plowright & Ferris, 1959).

Cultivation. *In eggs.* Multiplies in the embryo and chorioallantoic membrane, producing pocks (van Rooyen et al., 1959).

In tissue culture of embryonic calf and lamb kidney and in calf and lamb testis. CPE appears slowly in early cultures, but as early as 24 hours after adaptation spindle cells appear and later round up. Inclusions like those of sheep pox are formed (Plowright & Witcomb, 1959) but no syncytia.

Distribution. South and East Africa. First seen in northern Rhodesia and Madagascar in 1929; it appeared in the Transvaal in 1945; it was widespread there for a number of years before being recognized in Kenya.

Pathogenicity. *In cattle.* Fever with formation of multiple nodules in the skin, lesions in mucous membranes and viscera, adenitis; these seem constant features. The Allerton virus (see p. 345), a member of the herpesvirus group, was isolated from cattle supposed to be suffering from lumpy skin disease, but is not any longer thought to be causally related.

Experimentally, the cultivated virus produces fever and local reaction in cattle, and in rabbits a transient local reaction with some generalized lesions (Alexander et al., 1957).

Pathological lesions. Cytoplasmic inclusions like those of sheep pox are found in epithelial cells and histiocytes.

Ecology. The first outbreak in cattle in Kenya was apparently in association with a disease in sheep (Burdin & Prydie, 1959).

Control. The Isiolo or Kedong strains of 'sheep pox' have been given to cattle intradermally to produce immunity to Neethling virus (Capstick et al., 1959). These are, in fact, nearer to goat pox. True sheep pox does not protect as well (P. B. Capstick, personal communication, 1963. Neethling virus attenuated by egg passage forms a very efficient vaccine (K. E. Weiss, personal communication).

SUBGROUP IV: AVIAN POXES

FOWL POX

Synonyms: Bird-pox. Epithelioma contagiosum. Fowl diphtheria. *Poxvirus avium.*

Poxviruses affect numerous birds and are not clearly separable. Some strains are rather specific, others attack birds of various families. Canary pox will be described separately, but is not to be sharply distinguished from fowl pox.

Morphology and development. Elementary bodies or Borrel bodies have an estimated diameter of 332 × 284 nm by electron-microscopy (Boswell, 1947). Moreover, they are found in a cytoplasmic inclusion body, the Bollinger body, which consists of a matrix in which the virus particles are embedded. This matrix is osmiophilic, very resistant to enzymes, but dissolved by sodium lauryl sulphate. It probably contains both protein and lipid and is said to give a positive Feulgen reaction after previous extraction of lipids (Todd & Randall, 1958).

The developing virus appears first in the form of fairly undifferentiated particles, but soon acquires the characteristic pox virus appearance of a dumb-bell-like structure within an outer membrane (Morgan & Wyckoff, 1950; Eaves & Flewett, 1955).

Structures described as rodlets, tubules or filaments, are present in the walls of the virion (Herzberg et al., 1964; Hyde et al., 1965). These are core elements with a folded appearance. Arhelger & Randall (1964) have published pictures showing engulfment of a virion as it first enters a cell and release of new virions by budding from the cell-surface.

Chemical composition. DNA has a MW of 200–240 × 10^6 daltons and is probably linear and about 100 μm in length. GC content of the nucleoprotein 35 per cent. DNA is said to have been isolated and to be infectious only when tested on the CAM (Randall et al., 1966).

Physico-chemical characters. Virus is inactivated by 1 per cent caustic potash, but only after being freed from its matrix. It is resistant to ether, but sensitive to chloroform, and withstands 1 per cent phenol and 1:1,000 formalin for 9 days. Readily preserved by drying or freeze-drying for periods of several years. Inactivated by heating 30′ at 50°C or 8′ at 60°C.

Haemagglutination. A haemagglutinin, more readily separable from the virus bodies than with other poxes is described by Mayr (1956). Rones-Zakay (1966) and Garg et al. (1968) also describe haemagglutins.

Antigenic properties. There is some cross-neutralization among fowl, pigeon and canary pox. The bird poxes in fact form a family of viruses of varying degrees of cross-relationship. Fowl and pigeon poxes lie more closely together than either does to canary pox. Neutralization tests can be carried out by testing mixtures on scarified fowl combs or, more conveniently, by a method involving pock-counting on chorio-allantoic membranes (Burnet & Lush, 1936a). Precipitating antibodies can be revealed by gel diffusion (Wittman, 1958).

Cultivation. *In eggs.* Pocks are produced on chorioallantoic membranes of hens' and ducks' eggs, the lesions being first proliferative, later necrotic. Growth is best at 37°C, maximal titre being reached in 3 days on 8–12-day embryos (Haig, 1951).

In tissue culture. Both fowl and pigeon pox grow well and produce CPE in chick-embryo tissue cultures. Bollinger bodies may be found *in vitro*. Growth in chick fibroblasts was not accompanied by CPE or inclusion formation (Bang et al., 1951). A CPE was procured in cells of human amnion in culture, even though no multiplication occurred (Burnett & Frothingham, 1968).

Pathogenicity. Disease in fowls runs a course of 3 or 4 weeks. There are proliferative lesions followed by scabbing on the skin especially the head, sometimes on feet and vent. Involvement of the trachea is 'fowl diphtheria'. Various eye lesions. Caseous material may collect in the infra-orbital sinuses. Similar lesions in other birds.

High-range includes fowls, turkeys, pheasants, partridges, quail, grouse, pigeons, rhea, storks, ducks, sparrows and cormorants. A fatal epizootic among crows occurred in Slovakia in 1964 (Ursiny & Jakulik, 1966).

Experimentally, birds of practically any species can be infected by a variety of routes. A strain has been modified by intracerebral passage to give meningo-encephalitis in chicks (Buddingh, 1938). A mutant described by Goodpasture (1959) produced adenoma-like lesions on inoculation into kidneys. Otherwise the experimental disease is much like that occurring naturally. One of 4 strains of fowl pox inoculated into canaries produced necrotic lesions and deaths much like those caused by canary pox itself (Burnet, 1933). The virus may produce local lesions in mammals, e.g. in mouse lungs, but is not propagable in series (Nelson, 1941).

Pathology. The local lesions are caused by heaping up of epithelial cells continuing for some days or even weeks before necrosis sets in. The Bollinger bodies in cytoplasm already described (p. 394) attain a diameter larger than that of the nucleus.

Ecology. Transmission is by direct contact or through mechanical transfer by biting mosquitoes (*Culex* & *Aëdes* spp.). No multiplication in the mosquitoes is reported, but they may remain infectious for as long as 210 days (Bos, 1934). Epidemics usually occur in spring, summer or autumn. French & Reeves (1954) isolated 5 bird pox viruses from wild caught mosquitoes; these were related with various degrees of closeness to each other and to fowl pox. Fowl pox may be transmitted also by *Argas persicus*, and the virus may be transmitted transovarially in the ticks (Shirinov et al., 1969).

Control. Pigeon-pox or fowl-pox viruses attenuated by growth in eggs have been widely used as live vaccines by the feather follicle or 'stick' method of inoculation. There are varying reports as to efficacy; immunity is not as permanent as could be desired (Glover, 1939).

CANARY POX

This fatal disease of canaries described by Kikuth & Gollub (1932) was shown by Burnet (1933) to be closely related to fowl pox.

Pathogenicity. The fatal disease following intramuscular inoculation of canaries is at first sight unlike fowl pox, but (cf. p. 396) fowl pox itself may behave like this in canaries. There is local necrosis with oedema and exudates over serous membranes, and pneumonia with numerous cytoplasmic inclusions visible in bronchiolar epithelium.

Sparrows are readily infected; chicks and other birds usually not.

A pox infecting juncos in North America gave rise to both nuclear and cytoplasmic inclusions (Beaver & Cheatham, 1963). A sparrow pox is also described (Kato et al., 1965).

AVIAN ARTHRITIS

A virus causing arthritis in chicks is unrelated to fowl pox, but may be a poxvirus (Olson & Kerr, 1966). Its diameter is probably between 0·1 and 0·22 μm, it is ether- and chloroform-resistant and is not wholly inactivated after 8 to 10 hours at 60°. Taylor et al. (1966) cultivated it in the yolk sacs of fertile eggs. Inclusion bodies apparently containing DNA were present in cytoplasm of infected cells. Lesions resembling rheumatoid arthritis were produced by inoculation into the foot-pads of chicks.

SUBGROUP V. VIRUSES RELATED TO MYXOMA

MYXOMATOSIS

Reviews: Fenner (1965). Fenner & Ratcliffe (1965).

Morphology. Similar to vaccinia by electron-microscopy. Negative staining for electron-microscopy reveals superficial structures giving a beaded appearance to the surface. The tubules are said to be rather wider than with other poxviruses (Chapple & Westwood, 1963; Padgett et al., 1964). They are not visible in virions into which phosphotungstic acid has penetrated better.

Chemical composition. The cytoplasmic inclusions believed to consist largely of virus are Feulgen-positive (Kato & Cutting, 1959). A peculiar lipid was found in a centrifuged deposit of virus particles (Balls et al., 1940).

Physico-chemical characters. The specific gravity is estimated as 1·3.

Inactivated in 25′ at 55°C—rather more quickly than other poxviruses; 50° for 1 hour killed. Very stable in 50 per cent glycerol or when frozen and dried. It survives for many months in skins of affected rabbits at ordinary temperatures (Jacotot et al., 1955).

It is ether-sensitive—unlike most poxviruses—but, like them, is resistant to sodium desoxycholate (Andrewes & Horstmann, 1949). (With almost all other viruses sensitivity to ether and to bile salts are parallel.) 0·1 m potassium salicylate inactivates quickly. The action of various antiseptics was described by Moses (1911).

Haemagglutination. Not reported.

Antigenic properties. Neutralizing antibodies against myxoma can be demonstrated by tests in rabbit skins or on chorioallantoic membranes of eggs, but not always very readily. Soluble antigens demonstrable by precipitin tests in tubes (Smadel et al., 1940) or by gel diffusion (Mansi & Thomas, 1958) are present in tissues and sometimes serum of acutely ill rabbits. Walker, quoted by Fenner & Ratcliffe) (1965) identified 8 myxoma antigens by electrophoresis. Complement fixation can also be used (Fenner et al., 1953). Myxoma virus is very close antigenically to rabbit fibroma and to the Californian virus, but gel-diffusion tests reveal differences (Reisner et al., 1963, Fenner 1965).

Cultivation. *In developing eggs.* Pocks are formed on chorioallantoic membranes, as with many other poxviruses. Pock-counting can be used for virus titration, but the eggs are $2\frac{1}{2}$ times less sensitive than the rabbit's skin (Fenner & McIntyre, 1956).

In tissue culture. The virus grows and produces CPE not only in tissue cultures of susceptible species (rabbit, cotton-tail) but also in cultures of squirrel, rat, hamster, guinea-pig and even human tissues. After more than 40 passages in a cell-line derived from suckling-rabbit kidney the virus was sufficiently attenuated to be useful for immunization (McKercher & Saito, 1964). Plaques develop on monolayers. Tissues of some guinea-pigs are susceptible only after they have been cultivated for a few days outside the body. Only tissues of very young rats are susceptible (Chaproniere & Andrewes, 1957). Harisijades et al. (1967) reported attenuation after prolonged growth of human foetal tissues.

Distribution. The virus exists in the natural state in Uruguay and Brazil in the local wild rabbits (*Sylvilagus brasiliensis*). It has now been introduced into Australia, Chile and Europe where it is widespread amongst the rabbits, which belong to the genus *Oryctolagus*.

Pathogenicity. In *Sylvilagus* it produces only a local swelling; the disease is not fatal.

In wild and domestic *Oryctolagus* it produces a disease which, on first contact with the new host, is more than 99 per cent fatal. The first symptoms are those of blepharo-conjunctivitis leading to sealing of the eyelids with inspissated pus. Nose, muzzle, anal and genital orifice swell up. The 'myxomata' are subcutaneous gelatinous swellings. Rabbits become thin and apathetic and may die 2–5 days after symptoms appear. With severe infections the discrete tumours may not be seen. Disease of varying degrees of mildness are seen when either the virus has become attenuated or rabbits have become genetically resistant. The disease then runs a slower course and is often not fatal; local lesions differ widely in their character. Grades of virulence of strains are described by Fenner & Marshall (1957).

As already mentioned, *Sylvilagus* species are the natural hosts. Apart from *Oryctolagus*, natural transmission has been reported only—and very rarely—in hares (*Lepus*) in Britain and France. The virus has been propagated as an inapparent infection in brains of suckling mice (Andrewes & Harisijades, 1955), in a guinea-pig sarcoma and in homologous tissue grafts in guinea-pigs (Chaproniere & Andrewes, 1958). Before the virus was introduced into Australia, Bull & Dickinson (1937) tested the susceptibility of a wide range of mammals and birds and found all but rabbits to be refractory. Hurst (1937b) obtained an attenuated strain (neuromyxoma) by intracerebral passage in rabbits.

The face and eye swellings give typically affected rabbits a characteristic 'leonine' facies. The subcutaneous swellings are firm and whitish and ooze a serous fluid on section. Orchitis is common. The 'tumours' consist of proliferation of undifferentiated mesenchymal cells which assume a stellate form, but their bulk is made up of the sero-mucinous exudate. Rivers (1930) described masses of acidophilic granules in epithelium of skin and elsewhere. The pathological lesions in various organs are described by him, by Hurst (1937a) and by Ahlström (1940).

Ecology. *Transmission* in South America is probably through arthropod vectors, mainly mosquitoes. The same is true in Australia, the main vector being *Anopheles annulipes*, though many other insects can carry virus also. Transmission is apparently mechanical. In Britain the rabbit flea *Spilopsyllus cuniculi* is the main vector, though *Anopheles atroparvus* probably plays a role on the South Coast and also in France. Rabbit-to-rabbit spread occurs only across distances of a few inches.

Other aspects. This is not the place to describe the fascinating effects of the interaction of virus and rabbit in Australia, resulting within a few years in great reduction in severity of the disease, due partly to attenuation of virus but even more to increased genetic resistance in the rabbits—see Fenner's review (1965), also Marshall & Fenner (1958) and Fenner & Ratcliffe (1965). In Britain, the picture has been different doubtless because of transmission by fleas rather than mosquitoes (Andrewes et al., 1959; Fenner & Chapple, 1965). A possibly larger role played by mosquitoes in Britain is now under study (Service, 1971).

Control. *Vaccination.* Rabbits can be immunized by inoculation of fibroma virus, which produces only a local lesion and usually protects at least against fatal myxomatosis (Shope, 1938).

Other measures. Domestic rabbits can be protected by prevention of access of mosquitoes or fleas from without. The disease does not spread in a rabbit house where these are excluded.

FIBROMATOSIS OF RABBITS

Synonym: Shope fibroma.

Review: Shope (1932)—the original description.

Antigenic properties. Several workers have studied the immunological relation with myxoma, which is extremely close. Some find them almost identical, others that there are minor differences revealed in cross-complement-fixation or precipitation tests. Comparison of the 2 viruses by agar gel diffusion (Fayet et al., 1957; Mansi & Thomas, 1958) suggests that most of their antigens are common, but there may exist specific ones also. Ewton & Hodes (1967) identified a new cytoplasmic RNA associated with polyribosomes in cells infected with Shope fibroma.

Cultivation. *In developing eggs.* The virus (OA strain) can be propagated on the chorioallantoic membrane, but, in contrast to myxoma, no characteristic lesions were produced nor was the embryo invaded.

In tissue culture. Growth occurs in cultures of rabbit and cotton-tail rabbit tissues, with CPE, also in tissue cultures of guinea-pig, rat and man (Chaproniere & Andrewes, 1957). Foci of heaped-up cells appear on rabbit-kidney monolayers (Padgett et al., 1962). Successive changes in rabbit cells are described by Hinze & Walker (1964) and Israeli & Sachs (1964).

Pathogenicity. In cotton rabbits (*Sylvilagus*) the natural host, and in European rabbits (*Oryctolagus*) the virus causes subcutaneous swellings which normally remain localized; in nature they are commonly found on the foot. The lesions are soft and rubbery, appearing 3 to 5 days after intradermal inoculation and persisting for 10–15 days before necrosis and regression begin. Intratesticular inoculation leads to enlargement of the testis to several times its natural size; it may remain large up to 40 days. This description applies to the original (OA) strain.

Experimental infection. An acute disease with generalized lesions, sometimes fatal, is seen when rabbits are treated with tar at the time of inoculation (Ahlström & Andrewes, 1938), cortisone (Harel, 1956) or X-rays (Clemmesen, 1939) or if very young rabbits are injected (Duran-Reynals, 1945).

Fibroma virus has been propagated intracerebrally in 1-day-old mice (Dalmat, 1958).

Pathological lesions. Inoculation of virus into various tissues of rabbits leads to an immediate inflammatory reaction quickly followed by fibroblastic proliferation; this is the basis of production of the tumour-like swellings (Ahlström, 1938). A variant, the inflammatory (IA) strain was described by Andrewes (1936); in this the lesions are inflammatory throughout with intense accumulation of small mononuclear cells; fibroblastic proliferation is almost absent. Shope (1932) has described the occurrence in cotton-tails of proliferated epithelium with eosinophilic cutoplasmic inclusions in skin overlying fibromata.

Berry & Dedrick (1936) reported that heat-inactivated myxoma introduced into a rabbit along with fibroma virus led to production of fully virulent myxoma. This is more regularly demonstrable in tissue culture (Kilham, 1958) and is part of a more general phenomenon of mutual reactivation by poxviruses studied by Joklik et al. (1960). The properties of the transforming agent were studied by Kilham et al. (1958).

Ecology. *Transmission.* Experimentally *Aëdes and Culex* mosquitoes transmit the disease between cotton-tails or baby domestic rabbits—not readily between adult domestic rabbits (Kilham & Dalmat, 1955; Dalmat, 1959).

CALIFORNIAN RABBIT FIBROMATOSIS

This virus is like myxoma and Shope fibroma viruses morphologically. It is closely related to them antigenically, but is distinguishable in gel diffusion tests. It is attenuated after prolonged cultivation in baby rabbit-kidney cells.

It produces fibromata in its natural host in California, the brush rabbit, *Sylvilagus bachmani* (Marshall & Regnery, 1960). Experimentally it causes fibromata in other *Sylvilagus* species, but these are not transmissible in series (Fenner, 1964). In European rabbits it produces a rapidly fatal disease, but without the typical swellings of myxomatosis. The main natural vector is *Anopheles freeborni* (Grodhaus et al., 1963).

FIBROMATOSIS OF HARES

A virus, serologically related to myxoma and morphologically similar, is described as causing fibromata in hares in North Italy and southern France (Leinati et al., 1961). It was transmissible to rabbits and serially transmissible in suckling rabbits. It may be identical with the fibrosarcoma of hares described by Dungern & Coca (1903).

SQUIRREL FIBROMA

Multiple fibromata occur naturally in grey squirrels (*Sciurus carolinensis*) in North America and are caused by an agent serologically related to the rabbit fibroma virus (Kilham et al., 1953). This and the hare fibroma are less closely related antigenically to the 3 rabbit viruses than these are to each other (Fenner, 1965). This produces lesions in domestic rabbits for 1 passage only, but is serially transmissible in woodchucks (*Marmota monax*). In suckling squirrels the virus will produce not only generalized skin nodules (Kilham, 1955) but also lung adenomata (Kirschstein et al., 1958). Cytoplasmic inclusions are present in the fibroma cells.

A *fibroma of deer* probably belongs in the papovavirus group rather than among the poxviruses (see p. 297).

SUBGROUP VI. UNCLASSIFIED POXVIRUSES

HORSE POX

Synonym: Contagious pustular dermatitis (of the horse). Contagious pustular stomatitis. Grease or grease-heel is often considered as a synonym but may be caused by a different virus.

Antigenic properties. No reliable information as to crossing with other poxes. Numerous reports of cross-immunity with vaccinia and other poxes must be regarded with suspicion.

Distribution. The disease is now quite rare. In 1964 the F.A.O. only recorded it, at low incidence, from Mexico, Denmark, France, Norway, Hungary, Spain and Jordan (F.A.O., 1964).

Pathogenicity. Papular lesions developing into vesicles develop on the lips and buccal mucosa, sometimes in the nose. There is fever, drooling of saliva; a few die. Course 10–14 days; 3–4 weeks in severe cases. Possibly identical with coital exanthema of the horse. Finger lesions have been reported to occur in attendants (Bub, 1942).

Control. Skin lesions are milder than those of the mouth; and good results have been claimed from vaccinating on the skin.

In *Grease* vesicles turn into pustules and then crusts develop on flexor surfaces of lower parts of legs. Jenner thought this the origin of cow pox, but this is not generally believed. It may or may not be the same as horse pox: the latter disappeared from Britain some time before grease did (R. E. Glover—personal communication, 1963).

SWINE POX

Two pox-diseases occur in swine. One is probably vaccinia, the other a distinct entity. Manninger et al. (1940) suggest calling the former 'swine pox', the latter 'pseudo-swine-pox'. Most other authors refer to the true swine disease as 'swine pox' and call the other, what it is, 'vaccinia'. We shall follow this as the reasonable course.

Morphology. Reczko (1959) describes bodies 260 nm revealed by electron-microscopy; there are surrounding membranes. Also seen in the cytoplasm were 'crystalline' bodies 800 nm across, not seen with other poxes. Blakemore & Abdussalam (1956) and Datt (1964) also describe bodies rather larger, even larger than those of vaccinia.

Cultivation. *In eggs.* Not reported.

In tissure culture. The virus grows, with CPE, in pig kidney, testis and embryonic lung and brain. On monolayers very small plaques were produced (Kasza et al., 1960). No growth occurred in tissue cultures of cattle, sheep, mouse, roe-deer (Mayr, 1959).

Distribution. The disease is reported, usually in low incidence, from a few countries in every continent.

Pathogenicity. *In pigs* it affects chiefly very young animals with generalized pocks 1 cm or more across; these are followed by crusts which soon fall off.

Experimentally, the incubation period is 4–5 days from the time of inoculation. Guinea-pigs and baby mice are insusceptible (Mayr, 1959), so are calves, sheep and goats. Rabbits injected intradermally developed papular lesions, but these could not be passed more than 2 or 3 times in series (Datt, 1964).

Pathological lesions. Blakemore & Abdussalam (1956) describe cytoplasmic and intranuclear inclusions. Other workers (Lübke, 1960;

Kasza et al., 1960) refer to nuclear vacuolation or ballooning as well as cytoplasmic inclusions.

Ecology. *Transmission.* The pig louse (*Haematopinus suis*) is of major importance in transmitting the disease (Shope, 1940), but the disease may occur in the absence of lice.

Control. Vaccination is not considered necessary, but it is reported that glycerolated virus gives only local lesions (Schwartz, 1958).

Genital Papilloma of Pigs

Description: Parish (1961).

The virus probably belongs with the poxviruses rather than with the other papillomaviruses (Allison, 1965).

Morphology. The virus passed a filter with average pore-size 175 nm.

Antigenic properties. Neutralizing antibodies were detected in sera of hyperimmunized pigs and rabbits, not in those of recovered pigs. Conglutinating complement absorption tests were also positive; and antigen was detected in gel-diffusion tests at the stage of maximal growth (Parish, 1962). Recovered pigs were immune to challenge.

Pathogenicity. The natural lesions consisted of papillomata in the genital region of boars. Infection was transmitted by series by scarification or injection into the genital skin of adult pigs; the incubation period was about 8 weeks. Injections elsewhere into the skin were without effect, nor were a man, a calf, rabbits, guinea-pigs, mice or embryonated eggs susceptible. Cytoplasmic inclusions were observed in papillomatous lesions.

CAMEL POX

Synonym: Photo-Shootur.

This occurs in the Middle East, in North East Africa and Pakistan. Camel drivers may contract local lesions on hands and arms from contact with affected animals (Amanschulow et al., 1930; Leese, 1909). In Turmenia there are outbreaks every third to fifth year usually in late summer, and as many as 80,000 camels may be affected. The disease is usually mild, but is sometimes severe, causing abortions and deaths in young animals. Pustules are seen first round the lips and nose; other skin lesions follow. There may be keratitis followed by corneal opacity. Incubation period 4–15 days. (Borisovich & Orekhov, 1966).

RHINOCEROS POX

A pox-like disease in a captive rhinocerus has been described by Grunberg & Burtscher (1968) and is said to be an avian pox (Mayr & Mahnel, 1970). The virus could be cultivated in chick embryos.

MOLLUSCUM CONTAGIOSUM

Morphology and development. Elementary bodies have a size estimated by electron-microscopy as 302 × 226 nm (Boswell, 1947), but other reports give a maximum diameter of up to 360 or even 390 nm. Dourmashkin & Bernhard (1959) report an eccentric nucleoid and single or double membranes round the particles. The nucleoid is 50–100 nm across. They also describe stages referred to as 'viroplasm with double-leafed cleaving membranes' and 'free nucleoid particles'. Other reports describe a dense cortex and hollow interior; possibly representing an extra stage of development (Gaylord & Melnick, 1953). The particles are at first scattered, but soon come to lie within a protein matrix divided into locules by septa; tryptic digestion reduces this matrix to a gelatinous mass (Goodpasture & Woodruff, 1931). Particles 'negatively stained' for electron-microscopy may look like a 'ball of yarn', as does orf (Howatson, 1962).

Chemical composition. The studies of Rake & Blank (1950) suggest that the elementary bodies contain DNA.

Physico-chemical characters. Virus retains activity for a month in 50 per cent glycerol.

Antigenic properties. A heat-labile soluble antigen can be used for complement fixation, but most patients do not develop complement-fixing antibodies. There are no cross-reactions with other poxes (Mitchell, 1953). Neva (1962) reported that hyperimmune guinea-pig sera would neutralize the cytotoxic effects of the virus in tissue culture. The effects described by Postlethwaite (1964) were weakly neutralized by sera of patients.

Cultivation. The virus does not grow in eggs, but multiplication in tissue cultures of HeLa cells was reported by Dourmashkin & Febvre (1958). Neva (1962) and Chang & Weinstein (1961) described cytopathic effects in tissue cultures of human cells, while Postlethwaite (1964) found that extracts from lesions would induce in mouse-embryonic cells resistance against other viruses. None of these authors obtained evidence of indefinite propagation in series, even when organ cultures were used (Prose et al., 1969).

Distribution. World-wide.

Pathogenicity. Exclusively a human disease; lesions are confined to the skin. After an incubation period variously estimated at 14–50 days, pimples develop and increase to form nodules having a diameter of 2 mm. They become pearly white and may develop an opening revealing a white core. Lesions persist for months. There is proliferation, hyperplasia of epidermal cells, those nearest the surface containing the large molluscum body up to 24 μm across. The nature of these bodies is described above.

Ecology. *Transmission.* By direct contact and fomites. Swimming baths may be a source of cross-infection (Postlethwaite et al., 1967); the epidemic in Alaska affected 77 per cent of young boys exposed (Overfield & Brody, 1966).

YABA VIRUS

Virus of 'subcutaneous tumours in monkeys'.

Description. Niven et al. (1961).

Morphology. Closely resembles vaccinia. Noyes (1965) described long wavy tubules on the surface of the virion.

Physico-chemical characters. GC content 32·5 per cent. Density in CsCl 1·69 (Yohn & Gallagher, 1969). Tests for ether-resistance gave equivocal results. Exposure for 1 hour to 56° or pH 3 at room temperature inactivated.

Haemagglutination. None demonstrated.

Antigenic properties. Recovered rhesus monkeys are immune to reinoculation. There is no cross-immunity with vaccinia. Immunodiffusion tests revealed 12 lines of precipitation when culture fluids were used as antigens; with extracts of lesions 17 lines were seen. (Olsen & Yohn, 1970).

Cultivation. *In eggs.* Strandström, et al. (1966) cultivated the virus on the CAM, but only if 10–11-day-old eggs were used. Virus was also cultivated by inoculating the yolk sac and harvesting allantoic fluid.

In tissue culture. Cultivation has been reported in primary fetal human kidney, *Cercopithecus* kidney and in a monkey cell-line (MK2) (Levinthal & Schein, 1964; Kato et al., 1965; Noyes, 1965). Cytoplasmic inclusions appeared in infected cells. Yohn et al. (1966) have made detailed studies of the virus's development *in vitro*. Taylor et al. (1968) found an immunofluorescent technique better than recognition of CPE for titrating the virus in tissue cultures.

Distribution. The disease appeared in a colony of rhesus monkeys housed in the open at Yaba, Nigeria (Bearcroft & Jamieson, 1958). It spread to 20 or 35 rhesus in a few weeks and affected also 1 baboon (*Papio papio*).

Pathogenicity. The virus causes tumour-like growths particularly on heads and limbs of rhesus and cynomolgus monkeys. These reach a diameter of 25–45 mm and may project 25 mm above surrounding skin. They are liable to break down and ulcerate. Regression usually begins in 4–6 weeks and is complete in 6–12 weeks. In an inoculated *Macacus fuscatus* a 'tumour' was still present after 5 months. They are firm and white on section; local lymph nodes may be enlarged.

Lesions consist of accumulations of large polygonal cells probably derived from fibrocytes. Cytoplasmic inclusions consist of masses of virus particles.

In *Cercopithecus aethiops* monkey lesions appeared 11 days after inoculation and began to regress a week later. They remained flat and not bigger than 30 mm across. In inoculated cancer patients and in 1 accidental laboratory infection in man, the virus produced local skin nodules appearing after 5–7 days, reaching 2 cm in diameter and regressing after 3–4 weeks. Virus was recovered from the lesions (Grace et al., 1962). Wolfe et al. (1968) exposed rhesus and cynomolgus monkeys to a virus aerosol and thus obtained growths in lungs, nose and subcutaneous tissues. Behbehani et al. (1969) obtained a recombinant between Yaba and vaccinia viruses, having oncogenic properties in rhesus monkeys and antigenic components of both viruses.

Ecology. The spontaneous disease has only been seen where monkeys were kept out of doors in Africa; the natural host is probably therefore an African mammal, most probably a primate. Spread by arthropod has been neither demonstrated nor excluded.

TANAPOX VIRUS

What appears to be a distinct poxvirus has been studied by Downie et al. (1971). It caused localized skin lesions in children in Kenya. It seems from serological tests (Downie et al., 1971) to be identical with viruses causing outbreaks of disease in captive monkeys (Nicholas & McNulty, 1968; Crandell et al., 1969).

Morphology. Closely resembles vaccinia.

Haemagglutination. None demonstrated.

Antigenic characters. Related to Yaba virus (see p. 408) (Crandell et al., 1969), also to swine pox (Nicholas, 1970).

Cultivation. No growth in eggs. Multiplies in tissue cultures of monkey kidney, but not in those of species other than primates.

Pathogenicity. Causes single or multiple skin lesions in monkeys, but no systemic involvement. Monkey handlers have been infected. Other species apparently insusceptible (Downie et al., 1971).

Ecology. Presumably a disease of wild African monkeys. Possibly insect transmitted.

FIBROVASCULAR GROWTHS IN GUINEA-PIGS

Pox-like particles have been described as occurring in proliferative lesions in the thigh muscles of guinea-pigs (Hampton et al., 1968).

REFERENCES

Abdussalam, M. (1957a) *J. comp. Path.*, **67**, 217 and 307.
Abdussalam, M. (1957b) *Am. J. vet. Res.*, **18**, 614.
Abdussalam, M. (1958) *J. comp. Path.*, **68**, 23.
Abdussalam, M., & Coslett, V. E. (1957) *J. comp. Path.*, **67**, 145.
Ahlström, C. G. (1938) *J. Path. Bact.*, **46**, 461.
Ahlström, C. G. (1940) *Acta path. microbiol. scand.*, **17**, 377.
Ahlström, C. G., & Andrewes, C. H. (1938) *J. Path. Bact.*, **47**, 65.
Alexander, R. A., Plowright, W., & Haig, D. A. (1957) *Bull. épizoot. Dis. Afr.*, **5**, 489.
Allison, A. C. (1965) in *The comparative physiology and pathology of the skin*, p. 678. Oxford: Blackwell.
Amanschulow, S. A., Samarzew, A. A., & Arbusow, L. N. (1930) *Z. Infekt.-Kr. Haustiere*, **38**, 186.
Andrewes, C. H. (1936) *J. exp. Med.*, **63**, 157.
Andrewes, C. H., & Harisijades, S. (1955) *Br. J. exp. Path.*, **36**, 18.
Andrewes, C. H., & Hortsmann, D. M. (1949) *J. gen. Microbiol.*, **3**, 290.
Andrewes, C. H., Thompson, H. V., & Mansi, W. (1959) *Nature, Lond.*, **184**, 1179.
Angeloff, S., Panajotoff, P., Manolawa, N., & Nikoloff, P. (1956) *Arch. exp. vet. Med.*, **10**, 365.
Anthony, R. L., Taylor, D. L., Daniel, R. W., Cole, J. L., & McCrumb, F. R. (1970) *J. infect. Dis.*, **12**, 295.
Appleyard, G. (1961) *Nature, Lond.*, **190**, 465.
Arhelger, R. B., & Randall, C. C. (1964) *Virology*, **22**, 59.
Arita, I., & Henderson, D. A. (1968) *Bull. Wld Hlth Org.*, **38**, 277.
Aygün, S. T. (1955) *Arch. exp. vet. Med.*, **9**, 415.
Bablanian, R. (1968) *J. gen. Virol.*, **3**, 51.
Balls, A. K., Jansen, E. F., & Axelrod, B. (1940) *Enzymologia*, **8**, 267.
Bang, F. B., Levy, E., & Gey, G. O. (1951) *J. Immunol.*, **66**, 329.
Bauer, D. J. (1955) *Br. J. exp. Path.*, **36**, 105.
Bauer, D. J., & Sadler, P. W. (1961) *Nature, Lond.*, **190**, 1167.

Bauer, D. J., St. Vincent, L., Kempe, C. H., & Downie, A. W. (1963) *Lancet*, **2**, 494.
Bauer, D. J., St. Vincent, L., Kempe, C. H., Young, P. A., & Downie, A. W. (1969) *Am. J. Epidemiol.*, **90**, 130.
Baxby, D., & Hill, B. J. (1969) *Vet. Rec.*, 85, 315.
Bearcroft, W. G. C., & Jamieson, M. (1958) *Nature, Lond.*, **182**, 195.
Beaver, D. L., & Cheatham, W. J. (1963) *Am. J. Path.*, **42**, 23.
Bedson, H. S., & Dumbell, K. R. (1961) *J. Hyg., Camb.*, **59**, 457.
Bedson, H. S., & Dumbell, K. R. (1964) *J. Hyg., Camb.*, **62**, 147.
Bedson, H. S., & Dumbell, K. R. (1967) *Br. Med. Bull.*, **23**, 119.
Behbehani, A. M., Barrick, S., & Bolano, C. R. (1969) *Proc. Soc. exp. Biol. Med., U.S.A.*, **132**, 738.
Bennett, S. C. J., Horgan, E. S., & Hasseeb, M. A. (1944) *J. comp. Path.*, **54**, 131.
Bernkopf, W., Nishmi, M., & Rosin, A. (1959) *J. Immunol.*, **83**, 635.
Berry, G. P., & Dedrick, H. M. (1936) *J. Bact.*, **31**, 50.
Bhambani, B. D., & Murty, D. K. (1963) *J. comp. Path.*, **73**, 349.
Blakemore, F., & Abdussalam, M. (1956) *J. comp. Path.*, **66**, 373.
Borrel, A. (1903) *Annls Inst. Pasteur, Paris*, **17**, 123 and 732.
Borisovich, Y. F., & Orekhov, M. D. (1966) *Veterinariya, Moscow*, **3**, 50.
Borisovich, Y. F., Koval, G. L., Bagmet, L. G., & Pal'gov, A. A. (1966) *Trudy nauchno kontrol Inst. Vet. Preperatov*, **13**, 29.
Bos, A. (1934) *Z. Infekt.-Kr. Haustiere*, **46**, 195.
Boswell, F. W. (1947) *Br. J. exp. Path.*, **28**, 253.
Boué, A., Baltazard, M., & Vieuchange, J. (1957) *C.r. hebd. Séanc. Acad. Sci., Paris*, **244**, 1571.
Bras, G. (1952) *Documenta Med. geogr. trop.*, **4**, 303.
Briody, B. A. (1959) *Bact. Rev.*, **23**, 61.
Brown, A., Elsner, V., & Officer, J. E. (1960) *Proc. Soc. exp. Biol. Med., U.S.A.*, **104**, 605.
Bub, (1942) *Z. Veterinark.*, **54**, 137.
Buddingh, G. J. (1938) *J. exp. Med.*, **67**, 921 and 933.
Bull, L. B., & Dickinson, C. G. (1937) *J. Coun. scient. ind. Res., Aust.*, **10**, 291.
Burdin, M. L., & Prydie, J. (1959) *Nature, Lond.*, **183**, 949.
Burnet, F. M. (1933) *J. Path. Bact.*, **37**, 107.
Burnet, F. M., & Lush, D. (1936a) *Br. J. exp. Path.*, **17**, 302.
Burnet, F. M., and Lush, D. (1936b) *J. Path. Bact.*, **42**, 469.
Burnet, F. M., & Stone, J. D. (1946) *Aust. J. exp. Biol. med. Sci.*, **24**, 1.
Burnett, J. W., & Frothingham, T. E. (1968) *Arch. ges. Virusforsch.*, **24**, 137.
Cabasso, V., & Moore, I. F. (1957) *Proc. Soc. exp. Biol. Med., U.S.A.*, **95**, 605.
Capstick, P. B., Prydie, J., Coackley, W., & Burdin, M. L. (1959) *Vet. Rec.*, **71**, 422.
Carson, C. A., & Kerr, K. M. (1967) *J. Am. vet. med. Ass.*, **151**, 183.
Carson, C. A., Kerr, K. M., & Grumbles, L. C. (1968) *Am. J. vet. Res.*, **29**, 1783.
Castrucci, G., McKercher, D. G., Cilli, V., Arancia, G., & Nazionali, C. (1970) *Arch. ges. Virusforsch.*, **29**, 315.
Chang, T., & Weinstein, L. (1961) *J. invest. Dermat.*, **37**, 433.
Chapple, P. J., & Westwood, J. C. N. (1963) *Nature, Lond.*, **199**, 199.
Chaproniere, D. M., & Andrewes, C. H. (1957) *Virology*, **4**, 351.
Chaproniere, D. M., & Andrewes, C. H. (1958) *Virology*, **5**, 120.
Chu, C. M. (1948) *J. Hyg., Camb.*, **50**, 42.
Christensen, L. R., Bond, E. & Matanic, B. (1967) *Lab. Anim. Care*, **17**, 281.
Clark, E., & Nagler, F. P. O. (1943) *Aust. J. exp. Biol. med. Sci.*, **21**, 103.
Clemmesen, J. (1939) *Am. J. Cancer*, **35**, 378.
Cohen, G. H., & Wilcox, W. C. (1968) *J. Virol.*, **2**, 449.
Colby, C., & Duesberg, P. H. (1969) *Nature, Lond.*, **222**, 940.
Collier, L. H. (1954) *Bact. Rev.*, **18**, 74.
Collier, L. H. (1955) *J. Hyg., Camb.*, **53**, 76.
Councilman, W. T., Magrath, G. B., & Brinckerhoff, W. L. (1904) *J. med. Res.*, **11**, 12.
Craigie, J., & Wishart, F. O. (1934) *Br. J. exp. Path.*, **15**, 390.

Crandell, R. A., Casey, H. W., & Brumlow, W. B. (1969) *J. infect. Dis.*, **119**, 80.
Cross, R. M., Kaplan, C., & McClean, D. (1957) *Lancet*, **1**, 446.
Cutchins, E., Warren, J., & Jones, W. P. (1960) *J. Immunol.*, **85**, 275.
Dales, S. (1963) *J. cell. Biol.*, **18**, 51.
Dalmat, H. T. (1958) *Proc. Soc. exp. Biol. Med., U.S.A.*, **97**, 219.
Dalmat, H. T. (1959) *J. Hyg., Camb.*, **57**, 1.
Datt, N. S. (1964) *J. comp. Path.*, **74**, 62.
Deeker, C., Guir, J., & Kirn, A. (1969) *J. gen. Virol.*, **4**, 221.
Dourmashkin, R., & Bernhard, W. (1959) *J. Ultrastruct. Res.*, **3**, 11.
Dourmashkin, R., & Febvre, H. L. (1958) *C.r. hebd. Séanc. Acad. Sci., Paris*, **246**, 2308.
Douglas, S. R., Smith, W., & Price, L. R. W. (1929) *J. Path. Bact.*, **32**, 99.
do Valle, L. A. R., de Melo, P. R., Gomes, L. F. de S., & Provenca, L. M. (1965) *Lancet*, **2**, 976.
Downie, A. W. (1939) *Br. J. exp. Path.*, **20**, 158.
Downie, A. W. (1947) *J. Path. Bact.*, **48**, 361.
Downie, A. W. (1959) in *Viral and rickettsial infection of man*. Eds.: Rivers & Horsfall. 3rd ed. London: Pitman Medical Publ. Co.
Downie, A. W., Fedson, D. S., St. Vincent, L., Rao, A. R., & Kempe, C. H. (1969) *J. Hyg., Camb.*, **67**, 619.
Downie, A. W., & Haddock, D. W. (1952) *Lancet*, **1**, 1049.
Downie, A. W., & McCarthy, K. (1950) *Br. J. exp. Path.*, **31**, 789.
Downie, A. W., & Macdonald, A. (1950) *J. Path. Bact.*, **62**, 389.
Downie, A. W., Taylor Robinson, C. H., Caunt, A. E., Nelson, G. S., Manson-Bahr, P. E. C., & Matthews, T. C. H. (1971) *Br. med. J.*, **1**, 263.
Dungern von, E., & Coca, A. F. (1903) *Z. Immun.-Forsch.*, **2**, 391.
Dumbell, K. R., & Bedson, H. S. (1966) *J. Path. Bact.*, **91**, 459.
Dunham, C. G., & MacNeal, W. J. (1943) *J. Lab. clin. Med.*, **28**, 947.
Duran-Reynals, F. (1945) *Cancer Res.*, **5**, 25.
Easterbrook, K. B. (1966) *J. ultrastruct. Res.*, **14**, 484.
Easterbrook, K. B. (1967) *J. Virol.*, **1**, 643.
Eaves, G., & Flewett, T. F. (1955) *J. Hyg., Camb.*, **53**, 102.
Ewton, D., & Hodes, M. E. (1967) *Virology*, **33**, 77.
F.A.O. (1964) *Anim. Health Yearbook.*
Fayet, M. T., Mackowiak, C., Camand, R., & Leftheriotis, E. (1957) *Annls Inst. Pasteur, Paris*, **2**, 466.
Fenner, F. (1947a) *Aust. J. exp. Biol. med. Sci.*, **25**, 257.
Fenner, F. (1947b) *Aust. J. exp. Biol. med. Sci.*, **25**, 275 and 327.
Fenner, F. (1948a) *Br. J. exp. Path.*, **29**, 69.
Fenner, F. (1948b) *J. Hyg., Camb.*, **46**, 383.
Fenner, F. (1949) *J. Immunol.*, **63**, 341.
Fenner, F. (1958) *Virology*, **5**, 502.
Fenner, F. (1962) *Proc. roy. Soc. B.*, **156**, 288.
Fenner, F. (1964) in *Newcastle disease: an evolving pathogen*, p. 327. Madison: Univ. of Wisconsin Press.
Fenner, F. (1965) *Austral. J. exp. Biol. med. Sci.*, **43**, 143.
Fenner, F., & Chapple, P. J. (1965) *J. Hyg., Camb.*, **63**, 175.
Fenner, F., & McIntyre, G. A. (1956) *J. Hyg., Camb.*, **54**, 246.
Fenner, F., & Marshall, I. D. (1957) *J. Hyg., Camb.*, **55**, 149.
Fenner, F., Marshall, I. D., & Woodroofe, G. M. (1953) *J. Hyg., Camb.*, **51**, 225.
Fenner, F., & Ratcliffe, F. N. (1965) *Myxomatosis*. London/New York: Cambridge Univ. Press.
French, E. L., & Reeves, W. C. (1954) *J. Hyg., Camb.*, **52**, 551.
Friedman-Kien, A. E., Rowe, W. P., & Banfield, W. G. (1963) *Science*, **140**, 1335.
Garg, S. K., Sethi, M. S., & Negi, S. K. (1968) *Indian vet.*, **45**, 186.
Gaylord, W. H., & Melnick, J. L. (1953) *J. exp. Med.*, **98**, 157.
Gillen, A. L., Burr, M. M., & Nagler, F. P. O. (1950) *J. Immunol.*, **65**, 701.

Gispen, R. (1955) *J. Immunol.*, 75, 134.
Gispen, R., Verlinde, J. D., & Zwart, P. (1967) *Arch. ges. Virusforsch.*, **21**, 205.
Gledhill, A. W. (1962) *Nature, Lond.*, **196**, 298.
Glover, R. E. (1930) *Proc. 11th internat. vet. Congress.*
Glover, R. E. (1939) *J. comp. Path.*, **52**, 29.
Goodpasture, E. W. (1959) *Am. J. Path.*, **35**, 213.
Goodpasture, E. W., & Woodruff, C. E. (1931) *Am. J. Path.*, **7**, 1.
Gordon, M. H. (1925) Studies of the Viruses of Vaccinia and Variola, *Spec. Rep. Ser. med. Res. Coun., Lond.*, No. 98.
Grace, J. T., Mirand, E. A., Millian, S. J., & Metzger, R. S. (1962) *Fed. Proc.*, **21**, 32.
Grausgruber, W. (1964) *Zentbl. Bakt. ParasitKde, I. Abt. Orig.*, **195**, 175.
Greene, H. S. N. (1934) *J. exp. Med.*, **60**, 427.
Greenwood, M., Hill, A. B., Topley, W. W. C., & Wilson, J. (1936) *Spec. Rep. Ser. med. Res. Coun., Lond.*, No. 209, p. 64.
Grodhaus, G., Regnery, D. C., & Marshall, I. D. (1963) *Am. J. Hyg.*, **77**, 205.
Grünberg, W., & Burtscher, H. (1968) *Zentbl. VetMed.*, **15B**, 649.
Haddow, J. R., & Idnani, J. A. (1948) *Indian vet. J.*, **24**, 332.
Hahon, N., Ratner, M., & Kozikowski, M. (1958) *J. Bact.*, **75**, 707.
Haig, D. A. (1951) *Onderstepoort, J. vet. Res.*, **25**, 17.
Hamre, D., Bernstein, J., & Donovick, R. (1950) *Proc. Soc. exp. Biol. Med., U.S.A.*, **73**, 275.
Hampton, E. G., Bruce, M., & Jackson, F. L. (1968) *J. gen. Virol.*, **2**, 205.
Harel, J. (1956) *C.r. Séanc. Soc. Biol., Paris*, **150**, 351.
Harisijades, S. S., Harisijades, B., & Lukic, R. (1967) *Mikrobiologija, Belgrade*, **4**, 199.
Hart, L., Hayston, J. T., & Keat, J. C. (1949) *Aust. vet. J.*, **25**, 40.
Herzberg, K., Lang, D., Reuss, K., & Dahn, R. (1964) *Zentbl. Bakt. ParasitKde, I. Abt. Orig.*, **195**, 133.
Hinze, H. C., & Walker, D. L. (1964) *J. Bact.*, 88, 1185.
Holowczak, J. A., & Joklik, W. K. (1967) *Virology*, **33**, 717.
Horgan, S. E., & Hasseeb, M. A. (1939) *J. Hyg., Camb.*, **39**, 615.
Horgan, S. E., & Hasseeb, M. A. (1947) *J. comp. Path.*, **57**, 8.
Howatson, A. F. (1962) *Fed. Proc.*, **21**, 947.
Hurst, E. W. (1937a) *Br. J. exp. Path.*, **18**, 1.
Hurst, E. W. (1937b) *Br. J. exp. Path.*, **18**, 15.
Huygelen, C., Mortelmans, J., Thienport, D., Biche, Y., & Pinckers, F. (1958) *Zentbl. VetMed.*, **5**, 859.
Hyde, J. M., Gafford, L. G., & Randall, C. C. (1965) *J. Bact.*, **89**, 1557.
Israeli, E., & Sachs, L. (1964) *Virology*, **23**, 473.
Jacotot, H., Vallée, A., & Virat, N. (1955) *Annls Inst. Pasteur, Paris*, **89**, 290.
Jansen, J. (1941) *Zentbl. Bakt. ParasitKde, I. Abt. Orig.*, **148**, 65.
Jansen, J., & Kunst, H. (1959) *T. Diergeneesk.*, **84**, 947.
Ježić, J. A. (1932) *Z. Immunforsch.*, **75**, 456.
Joklik, W. K. (1962a) *J. molec. Biol.*, **5**, 265.
Joklik, W. K. (1962b) *Virology*, **18**, 9.
Joklik, W. K. (1966) *Bact. Rev.*, **30**, 33.
Joklik, W. K. (1968) *Ann. rev. Microbiol.*, **22**, 359.
Joklik, W. K., Abel, P., & Holmes, I. H. (1960) *Nature, Lond.*, **186**, 992.
Kaplan, C. (1958) *J. gen. Microbiol.*, **18**, 58.
Kameyama, S., Takahashi, M., Toyoshima, T., Kato, S., & Kamahora, J. (1959) *Biken's J.*, 2, 341.
Kasza, L., Bohl, E. H., & Jones, D. O. (1960) *Am. J. vet. Res.*, **21**, 269.
Kataria, R. S., & Singh, I. P. (1970) *Acta virol., Prague*, **14**, 307.
Kates, J., & Beeson, J. (1970) *J. molec. Biol.*, **50**, 1 and 19.
Kates, J., & McAusland, B. R. (1967) *Proc. natn. Acad. Sci. U.S.A.*, 58, 134.
Kato, S., & Cutting, W. (1959) *Stanf. med. Bull.*, **17**, 34.
Kato, K., Horiuchi, T., & Tsubahara, H. (1965) *Natn. Inst. Anim. Hlth Q., Tokyo*, **5**, 130.
Kato, S., Takahashi, M., Kameyama, S., & Kamahora, J. (1959) *Biken's J.*, **2**, 353.

Kaufman, H. E., Nesburn, A. B., & Maloney, E. D. (1962) *Virology*, **18**, 567.
Kempe, C. H., Dekking, F., St. Vincent, L., Rao, A. R., & Downie, A. V. (1969) *J. Hyg., Camb.*, **67**, 631.
Kikuth, W., & Gollub, H. (1932) *Zentbl. Bakt. ParasitKde, I. Abt. Orig.*, **125**, 313.
Kilham, L. (1955) *Am. J. Hyg.*, **61**, 55.
Kilham, L. (1958) *J. natn. Cancer Inst.*, **20**, 729.
Kilham, L., & Dalmat, H. T. (1955) *Am. J. Hyg.*, **61**, 45.
Kilham, L., Herman, C. M., & Fisher, E. R. (1953) *Proc. Soc. exp. Biol. Med., U.S.A.*, **82**, 298.
Kilham, L., Lerner, E., Hiatt, C., & Shack, J. (1958) *Proc. Soc. exp. Biol. Med., U.S.A.*, **98**, 689.
Kirschstein, R. L., Rabson, A. S., & Kilham, L. (1958) *Cancer Res.*, **18**, 1340.
Ledingham, J. C. C. (1924) *Br. J. exp. Path.*, **5**, 332.
Leese, A. S. (1909) *J. trop. vet. Sci.*, **4**, 1.
Leinati, L., Cilli, V., Mandelli, G., Castrucci, G., Carrara, O., & Scatozza, F. (1961) *Boll. Ist. sieroter. milan.*, **40**, 295.
Levaditi, C., & Nicolau, S. (1923) *Annls. Inst. Pasteur, Paris*, **37**, 1.
Levinthal, J. M., & Schein, H. M. (1964) *Virology*, **23**, 268.
Lillie, R. D. (1930) *Arch. Path.*, **10**, 241.
Lipschütz, B. (1920) *Arch. Derm. Syph.*, **127**, 193.
Loh, P. C., & Riggs, J. L. (1961) *J. exp. Med.*, **114**, 149.
Lübke, A. (1960) *Dtsch. tierärztl. Wschr.*, **67**, 113.
McCrea, J. F., Angerer, S., & O'Loughlin, J. (1962) *Virology*, **17**, 208.
McKercher, D. G., & Saito, J. K. (1964) *Nature, Lond.*, **202**, 933.
Magnus, P. von, Anderson, E. K., Petersen, K. B., & Birch-Andersen, A. (1959) *Acta path. microbiol. scand.*, **46**, 156.
Manninger, R., Csontos, J., & Sályi, G. (1940) *Arch. Tierheilk.*, **75**, 159.
Mansi, W., & Thomas, V. (1958) *J. comp. Path.*, **68**, 188.
Marennikova, S. S., Gurvich, E. B., & Yemasheva, M. A. (1959) *Probl. Virol.*, **4**, 70.
Marquardt, J., Falsen, E., & Lycke, E. (1967) *Arch. ges. Virusforsch.*, **20**, 109.
Marshall, I. D., & Fenner, F. (1958) *J. Hyg., Camb.*, **56**, 288.
Marshall, I. D., & Regnery, D. R. (1960) *Nature, Lond.*, **188**, 73.
Mason, J. H., & Neitz, W. O. (1940) *Onderstepoort J. vet. Res.*, **15**, 159.
Mayr, A. (1956) *Arch. ges. Virusforsch.*, **6**, 439.
Mayr, A. (1959) *Arch. ges. Virusforsch.*, **8**, 156.
Mayr, A. (1966) *Zentbl. Bzkt. ParasitKde, I. Abt. Orig.*, **199**, 145.
Mayr, A., & Mahnel, H. (1970) *Arch. ges. Virusforsch.*, **31**, 51.
Mitchell, J. C. (1953) *Br. J. exp. Path.*, **34**, 44.
Morgan, C., Ellison, S. A., Rose, H. M., & Moore, D. (1954) *J. exp. Med.*, **100**, 301.
Morgan, C., & Wyckoff, R. W. G. (1950) *J. Immunol.*, **65**, 285.
Moritsch, H. (1956) *Zentbl. Bakt. ParasitKde, I. Abt. Orig.*, **166**, 427.
Moses, A. (1911) *Mem. Inst. Oswaldo Cruz*, **3**, 46.
Munz, E. K., & Owen, N. C. (1966) *Onderstepoort J. vet. Res.*, **33**, 3.
Nagington, J. (1968) *Vet. Rec.*, **82**, 477.
Nagington, J., & Horne, R. W. (1962) *Virology*, **16**, 248.
Nagington, J., Lauder, I. M., & Smith, J. S. (1967) *Vet. Rec.*, **81**, 306.
Nagington, J., Newton, A. A., & Horne, R. W. (1964) *Virology*, **23**, 461.
Nagington, J., Tee, G. H., & Smith, J. S. (1965) *Nature, Lond.*, **208**, 505.
Nagler, F. P. O. (1942) *Med. J. Aust.*, **1**, 281.
Nelson, J. B. (1941) *J. exp. Med.*, **74**, 203.
Neva, F. A. (1962) *Arch. int. Med.*, **110**, 720.
Nicholas, A. H. (1970) *J. natn. Cancer Inst.*, **55**, 907.
Nicholas, A. H., & McNulty, W. P. (1968) *Nature, Lond.*, **216**, 745.
Nii, S. (1959) *Biken's J.*, **3**, 195.
Niven, J. S. F., Armstrong, J. A., Andrewes, C. H., Pereira, H. G., & Valentine, R. C. (1961) *J. Path. Bact.*, **81**, 1.

Noordaa, van, J., Dekking, F., Posthuma, J., & Beunders, B. J. W. (1967) *Arch. ges. Virusforsch.*, **22**, 210.
Noyes, W. F. (1953) *Proc. Soc. exp. Biol. Med., U.S.A.*, **83**, 426.
Noyes, W. F. (1962) *Virology*, **17**, 282.
Noyes, W. F. (1965) *Virology*, **25**, 666.
Olsen, R. G., & Yohn, D. S. (1970) *J. Virol.*, **5**, 212.
Olson, N. O., & Kerr, K. M. (1966) *Avian Dis.*, **10**, 470.
Overfield, T. M., & Brody, J. A. (1966) *J. Pediatrics*, **69**, 640.
Padgett, B. L., Moore, M. S., & Walker, D. L. (1962) *Virology*, **17**, 462.
Padgett, B. L., Wright, M. J., Jayne, C., & Walker, D. L. (1964) *J. Bact.*, **87**, 454.
Pandey, R., & Singh, I. P. (1970) *Res. vet. Sci.*, **11**, 195.
Parish, W. E. (1961) *J. Path. Bact.*, **81**, 331.
Parish, W. E. (1962) *J. Path. Bact.*, **83**, 429.
Peters, D., & Müller, G. (1963) *Virology*, **21**, 251.
Peters, D., Müller, G., & Büttner, D. (1964) *Virology*, **23**, 609.
Peters, J. C. (1966) *Tijdschr. Diergeneesk*, **91**, 387.
Plowright, W., & Ferris, R. D. (1958) *Br. J. exp. Path.*, **39**, 424.
Plowright, W., & Ferris, R. D. (1959) *Virology*, **7**, 357.
Plowright, W., & Witcomb, M. A. (1959) *J. Path. Bact.*, **78**, 397.
Porterfield, J. S., & Allison, A. C. (1960) *Virology*, **10**, 233.
Postlethwaite, R. (1964) *Virology*, **22**, 508.
Postlethwaite, R., Watt, J. A., Hawley, T. G., Simpson, I., & Adam, J. (1967) *J. Hyg., Camb.*, **65**, 281.
Prier, J. E., Sauer, R. M., Malsberger, R. E., & Sillaman, J. M. (1960) *Am. J. vet. Res.*, **21**, 381.
Prose, P. H., Friedman-Kien, A. E., & Vilček, J. (1969) *Am. J. Path.*, **55**, 249.
Rafyi, A., & Ramyan, H. (1959) *J. comp. Path.*, **69**, 141.
Rake, G., & Blank, H. O. (1950) *J. invest. Derm.*, **15**, 81.
Ramyar, H., & Hessami, M. (1967) *Zentbl. VetMed.*, **14B**, 516.
Randall, C. C., Gafford, L. G., Soehner, R. L., & Hyde, J. H. (1966) *J. Bact.*, **91**, 95.
Reczko, E. (1959) *Arch. ges. Virusforsch.*, **9**, 193.
Reisner, A. H., Sobey, W. R., & Conolly, D. (1963) *Virology*, **30**, 539.
Rivers, T. M. (1930) *J. exp. Med.*, **51**, 965.
Roberts, J. A. (1962) *Br. J. exp. Path.*, **43**, 451 and 462.
Roberts, J. A. (1964) *J. Immunol.*, **92**, 837.
Robinow, C. F. (1950) *J. gen. Microbiol.*, **4**, 242.
Rondle, C. J. M. (1969) *J. gen. Virol.*, **4**, 453.
Rondle, C. J. M., & Dumbell, K. R. (1962) *J. Hyg., Camb.*, **60**, 41.
Rones-Zakay, Z. (1966) *Israel J. med. Sci.*, **2**, 662.
Rouhandeh, H., Engler, R., Fouad, M. T. A., & Sells, L. L. (1967) *Arch. ges. Virusforsch.*, **20**, 363.
Sabban, M. S. (1957) *Am. J. vet. Res.*, **18**, 618.
Sabban, M. S. (1960) *Bull. Off. int. Épizoot.*, **53**, 1527.
Salaman, M. H., & Tomlinson, A. J. H. (1957) *J. Path. Bact.*, **74**, 17.
Sarov, I., & Beeker, Y. (1967) *Virology*, **33**, 369.
Schwartz, L. H. (1958) in *Diseases of swine*. Ed.: Dunne Ames: Iowa State Press.
Service, M. W. (1971) *J. Hyg., Camb.*, **69**, 105.
Shama, R. M., & Bhatia, H. M. (1959) *Indian J. vet. Sci.*, **28**, 205.
Sharp, D. G., Sadhukhan, P., & Galasso, G. J. (1964) *Proc. Soc. exp. Biol. Med., U.S.A.*, **115**, 811.
Shirinov, F. B., Fagaliev, I. A., Alekperov, Y. G., Ibraginova, A. A. (1969) *Veterinariya, Moscow*, **12**, 37.
Shope, R. E. (1932) *J. exp. Med.*, **56**, 803.
Shope, R. E. (1938) *Proc. Soc. exp. Biol. Med., U.S.A.*, 38, 86.
Shope, R. E. (1940) *Arch. ges. Virusforsch.*, **1**, 457.

Smadel, J. E., Rivers, T. M., & Hoagland, C. L. (1942) *Arch. Path.*, **34**, 275.
Smadel, J. E., Ward, S. M., & Rivers, T. M. (1940) *J. exp. Med.*, **72**, 129.
Strandström, H. V., Ambrus, J. L., & Owens, G. (1966) *Virology*, **28**, 479.
Subak-Sharpe, J. H., Timbury, M. C., & Williams, J. F. (1969) *Nature, Lond.*, **222**, 341.
Takahashi, M., Kameyama, S., Kato, S., & Kamahora, J. (1959) *Biken's J.*, **2**, 27.
Taylor, D. L., Olson, N. O., & Burrell, R. G. (1966) *Avian Dis.*, **10**, 462.
Taylor, D. O. N., Klauber, M. R., Lennette, E. H., & Wiener, A. (1968) *J. natn. Cancer Inst.*, **40**, 147.
Todd, W. M., & Randall, C. C. (1958) *Arch. Path.*, **66**, 150.
Trueblood, M. S. (1966) *Cornell Vet.*, **56**, 521.
Tunnicliff, E. A. (1949) *Am. J. vet. Res.*, **10**, 240.
Tunnicliff, E. A., & Matishek, P. H. (1941) *Science*, **94**, 283.
Turner, G. S., Squires, E. J., & Murray, H. G. S. (1970) *J. Hyg., Camb.*, **68**, 197.
Ursiny, J., & Jakulik, J. (1966) *Veterinarstvi*, **16**, 23.
Van Rooyen, P. J., Kümm, N. A. L., Weiss, K. E., & Alexander, R. A. (1959) *Bull. épizoot, Dis. Afr.*, **7**, 79.
Vieuchange, J. (1959) *Bull. Soc. Path. exot.*, **52**, 432.
Weiss, K. E. (1968) *Virology monogr.*, **3**, 111. Berlin: Springer.
Wenner, H. A., Macasaet, F. D., Kamitsuka, P. S., & Kidd, P. (1968) *Am. J. Epidemiol.*, **87**, 551.
Wesslén, T. (1956) *Arch. ges. Virusforsch.*, **6**, 430.
Westwood, J. C. N., Boulter, E. A., Bowen, E. T. W., & Maber, H. B. (1966) *Br. J. exp. Path.*, **47**, 453.
Westwood, J. C. N., Harris, W. J., Zwartouw, H. T., Titmuss, D. H. J., & Appleyard, G. (1964) *J. gen. Microbiol.*, **34**, 67.
Westwood, J. C. N., Phipps, P. H., & Boulter, E. A. (1957) *J. Hyg., Camb.*, **45**, 123.
Westwood, J. C. N., Zwartouw, H. T., Appleyard, G., & Titmuss, D. H. J. (1965) *J. gen. Microbiol.*, **38**, 47.
Wheeler, C. E., & Cawley, E. P. (1956) *Am. J. Path.*, **32**, 535.
WHO (1959). *Requirements for smallpox vaccine*. Technical report series no. 180.
WHO (1964) Technical report series no. 283.
WHO (1968) Technical report series no. 393.
WHO (1971) *WHO Chron.*, **25**, 370.
Wilcox, W. C. I., & Cohen, G. H. (1967) *J. Virol.*, **1**, 500.
Wilcox, W. C. J., & Cohen, G. H. (1969) *Curr. topics Microbiol. Immunol.*, **47**, 1.
Wilkinson, G. T., & Prydie, J. (1970) *Vet. Rec.*, **87**, 766.
Wittman, G. (1958) *Zentbl. VetMed.*, **5**, 769.
Wittman, G., & Matheka, H. D. (1958) *Mschr. Tierheilk.*, **10**, 161.
Wolfe, L. G., Griesemer, R. A., & Farrell, R. L. (1968) *J. natn. Cancer Inst.* **41**, 1715.
Woodroofe, G. M. (1960) *Virology*, **10**, 379.
Woodroofe, G. M., & Fenner, F. (1960) *Virology*, **12**, 272.
Woodroofe, G. M., & Fenner, F. (1962) *Virology*, **16**, 334.
Woodson, B. (1968) *Bact. Rev.*, **32**, 127.
Yohn, D. S., & Gallagher, J. F. (1969) *J. Virol.*, **3**, 114.
Yohn, D. S., Haendiges, V. A., & Grace, J. T. (1966) *J. Bact.*, **91**, 1977.
Yuan, C. J., Lee, P. C., & Cheng, Y. S. (1957) *Acta vet. zootechn. sinica*, **2**, 15.
Zakay-Roness, R. A., & Bernkopf, H. (1962) *J. Immunol.*, **88**, 185.
Zwartouw, H. T., Westwood, J. C. N., & Harris, W. J. (1965) *J. gen. Microbiol.*, **38**, 47.

17

Iridoviruses

The viruses described in this chapter have been provisionally grouped with the iridescent virus of craneflies (*Tipula*) and others infecting insects. All are relatively large DNA-containing viruses having icosahedral symmetry with 812, possibly more, capsomeres. Those infecting insects appear to be ether-resistant, while those of mammals are apparently ether-sensitive (Bellett, 1968).

AFRICAN SWINE FEVER

Synonym: Wart-hog disease.

Review: Maurer et al. (1958a). Scott (1965).

An extensive bibliography has been published (Anon, 1967).

Morphology. Diameter about 200 nm; some forms are hexagonal. An unstable icosahedral form may be stabilized by addition of outer membranes (Almeida et al., 1967). Breese & DeBoer (1966) describe regularly hexagonal particles 175 × 215 nm with a nucleoid 72 × 89 nm; growth was in cytoplasm.

Chemical composition. Adldinger et al. (1966) obtained an infectious DNA.

Physico-chemical characters. Inactivated in 30′ at 55°, 10′ at 60°—less heat-resistant than swine fever. Survives for years when dried at room temperature or frozen on skin or muscle. It is resistant to most disinfectants (cf. Scott, 1965), but is inactivated by 1 per cent formaldehyde in 6 days, by 2 per cent NaOH in 24 hours and by chloroform and ether.

Haemagglutination is not reported, but haemadsorption of pig RBCs is seen in cultures in pig marrow or 'buffy coat' (Malmquist & Hay, 1960). Coggins (1968) isolated in culture a variant which failed to show haemadsorption.

Antigenic properties. Quite distinct in cross-immunity tests from swine fever. Immunity even to the homologous strain seems to be transient. In Africa there are several serotypes, but possibly only 1 in Europe. Specific and group antigens have been revealed by means of complement fixation, gel diffusion and other tests. The CF test is group-reactive, but there is said to be a specific haemadsorption-inhibition test (Malmquist, 1963; Stone & Hess, 1965).

Cultivation. Grows in yolk sac of developing eggs, killing embryos in 6 or 7 days (McIntosh, 1952). It has been grown in cultures of pig bone marrow and buffy coat and shows haemadsorption in 24 hours, later CPE (Malmquist & Hay, 1960). It has also been adapted to grow in chick embryo (Hess et al., 1965) and pig-kidney tissue cultures. Only some strains produce CPE in the kidney cultures, but the virus has been adapted to produce plaques in such cultures (Parker & Plowright, 1968). Breese & Ozawa (1969) grew the virus in tissue cultures of monkey kidney; nuclear and cytoplasmic inclusions were seen.

Distribution. In East, South and West Africa, Reached Portugal and Spain in 1957, France in 1964 and Italy in 1967.

Pathogenicity. Very fatal amongst domestic pigs. The incubation period is 7–9 days in the field, 2 to 5 days experimentally. Symptoms include high fever, cough and diarrhoea and resemble those of classical swine fever. Growth of virus begins in tonsils and retropharyngeal glands, but there is soon generalization, especially to lymph nodes and spleen. There is leucopenia. Surviving pigs may have viraemia for months (De Tray, 1957). The disease probably naturally infects wart-hogs (*Phacochoerus*) and probably bush-pigs (*Potamochoerus*) and giant forest-hogs (*Hylochoerus*), and these too may be carriers.

Experimentally, the virus has been adapted to infect goats and rabbits (Mendes, 1962). After 100 passes it was less virulent for pigs, but did not immunize against a virulent strain.

Pathologically the lesions resemble those of classical swine fever, but are more severe with many haemorrhages and with severe karyorrhexis or lymphocytes (de Kock et al., 1940; Maurer et al., 1958b). Post mortems reveal fluid in body cavities, haemorrhagic lymph nodes, engorgement and oedema of the gall-bladder.

Ecology. Domestic pigs in Africa may contract infection from wart-hogs; these may liberate virus particularly during periods of farrowing and other stresses. Thereafter infection is by contact and through fomites. Infected premises remain dangerous for long periods. A new enzootic cycle has developed in Europe, where the disease has become less virulent, and chronically infected domestic pigs perpetuate the infection. The exact method of transmission is obscure (Scott, 1965).

Virus has been recovered from Argasid ticks (Botija, 1963); these have been readily found in burrows of wart-hogs in East Africa, and actual multiplication of the virus in ticks has been demonstrated (Plowright et al., 1969).

Control. Attenuated strains have been used as vaccines in Spain and Portugal, but vaccinated pigs may continue to carry the virus. Vaccinated pigs may develop high-titre antibodies, yet fail to develop good resistance (Stone & Hess, 1967).

FROG VIRUSES

Reviews: Lunger (1966). Granoff (1969).

DNA-containing viruses have been isolated by Granoff et al. (1966) and others from leopard-frogs (*Rana pipiens*) with and without kidney tumours (see p. 357); they are unlikely to be the cause of the tumours though they could conceivably act as 'helpers' of the Lucké herpesvirus. They have been given serial numbers FV 1, 2, etc. or LT 1, 2, etc., but all are apparently alike antigenically; so too is the tadpole oedema virus of bull frogs (*R. catesbiana*) (Wolf et al., 1968).

Morphology. Similar to *Tipula* virus. Diameter 120 × 130 nm. Growth in cytoplasm.

Chemical composition. GC ratio 53 per cent (Maes & Granoff, 1967) or 56–58 per cent. DNA 30·1 per cent, protein 55·8 per cent, lipid 14·2 per cent. MW of DNA 130×10^6 (Smith & McAusland, 1969).

Physico-chemical characters. Density of some particles 1·305, of others 1·28 (Morris et al., 1966).

Antigenic properties. Viruses, so far isolated, are similar according to neutralization, fluorescent antibody and precipitation tests. (Came et al., 1968; Morris & Roizman, 1967) and distinct from the Lucké herpesvirus and from African swine fever.

Cultivation. The viruses have been grown in tissue cultures of cells from fish, amphibia, reptiles, birds and mammals causing CPE. Growth is best at 30°; formation of plaques is described by Lehane et al. (1967). Multiplication viruses also in hens' eggs held at 30° (Came et al., 1968).

Pathogenicity. The viruses were lethal to tadpoles of several *Rana* or *Bufo* species (Came et al., 1968; Tweedell & Granoff, 1968). They would also multiply in immature or adult newts (*Triturus*), but not in adult frogs.

GECKO VIRUS

Morphologically similar viruses have been seen in the red blood cells of a reptile (*Gehyra variegata*) (Stehbens & Johnston, 1966). Their diameter was about 220 nm, and there was a core 120 to 160 nm across. Transmission to other geckos was accomplished, but there was no report of successful filtration.

LYMPHOCYSTIS

Review: Weissenberg (1965).

Morphology. Virions resemble those of the *Tipula* virus, having a diameter of 300; and icosahedral, with filaments 4 × 200–300 nm attached to vertices of the icosahedron. In thin sections 3 dark layers (stained by heavy metals) were seen, alternating with 2 clear layers (Zwillenberg & Wolf, 1968). Other workers failed to see capsomeres and estimated the diameter as 200 nm (Midlige & Malsberger, 1968). Walker (1962) describes in lymphocystis disease of the pike-perch (*Stizostedion*) bodies about 200 nm across, with a capsid 12 nm in thickness and a nucleoid 150 nm across consisting of a ball of 10 nm-wide threads.

Chemical composition. A DNA virus, but extracted DNA was not infectious (Wolf et al., 1966).

Physico-chemical characters. Virus survives 20 months at −20°, and 3 cycles of alternate freezing and thawing. It is ether-sensitive and, unlike most animal viruses, is not readily preserved in 50 per cent glycerol in the cold (Wolf, 1962). Readily lyophilized.

No *haemagglutination* nor haemadsorption could be demonstrated.

Antigenic properties. Recovered bluegills were partly but not wholly refractory to reinoculation (Wolf, 1962).

Cultivation. Grützner (1956) reported specific changes in cell cultures of *Lebistes* inoculated with virus. Wolf et al. (1966) isolated the virus in cultures of cyprinid fish cells. Lymphocystic cells developed slowly *in vitro* at 23°–25°.

Pathogenicity. Lymphocystis is a common chronic but rarely fatal disease affecting fish of several orders. Tumour-like masses of the skin and fins persist for long periods, but ultimately regress: secondary 'tumours' may appear elsewhere on the skin.

Experimentally, transmission has been accomplished by implantation, injection and other methods. Some workers have had difficulty in transmission between different species, but Wolf (1962) was more successful using several centrarchid fishes. The bluegill (*Lepomis macrochirus*) proved to be a convenient experimental animal which could be bred in hatcheries. Fish of unrelated families were refractory to Wolf's virus. Wolf & Carlson (1965) followed the course of development in bluegills.

Large cytoplasmic inclusions occur in 'tumour' cells, which may be up to 2000 nm across. They contain osmiophilic granules joined in a filamentous lattice (Weissenberg, 1956). Similar structures have been seen in flounder, perch and other fish. The growths themselves are of soft, jelly-like consistency.

Ecology. Outbreaks in nature occur mainly in summer months. Weissenberg (1945) reported spread to healthy fish, but Wolf (1962), who freed his fish from ectoparasites found no spread. Nigrelli (1952) suspected that parasites played a role. On the other hand, the huge lymphocystis cells burst in water, and this may represent the mode of spread to other fish.

REFERENCES

Adldinger, H. H., Stone, S. S., Hess, W. R., & Bachrach, H. L. (1966) *Virology*, **30**, 750.
Almeida, J. D., Waterson, A. P., & Plowright, W. (1967) *Arch. ges. Virusforsch.*, **20**, 392.
Anon (1967) *Bull. Off. int. Épizoot.*, **67**, 999.
Bellett, A. J. D. (1968) *Adv. virus Res.*, **13**, 225.
Botija, C. Sanchez (1963) *Bull. Off. int. Épizoot.*, **60**, 895.
Breese, S. S., & DeBoer, C. J. (1966) *Virology*, **28**, 420.
Breese, S. S., & Ozawa, Y. (1969) *J. Virol.*, **4**, 109.
Came, P. E., Geering, G., Old, L. J., & Boyse, E. A. (1968) *Virology*, **36**, 392.
Coggins, L. (1968) *Cornell Vet.*, **58**, 12.
de Kock, G., Robinson, E. M., & Keppel, J. J. G. (1940) *Onderstepoort J. vet. Sci.*, **14**, 31.
De Tray, D. E. (1957) *Am. J. vet. Res.*, **18**, 811.
Granoff, A. (1969) *Curr. topics Microbiol. Immunol.*, **50**, 107.
Granoff, A., Came, P. E., & Breeze, D. E. (1966) *Virology*, **29**, 133.
Grützner, L. (1956) *Zentbl. Bakt. ParasitKde, I. Abt. Orig.*, **165**, 81.
Hess, W. R., Cox, B. F., Henschele, W. P., & Stone, S. S. (1965) *Am J. vet. Res.*, **26**, 141.
Lehane, D. E., Clark, H. F., & Karzon, D. T. (1967) *Proc. Soc. exp. Biol. Med., U.S.A.*, **125**, 50.
Lunger, P. D. (1966) *Adv. Virus. Res.*, **12**, 1.
McIntosh, B. M. (1952) *Jl S. Afr. vet. med. Ass.*, **23**, 217.
Maes, R., & Granoff, A. (1967) *Virology*, **33**, 491.
Malmquist, W. A. (1963) *Am. J. vet. Res.*, **24**, 450.
Malmquist, W. A., & Hay, D. (1960) *Am. J. vet. Res.*, **21**, 106.
Maurer, F. D., Griesemer, R. A., & Jones, T. C. (1958a) in *Diseases of swine.* Ed.: Dunne. Ames: Iowa State College Press.
Maurer, F. D., Griesemer, R. A., & Jones, T. C. (1958b) *Am. J. vet. Res.*, **19**, 517.
Mendes, A. M. (1962) *Bull. Off. int. Épizoot.*, **58**, 699.
Midlige, F. H., & Malsberger, R. G. (1968) *J. Virol.*, **2**, 830.
Morris, V. L., & Roizman, B. (1967) *Proc. Soc. exp. Biol. Med., U.S.A.*, **124**, 507.
Morris, V. L., Spear, P. G., & Roizman, B. (1966) *Proc. natn. Acad. Sci., U.S.A.*, **56**, 1155.

Parker, J., & Plowright, W. (1968) *Nature, Lond.*, **219**, 524.
Plowright, W., Parker, J., & Peirce, M. A. (1969) *Nature, Lond.*, **221**, 1071.
Scott, G. R. (1965) *Vet. Rec.*, **77**, 1421.
Smith, W. R., & McAusland, B. R. (1969) *J. Virol.*, **4**, 339.
Stehbens, W. E., & Johnston, M. R. L. (1966) *J. ultrastruct. res.*, **15**, 543.
Stone, S. S., & Hess, W. R. (1965) *Virology*, **26**, 622.
Stone, S. S., & Hess, W. R. (1967) *Am. J. vet. Res.*, **28**, 475.
Tweedell, K., & Granoff, A. (1968) *J. natn. Cancer Inst.*, **40**, 407.
Weissenberg, R. (1945) *Zoologica*, **30** (Part 4), 169.
Weissenberg, R. (1956) *Arch. ges. Virusforsch.*, **7**, 1.
Weissenberg, R. (1965) *Ann. N.Y. Acad. Sci.*, **126**, 362.
Wolf, K. (1962) *Virology*, **18**, 249.
Wolf, K., Bullock, G. L., Dunbar, C. A., & Quimby, M. C. (1968) *J. infect. dis.*, **118**, 253.
Wolf, K., & Carbon, C. P. (1965) *Ann. N.Y. Acad. Sci.*, **126**, 414.
Zwillenberg, L. O., & Wolf, K. (1968) *J. Virol.*, **2**, 393.

PART III

Uncharacterized Viruses

18

Uncharacterized Viruses

HEPATITIS IN MAN

Review: WHO (1970). Havens (1967). MacCallum (1955). Ward & Krugman (1962). Zuckerman (1970).

Two clinically similar forms of hepatitis in man are known. One, caused by virus A, is known as infectious hepatitis, can occur in epidemic form, particularly among children, and has a relatively short incubation period, usually less than 40 days; virus is present in faeces as well as in blood and perhaps in nasopharynx also. Virus B causes serum hepatitis; it is commonly transmitted by transfusions of blood or blood products directly or indirectly through use of contaminated needles. It affects all ages, has an incubation period of usually more than 40 days; virus is normally absent from stools.

It has recently been realised that none of the above criteria is satisfactory. Virus B may, though not readily, be transmitted by normal contact and the incubation periods of A and B infections overlap. There is often no difficulty in deciding which one is dealing with, but in the absence of specific diagnostic criteria, many conclusions have to be accepted with reserve. According to Krugman et al. (1970) heating for 1 minute to 98° destroys the antigenicity of virus A, not that of B.

Infectious Hepatitis

Synonyms: Infective hepatitis (hepatitis virus A). (IH). Epidemic jaundice. Catarrhal jaundice.

Reviews: Havens (1954, 1963). MacCallum (1953, 1955). Ward & Krugman (1962).

Morphology. There is no agreement as to the virus's size. Various workers have described bodies with diameters of 40 × 60 nm (Gueft, 1961), 42 × 58 nm (Braunsteiner et al., 1957) or—from tissue culture—12–18 nm (Rightsel et al., 1961).

Physico-chemical characters. Survives 30′ at 56°; also $1\frac{1}{2}$ years at −10 to 20°. Ether-stable. Residual chlorine 1 part/million inactivates in the absence of impurities.

Haemagglutination. Sera from infected persons agglutinate rhesus RBCs more frequently than do others, but the test is not specific. Schmidt & Lennette (1961) have reviewed the reported HA-tests.

Antigenic properties. Challenge of recovered volunteers with homologous virus shows that there is immunity to that, but apparently none to serum hepatitis. Pools of gamma globulin from randomly obtained human sera show protective action. Most workers find that Australia antigen (see p. 426) is absent from patients with IH, but that regard must be paid to difficulty in diagnosis. However, Prete et al. (1970) have described a serum from a repeatedly transfused patient in Italy which reacts not only with Australia antigen but also with an antigen present in sera of numerous patients with IH; these were affected in 3 different epidemics. Ninety per cent of patients' sera contained the antigen if tested during the first week of symptoms. This epidemic-hepatitis-associated-antigen (EHAA) was labile at −20° and titres gradually fell on storage.

Cultivation. There have been several claims to have cultivated a virus. That of Rightsel et al. (1961) produced CPE in a particular line of human cancer cells (Detroit 6). The virus was said to be ether-resistant, 12–18 nm across, not inactivated in 30′ at 60°; there were 3 serotypes. More than a hundred cases are on record of hepatitis in man, apparently contracted from chimpanzees and other apes (Davenport et al., 1966). There are discrepancies between the accounts of different workers and no results from any laboratory have been regularly reproducible elsewhere. According to one report (WHO, 1964) claims for 18 viruses or groups of viruses are competing for recognition.

Distribution. Distribution is world-wide; the disease is particularly prevalent in the Middle East and around the Mediterranean.

Pathogenicity. Onset in man may be acute or insidious with anorexia, nausea and abdominal tenderness. There may be a period of improvement before jaundice appears, but many patients, especially children, are anicteric. Mortality is low, but cirrhosis or liver atrophy may follow in a few cases. The incubation period is from 15 to 40 days, average 25. A milder form occurring amongst children had a longer incubation—35 to 56 days (Ward & Krugman, 1962).

Experimentally, volunteers have been infected *per os* or parenterally. Virus is present in blood, urine and faeces. Virus has been recovered from urine a week before the appearance of jaundice (Giles et al., 1964). Many of the infected persons have had subicteric infection detected by finding abnormalities in liver-function tests. There have been claims to have reproduced the disease in African monkeys, marmosets, rats, guinea-pigs, pigs, ducks, canaries and other species; none has been generally confirmed.

Pathological lesions. Focal mid-zonal necrosis of liver parenchyma and portal infiltration have been seen in biopsy specimens. In fulminant cases there is acute yellow atrophy of the liver with multiple haemorrhages (Lucké, 1944). Regeneration of liver is rapid in convalescence. Sheehey et al. (1964) described round-cell infiltration and atrophy of villi in jejunum in biopsy specimens. It is likely that there is a general infection and that liver damage, often its most striking manifestation, may be absent.

Ecology. Infection commonly occurs through faecal contamination of food or water; infected oysters have caused some serious outbreaks. Opinions differ as to occurrence of virus in the nasopharynx and the importance of infection by the respiratory route. Virus may be present in blood and faeces during the incubation period and for months after recovery; this virus and not only homologous serum hepatitis can thus be transmitted by transfusions.

Control. Improved hygiene should be aimed at. Pooled γ globulin from normal adults, given during the incubation period, is believed to be effective in preventing illness if not infection (Krugman & Ward, 1961–62). The unexpectedly long duration of its efficacy has suggested that it operates as a passive-active immunizing agent, permitting an immunizing subclinical infection in those exposed before its effects have quite worn off (Stokes et al., 1951).

Serum Hepatitis

Synonyms: Hepatitis (virus B). Homologous serum hepatitis (SH).

Review: Blumberg et al. (1970).

Morphology. Passed a 52 nm gradocol membrane (McCollum, 1952) and diameter therefore likely to be 30 nm or less. Particles described under 'Antigenic Properties' may or may not represent the virus itself. Recently, 2 groups of workers have reported that RNA is present at least in the larger particles (Jokelainen et al., 1970) or perhaps in all (Józwiak et al., 1971); in the latter report the GC content is said to be about 48 per cent.

Physico-chemical characters. Survives heating at 60° for 4 not 10 hours, also, in serum or plasma, at least 6 months at room temperature and for years when frozen or dried. Withstands 0·25 per cent phenol for over a year. Ether-resistant. UV radiation is unreliable as a method of inactivation.

Antigenic properties. No cross-immunity with hepatitis virus A, but there is some evidence of homologous immunity (Neefe et al., 1947). γ globulin from pooled human sera does not protect. Interest in the subject has suddenly awakened since the discovery of the Australia or SH antigen (Blumberg et al., 1967; Prince, 1968). Sera of people who have been repeatedly transfused may develop antibodies reacting in double-diffusion tests in agar with an antigen present particularly in sera from patients with SH. A complement-fixation test may be more sensitive (Purcell et al., 1969). Electron-microscopy reveals in such sera abundant particles 16–50 nm, but mainly 20–25 nm in diameter, also tubules 20 nm in width (Almeida & Waterson, 1969).

Particles had a buoyant density of 1·20/cm^3 in CsCl and were not degraded by ether nor overnight heating at 56° (Gerin et al., 1969). Some sera contain what appear to be antigen-antibody complexes, and their presence may account for the fact that many hepatitis sera are anticomplementary. Particles are said to contain no nucleic acid, but some lipids may be present. Tests with fluorescent antibody reveal presence of the antigen in liver cells, mainly in nuclei (Millman et al., 1970; Coyne et al., 1970). Antigen may be only transiently present in sera or may persist for many years in sera of known carriers of infectivity (Zuckerman & Taylor, 1969).

Numerous papers have been published on presence of antigen in various kinds of liver disease. Most workers, as already mentioned (p. 424) find it absent from IH sera. It has been found in sera of chimpanzees, spider monkeys, vervets and marmosets; these might have been infected in captivity (Blumberg et al., 1970).

Because the particles in sera are not uniform in size and may lack nucleic acid, the question of whether they represent the virus itself remains open. Some workers (Zuckerman et al., 1970) have described much larger particles, not unlike coronaviruses, in sera; their nature is quite doubtful.

Cultivation: Many unconvincing reports.

Pathogenicity. Symptoms in man are not clearly separable from those caused by hepatitis virus A. Persons of all ages are susceptible (A virus usually affects younger people), but this difference may reflect the different method of transmission. Some subjects carry virus in the blood for as long as 5 years; they may show abnormalities in liver function tests. Incubation commonly 40–150 days. Various primates seem to be susceptible to infection; Australia antigen has been found in their sera after inoculation, though absent previously.

Ecology. Virus is present in the blood of some normal persons, apparently 0·78 per cent in 1 survey (Lehane et al., 1949). In Southeast Asia, Oceania and other tropical countries, incidence may be as high as 20 per cent (Blumberg et al., 1968), usually in the absence of symptoms. Arthropod transmission may play a part.

Infection is commonly transmitted by transfusion or by any other injection of human blood, serum or plasma or when several people are injected without proper sterilization of needles or syringes in between. Thousands of cases occurred during the Second World War as a result of injecting yellow-fever vaccine containing human serum. Contact infection may occur also: patients with Downs' syndrome living in institutions, but not those at home, have a high incidence of antigen in their sera and often chronic liver damage. There have been numerous outbreaks in renal dialysis units, not only in patients but in staff who have contact with the blood.

Control. Many infections can be avoided if blood or plasma for transfusions is not pooled and if syringes and needles are sterilized by autoclaving or boiling for 10 minutes between any 2 injections. Immersion in alcohol is useless. No satisfactory method of inactivating virus in blood or plasma without destroying their useful properties has been established. Fractions from blood, such as serum albumin, and more especially γ globulin, can, however, be treated so as to render them safe. Tests for Australia antigen are not yet sufficiently sensitive to permit detection of dangerous donors, but perhaps 20 per cent can be thus detected (Levene & Blumberg, 1969).

HEPATITIS IN MARMOSETS

Deinhardt et al. (1967) observed hepatitis in marmosets (*Sanguinus*) inoculated with serum or plasma from early cases of IH in man. It seemed possible that the human infection had been transmitted, but Parks & Melnick (1969) obtained what seemed to be the same agent from uninoculated marmosets. The virus was ether-sensitive, passed a 50-nm membrane and was not associated with presence of Australia antigen.

HEPATITIS IN HORSES

There are reports of hepatitis in horses following injection of homologous serum or tissues (Findlay & MacCallum, 1938). The disease has followed attempts at immunizing against horse sickness in South Africa, against some equine encephalomyelitis in North America (Marsh, 1937) and against *Clostridia* in Scotland. In Marsh's cases the incubation period was 40–72 days in 90 per cent, and the mortality in horses developing symptoms was about 90 per cent. Besides hepatitis, there were diffuse enteritis and renal damage. There are no reports of attempts to recover a transmissible agent.

EPIDEMIC VIRAL GASTRO-ENTERITIS OF MAN

Synonyms: Acute infectious gastro-enteritis. Winter vomiting disease. Viral dysentery. Infantile diarrhoea (in part) —and various other names.

Review: Reimann (1963). Cheever (1967).

There is obscurity as to the relationships between neonatal diarrhoea and the diarrhoeas of older infants and of adults.

A number of viruses are probably concerned; some epidemics, especially in children, may be caused by enteroviruses (see p. 12). There is a little information about some others, largely obtained as a result of transmission experiments in human volunteers. An agent causing winter vomiting disease is probably one of the smaller ether-resistant viruses (D. A. J. Tyrell, personal communication 1971).

Marcy Virus

Stool-filtrates obtained from adults suffering from epidemic diarrhoea were fed to volunteers, most of whom developed severe watery diarrhoea with little or no fever; some had vomiting also. The incubation period was 1 to 5 days, average 3 (Gordon et al., 1947). The infection was transmissible in series. Sufferers reinoculated after 2 weeks proved to be immune, but 2 subjects tested after 10 months were susceptible. Immunity is therefore transient. A virus from Japan is probably identical with Gordon's from the United States (Fukumi et al., 1957).

Virtually nothing is known of the properties of this virus. It could not be cultivated in fertile eggs, nor are there definite reports of growth in tissue culture.

FS Virus

Another agent, obtained from bacteria-free preparations of stools of adults with diarrhoea, caused rather different symptoms with some fever, more constitutional symptoms, often abdominal pain without diarrhoea (Jordan et al., 1953; Badger et al., 1956). There was no cross-immunity with the Marcy virus.

Buddingh & Dodd's (1944) virus associated with stomatitis and diarrhoea may have been an atypical herpes simplex virus.

ROSEOLA INFANTUM

Synonyms: Exanthema subitum. Sixth disease.

A fever affecting small children, usually sporadic in distribution. Incubation period 10 days. Fever occurs with leucopenia and sometimes lymphoadenopathy; the rash usually appears as the temperature falls. Evidence that it is a virus disease rests on transmission to other children with bacteria-free serum. Throat-washings produced fever, leucopenia but no rash in monkeys (Kempe et al., 1950; Hellstrom & Vahlquist, 1951). A claim to have propagated the virus in mice (Nagayama, 1960) awaits confirmation.

Another exanthem affecting children, *erythema infectiosum*, or fourth disease, is believed to be caused by a virus, but definite evidence is lacking.

CAT-SCRATCH DISEASE

Review: Warwick (1967).

There is no firm evidence that this is a viral disease. There have been claims by some, denied by others, that it is a chlamydial infection. The herpes-like particles seen by Kalter et al. (1969) may have been only passengers. The symptoms include fever, malaise and generalized lymphadenopathy and rashes. Not every case gives a history of cat-scratch or cat-bite. Warwick (1967) considers it as a syndrome possibly associated with a variety of agents.

LETHAL INTESTINAL VIRUS OF INFANT MICE

Description: Kraft (1962).

Review: Kraft (1966).

Morphology. Virus readily passes a 220 nm membrane.

Physico-chemical characters. Inactivated in 30 minutes at 50°. Ether-sensitive.

Haemagglutination. None demonstrated.

Antigenic properties. Antibodies are found in some mice. No relation to epidemic diarrhoea of suckling mice (p. 66).

Cultivation. Attempts at propagation in tissue culture have been unsuccessful.

Pathogenicity. Very young mice do not take milk and die in a few days without diarrhoea. Diarrhoea may be seen in mice 7–20 days old. Growth may be stunted in survivors.

In baby mice the intestinal wall is ulcerated. The mucous layer is thinned and villi almost disappear. There are many syncytia and balloon cells. Changes are less severe in older mice, and the damage can be repaired (Biggers et al., 1964).

Control. A strict culling programme apparently eliminated the disease from a colony (Kraft, 1966).

MOUSE THYMIC VIRUS

Description: Rowe & Capps (1961).

Morphology. Diameter 75–100 nm by electron-microscopy. Body made up of granules 19–36 nm across with an outer wall 10–27 nm in diameter. Particles in the cytoplasm were 119–132 nm across. Some filaments were present in nuclei. May well prove to be a herpesvirus.

Physico-chemical characters. Ether-sensitive. Inactivated in 30 minutes at 50°C. Survived well at −60°C and after shell freezing.

Antigenic properties. Very poor antibody formation in mice.

Habitat. Probably enzootic in some stocks of laboratory mice and in wild mice (*Mus musculus*).

Pathogenicity. Produces a nonfatal infection in baby mice with massive necrosis of the thymus after 12–14 days, going on to scarring. The virus may persist in the mice for months. Nuclear inclusions are present in affected tissues.

MOUSE PAPULE AGENT

Little is known of this papular disease described by Kraft & Moore (1961).

VIRUS DISEASE OF SQUIRRELS

Vizoso et al. (1966) isolated a virus from red squirrels (*Sciurus vulgaris*) and grey squirrels (*S. carolinensis*), both normal and sick. The virus, ether-sensitive, killed chick embryos when inoculated amniotically and multiplied in hamster, mouse, human, and rabbit-cell cultures. It produced syncytia, cytoplasmic inclusions and caused cell destruction. In L (mouse) cells haemagglutinins and haemolysins appeared. In hamster cells fibroblasts behaved as if 'transformed' and caused transplantable tumours after injection into hamsters.

M-P VIRUS

A virus, isolated from an Ehrlich mouse carcinoma; it caused lymphocytopenia and hepatitis with increases in pleural and peritoneal fluids; 20–30 per cent of inoculated mice died. Reported to be ether-resistant, to pass a 220 nm but not a 150 nm membrane and not to agglutinate bovine or chick RBCs. Distinct from mouse hepatitis, reovirus 3 and mouse leukaemia viruses. It was oncolytic for several mouse tumours (Molomut et al., 1965). The virus described by Mouriquand et al. (1965) may be the same.

SKIN-HETEROGENIZING VIRUS OF MICE

Svet-Moldavsky et al. (1968; 1970) have described a virus, present in many transplantable tumours of mice; its only known effect is to induce formation of strong transplantation antigens in the skin. It seems to be very small, ether-resistant and active at very high dilutions.

REGIONAL ILEITIS IN HAMSTERS

A virus isolated from this condition by Tomita & Jonas (1968) was ether-sensitive, between 100–200 nm in diameter, cultivable in hamster-embryo fibroblasts, lethal to 7-day-old hamsters; it could perhaps be a herpes virus.

MISCELLANEOUS VIRUSES INFECTING DOGS

Contagious rhinotonsillitis (Fontaine et al., 1957; Goret et al., 1959) was transmissible to foxes, but not other species; no report is available of any comparison with the canine herpesvirus (p. 349).

A *malignant canine lymphocytopenia* with intracytoplasmic inclusion bodies was described by Reihart et al. (1952).

The contagious venereal sarcoma of dogs has not yet been shown to have a viral cause, though C-type particles have been observed budding from cell surfaces (Sapp & Adams, 1970). A claim to have infected new-born puppies with cell-free material (Kakuk et al., 1968) awaits confirmation.

ALEUTIAN DISEASE OF MINK

Review: Karstad (1967).

This disease occurs in mink, mainly those homozygous for the recessive Aleutian gene which determines a desirable coat colour.

Morphology. Particles 25 nm in diameter were in crystalline array in cytoplasm of endothelial cells of arteries in the kidney (Tsai et al., 1969).

Chemical composition. DNA from affected spleens reproduced the disease (Karstad, 1967).

Physico-chemical characters. Ether-resistant. Not unusually resistant to formalin, but more heat-resistant than most viruses: 30 minutes at 80° did not wholly inactivate (Eklund et al., 1968). Readily inactivated by UV radiation (Goudas et al., 1970). Reports of passage through dialysing membranes not confirmed.

Cultivation. Not certainly achieved.

Pathogenicity. Symptoms include loss of weight, anorexia and diarrhoea; death follows after some months. Virus is present in saliva, blood, faeces and urine. Infected animals have greatly increased γ globulin of the 6·4 S variety in the blood and there are lesions of blood vessels including those of glomeruli (Porter et al., 1965); the renal lesions are probably the result of an accumulation of antigen-antibody complexes in kidneys (Porter et al., 1969; Henson et al., 1969). Virus multiplication occurs rapidly after infection, though symptoms only develop after months (Porter et al., 1969). Viraemia may persist for months or even for life.

The condition has been recognised in ferrets housed in contact with infected mink, but a similar disease may occur naturally in ferrets (Kenyon et al., 1967).

Ecology. Padgett et al. (1967) have studied genetic aspects of susceptibility. This influences course and duration, though transplacental transmission also occurs.

PUFFINOSIS

A disease affecting young Manx shearwaters.

Description: Stoker & Miles (1953).

Morphology. A small virus, passing a 48 nm but not a 31 nm gradocol membrane.

Haemagglutination. Not demonstrable.

Antigenic properites. Neutralized by convalescent shearwater serum; complement fixation could also be shown.

Cultivation was possible on the CAM of fertile eggs. Small pocks were produced after 12 passages. Three passages were also achieved in the allantoic cavity.

Pathogenicity. A disease affecting young Manx shearwaters (*Puffinus puffinus*) after desertion by their parents 2 months after hatching. There were blisters on the feet and patchy lung consolidation. On transmission the disease was fatal to young shearwaters; and lesions were also produced on the webbed feet of ducklings, which often died. Vesicles were also produced on the breast of a pigeon, but chickens and mammals were unsusceptible. Virus was also obtained from a young Herring gull (*Larus argentatus*). The disease also affects fulmars (Macdonald et al., 1967).

Dane et al. (1953) suggest that the shearwaters may have been infected from gulls. Comparison with the Hughes virus (Hughes et al., 1964), an arbovirus affecting sea-birds (p. 131), would be interesting.

OTHER AVIAN DISEASES

Hepatitis of Turkeys

This virus, lethal to 1-day-old turkey poults but not chicks, was described by Snoeyenbos et al. (1959) and Mongeau et al. (1959).

Hepatitis of Pheasants

The virus was lethal only to pheasants (Rosen et al., 1965).

Disease of Muscovy Ducks

A fatal disease causing diarrhoea and lesions of liver and spleen has been described in Muscovy ducks (Kaschula, 1950).

VIRUSES OF FISH

Fish viruses about which very little is known are infectious dropsy of carp (Roegner–Aust & Schleich, 1951), and a virus causing contagious stomatitis of fish in South America (Pacheco, 1935). Clem et al. (1965) isolated an ether-sensitive 'orphan' virus in tissue cultures of a marine fish (the grunt, *Haemulon*). A virus causation has been suspected also for a number of tumours in fish.

Other fish viruses referred to are those of infectious pancreatic necrosis of trout (p. 67), haemorrhagic septicemia of salmonids (Egtved virus, p. 200), sock-eye and Chinook salmon viruses (p. 200), carp pox (p. 359) and lymphocystis (p. 418).

ANOMALOUS AGENTS

We consider next 4 'slow virus infections'. The infective agents have such remarkable properties that doubt has been thrown on whether they should be regarded as viruses.

SCRAPIE

Synonyms: Tremblant du Mouton. Rida (Iceland).

Review: Stamp (1967).

Morphology. Unknown. Diameter less than 100 nm and possibly about 27 nm (Gibbs, 1967). Claims that the agent will pass a dialysing membrane are not confirmed.

Physico-chemical characters. An extraordinarily stable virus which withstands boiling for 3 hours, also exposure to 20 per cent formalin for 18 hours at 37° (Pattison, 1965), and 0·35 per cent formalin for 3 months. It survives well when dried, and for several years at −40°C (Stamp et al., 1959). Treatment with ether leads to modulate reduction in titre.

Density in CsCl 1·32 (Gibbs, 1967). Stable in mouse-brain tissue for 24 hours at pH range of 2·1 to 10·5. Unstable in 2·5 M CsCl. Unaffected by β-propiolactone. Sensitive in some circumstances to trypsin (Hunter & Millson, 1967).

Alper et al. (1967) and Haig et al. (1969) found the agent almost completely resistant to UV radiation of a wavelength, 254 nm, specifically absorbed by nucleic acids. They suggested that it might lack any nucleic acid. Latarjet et al. (1970) found later that there was maximal inactivation by radiation of a wavelength of 237 nm. R. Markham has pointed out (personal communication, 1970) that suggestions such as those of Alper et al. are only valid if particles are dispersed as monomers; further, inactivation by radiation is not a necessary consequence of absorption of a particular wavelength.

In view of the agent's remarkable properties, various suggestions have been put forward as to what it might be: a nucleic acid surrounded by an unusually stable protein capsid or perhaps a defective virus (Gajdusek, 1967); a small NA core surrounded by a mucopolysaccharide or polysaccharide coat (Adams & Caspary, 1967); a basic protein (Pattison & Jones, 1967); a replicable change in the structure of a cellular membrane (Gibbons & Hunter, 1967); a linkage substance between a DNA-polysaccharide 'subvirus' and cell membranes (Adams & Field, 1968).

Antigenic properties. In spite of intensive effort, no specific serological reactions have been demonstrated. Does not induce interferon nor is it affected by it (Katz & Koprowski, 1968).

Cultivation. The virus has multiplied in cultures of the brain of an infected mouse. The forty-first subculture of the cell-line was still infective (Clarke & Haig, 1970). No report of transmission to normal cells.

Distribution. The disease is prevalent in Europe, particularly so in Scotland and has been introduced in infected sheep to Canada, thence to the United States, also to New Zealand and Australia whence it was soon eradicated; it has also been reported to occur in India. There has been a report of isolation of the virus from normal mice, but no such recovery has been reported from a laboratory where scrapie was not already being studied.

Pathogenicity. In one form, affected sheep have intense pruritus, so that they rub off their wool on fences and other objects and bite the affected area. There soon follow tremors, incoordination and paralysis, but death does not occur for several weeks from the first appearance of symptoms. The pruritus may be absent. The natural incubation period is from 1 to 4 years, but after inoculation it may be 6 months to 2 years. Sheep are the only naturally affected species, though infection may pass from sheep to goats by contact. Virus is present in various tissues, but probably in highest concentration in nervous tissues, and these have been chiefly used for passage. Sheep of different breeds show very different susceptibility to inoculation, and their susceptibility seems to depend in part on genetic factors, though few agree with the view (Parry, 1960) that it is a hereditary disease. Goats can be experimentally infected by inoculation by various routes and, unlike sheep, are almost 100 per cent susceptible. The incubation period is longer at first, but decreases after goat-to-goat passage (Pattison et al., 1959). Virus has been segregated on goats into itching and sleepy strains which breed fairly true (Pattison & Millson, 1961). Chandler (1961; 1962) has transmitted an agent from the sleepy strain to mice. Rather indefinite symptoms involving the CNS appeared in 10 months, but the agent was serially transmissible, and the incubation period fell, after passage, to 3 or 4 months. Multiplication occurs first in lymphoid tissues, and virus is then apparently accumulated in the spleen, where it may be detected after 18 days, reaching maximal levels by 42–52 days after inoculation. It reaches the brain later, having possibly been first modified (Eklund et al., 1967; Mould et al., 1970). Transmission to rats has been reported (Chandler & Fisher, 1963); also to golden hamsters (Zlotnik & Rennie, 1965). Hanson et al. (1971) succeeded in infecting mink with virus from naturally infected sheep.

Pathological lesions are unexpectedly few. Vacuolation of neurones, which is scanty in normal sheep and goat brains, occurs much more abundantly in scrapie brains, especially in brain stem and medulla. Much more prominent is a status spongiosus of the grey matter (Holman & Pattison, 1943). Lesions in the mice were similar and could be detected in as little as 2 months. Chandler (1967) and Field & Raine (1966) employed electron-microscopy to study changes in nervous tissue.

Rida, a chronic encephalopathy of sheep in Iceland, is probably scrapie though scratching is little in evidence. Reports suggesting a different pathology (Sigurdsson, 1954) did not take account of confusion between this and a subsequently recognized nervous disease of sheep, Visna.

Ecology. Despite earlier doubts, there is no doubt that infection can be transmitted orally and by contact between sheep and goats (Brotherston et al., 1968).Virus has also spread by contact among mice (Dickinson et al., 1964; Morris et al., 1965; Zlotnik, 1968). Greig (1940) believed that infection could spread from infected pastures. The long incubation period, the confused and confusing role of heredity and the remarkable stability of the agent have combined to baffle investigators.

Control by slaughter has been practised in countries to which the disease has been introduced.

Encephalopathy of Mink

Physico-chemical characters. The virus of this disease resembles that of scrapie. It is less than 50 nm in diameter, resistant to 10 per cent formalin and to UV radiation; it withstands boiling for 15 minutes. Some sensitivity to ether. Susceptible to digestion with pronase (Hartsough & Burger, 1965; Marsh & Hanson, 1969).

Antigenic properties and cultivation. Not reported.

Pathogenicity. Infected mink show excitability or somnolence and other symptoms of nervous origin. It is not readily infectious, but may spread by cannibalization. It is transmissible to other mink by injection or *per os* with incubation periods of 5 and 8 months respectively. Infection has been transmitted to goats (Zlotnik & Barlow, 1967), also to mice, ferrets and hamsters (Marsh et al., 1969; Barlow & Rennie, 1970).

Histological changes include vacuolization of neuroglia.

It does not seem that Aleutian disease of mink (see p. 431) has similar properties.

Kuru

A disease of the CNS, commonly fatal, restricted to members of the Fore people, a small cultural and linguistic group of Melanesians in the highlands of New Guinea.

Physico-chemical characters. The infective agent has passed a 220-nm filter. It survives heating for 30 minutes at 85°, lyophilization and storage for over 5 years at −70° (Gibbs & Gajdusek, 1970).

Antigenic properties and cultivation. Not reported.

Pathogenicity. The disease begins with ataxia, tremors, emotional instability, progressing to complete incapacity with aphonia and dysphagia. The disease has been transmitted to chimpanzees (Gajdusek et al., 1966; 1967). On passage to a second chimpanzee in series the incubation period fell from $1\frac{1}{2}$ to $2\frac{1}{2}$ years down to 1 year. The virus was injected IC or peripherally and was recovered from viscera as well as from brain (Gibbs & Gajdusek, 1970). Transmission to spider monkeys (*Ateles*) has also been successful: ataxia and tremor appeared in 23–26 months. Other monkeys tested were apparently insusceptible (Gajdusek et al., 1968). Pathological changes consist of widespread neuronal degeneration, myelin loss, microglial proliferation and perivascular cuffing.

Ecology. The ecology has been studied by Matthews (1967). Susceptibility apparently depends on possession of a single gene, and the facts are consistent with the view that homozygotes of both sexes die in childhood, while female, but only rarely male, heterozygotes develop the disease and die in adult life.

Spongiform Encephalopathy

Synonym: Creutzfeldt-Jakob disease.

In this disease patients have dementia, severe visual disturbances and myoclonic jerking.

Infection was passed from brain biopsy material to chimpanzees, and at least once in series. The incubation period was 12 to 14 months after IC injection, and symptoms were much as in man.

Other injected species remained normal (Gibbs & Gadjusek, 1969).

The particles seen in the brain by Vernon et al. (1970) were some 65–85 nm in diameter, some smaller; they suggest a similarity to subacute sclerosing panencephalitis (see p. 241) rather than to scrapie or kuru.

REFERENCES

Adams, D. H., & Caspary, E. A. (1967) *Br. med. J.*, **3**, 173.

Adams, D. H., & Field, E. J. (1968) *Lancet*, **2**, 714.

Almeida, J. D., & Waterson, A. P. (1969) *Lancet*, **2**, 983.

Alper, T., Cramp, W. A., Haig, D. A., & Clarke, M. C. (1967) *Nature, Lond.*, **214**, 764.

Badger, G. F., McCorkle, L. P., Curtiss, G., Dingle, J. H., Hodges, R. G., & Jordan, W. S. (1956) *Am. J. Hyg.*, **64**, 376.

Barlow, R. M., & Rennie, J. C. (1970) *J. comp. Path.*, **80**, 75.

Biggers, D. C., Kraft, L. M., & Spiez, H. (1964) *Am. J. Path.*, **45**, 413.

Blumberg, B. S., Gertsley, B. J. S., Hungerford, D. A., London, T., & Sutnick, A. I. (1967) *Ann. int. Med.*, **66**, 924.

Blumberg, B. S., Sutnick, A. I., & London, W. T. (1968) *Bull. N.Y. Acad. Med.*, **44**, 1566.

Blumberg, B. S., Sutnick, A. I., London, W. T., & Millman, J. (1970) *New Engl. J. Med.*, **283**, 349.

Braunsteiner, H., Fellinger, K., Beyreder, T., Grabner, G., & Neumayr, A. (1957) *Wien Z. inn. Med.*, **38**, 231.

Brotherston, J. G., Renwick, C. C., Stamp, J. T., Zlotnik, I., & Pattison, I. H. (1968) *J. comp. Path.*, **78**, 9.

Chandler, R. L. (1961) *Lancet*, **1**, 1378.

Chandler, R. L. (1962) *Lancet*, **1**, 107.

Chandler, R. L. (1967) *Res. vet. Sci.*, **8**, 98 and 166.

Chandler, R. L., & Fisher, J. (1963) *Lancet*, **2**, 1165.

Cheever, F. S. (1967) *Med. Clinics. N. America*, **51**, 637.

Clarke, M. C., & Haig, D. A. (1970) *Nature, Lond.*, **225**, 100.

Clem, L. W., Sigel, M. M., & Früs, R. R. (1965) *Ann. N.Y. Acad. Sci.*, **126**, 343.

Coyne, V. E., Millman, I., Cerda, J., & Gerstley, J. S. (1970) *J. exp. Med.*, **131**, 307.

Dane, D. S., Miles, J. A. R., & Stoker, M. G. P. (1953) *J. anim. Ecol.*, **22**, 123.

Davenport, F. M., Hennessy, A. V., Christopher, N., & Smith, K. C. (1966) *Am. J. Epidemiol.*, **83**, 146.

Deinhardt, F., Holmes, A. W., Capps, R. B., & Popper, H. (1967) *J. exp. Med.*, **125**, 673.

Dickinson, A. G., Mackay, J. M. K., & Zlotnik, I. (1964) *J. comp. Path.*, **74**, 250.

Eklund, C. M., Hadlow, W. J., Kennedy, R. C., Boyle, C. C., & Jackson, T. A. (1968) *J. infect. Dis.*, **118**, 510.

Eklund, C. M., Kennedy, R. C., & Hadlow, W. J. (1967) *J. infect. Dis.*, **117**, 15.

Field, E. J., & Raine, C. S. (1966) *Res. vet. Sci.*, **7**, 292.

Findlay, G. M., & MacCallum, F. O. (1938) *Proc. r. Soc. Med.*, **31**, 799.

Fontaine, M., Ricq, A., Brion, A., & Goret, P. (1957) *C.r. hebd. Seánc. Acad. Sci., Paris*, **245**, 122.

Fukumi, H., Nakaya, R., Hatha, S., Noriki, H., Yunoki, H., Akagi, K., Saito, T., Uchiyama, K., Kobari, K., & Nakamishi, R. (1957) *Jap. J. med. Sci.*, **10**, 1.

Gajdusek, C. (1967) *Curr. topics Microbiol. Immunol.*, **40**, 59.

Gajdusek, C., Gibbs, C. J., & Alpers, M. (1966) *Nature, Lond.*, **209**, 794.

Gajdusek, C., Gibbs, C. J., & Alpers, M. (1967) *Science*, **155**, 212.

Gajdusek, C., Gibbs, C. J., Asher, D. M., & David, E. (1968) *Science*, **162**, 693.
Gerin, J. L., Purcell, R. H., Hoggan, M. D., Holland, P. V., & Chanock, R. M. (1969) *J. Virol.*, **4**, 763.
Gibbons, R. A., & Hunter, G. D. (1967) *Nature, Lond.*, **215**, 1041.
Gibbs, C. J. (1967) *Curr. topics Microbiol. Immunol.*, **40**, 44.
Gibbs, C. J., & Gajdusek, D. C. (1969) *Science*, **165**, 1023.
Gibbs, C. J., & Gajdusek, D. C. (1970) *Am. J. trop. Med. Hyg.*, **19**, 138.
Giles, J. P., Liebhaber, H., Krugman, S., & Lattimer, C. (1964) *Virology*, **24**, 107.
Gordon, I., Ingraham, H. S., & Korns, R. F. (1947) *J. exp. Med.*, **86**, 409.
Goret, P., Brion, A., Fontaine, M., Pilet, C., Girard, M., & Girard, M. (1959) *Bull. Acad. vet. Fr.*, **32**, 623 and 635.
Goudas, P., Karstad, L., & Tabel, H. (1970) *Can. J. comp. Med.*, **34**, 118.
Greig, J. R. (1940) *Vet. J.*, **96**, 203.
Gueft, B. (1961) *Arch. Path.*, **72**, 61.
Haig, D. A., Clarke, M. C., Blum, E., & Alper, T. (1969) *J. gen. Virol.*, **5**, 455.
Hartsough, G. R., & Burger, D. (1965) *J. infect. Dis.*, **115**, 387.
Havens, W. P. (1954) *Ann. rev. Microbiol.*, **8**, 289.
Havens, W. P. (1963) *Medicine*, **14**, 57.
Havens, W. P. (1967) *Med. clinics N. America*, **51**, 653.
Hellström, B., & Vahlquist, B. (1951) *Acta paediat., Stockh.*, **40**, 189.
Henson, J. B., Gorham, J. R., Padgett, G. A., & Davis, W. C. (1969) *Arch. Path.*, **87**, 21.
Holman, H. H., & Pattison, I. H. (1940) *J. comp. Path.*, **53**, 231.
Hughes, L. E., Clifford, C. M., Thomas, L. A., Denmark, H. A., & Philip, C. B. (1964) *Am. J. trop. Med. Hyg.*, **13**, 118.
Hunter, G. D., & Millson, G. C. (1967) *J. comp. Path.*, **77**, 301.
Jokelainen, P. T., Krohn, K., Prince, A. M., & Finlayson, N. D. C. (1970) *J. virol.*, **6**, 6.
Jordan, W. S., Gordon, I., Dorrance, L. R. (1953) *J. exp. Med.*, **98**, 461.
Józwiak, W., Košcielak, J., Madalínski, K., Brzosko, W. J. (1971) *Nature New Biol.*, **229**, 92.
Kakuk, T. J., Hinz, R. W., Langham, R. F., & Conner, G. H. (1968) *Cancer Res.*, **28**, 716.
Kalter, S. S., Kim, C. S., & Heberling, R. L. (1969) *Nature, Lond.*, **224**, 190.
Karstad, L. (1967) *Curr. topics Microbiol. Immunol.*, **40**, 9.
Kaschula, V. R. (1950) *J. S. Afr. vet. med. Ass.*, **21**, 18.
Katz, M., & Koprowski, H. (1968) *Nature, Lond.*, **219**, 639.
Kempe, C. H., Shaw, E. B., Jackson, J. R., & Silver, H. K. (1950) *J. Pediat.*, **37**, 561.
Kenyon, A. J., Howard, E., & Buko, L. (1967) *Am. J. vet. Res.*, **28**, 1167.
Kraft, L. M. (1962) *Science*, **137**, 282.
Kraft, L. M. (1966) *The viruses of laboratory rodents.* Natn. Cancer Inst. Monogr. no. 20, p. 55.
Kraft, L. M., Moore, A. E. (1961) *Z. Versuchstierk*, **1**, 66.
Krugman, S., Giles, J. P., & Hammond, J. (1970) *J. infect. Dis.*, **122**, 432.
Krugman, S., & Ward, R. (1961–62) *Yale J. Biol. Med.*, **34**, 169.
Latarjet, B., Muel, B., Haig, D., Clarke, M. C., & Alper, T. (1970) *Nature, Lond.*, **227**, 1341.
Lehane, O., Kwantes, C. M. S., Upward, M. G., & Thompson, D. R. (1949) *Br. med. J.*, **2**, 572.
Levene, C., & Blumberg, B. S. (1969) *Nature, Lond.*, **221**, 195.
Lucké, B. (1944) *Am. J. Path.*, **20**, 471.
MacCallum, F. O. (1953) *Br. med. Bull.*, **9**, 221.
MacCallum, F. O. (1955) in *Virus and rickettsial diseases.* Eds.: Bedson, Downie, MacCallum, Stuart-Harris, 2d Ed. London: Arnold.
McCollum, R. W. (1952) *Proc. Soc. exp. Biol. Med., U.S.A.*, **81**, 157.
Macdonald, J. W., McMartin, D. A., Walker, K. G., Cairns, M., & Dennis, R. H. (1967) *Br. Birds*, **60**, 356.
Marsh, R. F., Burger, D., Eckroade, R., Zu Rhein, G. M., & Hanson, R. P. (1969) *J. infect. Dis.*, **120**, 713.
Marsh, R. F., & Hanson, R. P. (1969) *J. Virol.*, **3**, 176.

Matthews, J. C. (1967) *Lancet*, **1**, 821.
Millman, I., Loeb, L. A., Bayer, M. E., & Blumberg, B. S. (1970) *J. exp. Med.*, **131**, 1190.
Molomut, N., Padnos, M., & Smith, W. (1965) *J. natn. Cancer Inst.*, **34**, 403.
Mongeau, J. D., Truscott, R. B., Ferguson, A. E., & Connell, M. C. (1959) *Avian Dis.*, **3**, 338.
Morris, J. A., Gajdusek, D. C., & Gibbs, C. J. (1965) *Proc. Soc. exp. Biol. Med., U.S.A.*, **120**, 108.
Mould, D. L., Dawson, A. McL., & Rennie, J. C. (1970) *Nature, Lond.*, **228**, 779.
Mouriquand, J., Mouriquand, C., Darmault, J., & Grivet, M. N. (1965) *C.r. hebd. Séanc. Acad. Sci., Paris*, **260**, 2952.
Nagayama, T. (1960) *Acta med. univ. kagoshima*, **3**, 41.
Neefe, J. R., Baty, J. B., Reinhold, J. G., & Stokes, J. (1947) *Am. J. publ. Hlth*, **37**, 365.
Pacheco, G. (1935) *Mem. Inst. Oswaldo Cruz*, **30**, 349.
Padgett, G. A., Gorham, J. R., & Henson, J. B. (1967) *J. infect. Dis.*, **117**, 35.
Parks, W. P., & Melnick, J. L. (1969) *J. infect. Dis.*, **120**, 539.
Parry, H. B. (1960) *Nature, Lond.*, **185**, 441.
Pattison, I. H. (1965) *J. comp. Path.*, **75**, 159.
Pattison, I. H., Gordon, W. S., & Millson, G. C. (1959) *J. comp. Path.*, **69**, 300.
Pattison, I. H., & Jones, K. M. (1967) *Vet. Rec.*, **80**, 2.
Pattison, I. H., & Millson, G. C. (1961) *J. comp. Path.*, **70**, 182.
Porter, D. D., Dixon, F. J., & Larsen, A. E. (1965) *J. exp. Med.*, **121**, 889.
Porter, D. D., Larsen, A. E., & Porter, H. G. (1969) *J. exp. Med.*, **130**, 575.
Prete, del, S., Constantino, D., Doglia, M., Graziina, M., Ajdukiewicz, A., Dudley, F. J., Fox, R. A., & Sherlock, S. (1970) *Lancet*, **2**, 579.
Prince, A. M. (1968) *Proc. natn. Acad. Sci. U.S.A.*, **60**, 814.
Purcell, R. H., Holland, P. V., Walsh, J. H., Wong, D. C., Morrow, A. G., & Chanock, R. M. (1969) *J. infect. Dis.*, **120**, 383.
Reihart, O. F., Reihart, H. W., & Schenker, J. R. (1952) *N. Am. Vet.*, **33**, 174.
Reimann, H. A. (1963) *Am. J. med. Sci.*, **246**, 404.
Rightsel, W. A., Keltsch, R. A., Taylor, A. R., Boggs, J. B., & McLean, I. W. (1961) *J. Am. med. Ass.*, **177**, 671.
Roegner-Aust, S., & Schleich, F. (1951) *Z. Naturforsch.*, **6b**, 448.
Rosen, M. N., Hunter, B. F., & Brunetti, O. A. (1965) *Avian Dis.*, **9**, 382.
Rowe, W. P., & Capps, W. I. (1961) *J. exp. Med.*, **113**, 831.
Sapp, W. J., & Adams, E. W. (1970) *Am. J. vet. Res.*, **31**, 1321.
Schmidt, N. J., & Lennette, E. H. (1961) *Progr. med. Virol.*, **3**, 32.
Sheehey, T. W., Artenstein, M. S., & Green, R. W. (1964) *J. Am. med. Ass.*, **190**, 1023.
Sigurdsson, B. (1954) *Br. vet. J.*, **110**, 341.
Snoeyenbos, G. H., Basch, H. I., & Sevoian, M. (1959) *Avian Dis.*, **3**, 377.
Stamp, J. T. (1967) *Br. med. Bull.*, **23**, 133.
Stamp, J. T., Brotherston, J. G., Zlotnik, I., Mackay, J. M. K., & Smith, W. (1959) *J. comp. Path.*, **69**, 268.
Stoker, M. G. P., & Miles, J. A. R. (1953) *J. Hyg., Camb.*, **51**, 195.
Stokes, J., Farquhar, J. A., Drake, M. E., Capps, R. B., Ward, C. S., Mills, O. W., & Kitts, A. W. (1951) *J. Am. med. Ass.*, **147**, 714.
Svet-Moldavsky, G. J., Mkheidze, D. M., Liozner, A. L., & Bykovsky, A. P. (1968) *Nature, Lond.*, **217**, 102.
Svet-Moldavsky, G. J., Liozner, A. L., Mkheidze, D. M., Sokolov, P. P., & Bykovsky, A. P. (1970) *J. natn. Cancer Inst.*, **45**, 475.
Tomita, Y., & Jonas, A. M. (1968) *Am. J. vet. Res.*, **29**, 445.
Tsai, K. S., Grinyer, I., Pan, I. C., & Karstad, L. (1969) *Can. J. Microbiol.*, **15**, 138.
Vernon, M. L. et al. (1970) *Lancet*, **1**, 964.
Vizoso, A. D., Hay, R., & Battersby, T. (1966) *Nature, Lond.*, **209**, 1263.
Ward, R., & Krugman, S. (1962) *Progr. med. Virol.*, **4**, 87.
Warwick, W. J. (1967) *Progr. med. Virol.*, **9**, 256.
WHO (1970) *Bull. Wld Hlth org.*, **42**, 957.

WHO (1964) *Expert committee on hepatitis.* Technical report series, no. 285.
Wolf, K. (1966) *Adv. Virus Res.*, 12, 35.
Zlotnik, I. (1968) *J. comp. Path.*, 78, 19.
Zlotnik, I., & Barlow, R. M. (1967) *Vet. Rec.*, 81, 55.
Zlotnik, I., & Rennie, J. L. (1965) *J. comp. Path.*, 75, 147.
Zuckerman, A. J. (1970) *Virus diseases of the liver.* London: Butterworth.
Zuckerman, A. J., & Taylor, P. E. (1969) *Nature, Lond.*, 223, 81.
Zuckerman, A. J., & Taylor, P. E., & Almeida, J. D. (1970) *Br. med. J.*, 1, 262.

Index